*Perspectives
on Cognitive
Neuroscience*

Perspectives on Cognitive Neuroscience

edited by

RICHARD G. LISTER

HERBERT J. WEINGARTNER

National Institute on Aging
The National Institute on Alcohol Abuse and Alcoholism

New York   Oxford
OXFORD UNIVERSITY PRESS
1991

We dedicate this book to our families,
Elizabeth, Eric, Thelma (Feathers), Gordon, Michael and Robert

QP
360
.P462
1992

Oxford University Press

Oxford   New York   Toronto
Delhi   Bombay   Calcutta   Madras   Karachi
Petaling Jaya   Singapore   Hong Kong   Tokyo
Nairobi   Dar es Salaam   Cape Town
Melbourne   Auckland

and associated companies in
Berlin   Ibadan

Library of Congress Cataloging-in-Publication Date
Perspectives on cognitive neuroscience /
edited by Richard G. Lister and Herbert J. Weingartner.
p.   cm.   Includes bibliographical references and index.
1. Neuropsychology.
2. Cognitive psychology.
3. Cognitive science.
I. Weingartner, Herbert.   II. Lister, Richard G.
[DNLM: 1. Brain—physiology.
2. Cognition—physiology.
3. Cognition—Disorders.
4. Memory—physiology.
5. Psychophysiology.
WL 102 P467]
QP360.P462   1992   152—dc20   91-1958
ISBN 0-19-506151-9

9 8 7 6 5 4 3 2 1

Printed in the United States of America
on acid-free paper

Preface

We are in the midst of a brain science revolution. Neuronal events are being described with increasing precision, and during the past two decades our understanding of the neuroanatomy of the central nervous system has grown tremendously. In addition, we have learned a great deal about how neurons communicate with one another, and the dynamic neurochemical and neurophysiological processes involved in information processing. All of this information would be lifeless if it were not possible to relate those neurobiological events to behavior. Our ultimate goal in cognitive neuroscience is to develop as complete a description as possible of the processes of the mind. A cognitive model that ignores biology would have to be considered incomplete. Likewise, a complete description of brain without a consideration of behavior would also be inadequate. Ideally, we would like to describe cognition from the molecular to the behavioral level, and clearly this would involve an analysis of brain and behavior from many different perspectives. It is the aim of this volume to represent some of these perspectives and to relate them to one another.

Where does our current body of knowledge about brain and behavior come from? Several scientific cultures have contributed to a psychobiological understanding of behavior. From about the mid-nineteenth century one can identify five relatively separate traditions that have influenced contemporary cognitive neuroscience. The first has its roots in the study of clinical phenomena, which involves an examination of the effects of circumscribed changes in higher mental function and which relates such changes to a localized neuropathology. A clear example of this tradition comes from the work of Broca on the localization of language and lateralization. Neuropsychological studies based on the dramatic accidents of nature continue to make a very significant contribution to cognitive science. In this volume examples of this approach are represented in the chapters by Schacter and Nadel and by Rapp and Caramazza.

A second tradition emphasizes the scientific study of behavior, focusing on the development of methods for quantifying behavior and of theoretical positions for organizing the data obtained. The focus for this research has been the study of normal as opposed to impaired behavior. Although psychologists such as James, Watson, Skinner, and the current cognitive theorists thought about higher mental functions very differently, they share a commitment to psychological explanations of higher mental functions. During the past several decades experimental psychologists have developed many models and theories which describe higher mental functions in psychological terms. The chapters by Mackintosh and Watkins illustrate this approach.

A third tradition is based upon a biological approach to the study of behavior in both animals and humans. The primary aim of the investigators from this tradition is to understand the biological mechanisms underlying behavior. The long history of physiological and comparative psychology provides numerous examples. In this volume examples of this approach are provided by Thompson and Gluck and by Olton, Wible, and Markowska.

A fourth tradition has emerged from those interested in developing a detailed knowledge of central nervous function almost independent of behavior. Much of contemporary neuroscience is concerned with this theme, and our current understanding of cognitive neuroscience has been critically dependent on this knowledge base. In this volume the contribution of Bowen illustrates this tradition.

A fifth and final area of study that is relevant to cognitive neuroscience can be loosely defined by the domain of research on artificial intelligence. Workers in this area include engineers and mathematicians as well as cognitive scientists. The goal of these researchers is to develop machines and programs that can perform cognitive operations. Ideas from this area have been used by a number of neuroscientists in developing models of brain function. The chapter by McNaughton and Smolensky provides an example of this approach.

As evident in these five approaches, researchers in cognitive neuroscience are confronted with relevant information that comes from many different fields. A full appreciation of this information appears to require a detailed knowledge which can be obtained only from years of experience. It is clearly impossible to achieve this knowledge in all areas. Nevertheless, it should be possible to appreciate the conceptual logic of questions being asked and the major findings that have emerged from the different fields that make up cognitive neuroscience. This book has been developed so that each of the themes presented by the contributors is understandable to readers with diverse backgrounds. We hope that cognitive scientists with no biological background will find discussions of the neurobiology of cognition helpful in developing their research programs. Similarly, we hope that basic neuroscientists will find discussions of the subtleties of behavior interesting. We also hope that clinicians will find the contributions that are concerned with basic cognitive neuroscience helpful for developing more effective methods of diagnosing and treating cognitive disorders. Finally, and perhaps most important, we hope that the volume is of value to those students whose interest in cognitive neuroscience has just been awakened. Although we hesitate to suggest how the book should be read, we nevertheless recommend that each contribution be seen as representing a strategy for considering the determinants of cognitive behavior rather than as a detailed summary of what is known in each area.

The contributions have been organized into four parts, each of which is tied together by a common theme. Part I addresses the issue of how to examine cognitive problems from different levels of analysis. The contributors in this section are from disciplines as diverse as molecular biology, genetics, artificial intelligence, experimental psychology, and functional neuroanatomy. They discuss approaches to the study of learning and memory that range from the molar (global) to the molecular.

The contributions in Part II focus on the dissociation of different types of cognitive processes and the development of cognitive models based on these dissociations. Some of the questions that are addressed include whether there are different types of memory and what constitutes evidence that memory systems are psychobiologically distinct. A diverse set of strategies are covered by the various contributors.

Part III focuses on factors that modulate cognitive processes. Emotional state, arousal, and reinforcement contingencies all can have a major impact on cognition, although they have often been ignored in cognitive research. The contributors to Part III discuss the mechanisms involved in the modulation of cognition.

In Part IV the contributors discuss the various approaches to the study of cognitive failures. They analyze cognitive failures from both behavioral and neurobiological perspectives, and the research discussed is clinically important in that it provides a framework for developing methods of treating cognitive disorders. However, it also casts considerable light on basic cognitive mechanisms.

Preceding each of these parts is a short introduction with a general framework to help the reader integrate the various contributions. Immediately following each part of the book we discuss in more detail some of the points that emerge from each chapter. We begin by highlighting the basic ideas and issues that are addressed by each contributor. We then attempt to integrate the chapters within each part, showing how they relate to contributions elsewhere in the volume. We conclude the book with a short agenda for future research in cognitive neuroscience.

We are most grateful to the contributors not only for their chapters but also for the careful reviews of our introductions, summaries, and commentaries. Their many helpful suggestions were incorporated into the final manuscript and have strengthened our discussions and made the book all the more cohesive.

The views expressed are those of the authors and do not necessarily reflect the views of the institutes with which the authors are affiliated.

My dear friend, collaborator, and colleague Richard Lister died on July 8, 1991, at the age of 33. The field of cognitive neuroscience has lost a creative and brilliant young scientist. In a short time, Richard had already made a significant contribution to our understanding of the neurobiology of cognition, and more important, was an ideal role model, expressing the very best aspirations and goals that should be part of the process of carrying out science. He was a superb scholar but also an extraordinary human being. Great fun and excitement accompanied all of our research projects. We both loved the world of brain-behavior research, and what can be more exciting than to develop a picture of the mechanisms that define the human experience? That is what we saw, what I have always seen as the tapestry of cognitive neuroscience. We approached all of our research projects much like children at play. It gave us enormous pleasure to think through issues, ideas, theory, data, the plans for experiments, mixing humor and flights of the imagination with disciplined, systematic hard work. It was a fulfilling atmosphere for research. I will never forget Richard and will miss him very much.

Richard and I worked closely together for about three years in completing this book. That may seem like a long time for an edited volume; however, the reader should understand that we did not view this book as a series of disconnected themes but as an integration of contributions that could bridge all of cognitive neuroscience. It was not an easy task to accomplish that goal. We encouraged the contributors not only to consider their specialized areas of research but also to realize that they would have important things to say to researchers who were often unaware of their work. The contributors provided us with superb chapters, and it was up to us to develop the integrative broad perspective that would relate these many distinct areas of research to one another. I think we achieved our goal of providing an organized picture of cognitive neuroscience. It was enormously fulfilling, and we took much pleasure in working on this book together.

August 1991   R.G.L.
Bethesda, Maryland   H.J.W.

Contents

III. Modulation of Cognition

IV. Clinical Perspectives

Contributors

Marilyn Albert
  Department of Psychiatry
  Massachusetts General Hospital,
    CNY-6
  Boston, MA 02114

José Ambros-Ingerson
  Bonney Center for the Neurobiology
    of Learning and Memory
  University of California
  Irvine, CA 92717

Craig H. Bailey
  Center for Neurobiology and
    Behavior
  College of Physicians and Surgeons
  Columbia University
  722 West 168th Street
  New York, NY 10032

Linda M. Bierer
  Psychiatry Service
  Bronx VA Medical Center
  130 W. Kingsbridge Road
  Bronx, NY 10468

David M. Bowen
  Department of Neurochemistry
  Institute of Neurology
  National Hospital for Nervous
    Diseases
  Queens Square
  1 Wakefield Street
  London, WC1N 1PJ
  ENGLAND

Alfonso Caramazza
  Johns Hopkins University
  Cognitive Science Center
  Ames Hall
  Baltimore, MD 21218

Patricia S. Churchland
  Department of Philosophy
  University of California-San Diego
  La Jolla, CA 92093

Kenneth L. Davis
  Psychiatry Service
  Bronx VA Medical Center
  130 W. Kingsbridge Road
  Bronx, NY 10468

Michael W. Eysenck
  Department of Psychology
  Royal Holloway and Bedford New
    College
  Egham Hill
  Egham, Surrey, TW20 OEX
  ENGLAND

Mark A. Gluck
  Department of Psychology
  Stanford University
  Stanford, CA 94305

Richard Granger
  Bonney Center for the Neurobiology
    of Learning and Memory
  University of California
  Irvine, CA 92717

Ruben C. Gur
Brain Behavior Laboratory
Neuropsychiatry Program
Departments of Psychiatry and
Neurology
University of Pennsylvania
Philadelphia, PA 19104

Raquel E. Gur
Brain Behavior Laboratory
Neuropsychiatry Program
Departments of Psychiatry and
Neurology
University of Pennsylvania
Philadelphia, PA 19104

Vahram Haroutunian
Psychiatry Service
Bronx VA Medical Center
130 W. Kingsbridge Road
Bronx, NY 10468

William Hirst
New School for Social Research
Department of Psychology
65 Fifth Avenue
New York, NY 10003

Marcia K. Johnson
Department of Psychology
Princeton University
Princeton, NJ 08544-1010

Raymond P. Kesner
Department of Psychology
University of Utah
Salt Lake City, UT 84112

George F. Koob
Department of Neuropharmacology
Research Institute of Scripps Clinic
La Jolla, CA 92037

John Larson
Bonney Center for the Neurobiology
of Learning and Memory
University of California
Irvine, CA 92717

Howard Leventhal
30 College Avenue
Institute for Health, Health Care
Policy, and Aging Research
Rutgers University
New Brunswick, NJ 08903

Muriel D. Lezak
Department of Neurology
Oregon Health Sciences University
3181 S.W. Sam Jackson Park Road
Portland, OR 97201

Gary Lynch
Bonney Center for the Neurobiology
of Learning and Memory
University of California
Irvine, CA 92717

N. J. Mackintosh
Department of Experimental
Psychology
University of Cambridge
Downing Street
Cambridge, CB2 3EB
ENGLAND

Alicja L. Markowska
Department of Psychology
Johns Hopkins University
Baltimore, MD 21218

James L. McGaugh
Center for the Neurobiology of
Learning and Memory
University of California
Irvine, CA 92717

Bruce L. McNaughton
Department of Psychology
University of Arizona
Tucson, AZ 85721

William Milberg
  GRECC
  VA Medical Center
  1400 VFW Parkway
  West Roxbury, MA 02132

Morris Moscovitch
  Department of Psychology
  Erindale College
  University of Toronto
  Mississauga Road
  Mississauga, Ontario, L5L 1C6
  CANADA

  The Rotman Institute of Baycrest
    Centre for Geriatric Care
  Toronto, Ontario
  CANADA

Lynn Nadel
  Department of Psychology
  Cognitive Science Program
  University of Arizona
  Tucson, AZ 85721

David S. Olton
  Department of Psychology
  Johns Hopkins University
  Baltimore, MD 21218

Alan M. Palmer
  Departments of Psychiatry and
    Neuroscience
  Western Psychiatric Institute and
    Clinic
  University of Pittsburgh
  3811 O'Hara Street
  Pittsburgh, PA 15213

Brenda C. Rapp
  Johns Hopkins University
  Cognitive Science Center
  Ames Hall
  Baltimore, MD 21218

Anthony C. Santucci
  Department of Psychology
  Manhattanville College
  25 Purchase Street
  Purchase, NY 10577

Daniel L. Schacter
  Department of Psychology
  Harvard University
  Cambridge, MA 02138

Terrence J. Sejnowski
  The Salk Institute
  P.O. Box 85800
  San Diego, CA 92138-9216

Shepard Siegel
  McMaster University
  Department of Psychology
  1280 Main Street West
  Hamilton, Ontario, L8S 4K1
  CANADA

Paul Smolensky
  Department of Computer Science
  Institute of Cognitive Science
  University of Colorado
  Boulder, CO 80309

Ursula Staubli
  Bonney Center for the Neurobiology
    of Learning and Memory
  University of California
  Irvine, CA 92717

Richard F. Thompson
  Department of Psychology and
    Biological Sciences
  University of Southern California
  University Park
  Los Angeles, CA 90089-1061

Tim Tully
  Department of Biology
  Brandeis University
  Waltham, MA 02254

Carlo Umilta
  Dipartimento di Psicologia Generale
  Universita di Padova
  Piazza Capataniato, 3
  35139 Padova
  ITALY

Michael J. Watkins
  Department of Psychology
  Rice University
  Houston, TX 77251

L. Weiskrantz
  Department of Experimental
    Psychology
  University of Oxford
  South Parks Road
  Oxford, 0X1 3UD
  ENGLAND

Cynthia G. Wible
  Department of Psychology
  Johns Hopkins University
  Baltimore, MD 21218

I

Levels of Analysis

One of the major themes of this volume concerns the value of communication among scientists working in different disciplines. The contributions in this first section have been assembled with the intention of highlighting this theme.

Cognitive scientists argue that explanations in cognitive science are cast at a functional level and that neurobiological processes can be ignored for the same reason that electrical and engineering factors are separated from programming considerations in computer science (Pylyshyn, 1983). An example of how similar biological and engineering problems have been solved independently comes from the invention of radar and the discovery of echolocation in bats (see Griffin, 1958). It is clear that neuroscience and cognitive science can proceed (and historically have proceeded) independently. The issue we wish to address, however, is whether any mutual benefit can be gained from their interaction. That is, can cognitive scientists use any ideas generated from neuroscientific research to further their own research agenda, and vice versa? Had echolocation in bats been discovered earlier, might the engineers have used this knowledge base in the development of radar? In the following chapters, therefore, we wish the reader to consider the value of any one approach in isolation, and whether research in any one domain is significantly enhanced by the consideration of another.

Part I opens with a chapter by Churchland and Sejnowski that gives a general overview of cognitive neuroscience. The authors introduce the idea of *levels of analysis* and discuss some of the techniques that are used to describe cognitive phenomena at these different levels.

The next few chapters illustrate the different levels at which associative learning can be studied. The contributors represent such diverse disciplines as functional neuroanatomy, molecular biology, behavioral genetics, experimental psychology, and computational theory. Can these different approaches to the study of associative learning be integrated so that the whole is greater than the sum of the parts?

1

The second chapter, by Thompson and Gluck, begins with a behavioral description of classical conditioning, then describes the results of an extensive series of studies mapping the neuronal circuits that are responsible for simple associative learning, and concludes with a discussion of the importance of this research in the development of algorithms for classical conditioning.

The following two chapters, by Tully and by Mackintosh, view classical conditioning from two very different perspectives. Tully describes behavioral genetic research, looking for molecular mechanisms underlying components of classical conditioning in the fruit fly. Mackintosh, in contrast, describes classical conditioning from the perspective of an experimental psychologist and emphasizes some behavioral considerations that are often ignored by those working on simple systems.

Next, Bailey discusses the fascinating morphological changes that have been observed to accompany memory formation in one such simple system, *Aplysia*.

The chapters by McNaughton and Smolensky, and by Lynch et al., like that by Thompson and Gluck, both emphasize the value of interaction between neurobiologists and computationalists in the study of memory.

Watkins concludes the section with a chapter expressing an experimental psychologist's view of cognitive science. He questions the ability of neuroscientists to explain psychological phenomena, and his position presents a challenge to cognitive neuroscience.

References

Griffin, D. R. (1958). *Listening in the dark: The acoustic orientation of bats and men*. New Haven, CT: Yale University Press.

Pylyshyn, Z. W. (1983). Information science. In F. Machlup & U. Mansfield (Eds.), *The study of information* (pp. 63–74). New York: Wiley.

1

Perspectives on Cognitive Neuroscience

PATRICIA S. CHURCHLAND / TERRENCE J. SEJNOWSKI

Remarkable developments in the last two decades in cognitive science, computational theory, and neuroscience have engendered a new, if cautious, optimism for achieving some measure of integration and explanatory unification of the various levels of organization in nervous systems (Churchland, 1986; Mountcastle, 1986). Until quite recently, the immediate goals of neuroscientists and cognitive scientists were sufficiently distant from each other that each group's discoveries often seemed of merely academic significance to the other. The dominant model of computation in cognitive science, based on symbol processing, was inspired by the Turing machine and the von Neumann architecture of the digital computer. Symbol-processing models were unpromising as a means to bridge the gap because they did not relate to what was known about nervous systems at the level of signal processing. All this has begun to change quite dramatically, and there is gathering conviction among scientists that the time is right for a fruitful convergence of research from hitherto isolated fields.

In particular, important new techniques for investigating the functions of the brain have been invented in the last several decades that make possible a much better global picture of processing in the brain and a more detailed structural and functional description of the brain than was previously available. In addition, we have a much better understanding of the components and dimensions of cognitive abilities and more systematic and sophisticated methods for determining the behavioral parameters of psychological phenomena (Kosslyn, 1988; LeDoux & Hirst, 1986; Posner, 1989). Finally, the vast increase in the speed and availability of computers, together with new computational approaches to modeling in neuroscience (Schwartz, 1989; Sejnowski, Koch, & Churchland, 1988), has allowed some problems of information processing in the nervous system to be approached effectively and has suggested new ways for thinking about cognitive operations (Ballard, 1986; Grossberg & Kuperstein, 1986; Kohonen, 1984; Rumelhart & McClelland, 1986; Skarda & Freeman, 1987). These developments appear to favor genuine progress in generating theories that honor neurobiological and psychological constraints to explain how the brain actually works.

The Coevolutionary Research Strategy

With minimal caricature, one can describe two radically different research strategies for addressing complex phenomena such as perception, consciousness, and learning. The first is purely top-down, recommending that theories of cognition be determined solely from carefully garnered behavioral data and from computational insights (Fodor 1981; Fodor & Pylyshyn, 1988; Pylyshyn, 1980). The details of neuronal response properties and patterns of connectivity are viewed as irrelevant because the implementation of the computer that runs cognitive programs could have been designed in many different ways. The other strategy is purely bottom-up. It recommends that all of the details at the cellular level should be discovered first, and once these details are understood, the means whereby complex effects are produced will be more or less evident. The cold shoulder sometimes given to theory building in neuroscience was frequently justified by a version of this rationale. Fortunately, there has been a demonstrable softening of these two extremes in the past decade. The recognition that there are many levels of organization in the brain and that complex effects can emerge from the interaction of cells has underscored the importance of using the appropriate techniques for accessing different levels of organization. Accordingly, the preferred strategy is neither exclusively top-down nor exclusively bottom-up. Rather, it is a coevolutionary strategy, typified by the interaction between research domains, where research at one level provides constraints, corrections, and inspiration for research at other levels (Churchland, 1986).

Experimental psychology on both human subjects and animals is an essential part of the enterprise, for the obvious reason that accurate characterizations of psychological phenomena are necssary to guide the search for explanations and mechanisms. Trying to find a mechanism when the phenomenon is misdescribed or underdescribed is likely to be quixotic. Neurology is an essential part of the enterprise because it provides both important behavioral data on human subjects and hypothesizes connections between specific brain structures and behavior. Neuroscience is essential both to discover the functional capacities of neural components and because reverse engineering is an important strategy for figuring out how a novel device works. That is, when a new camera or chip appears on the market, competitors will take it apart to find out how it works. Typically, of course, they already know about devices of that general kind, so the problem can be manageable. Although we have to use reverse engineering to study the brain, our starting point is much further back, because we know so little about devices of that general kind. From our vantage point, the brain is a bit of alien technology, and hence it is especially difficult to know, among the facts available to us, which are theoretically important and which are theoretically uninteresting. Thus it looks productive to use a mixed-level strategy, which invites the coevolution of theory and experiment, of low-level theory and high-level theory, of realistic models that try to mimic the relevant range of neurobiological data and simplifying models that abstract general principles from a great deal of data (Sejnowski, Koch, & Churchland, 1988).

Levels

Concepts of levels are typically invoked in explaining what motivates different domains of research and how research domains are related to one another. However, there are in circulation at least three different notions of levels, each carving the landscape in a different way—levels of analysis, levels of organization, and processing levels. The first concerns the conceptual division of a given phenomenon in terms of different classes of questions that can be asked about it. The second refers to structural organization at different physical scales, ranging from molecules and neurons to circuits and systems. The third notion, that of processing levels, refers to an operation's "synaptic distance" from the sensory periphery, and serves as an index of its position in the information-processing sequence.

Levels of Analysis

A framework articulated by Marr and Poggio (1977) provided an important and influential background for thinking about levels in the context of computation by nervous structures.[1] This framework drew upon the conception of levels in computer science, and accordingly they identified three levels: (1) the *computational* level of abstract problem analysis, which decomposes the task into its main constituents (e.g., determining the three-dimensional structure of a moving object from successive views), (2) the level of the *algorithm*, which specifies a formal procedure to perform the task by providing the correct output for a given input; and (3) the level of physical *implementation*. An important element in Marr's (1982) view was that a higher level was largely independent of the levels below it, and hence that computational problems of the highest level could be analyzed independently of understanding the algorithm that performs the computation. Similarly, the algorithmic problem of the second level was thought to be solvable independently of understanding its physical implementation.

Unfortunately, two very different issues were confused in this doctrine of independence. One concerns whether, as a matter of discovery, one can ever figure out the relevant algorithm and the problem analysis independently of facts about implementation. The other concerns whether, as a matter of formal theory, a given algorithm that is already known to perform a task in a given machine (e.g., the brain) can be implemented in some other machine that has a different architecture (Churchland, 1985). So far as the latter is concerned, what computational theory tells us is that algorithms can be run on different machines, and in that sense and that sense alone, the algorithm is independent of the implementation. The formal point is straightforward: Because an algorithm is formal, no specific physical parameters (e.g., "vacuum tubes," "Ca^{++}") are part of the algorithm. That said, it is important to see that the purely formal point cannot speak to the issue of how best to *discover* the algorithm in fact used by a given machine, nor how best to arrive at the neurobiologically adequate task analysis. Certainly it cannot tell us that the discovery of the algorithms relevant to cognitive functions will be independent of a detailed understanding of the nervous system. Moreover, it does not tell us that any implementation is as good as any other. And it had better not, because different implementations display enormous differences in speed, size, efficiency, elegance, and so forth. The formal independence of algorithm from architecture is something we can exploit to build other machines once we know how the brain works, but it is no guide to discovery when we do not yet know how the brain works.

Current research suggests that considerations of implementation play a vital role in the kinds of algorithms we devise and in the kinds of computational insights available to us. Knowledge of brain architecture, so far from being irrelevant to the project, can be the essential basis and invaluable catalyst for devising likely and powerful algorithms—algorithms that have a reasonable shot at explaining how in fact the brain does the job.

Levels of Organization

How do the three levels of analysis map onto the nervous system? To begin with, there is organized structure at different scales: molecules, synapses, neurons, networks, layers, maps and systems (Churchland & Sejnowski, 1988; Crick, 1979; Grobstein, 1988; Shepherd, 1972) (see Figure 1.1). At *each* structurally specified stratum, we can raise the computational questions, What does that organization of elements do? What does it contribute to the wider, computational organization of the brain? In addition, there are physiological levels: ion movement, channel configurations, excitatory and inhibitory postsynaptic potentials, action potentials, evoked response potentials, and probably other intervening levels that we have yet to learn about and that involve effects at higher anatomical levels such as networks or systems. The range of structural organization implies, therefore, that there are many levels of implementation and that each has its companion task description. But if there are as many types of task descriptions as there are levels of structural organization, this diversity will probably be reflected in a multiplicity of *algorithms* that characterize how the tasks are accomplished. This, in turn, means that the notion of *the* algorithmic level is as oversimplified as is the notion of *the* implementation level. Structure at every scale in the nervous system—molecules, synapses, neurons, networks, layers, maps, and systems (Figure 1.1)—is separable conceptually but not detachable physically. What is picked out as a level is actually a boundary

Levels of Investigation

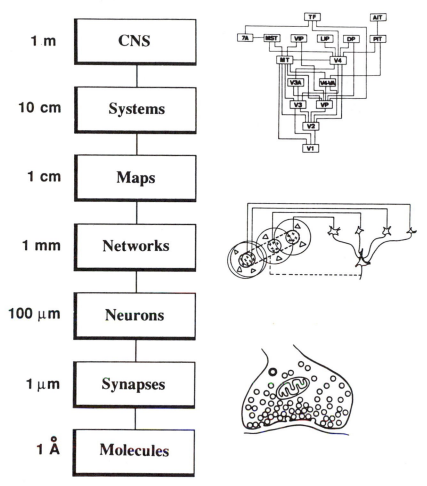

1 m	CNS
10 cm	Systems
1 cm	Maps
1 mm	Networks
100 μm	Neurons
1 μm	Synapses
1 Å	Molecules

Figure 1.1 Structural levels of organization in the nervous system. The spatial scale at which anatomical organizations can be identified varies over many orders of magnitude. Schematic diagrams to the right (top) illustrate a processing hierarchy of visual areas in monkey visual vortex (Maunsell & Newsome, 1987). (Center) A small network model for the synthesis of oriented receptive fields of simple cells in visual cortex (Hubel & Wiesel, 1962) and the structure of a chemical synapse (Kandel & Schwartz, 1984). Relatively little is known about the properties at the network level in comparison with the detailed knowledge we have of synapses and the general organization of pathways in sensory and motor systems.

imposed on the structure that depends on the techniques available to understand the phenomenon at hand. But in the brain, they are all part of one, integrated, unified biological machine.

It may be assumed that psychological phenomenon are the outcome of events at the highest level of organization, but this is not known. Some perceptual states, such as the "raw" pain of a toothache, might, for all we know now, be primarily a lower-level effect. But it may turn out that a particular aspect of attention depends on a variety of mechanisms, some of which can be found at the level of local neural networks, and others at the level of larger neural systems that reside in many different locations within the brain. Cognitive phenomena, such as planning and problem solving, may involve a complex interplay between several neural systems at many interacting levels.

Although Figure 1.1 specifies a set of structural levels, we must caution that this list is provisional and eminently revisable. Which structures constitute a theoretically significant level of organization in the nervous system, in the sense that its operations will figure in a more complete theory of how the brain works, is an empirical, not an a priori matter. We cannot tell, in advance of further study of the nervous system, how many levels there are, nor what is the nature of the structural and functional features of any given level. The count is imprecise for several reasons. Further research may lead to the subdivision of some categories, such as *systems*, into finer grained categories, and some categories may be profoundly misdrawn and may need to be completely reconfigured. As we come to understand more about the brain and how it works, new levels of organization may be postulated. This is especially likely at higher levels, where much less is known than at the lower levels.

Processing Levels

The third concept of levels seeks to specify where an operation is in an information-processing sequence in terms of the relative location of the structure carrying it out. Since signals are thought to undergo processing at every synaptic stage, the simple rule would be "the greater the distance from cells responding to sensory input, the higher the degree of information processing." Thus the level assigned is a function of synaptic distance from the periphery. On this measure, cells in the primary visual area of the neocortex that respond to oriented bars of light are at a higher level than are cells in the lateral geniculate nucleus (LGN), and these in turn are at higher levels than are retinal ganglion cells.

Once the sensory information reaches cerebral cortex, it fans out through corticocortical projections into a multitude of parallel streams of processing. In the primate visual system 24 visual areas have been identified (De Yoe & Van Essen, 1988). For every forward projection, there is a backprojection, and there are even massive feedback projections from primary visual cortex to the LGN. Given these reciprocal projections, it might seem that the processing levels do not really form a hierarchy, but there is a way to order the information flow by examining the layer of cortex into which fibers project. Forward projections always terminate in the middle layers of cortex, and feedback projections terminate in the upper and lower layers (Maunsell & Van Essen, 1983). However, we do not yet understand the function of these feedback pathways. If higher areas can affect the flow of information through lower areas, then the concept of sequential processing must be modified.

There are already indications that systems of strong reciprocal projections are found at the level of parietal and prefrontal cortex and that these areas of cortex are organized not in terms of a predominant hierarchy, but rather as a parallel distributed network of processing centers (Goldman-Rakic, 1988). Decisions to act and the execution of plans and choices could be the outcome of a system with distributed control rather than a single control center. Coming to grips with systems having distributed control will require both new experimental techniques and new conceptual advances. Perhaps more appropriate metaphors for this type of processing will emerge from studying models of interacting networks of neurons.

Color Vision: A Case Study

As an illustration of fruitful interactions between psychology and physiology on a problem in perception, we have chosen several examples from color vision. Similar examples can also be found in the areas of learning (see Kandel et al., 1987; Squire, 1987) and sensorimotor integration (see Wise & Desimone, 1988; Lisberger, 1988). Newton's ingenious prism experiment demonstrated that white light can be decomposed into a mixture of wavelengths and recombined to recover the white light. This physical description of color, however, did not satisfy artists, who were well aware that the perception of color involved complex spatial and temporal effects. As Goethe pointed out in *Zür Farbenlehre*, dark shadows often appear blue. The physical description of color and the psychological description of color perception are at two different levels: The link between them is at the heart of the problem of relating brain to cognition. Three examples will be given to illustrate how such links are being made in color vision.

The knowledge that mixtures of only three wavelengths of light are needed to match any color led Young to propose in 1802 that there are only three types of photoreceptors. Quite a different theory of color vision was later proposed by Hering, who suggested that color perception was based on a system of color opponents, one for yellow versus blue, one for red versus green, and a separate system for black versus white. Convincing experiments and impressive arguments were marshalled by supporters of these two rival theories for nearly a century. The debate was finally settled by physiological studies proving that both theories were right—in different parts of the brain. At the level of the retina three different types of color-sensitive photoreceptors were found as predicted by Young, and at the level of the thalamus and visual cortex there are neurons that respond to Hering's color opponents.[2] Evidently, even at this early stage of visual processing the complexity of the brain may lead to puzzles that can only be settled by knowing how the brain is constructed (Barlow, 1985).

The problem of color constancy is a more recent example. Red apples look red under a wide range of illumination even though the physical wavelengths impinging on the retina vary dramatically from daylight to interior lighting. Insights into color constancy have come from artists, who manipulate color contrasts in paintings; psychophysicists, who have quantified simultaneous contrast effects (Jameson & Hurvich, 1961); and theorists, who have modeled them (Hurlbert and Poggio, 1988; Land, 1986). Color constancy depends on being able to compute the intrinsic reflectance of a surface independently of the incident light. The reflectance of a patch of surface can be approximately computed by comparing the energy in wavelength bands coming from the patch of surface to the average energy in these bands from neighboring and distant regions of the visual field. The signature for a neuron that was performing this computation would be a long-range suppressive influence from regions of the visual field outside the classical receptive field. Neurons with such nonclassical color-selective surrounds have now been reported in visual cortex area V4 (Desimone, Schein, & Ungerleider, 1985; Zeki, 1983); nonclassical surrounds were first found for motion-selective cells in area MT by Allman, Miezen, & McGuiness (1985). If these neurons are necessary for color constancy, then their loss should result in impairments of color vision. Bilateral lesions of certain extrastriate visual areas in humans do produce achromatopsia—a total loss of color perception (Damasio, 1985)—though this condition is usually found together with other deficits and the damaged areas may not be homologous with area V4 in monkeys.

The third example comes from research on how form, motion, and color perception are processed in the visual system. If different parts of the system are specialized for different tasks (e.g., motion or color), then there should be conditions under which these specializations are revealed. Suppose that the color system is good at distinguishing colors, but not much else, and in particular, is poor at determining shape, depth, and motion, whereas the shape system is sensitive not to color differences but to brightness differences. When boundaries are marked only by color differences (all differences in brightness are experimentally removed) shape detection should be impaired. Psychophysical research has shown that this is indeed the case: The perceived motion of equiluminant contours is degraded (Ramachandran & Gregory, 1978), form cues such as shape-from-shading are difficult to interpret (Cavanagh & Leclerc, 1985), and perceived depth in random-dot stereograms collapses (Lu & Fender, 1972). Physiological and anatomical research has begun to uncover a possible explanation for these phenomena (Livingstone & Hubel, 1988). The separate processing streams in cerebral cortex mentioned earlier carry visual information about different properties of objects (De Yoe & Van Essen, 1988; Zeki, 1988). In particular, the predominant pathway for color information diverges from those carrying information on motion and depth (Figure 1.2). The separation is not perfect; however, but equiluminant stimuli are providing physiologists with a visual "scalpel" for tracking down the correlates of perceptual coherence in different visual areas.

The lessons learned from color perception may have general significance for studying other cognitive domains. So far as we know, only a small fraction of the neurons in the visual system appear to respond in a way that is similar to our perceptual report of color. The locations in the brain where links between physiological states and perceptual states can be found vary from the retina to deep in the visual system for different aspects of color perception (Teller & Pugh, 1983). New experimental techniques will be needed to study these links when the information is encoded in a large population of interacting neurons (Sejnowski, 1988; Wise & Desimone, 1988).

Techniques and Research Strategies

Color vision is a problem that has been studied for hundreds of years; we know much less about the biological basis of other perceptual and cognitive states. This will require recording from neurons in awake and behaving animals. However, this strategy cannot be used to study some cognitive abilities that are uniquely human, such as language processing. Fortunately, new techniques are becoming available for noninvasively measuring brain activity in humans. With these techniques the large-scale pattern of "what" is happening "where" and "when" in the brain can be determined; later, as techniques with higher resolution are developed, they can be focused on the relevant areas to ask "how" the processing is accomplished.

Regional blood flow in the brain can be used to monitor variable metabolic demands due to electrical activity either by using xenon-133 injected into the carotid artery and monitored with external radiation detectors or by noting changes in blood volume with positron emission tomography (PET) following the injection of oxygen-15 labeled water into the blood (Ingvar & Schwartz, 1974; Raichle, 1986; Roland, 1984a). One great advantage of these methods is that several different conditions can be studied in a single session because the clearance times and halflives are only several minutes. These techniques have been used to study voluntary motor activity (Fox, Fox, & Raichle, 1985) and selective attention to somatosensory stimuli (Roland, 1984b). PET recording offers significant opportunity to investigate the localization of higher functions, including language abilities in

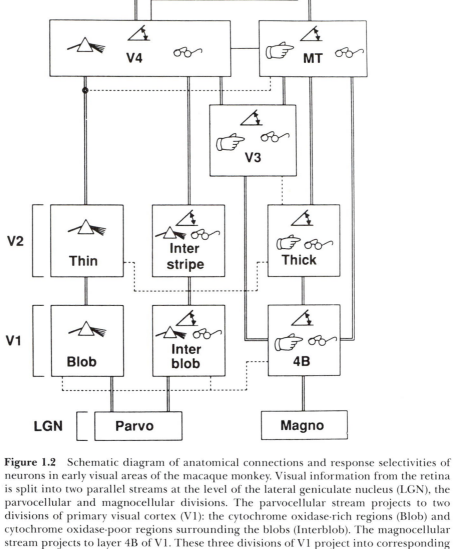

Figure 1.2   Schematic diagram of anatomical connections and response selectivities of neurons in early visual areas of the macaque monkey. Visual information from the retina is split into two parallel streams at the level of the lateral geniculate nucleus (LGN), the parvocellular and magnocellular divisions. The parvocellular stream projects to two divisions of primary visual cortex (V1): the cytochrome oxidase-rich regions (Blob) and cytochrome oxidase-poor regions surrounding the blobs (Interblob). The magnocellular stream projects to layer 4B of V1. These three divisions of V1 project into corresponding areas of V2: The Thin stripe, Interstripe, and Thick stripes of cytochrome oxidase-rich and oxidase-poor regions in V2. These areas in turn project to visual areas V3, V4 and MT (middle temporal area; also called V5). Heavy lines indicate robust primary connections, and thin lines indicate weaker, more variable connections. Dotted lines indicate connections that require additional verification. Not all projections from these areas to other brain areas are represented. The neurons in each visual area respond preferentially to particular properties of visual stimuli, as indicated by the icons: Tuned and/or opponent wavelength selectivity (Rainbow); Orientation selectivity (Angle symbol); Binocular disparity selectivity and/or strong binocular interactions (Spectacles); and Direction of motion selectivity (Pointing hand). (Courtesy of DeYoe & Van Essen, 1988.)

humans. For example, cognitive tasks such as reading single words and semantic tasks have been studied using a subtractive technique to localize individual mental operations (Petersen, Fox, Posner, Mintun, & Raichle, 1988; Posner, Petersen, Fox, & Raichle, 1988). The current spatial resolution of PET scanning is around 10 mm and the ultimate resolution, limited by the range of positrons, is estimated to be 2 to 3 mm. However, using an averaging technique that is applicable to point sources, it has been possible to map the visual field in the primary visual cortex of humans with a resolution of 1 mm (Fox, Miezing, Allman, Van Essen, & Raichle, 1987).

Magnetic resonance imaging (MRI) of hydrogen density has a much higher spatial resolution (about 0.1 mm in the plane of section, which is good enough to distinguish the line of Gennari, layer 4 of striate cortex). MRI is very useful for localizing lesions, tumors, and developmental abnormalities, but thus far is limited in not being able to assess dynamic brain activity. It is also possible to map other chemical elements in the brain using MRI—especially elements, such as sodium and phosphorus, whose concentrations vary with the functional state of the brain. However, the concentration of these chemical elements in living tissue is much less than that of hydrogen, and hence the signal-to-noise ratio and the resolution of the map using MRI is also much less. This will improve as higher magnetic fields are used, and it is likely that someday we will be able to image chemical reactions within intact brain tissue (Hsieh & Balaban, 1987; Luyten, 1990).

Another valuable source of information about brain function can be found by carefully studying humans with damage in localized regions of the brain. Despite limitations due to the variability of the damage from patient to patient and the difficulty of interpreting the abnormal function, valuable insights about the general organization of memory and perception have been gained from these studies. For example, bilateral lesions of the medial temporal lobe typically result in anterograde amnesia (Milner, 1966; Squire, Shimmamura, & Amaral, 1988) and lesions of the posterior parietal cortex have been implicated in loss of capacity to attend to the opposite side of the body and to the opposite hemispace (Mesulam, 1985). Some lesions produce surprisingly specific perceptual and linguistic deficits (Damasio, 1985; McCarthy & Warrington, 1988). The disconnection effects discovered in split-brain patients (Sperry & Gazzaniga, 1967) demonstrated fragmentation of experience and awareness, and confirmed the lateralization of certain functions, particularly speech production and spatioconstructive tasks, in some patients. Studies on split-brain patients can provide a unique source of information about the global organization of processing in the brain for perceptual and cognitive phenomena such as color constancy (Land, Hubel, Livingstone, Perry, & Burns, 1983) and mental imagery (Kosslyn, Holtzman, Gazzaniga, & Farrah, 1985).

Virtually everything we know about the microorganization of nervous systems has derived from work on animal brains, and such research is

Spatial and Temporal Resolution of Techniques for Studying Brain Function

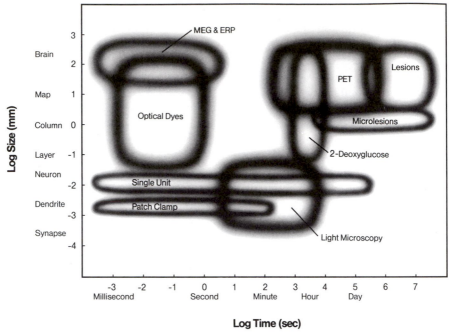

Figure 1.3 Schematic illustration of the ranges of spatial and temporal resolution of various experimental techniques for studying the function of the brain. The vertical axis represents the spatial extent of the technique, with the boundaries indicating the largest and smallest sizes of the region from which the technique can provide useful information. Thus, single-unit recording can provide information only from a small region of space, typically 10–15 μm on a side. The horizontal axis represents the minimum and maximum time interval over which information can be collected with the technique. Thus, action potentials from a single neuron can be recorded with millisecond accuracy over many hours. Patch recording allows the ionic currents through single ionic channels to be recorded. Optical dyes that are sensitive to membrane potential and ionic concentrations have been used with cellular resolution in tissue culture, where it is possible to obtain a clear view of single cells (see Smith, 1988). However, recordings from the central nervous system are limited in resolution by the optical properties of nervous tissue, and only about 0.1 mm resolution has been achieved (Blasdel & Salama, 1986). Confocal microscopy is a recent improvement of the light microscope that could be used for improving the resolution of the technique for three-dimensional specimens (Fine et al., 1988). ERP (Evoked Response Potential) and MEG (Magneto-encephalography) record the average electrical and magnetic activity over large brain regions and are limited to events that take place over about 1 (Hillyard & Picton, 1987; Williamson & Kauffman, 1988). The temporal resolution of PET (Positron Emission Tomography) depends on the lifetime of the isotope being used, which ranges from minutes to an hour. It may be possible to achieve a temporal resolution of seconds with ^{15}O to study fast changes in blood flow using temporal binning of the gamma ray events (equivalent to the poststimulus time histogram for action potentials) (Raichle, 1986). The 2-deoxyglucose technique has a time resolution of about 45 min and a spatial resolution of 0.1 mm with large pieces of tissue and 1μm with small pieces of tissue (Sokoloff, 1984). The 2-DG technique can also be applied to humans with PET (Phelps & Mazziotta, 1985). Lesions

14

absolutely indispensable if we are to have any hope of understanding the human brain. Of course there are limitations, inasmuch as there are non-trivial differences between the brains of different species, and we cannot blithely generalize from cat and monkey brains to a human brain. Even the problem of identifying homologous structures in different species can be vexing (Campbell & Hodos, 1970). Nevertheless, it may be that fundamental principles can be discovered in animal models, and that knowing these will provide the scaffolding for answering questions concerning those aspects of the human brain that make it unique.

Techniques for recording from populations of neurons are just now being developed. Multiple-electrode recording techniques are being used to search for coherent firing patterns in neural populations (Eckhorn et al., 1988; Llinas, 1985; Ts'o, Gilbert, & Wiesel, 1986). A promising approach for recording from large numbers of neurons simultaneously is the optical recording of electrical and ionic changes in the brain by direct observation. New optical dyes have been developed for noninvasive monitoring of changes in the membrane potential of neurons (Grinvald, 1985; Salzberg, Obaid, Senseman, & Gainer, 1983). This technique has recently been used to visualize the ocular dominance and orientation columns in visual cortex (Blasdel & Salama, 1986; see Figure 14). It also appears that small changes in the absorption of red light in visual cortex can be recorded, and that these changes are correlated with electrical responses of neurons even in the absence of dyes (Grinvald, Lieke, Frostig, Gilbert, & Wiesel, 1986). Ion-sensitive fluorescent dyes, such as the calcium-sensitive dye Fura-2, have also been developed that can monitor the change in intracellular ion concentration (Connor, Tseng, & Hockenberger, 1987; Smith, 1988; Tsien, 1988). These optical techniques could be used with confocal microscopy to produce three-dimensional images of physiological activity in vivo (Fine, Amos, Durbin, & McNaughton, 1988).

A useful way to get an overview of the assorted techniques is to graph them with respect to temporal and spatial resolution. This permits us to spot areas where techniques are not yet adequate to gain access to levels of organization at specific spatiotemporal resolutions, and to compare their strengths and weaknesses (see Figure 1.3). For example, it is apparent that we lack detailed information about processing in neural networks within

---

allow the interruption of function to be studied immediately following ablation of the tissue and for a long period of time following the ablation. Microlesion techniques make selective modifications using substances such as ibotenic acid, which destroys neurons but not fibers of passage, and 4-amino-phosphonobutyric acid, which selectively and reversibly block a class of glutamate receptors (Horton & Sherk, 1984; Schiller, 1982). Video-enhanced light microscopy has opened a window onto dynamical activity within neurons, such as the recent visualization of axonal transport of organelles on microtubules (Schnapp & Reese, 1986). All of the boundaries drawn here show rough regions of the spatio-temporal plane, where these techniques have been used, and are not meant to indicate fundamental limitations.

cortical layers and columns over a wide range of time scales, from milliseconds to hours. There is also a pressing need for experimental techniques designed to address the dendritic and synaptic levels of investigation in cerebral cortex. Without these data, it will not be possible to develop realistic models of information processing in cortical circuits. Another important source of insight into brain processing can come through the study of neural development (Changeux, 1986; Edelman, 1987).

Models

Although we need experimental data concerning the properties of neurons and behavioral data about psychological capacities, we also need to find models that explain how patterns of activity in neurons manage to represent surfaces, optical flow, and objects, and how networks learn, store, and retrieve information, how they accomplish sensorimotor integration, and so forth. Ideally, modeling and experimental research will have a symbiotic relationship, such that each informs, corrects, and inspires the other. To treat a mechanism as a black box is equivalent to deciding to ignore certain details of the mechanism in order to find the properties that play a prominent role in the function. Such simplifying and idealizing is essential to theorizing in science in general, although it is notoriously difficult to determine which properties one can safely ignore in constructing the theory of the function, and certainly no decision procedure exists for that problem.

We could treat the brain wholly as a black box, but the trouble with this strategy is that there are indefinitely many computational models one might dream up, and it might be that none are even close to how the brain in fact achieves solutions to difficult computational problems. The more profitable strategy will be to treat the brain as a lot of little black boxes. For example, one could take the individual neurons as a black box, thereby deciding not to worry for the nonce about such matters as types of single channels in neuronal membranes, the details of dendritic responsivity, the role of microtubules, and so on. One could choose a slightly larger size of black box—taking, say, cortical columns as units—where the details of connectivity within the column are ignored in the model, but intercolumn connectivity is analyzed and incorporated. Or, on an even larger scale, one could treat the visual areas as wholes, or the hippocampus or the cerebellar cortex as the level of organization one will try to address.

Although many diverse kinds of things are presented as models for some part of the nervous system, it is useful to distinguish between *realistic models*, which are genuinely and strongly predictive of some aspect of nervous system dynamics or anatomy, and *simplifying models*, which, though not so predictive, demonstrate that the nervous system could be governed by the principles specified in them. Connectionist network models (Arbib, 1987; Grossberg & Kuperstein, 1986; Hopfield & Tank, 1986; Kohonen, 1988;

Rumelhart & McClelland, 1986) are typically motivated by cognitive phenomena and are governed primarily by computational constraints; figuring
in the background are very general neurobiological constraints such as
number of processing units and time required to perform a task. Accordingly, these models are more properly considered demonstrations of what
could be possible, and sometimes of what is not possible. Realistic models of
real neural networks, by contrast, are primarily motivated by biological
constraints, such as the physiological and anatomical properties of specific
cell types (Koch & Segev, 1989; Sejnowski, Koch, & Churchland, 1988).
Despite their different origins and sources of dominant constraints, simplifying models and realistic neural models are both based on the mathematics of nonlinear dynamic systems in high-dimensional spaces (Abraham &
Shaw, 1982). The common conceptual and technical tools used in these
models should provide links between two rich sources of experimental data
and, consequently, connectionist and neural models have the potential to
coevolve toward an integrated, coherent account of information processing
in the mind–brain.

The ultimate goal of a unified account does not entail that it must be a
single model that will span all the levels of organization seen in nervous
systems, or that the highest levels will be explained directly in terms of
events at the molecular level. Instead, it is more probable that the integration will consist of a chain of theories and models linking adjacent levels.
The unifying connections will derive, therefore, from the chain of interlock-
ing theories in virtue of which phenomena at one level are explained in
terms of phenomena at a lower level, and those, in turn, by phenomena at
yet lower levels. Notice also that should one level be explained in terms of a
lower level, this does not mean that the higher level theory is no longer
useful, or that the phenomenon at that level no longer exist or that if they
do, they are no longer worth studying. On the contrary, such levels and the
theories pertinent to them will persist. As in chemistry and physics, genetics
and embryology, explanations coexist at all levels, from the molecular to the
systems levels.

Conclusions

It would be convenient if we could understand the nature of cognition
without understanding the nature of the brain itself. Unfortunately, it is
difficult, if not impossible, to theorize effectively on these matters in the
absence of neurobiological constraints. The primary reason is that computational space is consummately vast, and there are many conceivable solutions
to the problem of how a cognitive operation could be accomplished. Neurobiological data provide essential constraints on computational theories,
and they consequently provide an efficient means for narrowing the search
space. Equally important, the data are also richly suggestive in hints concern-

ing what might really be going on and what computational strategies evolution might have chanced upon. Moreover, it is by no means clear or settled yet what exactly are the functional categories at the congitive levels, and theories of lower level function may well be crucial to the discovery of the nature of higher level organization. Accordingly, despite the fact that the brain is experimentally demanding, neurobiology is indispensable in the task of discovering the theories that explain how we perform such activities as seeing, thinking, and being aware.

On the other hand, the possibility that cognition will be an open book once we understand the details of each and every neuron and its development, connectivity, response properties, and so forth, is likewise misconceived. Even if we could simulate, synapse for synapse, our entire nervous system, that accomplishment, by itself, would not be the same as understanding how it works. The simulation might be just as much of a mystery as the function of the brain currently is, for it might reveal nothing about the network and systems properties that hold the key to cognitive effects. Even simulations of small network models have capabilities that are difficult to understand (Andersen & Zipser, 1988; Edelman, 1987; Lehky & Sejnowski, 1988). Genuine theorizing about the nature of neurocomputation is therefore essential.

Both cognitive science and neuroscience share, at different levels, the goal of trying to understand how the mind–brain works. The social and institutional background for studying the relationship between mind and brain is rapidly evolving. Cross-disciplinary contact and collaboration are increasing; there are meetings on such topics as computational neuroscience and biological cognition; new journals for cognitive neuroscience, neural computation, and neural networks are being founded; departments of cognitive science and institutes for the study of the mind–brain are coming into existence. These developments are responses to the need for new ways to further cross-disciplinary research, and they are enormously important if the integrative program is to succeed. For it is not enough that researchers in one field can read the journals and attend the meetings of the other field. Rather, it will be through working together on common projects that the major breakthroughs are likely to come.

Acknowledgments

We are indebted to Dr. Francis Crick whose insights and critical judgments were a major resource in writing this article. We are also grateful to Drs. Paul Churchland, Richard Cone, and Donald MacLeod for helpful suggestions. TJS is supported by grants from the National Science Foundation, the Howard Hughes Medical Institute, the Office of Naval Research, and the Drown Foundation; PSC is supported by grants from the National Science Foundation and the James S. McDonnell Foundation.

Notes

1. The original conception of levels of analysis can be found in Marr and Poggio (1977). While Marr (1982) emphasized the importance of the computational level, the notion of a hierarchy of levels grew out of earlier work by Reichardt and Poggio (1976) on the visual control of orientation in the fly. In a sense, the current view on the interaction between levels is not so much a departure from the earlier views as it is a return to the practice that was previously established by Reichardt, Poggio, and even Marr himself, who published a series of papers on neural network models of the cerebellar cortex and cerebral cortex (see, for example, Marr, 1969, 1970). The emphasis on the computational level has nonetheless had an important influence on the problems and issues that concern the current generation of neural and connectionist models (Sejnowski, et al., 1988).

2. The possibility of a two-stage analysis for color vision was suggested as early as 1881, but no progress was made until physiological techniques became available for testing the hypothesis. There are still some issues that have not yet been fully resolved by this theory. The genes for the three cone photopigments have recently been sequenced (Nathans et al., 1986). The arrangement of the red and green pigment genes on the chromosome and unequal crossing over during homologous recombination can account for the incidence of various forms of color blindness (Vollrath, Nathans, & Davis, 1988).

References

Abraham, R. F., & Shaw, C. D. (1982). *Dynamics, the geometry of behavior*. Santa Cruz, CA: Aerial Press.

Allman, J., Miezin, F., & McGuiness, E. (1985). Stimulus specific response from beyond the classical receptive field: Neurophysiological mechanisms for local-global comparisons in visual neurons. In W. M. Cowan, E. M. Shooter, C. F. Stevens, & R. F. Thompson (Eds.), *Annual review of neuroscience* (pp. 407–430). Palo Alto, CA: Annual Reviews, Inc.

Andersen, R. A., & Zipser, D. (1988). The role of the posterior parietal cortex in coordinate transformations for visual-motor integration. *Canadian Journal of Physiology and Pharmacology, 66*(4), 488–501.

Arbib, M. (1987). *Brains, machines, and mathematics*. New York: Springer-Verlag.

Ballard, D. (1986). Cortical connections and parallel processing: Structure and function. *Behavioral and Brain Sciences, 9*, 67–120.

Barlow, H. B. (1985). The 12th Bartlett Memorial Lecture: The role of single neurons in the psychology of perception. *Quarterly Journal of Experimental Psychology, 37*, 121.

Blasdel, G., & Salama, G. (1986). Voltage-sensitive dyes reveal a modular organization of monkey striate cortex. *Nature, 321*, 579–585.

Brown, T. H., Chapman, P. F., Kairiss, E. W., & Keenan, C. L. (1988). Long-term synaptic potentiation. *Science, 242*, 724–728.

Campbell, C., & Hodes, W. (1970). The concept of homology and the evolution of the nervous system. *Brain, Behavior and Evolution, 3*, 353–367.

Cavanagh, P., & Leclerc, Y. (1985). Shadow constraints. *Investigative Ophthalmology, Supplement, 26*, 282.

Changeux, J. (1986). *Neuronal man*. Oxford, England: Oxford University Press.

Churchland, P. M. (1982). Is "Thinker" a natural kind? *Dialogue, 21*, 223.

Churchland, P. M. (1985). Conceptual progress and word-world relations: In search of the essence of natural kinds. *Canadian Journal of Philosophy, 15*, 1–17.

Churchland, P. S. (1986). *Neurophilosophy: Toward a unified science of the mind-brain*. Cambridge, MA: MIT Press.

Churchland, P. S., & Sejnowski, T. J. (1988). Neural representation and neural computation. In L. Nadel (Ed.), *Neural connections and mental computation*. Cambridge, MA:
MIT Press.

Crick, F. H. C. (1979). Thinking about the brain. *Scientific American, 240*, 219–232.

Damasio, A. (1985). Disorders of complex visual processing: Agnosias, achromotopsia,
Balint's syndrome, and related difficulties of orientation and construction. In M.-
M. Mesulam (Ed.), *Principles of behavioral neurology* (pp. 259–288). Philadelphia: F. A.
Davis Company.

Desimone, R., Schein, S. J., & Ungerleider, L. G. (1983). Contour, colour and shape
analysis beyond the striate cortex. *Vision Research, 25*, 441.

DeYoe, E. A., & Van Essen, D. C. (1988). Concurrent processing streams in monkey visual
cortex. *Trends in Neurosciences, 11*, 219–226.

Dodd, J., & Jessel, T. M. (1988). Axon guidance and the patterning of neuronal projections
in vertebrates. *Science, 242*, 692–699.

Edelman, G. M. (1987). *Neural Darwinism*. New York: Basic Books.

Fine, A., Amos, W. B., Durbin, R. M., & McNaughton, P. A. (1988). Confocal microscopy:
Applications in neurobiology. *Trends in Neurosciences, 11*, 346–351.

Goldman-Rakic, P. S. (1988). Topography of cognition: Parallel distributed networks in
primate association cortex. *Annual Review of Neuroscience, 11*, 137–156.

Grinvald, A., Lieke, E., Frostig, R., Gilbert, C., & Wiesel, T. (1986). Functional architecture
of cortex revealed by optical imaging of intrinsic signals. *Nature, 324*, 361–364.

Grobstein, P. (1988). Between the retinotectal projection and directed movement: Topography of a sensorimotor interface. *Brain Behavior and Evolution, 31* (1), 34–48.

Grossberg, S., & Kuperstein, M. (1986). *Neural dynamics of adaptive sensory-motor control*.
Amsterdam: North Holland.

Harrelson, A. L., & Goodman, C. S. (1988). Growth cone guidance in insects: Fasciclin II is
a member of the immunoglobulin superfamily. *Science, 242*, 700–708.

Hering, E. (1878). *Zur Lehre vom Lichtsinn* (Berlin).

Hillyard, S., & Picton, T. (1987). Electrophysiology of cognition. In F. Plum (Ed.), *Handbook
of physiology section 1: Neurophysiology* (pp. 519–584). New York: American Physiological Society.

Hopfield, J. J., & Tank, D. W. (1986). Computing with neural circuits: A model. *Science, 233*,
625–633.

Horton, J., & Sherk, H. (1984). Receptive field properties in the cat's lateral geniculate
nucleus in the absence of on-center retinal input. *Journal of Neuroscience, 4*, 374–380.

Hubel, D., & Wiesel, T. (1962). Receptive fields, binocular interaction and functional
architecture in the cat's visual cortex. *Journal of Physiology (London), 160*, 106–154.

Hurlbert, A. C., & Poggio, T. A. (1988). Synthesizing a color algorithm from examples.
Science, 239, 482–485.

Jameson, D., & Hurvich, L. M. (1959). Perceived color and its dependence on focal,
surrounding, and preceding stimulus variables. *Journal of the Optical Society of
America, 49*, 890–898.

Kandel, E. R., Klein, M., Hochner, B., Shuster, M., Siegelbaum, S. A., Hawkins, R. D.,
Glanzman, D. L., & Castellucci, V. F. (1987). Synaptic modulation and learning:
New insights into synaptic transmission from the study of behavior. In G. M.
Edelman, W. E. Gall, & W. M. Cowan (Eds.), *Synaptic function* (pp. 471–518). New
York: Wiley.

Kandel, E., & Schwartz, J. (1984). *Principles of neural science* (2nd ed.). New York: Elsevier.

Koch, C., & Segev, I. (1989). *Methods in neuronal modeling: From synapse to networks*.
Cambridge, MA: MIT Press.

Kohonen, T. (1984). *Self-organization and associative memory*. New York: Springer-Verlag.

Koob, G. F., & Bloom, F. E. (1988). Cellular and molecular mechanisms of drug dependence. *Science, 242*, 715–723.

Kosslyn, S., Holtzman, J., Gazzaniga, M., & Farrah, M. (1985). A computational analysis of mental imagery generation: Evidence for functional dissociation in split brain patients. *Journal of Experimental Psychology, General, 114,* 311–341.

Kosslyn, S. M. (1988). Aspects of a cognitive neuroscience of mental imagery. *Science, 240,* 1621–1626.

Land, E. H. (1986). An alternative technique for the computation of the designator in the retinex theory of color vision. *Proceedings of the National Academy of Sciences (U.S.A.), 83,* 3078–3080.

LeDoux, J., & Hirst, W. (1986). *Mind and brain: Dialogues in cognitive neuroscience.* Cambridge, England: Cambridge University Press.

Lehky, S. R., & Sejnowski, T. J. (1988). Network model of shape-from-shading: Neural function arises from both receptive and projective fields. *Nature, 333,* 452–454.

Lisberger, S. G. (1988). The neural basis for learning of simple motor skills. *Science, 242,* 728–735.

Livingstone, M. S., & Hubel, D. H. (1988). Segregation of form, color, movement, and depth: Anatomy, physiology, and perception. *Science, 240,* 740–749.

Llinas, R. (1985). Electronic transmission in the mammalian central nervous system. In M. Bennett & D. Spray (Eds.), *Gap junctions* (pp. 337–353). Cold Spring Harbor, NY: Cold Spring Harbor Publishing.

Lu, C., & Fender, D. H. (1972). The interaction of color and luminance in stereoscopic vision. *Investigative Ophthalmology, 11,* 482–490.

Luyten, P. (1990). Humans under the spectrometer. *Physics World,* September, pp. 36–40.

Marr, D. (1969). A theory of cerebellar cortex. *Journal of Physiology (London), 202,* 437–470.

Marr, D. (1970). A theory for cerebral neocortex. *Proceedings of the Royal Society, London, B 176,* 161–234.

Marr, D. (1982). *Vision.* San Francisco: Freeman.

Marr, D., & Poggio, T. (1977). From understanding computation to understanding neural circuitry. *Neurosciences Research Progress Bulletin, 15,* 470–488.

Maunsell, J., & Newsome, W. (1987). Visual processing in monkey extrastriate cortex. *Annual Review of Neuroscience, 10,* 363–401.

Maunsell, J., & Van Essen, D. (1983). The connections of the middle temporal visual area (MT) and their relationship to a cortical hierarchy in macaque monkey. *Journal of Neuroscience, 3,* 2563–2586.

McCarthy, R. A., & Warrington, E. K. (1988). Evidence for modality-specific meaning systems in the brain. *Nature, 334,* 428–430.

Mesulam, M., & Mesulam, M. (1985). Attention, confusional states, and neglect. In M. Mesulam (Ed.), *Principles of behavioral neurobiology.* Philadelphia: F. A. Davis Company.

Milner, B., (1966). In C. Whitty & O. Zangwill, (Eds.), *Amnesia following operation on the temporal lobes,* (pp. 109–133). London: Butterworth.

Mishkin, M., & Appenzeller, T. (1987). The anatomy of memory. *Scientific American, 256,* 80–89.

Mountcastle, V. B. (1986). The neural mechanisms of cognitive functions can now be studied directly. *Trends in Neurosciences, 9,* 505–508.

Nathans, J., Piantanida, T. P., Eddy, R. L., Shows, T. B., & Hogness, D. S. (1986). Molecular genetics of inherited variation in human color vision. *Science, 232,* 203–210.

Phelps, M., & Mazziotta, J. (1985). Positron emission tomography: Human brain function and biochemistry. *Science, 228,* 799–809.

Posner, M. I., Petersen, S. E., Fox, P. T., & Raichle, M. E. (1988). Localization of cognitive operations in the human brain. *Science, 240,* 1627–1631.

Pylyshyn, Z. (1980). Cognition and computation: Issues in the foundations of cognitive science. *Behavioral and Brain Sciences, 3,* 111–132.

Raichle, M. (1986). Neuroimaging. *Trends in Neurosciences, 9,* 525–529.

Ramachandran, V. S., & Gregory, R. L. (1978). Does color provide an input to human motion perception? *Nature, 275,* 55–56.

Reichardt, W., & Poggio, T. (1976). Visual control of orientation behavior in the fly. *Quarterly Reviews of Biophysics, 9,* 311–375.

Roland, P. (1984a). Organization of motor control by the normal human brain. *Human Neurobiology, 2,* 205–216.

Roland, P. (1984b). Somatotopic tuning of postcentral gyrus during focal attention in man. *Journal of Neurophysiology, 46,* 744–754.

Rumelhart, D. E., & McClelland, J. L. (1986). *Parallel distributed processing: Explorations in the microstructure of cognition. Vol. 1: Foundations.* Cambridge, MA: MIT Press.

Sacks, O., Wasserman, R. L., Zeki, S., & Siegel, R. M. (1988). Sudden color-blindness of cerebral origin. *Society of Neuroscience, Abstract, 14,* 1251.

Salzberg, B., Obaid, A., Senseman, D., & Gainer, H. (1983). Optical recording of action potentials from vertebrate nerve terminals using potentiometric probes provides evidence for sodium and calcium components. *Nature, 306,* 36–40.

Schiller, P. (1982). The central connections of the retinal on and off pathways. *Nature, 297,* 580–583.

Schnapp, B. J., & Reese, T. S. (1986). New developments in understanding rapid axonal transport. *Trends in Neurosciences, 9,* 155–162.

Sejnowski, T. J. (1988). Neural populations revealed. *Nature, 332,* 308.

Sejnowski, T. J., Koch, C., & Churchland, P. S. (1988). Computational neuroscience. *Science, 241,* 1299–1306.

Shepherd, G. M. (1972). The neuron doctrine: A revision of functional concepts. *Yale Journal of Biological Medicine, 45,* 584–599.

Skarda, C. A., & Freeman, W. J. (1990). Chaos and the new science of the brain. *Concepts in Neuroscience, 1,* 275–286.

Smith, S. J. (1988). Neuronal cytomechanics: The actin-based motility of growth cones. *Science, 242,* 708–715.

Sokoloff, L. (1984). *Metabolic probes of central nervous system activity in experimental animals and man.* Sunderland, MA: Sinauer Associates.

Sperry, R., Gazzaniga, M., Millikan, C., & Darley, F. (1967). Language following surgical disconnection of the hemispheres. In C. H. Millikan & F. L. Darley (Eds.), *Brain mechanisms underlying speech and language* (pp. 108–115). New York: Grune & Stratton.

Squire, L. (1987). *Memory and brain.* Oxford, England: Oxford University Press.

Squire, L., Shimamura, A., & Amaral, D. (1988). Memory and the hippocampus. In J. Byrne & W. Berry (Eds.), *Neural models of plasticity.* New York: Academic Press.

Teller, D. Y., & Pugh, E. N., Jr. (1983). Linking propositions in color vision. In J. D. Mollon & L. T. Sharpe (Ed.), *Color vision: Physiology and psychophysics* (pp. 577–589). New York: Academic Press.

Tsien, R. Y. (1988). Fluorescence measurement and photochemical manipulation of cytosolic free calcium. *Trends in Neurosciences, 11,* 419–424.

Ts'o, D., Gilbert, C., & Wiesel, T. (1986). Relationship between horizontal interactions and functional architecture in cat striate cortex as revealed by cross-correlation analysis. *Journal of Neuroscience, 6,* 1160–1170.

Vesalius, A. (1543). *De Humani Corporis.* Brussels.

Vollrath, D., Nathans, J., & Davis, R. W. (1988). Tandem array of human visual pigment genes at χq28. *Science, 240,* 1669–1672.

Williamson, S., Romani, G., Kaufman, L., & Modena, I. (1983). *Biomagnetism: An interdisciplinary approach.* New York: Plenum Press.

Williamson, S., & Kaufman, L. (1988). Neuromagnetism: A window on the brain. *ELAN (Japan).*

Wise, S. P., & Desimone, R. (1988). Behavioral neurophysiology: Insights into seeing and grasping. *Science, 242,* 736–741.

Young, T. (1802). On the theory of light and colours. *Philosophical Transactions of the Royal Society, London, 92,* 12–48.

Zeki, S. (1983). Colour coding in the cerebral cortex: The reaction of cells in monkey visual cortex to wavelengths and colours. *Neuroscience, 9,* 741–765.

Zeki, S., & Shipp, S. (1988). The functional logic of cortical connections. *Nature, 335,* 311–317.

2

Brain Substrates of Basic Associative Learning and Memory

RICHARD F. THOMPSON / MARK A. GLUCK

Recently there has been a great surge of interest across the disciplines of cognitive psychology, computer science, and neurobiology in understanding the information-processing capabilities of "adaptive networks" consisting of interconnected, neuronlike computing elements. These connectionist and parallel-distributed processing models are notable for their computational power and resemblance to psychological and neurobiological processes (e.g., Ackley, Hinton, & Sejnowski, 1985; Gluck & Bower, 1988a, 1988b; Hinton & Anderson, 1981; Rosenberg & Sejnowski, 1986; Rumelhart & McClelland, 1986). In essence, this represents a move toward understanding complex human learning and memory phenomena as emerging from complex configurations of the elementary associative processes of classical or Pavlovian conditioning that formed the core of classical learning theory in psychology. The basic processes and phenomena of classical conditioning appear to have reemerged as critical for understanding the more complex phenomena of learning and memory (Rescorla, 1988a, 1988b).

To date, the most productive research strategy for identifying and analyzing the neuronal substrates of behavior has been the model system approach: selecting a behavioral paradigm and organism exhibiting the behavioral phenomena at issue where sufficient experimental control can be exerted and where the essential neuronal circuits can to some degree be identified and analyzed. In the context of brain substrates of learning and memory, classical or Pavlovian conditioning has proved to be the Rosetta stone for both invertebrate and vertebrate preparations (Kandel, 1976; Thompson et al., 1976).

Classical Conditioning

Pavlovian (or classical) conditioning is most generally defined as a procedure by which an experimenter presents subjects with stimuli that occur in some prearranged relationship and measures changes in response to one of them. Typically, the experimenter arranges for one of the stimuli, the conditioned stimulus (CS), to be relatively neutral, and for a second stim-

24

ulus, the unconditioned stimulus (US), to reliably elicit a readily measured response. (The fact that the CS and US occur independently of the subjects' behavior is a defining feature of the procedure.) Changes in subjects' behavior in response to the CS over the course of training are said to reflect associative learning when it can be shown that the change is due only to the relationship between the CS and US, as opposed to habituation or sensitization processes produced by mere exposure to the two events—for example, when the correlation between CS and US occurrence is zero (e.g., Rescorla, 1968; Rescorla & Wagner, 1972; Wagner, 1969). When the CS and US are arranged to occur such that the CS onset shortly precedes US onset, the CS comes to elicit conditioned responses (CRs) that, in many instances, mimic the unconditioned response (UR) to the US (see Donegan & Wagner, 1987; Mackintosh, 1983). For example, when the CS is a tone and the US is food in the mouth, the CS comes to elicit salivation (Pavlov, 1927); when the US is an airpuff to the eye, the CS comes to elicit eyelid closure (Gormezano, 1972). However, there are instances where the CR differs from the UR. In heart-rate conditioning in the rabbit, the shock US elicits an increase in heart rate (the UR), but the response that is learned to a tone or light CS paired with the shock US is a *decrease* in heart rate. In conditioned response suppression, where an animal's ongoing behavior (lever pressing, licking) is used as the behavioral index, a shock US causes increases in behavioral responding. However, after classical conditioning training with a tone or light CS paired with shock, the CS causes suppression of responding in proportion to the strength of the association. Such conditioned behaviors, whether they mimic the UR or not, are said to reflect the development of excitatory association between subjects' representations of the CS and US or the CS and response to the US. In biological terms, CRs can be described as adaptive responses—responses that are well suited to cope with the US.

Some degree of consistent temporal relationship between the occurrence of the CS and the US—that is, some degree of contiguity—is necessary for excitatory conditioning to occur. The CS must precede the US in time, but there is no single optimal interval. Depending on the behavioral paradigm, the optimal interval can range from several hundred milliseconds (eyelid conditioning) to many seconds (heart-rate conditioning) to many minutes (taste aversion learning). Interestingly, the function relating degree of learning to CS–US onset interval has the same form in all these paradigms, rising rapidly to an optimum and then decaying more slowly as the interval is increased (Rescorla, 1988a). Regardless of the value of the optimal interval, in all these situations the organism learns a "causal" relationship between occurrence of the CS and the US.

Although contiguity between CS and US is necessary for conditioning, it is not sufficient, as Rescorla (1968) showed some years ago in classic studies. If two groups of animals are given the same number of paired CS–US trials but one group is given US-alone presentations as well, learning will be poorer. Indeed, learning develops in proportion to the degree of contingency between CS and US. Contingency is a much more sophisticated computation of causal relationships than is simple contiguity. A powerful implication of this discovery is that organisms will also learn the absence of a relationship between CS and US. Explicitly unpaired presentations of CS and US prior to paired training result in slower learning than is the case if the CS and US are first presented at random intervals. Such findings led to the notion of inhibition; the CS is said to have acquired inhibitory properties (Rescorla & Holland, 1982). Indeed, Pavlov (1927) first demonstrated conditioned inhibition: If CS_1, is paired with the US and $CS_1 + CS_2$ trials not paired with the US are also given, conditioned responding to CS_1 is much stronger than conditioned responding to $CS_1 + CS_2$. CS_2 has acquired inhibitory properties. Much current research in Pavlovian conditioning focuses on the ways in which temporal, logical, and qualitative relationships among stimuli influence conditioning (see Rescorla, 1988b).

Pavlovian conditioning is a very powerful tool for behavioral and neurobiological analysis of mechanisms of learning and memory. Thus, the occurrence of the CS and US is determined by the experimenter, not by the subjects' behavior. Of particular importance is the fact that the conditioned response is time-locked to the CS. Therefore, neural events can be analyzed relative to known temporal referents. This feature is a great advantage when trying to detect correlations between changes in neural events, say through electrophysiological recording, and changes in behavior.

In addition, the effects of experimental manipulations on learning rather than performance can be more easily evaluated than in instrumental procedures. The problem of learning versus performance has plagued the study of brain substrates of learning from the beginning. For example, does a brain lesion or the administration of a pharmacological agent impair a learned behavior because it affects the memory trace or because it alters the animal's ability or motivation to respond? By using Pavlovian procedures in preparations in which the CR mimics the UR, one can estimate the relative effects of such manipulation on learning and performance by comparing the subject's ability to generate the CR and UR before and after making a lesion or administering a drug.

In more general terms, Pavlovian conditioning may be the fundamental process by which organisms learn about the causal textures of their worlds. In Rescorla's words (1988b):

[Many textbook descriptions] of conditioning . . . come from a long and honorable tradition in physiology, the reflex tradition in which Pavlov worked and within which many early behaviorists thought. This tradition sees conditioning as a kind of low-level mechanical process in which the control over a response is passed from one stimulus to another. Much modern thinking about conditioning instead derives largely from the associative tradition originating in philosophy. It sees conditioning as the learning that results from exposure to relations among events in the environment. Such learning is a primary means by which the organism represents the structure of its world. Consequently, Pavlovian conditioning must have considerable richness, both in the relations it represents and in the ways its representation influences behavior, a richness that was not envisioned within the reflex tradition. (p. 152)

There has been a considerable resurgence of interest in Pavlovian conditioning over the past twenty years, due in large part to the key role it has played in both neurobiological and cognitive approaches to mechanisms of learning and memory. It is no accident that the analysis of neurobiological mechanisms of learning and memory in both invertebrate and vertebrate nervous systems has made its greatest progress using Pavlovian conditioning; the advantages have already been noted (see Kandel, 1976; Thompson et al., 1976; and below). In cognitive science, there is much current interest in parallel-distributed processing or connectionist models. To quote Rescorla again (1988b):

Connectionistic theories . . . bear an obvious resemblance to theories of Pavlovian conditioning. Both view the organism as using multiple associations to build an overall representation, and both view the organism as adjusting its representation to bring it into line with the world, striving to reduce any discrepancies. Indeed, it is striking that often such complex models are built on elements that are tied quite closely to Pavlovian associations. For instance, one of the learning principles most frequently adopted within these models, the so-called delta rule, is virtually identical to one popular theory in Pavlovian conditioning, the Rescorla–Wagner model. Both are error-correction rules, in which the animal uses evidence from all available stimuli and adjusts the strength of each stimulus based on the total error. Here, then, is a striking point of contact between Pavlovian conditioning and a portion of cognitive science. (p. 159)

Memory Traces and Memory Trace Circuits

The problem of localizing the neuronal substrates of learning and memory, first explored in depth by Lashley (1929) and later by Hebb (1949), has been the greatest barrier to progress and remains fundamental to all work on the biological basis of learning and memory. Mechanisms of memory storage cannot be analyzed until they have been localized. It is useful to distinguish between the neural circuitry essential for the development and expression of a particular form of learning—the essential (necessary and sufficient) memory trace circuit—and the subset of neural elements in this circuit that exhibit the training-induced plasticity necessary for the development of such behavior—the memory trace itself. The premise that memory traces are localized does not necessarily imply that a particular trace has a single anatomical location. Instead, the memory trace circuit might involve a number of loci, parallel circuits, and feedback loops. It may be said that for a given form of learning, there is a discrete set of loci whose neuronal elements exhibit the essential neuronal plasticity defining the memory trace. Evidence detailed here strongly supports this view.

To date, the focus of research on mammalian brain substrates of classical conditioning has been on the necessary first step: identification of the essential memory trace circuits, using a wide range of techniques—lesions, electrical microstimulation, anatomical pathway tracing, electrophysiological recording, microinfusion of pharmacological agents, 2-deoxyglucose, and so on—to establish the neuroanatomy of conditioned responses (see Thompson, Donegan, & Lavond, 1988). Once this has been done, the next step is to localize the sites of synaptic plasticity—the memory traces. Electrophysiological recording of the patterns of neural unit activity in each synaptic relay structure in the memory trace circuit over the course of learning is a powerful tool for localization of sites of memory storage.

When this electrophysiological analysis has been completed for each major synaptic relay in the essential memory trace circuit, the *functional* neuroanatomy of the conditioned response circuit will have been characterized and the sites of plasticity in all likelihood localized. The cellular-synaptic mechanisms of memory trace formation can then be analyzed. Once this reductionistic odyssey has been completed, it is necessary to show that the identified essential memory trace circuit and the mechanisms of synaptic plasticity embedded within it can in fact generate all the emergent behavioral phenomena of learning and memory, in this case the phenomena of classical conditioning. This can only be done by developing realistic mathematical-computational models of the identified circuits. At this point it is not merely interesting to develop computational models of the empirically identified and characterized neural memory networks and circuits, it is mandatory to do so; verbal-logical analysis at a quantitative level of the functioning of complex neural circuits is impossible (Gluck & Thompson, 1987).

Identification and analysis of essential memory trace circuits for certain forms of classical conditioning are rapidly approaching the point where computational modeling becomes essential. It would seem that classical conditioning will form the natural bridge between empirical neurobiological analysis of memory trace circuits and computational modeling of learning networks. The fact that the computational models will be of real neural networks rather than "neural-like" networks is perhaps novel and may well add an important new dimension to the development of computational modeling as well as to our understanding of the operations of real neural circuits in the brain (Donegan, Gluck, & Thompson, 1989).

The greatest progress to date in the identification and characterization of learning and memory circuits in the mammalian brain has been made in two classical conditioning paradigms: (1) conditioned fear and an autonomic index of fear, conditioned heart rate, and (2) conditioning of discrete behavioral responses (e.g., eyelid closure, limb flexion). We present brief overviews of these two learning and memory neural networks here.

The Neural Circuit for Conditioned Fear

Michael Davis and his colleagues at Yale have made impressive progress in identifying the essential memory trace circuit for conditioned fear (Davis, 1986). They use potentiation of the acoustic startle response in rats as the basic paradigm. The startle response in rats is elicited by a sudden, loud sound and can be measured easily as body movement ("jump"). The startle response exhibits habituation with repeated presentations of the same startle stimulus, and sensitization if another strong stimulus (e.g., loud continuous background white noise) is added (Davis, Gendelman, Tischler, & Gendelman, 1982). In a classic study, Brown, Kalish, and Farber (1951) utilized the startle response to study the conditioned emotional response (CER) or learned fear. They paired light and inescapable shock in an environment other than the startle apparatus—classically conditioned fear

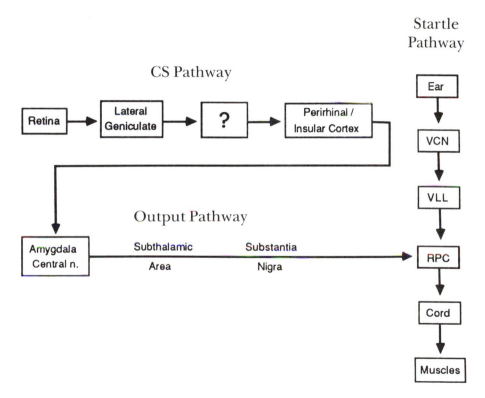

Figure 2.1   Schematic of the essential memory trace circuit hypothesized by Davis and associates for classical conditioning of potentiation of startle (fear). The conditioned fear circuit for a visual CS acts on the acoustic startle circuit at the level of the reticular formation (see text for details).
Abbreviations: RPC, nucleus reticularis pontis caudalis; VCN, ventral cochlear nucleus; VLL, ventral nucleus of the lateral lemniscus. (Courtesy of Michael Davis, Janice Hitchcock, and Jeffrey Rosen, Department of Psychiatry, Yale University.)

to the light CS. They then compared startle amplitude to an acoustic stimulus either alone or preceded by and paired with the light fear CS. Startle amplitude was significantly greater when the light CS was present, the CER.

Davis and colleagues first identified the essential startle circuit. The circuit consists of ventral cochlear nucleus, nuclei of the dorsal and ventral lateral lemniscus, (auditory relay nuclei), nucleus reticularis pontis caudalis (RPC—an efferent relay nucleus to spinal cord), and spinal cord (Davis et al., 1982). They then utilized the Brown et al. CER paradigm, and, using many of the techniques noted earlier, have identified many of the relays in essential CER circuit using a visual CS: lateral geniculate body, insular region of cerebral cortex, central nucleus of the amygdala and nuclei of the lateral lemniscus (Figure 2.1). The CER memory trace itself has not yet been localized, but the insular cortex and the amygdala are candidates.

Conditioned change in heart rate is generally viewed as an autonomic index of conditioned fear. The essential circuit for conditioned tachycardia to a visual CS in pigeons, where it has been most fully identified, by Cohen and associates (Cohen, 1980), is closely analogous to the CER circuit. Although less fully explored, work to date on conditioned heart rate in mammals is consistent (Kapp, Gallagher, Applegate, & Frysinger, 1982; Schneiderman et al., 1987). In particular, the central nucleus of the amygdala is critical. Electrophysiological recordings in the pigeon-heart-rate paradigm (Cohen, 1980) indicate that learning-related changes in neuronal activity to the visual CS do not occur in retina or in optic nerve fibers but are present in the visual thalamus and in later relays in the circuit. Cohen suggests that the memory traces may be distributed over several synaptic relays in the circuit, beginning at the thalamic level.

The Neural Circuit for Classical Conditioning of Discrete Behavioral Responses

Our work overwhelmingly argues that the cerebellum and its associated brain stem circuitry is essential—both necessary and sufficient—for classical conditioning of eyelid closure, limb flexion, and other discrete responses learned to deal with aversive stimuli (Thompson, 1986, 1988). Using the range of techniques already noted, we have shown that a very localized region of the cerebellar interpositus nucleus ipsilateral to the trained eye (eyelid conditioning) is critical. Thus, small electrolytic lesions of this region completely and permanently abolish the ipsilateral eyelid CR, have no effect on the reflex eyelid closure to the corneal airpuff US, and do not impair learning by the contralateral eye. If the lesion is made before training, the ipsilateral eye is unable to learn (Lincoln, McCormick, & Thompson, 1982; McCormick et al., 1981; McCormick & Thompson, 1984a). Kainic acid lesions of the interpositus nucleus, which spare fibers of passage, abolish the conditioned response with no accompanying degeneration in the inferior olive (Lavond, Hembre, & Thompson, 1985). Although most of our work has used the conditioned eyelid response, our evidence indicates that the cerebellar interpositus nucleus is necessary for the learning of any discrete behavioral response. Thus, the region of the interpositus nucleus essential for conditioning of the hind limb flexion reflex is medial to the region essential for the learned eyelid response (Donegan, Lowry, & Thompson, 1983). It appears that a complete set of motor "programs" are present and hardwired from interpositus to motor nuclei (see below), and are represented in a somatotopic organization consistent with the patterns of projection of somatic sensory information to the nucleus. Over the course of learning, the region of the interpositus appropriate for the particular unconditioned stimulus used (e.g., corneal airpuff, paw shock) becomes activated by the conditioned stimulus (see Figure 2.2).

An article recently appeared in the *Journal of Physiology* (London) in which our basic cerebellar lesion studies in rabbit have been replicated exactly in a human patient (Lye, O'Boyle, Ramsden, & Schady, 1988). In brief, the patient suffered a right cerebellar hemisphere infarction; brain scans verified that damage was limited to the right hemisphere. In essence, the patient's right eye was unable to learn the conditioned eyeblink response with repeated training sessions to each eye, but the left eye learned rapidly and robustly. Reflex responses of both eyes and sensory sensitivity to US and CS were normal. It is very gratifying to see that results of research on an animal model of basic associative learning and memory apply exactly to the human condition. We can now feel considerable confidence that the basic mechanisms of memory storage we elucidate in the rabbit eyelid preparation will apply in detail to the human brain.

Electrophysiological analysis shows several localized regions of cerebellar cortex and the lateral interpositus nucleus where neurons, including identified Purkinje cells, develop patterned changes in discharge frequency that precede and predict the occurrence and form of the learned response within trials and predict the development of learning over training trials (see Figure 2.3 and McCormick & Thompson, 1984b).

The essential efferent CR pathway consists of fibers that exit the interpositus nucleus ipsilateral to the trained side of the body in the superior cerebellar peduncle, cross to relay in the contralateral magnocellular division of the red nucleus, and then cross back to descend in the rubral pathway to act ultimately on motor neurons (Thompson, 1986; see also Figure 2.2). We do not know whether other efferent systems are involved in controlling the CR, although descending systems originating rostral to the midbrain are not required for learning or retention of the CR (Mauk & Thompson, 1987; Norman, Buchwald, & Villablanca, 1977).

Our recent research using lesion and microstimulation indicates that the essential reinforcement pathway for the US is comprised of climbing fibers from the dorsal accessory olive (DAO) portion of the inferior olive, activated from the trigeminal nucleus, that project through the inferior cerebellar

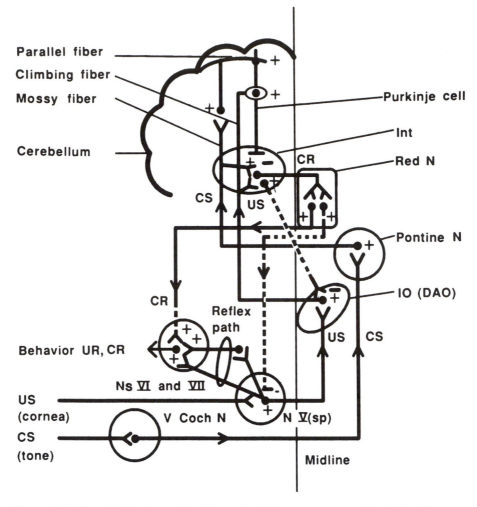

Figure 2.2 Simplified schematic of hypothetical memory trace circuit for discrete behavioral responses learned as adaptation to aversive events. The US (corneal airpuff) pathway seems to consist of somatosensory projections to the dorsal accessory portion of the inferior olive (DAO) and its climbing fiber projections to the cerebellum. The tone CS pathway seems to consist of auditory projections to the cerebellum. The efferent (eyelid closure) CR pathway projects from the interpositus nucleus (Int) of the cerebellum to the red nucleus (Red N) and via the descending rubral pathway to act ultimately on motor neurons. The red nucleus may also exert inhibitory control over the transmission of somatic sensory information about the US to the inferior olive (IO), so that when a CR occurs (eyelid closes), the red nucleus dampens US activation of climbing fibers. Evidence to date is most consistent with storage of the memory traces in localized regions of cerebellar cortex and possibly interpositus nucleus as well. Pluses indicate excitatory and minuses inhibitory synaptic action. Additional abbreviations: N V (sp), spinal fifth cranial nucleus; N VI, sixth cranial nucleus; N VII, seventh cranial nucleus; V Coch N, ventral cochlear nucleus. (From Thompson, 1986; reprinted by permission of *Science*, 1986.)

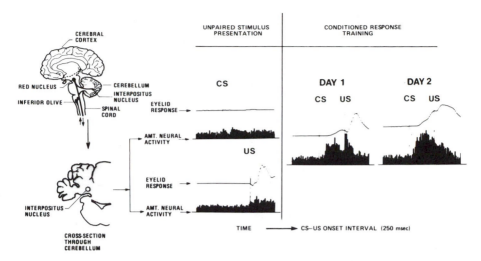

Figure 2.3   Neuronal unit activity recorded from the lateral interpositus nucleus during unpaired and paired presentations of the training stimuli. The animal was first given pseudorandomly unpaired presentations of the tone and corneal airpuff, in which the neurons responded very little to either stimulus. However, when the stimuli were paired together in time, the cells began responding within the CS period as the animal learned the eyeblink response. The onset of this unit activity preceded the behavioral NM response within a trial by 36 to 58 ms. Stimulation through this recording site yielded ipsilateral eyelid closure and NM extension. Each histogram bar is 9 ms in duration. The upper trace of each histogram represents the movement of the NM with up being extension across the eyeball (From McCormick and Thompson, 1984b.)

peduncle to the cerebellum (see Figure 2.2). Thus, lesions of the appropriate region of the DAO prevent acquisition and result in normal extinction of the behavioral CR with continued paired training in animals that already have been trained (McCormick, Steinmetz, & Thompson, 1985). Electrical microstimulation of this same region produces behavioral responses and functions as an effective US for normal learning of behavioral CRs; the precise behavioral response produced by DAO stimulation is learned as a normal CR to a tone or light CS (Mauk, Steinmetz, & Thompson, 1986.)

Preliminary evidence from our laboratory suggests that the auditory CS pathway projects from the cochlear nucleus to the contralateral lateral pontine nuclear region and from there to the cerebellum via mossy fibers in the middle cerebellar peduncle (Figure 2.2). Thus, large lesions of the middle cerebellar peduncle prevent acquisition and immediately abolish retention of the eyelid CR to all CS modalities (Solomon, Lewis, LoTurco, Steinmetz, & Thompson, 1986). However, lesions in the pontine nuclear region can selectively abolish the eyelid CR to an acoustic CS with no impairment of responding to a light CS (Steinmetz et al., 1987). If animals are trained with left pontine nuclear stimulation as the CS and subsequently tested for transfer to right pontine stimulation, transfer takes place immediately (in one trial) if the two electrodes have similar locations on the two sides. This suggests that under these conditions the traces are not formed in the pontine nuclei but instead are formed beyond the mossy fiber terminals in the cerebellum—that is, the pontine nuclei are afferent to the memory trace (Steinmetz, Rosen, Woodruff-Pak, Lavond, & Thompson, 1986).

Our results to date are strikingly consistent with the classical theoretical and computational models of the role of the cerebellum in motor learning (Albus, 1971; Eccles, 1977; Ito, 1974; Marr, 1969).

The interpositus nucleus formed an initial focus for electrophysiological analysis in classical conditioning of discrete behavioral responses, par- ticularly conditioned eyelid closure. Unit recordings from the region of interpositus—a region that, when destroyed, completely prevents learning and memory of the eyelid response—show the development of a pattern of increased frequency of unit discharges that models the amplitude–time course of the learned response, the conditioned eyelid closure, precedes it in time by as much as 60 ms or more within trials, develops in close association with the development of the eyelid CR and the trials of training, and does not model or relate to the reflex eyelid closure UR (see Figure 2.3; McCormick & Thompson 1984a, 1984b). This learning-induced unit re- sponse in the interpositus is completely predictive of the learned behavioral response both within trials and over the trials of training; it is isomorphic and temporally predictive. This, coupled with the interpositus lesion aboli- tion of the CR, argues strongly that the memory trace must be in the interpositus or in structures afferent to the interpositus. This hypothesis is strongly supported by other evidence as well (Chapman, Steinmetz, & Thompson, 1988; Thompson, 1986).

In an extensive signal-detection analysis of unit activity in the ventral cochlear nucleus, we showed definitively that there are no learning-induced or learning-related changes in unit activity in eyelid conditioning—units faithfully respond in relation to the acoustic properties of the auditory CS and show no changes as a result of learning (Kettner & Thompson, 1985). Preliminary unit recordings from the putative pontine relay in the CS pathway (relaying acoustic CS information from the cochlear nuclei to the cerebellum as mossy fibers; see Figure 2.2) similarly indicate that no learning-related changes occur; units typically show "on" response to audi- tory CS onset (and do not respond to visual CSs in trained animals). If these preliminary findings are representative, then it can be argued strongly that the memory trace is efferent from the pons.

Taken together, the interpositus, cochlear nucleus, and pontine record- ings argue that the memory trace must be in the cerebellum. Indeed, recordings from identified Purkinje cells in cerebellar cortex *do* show dramatic, learning-related changes in simple spike discharges (responses to mossy-parallel fiber activation). Prior to training the great majority of Purkinje cells that are influenced by the tone CS (in H VI, Crus I, Crus II) show variable increases in simple spike discharge frequency. After training, the majority show CS-evoked decreases in frequency of discharge that precede and "predict" the occurrence of the behavioral CR within trials. (Donegan, Foy, & Thompson, 1985; Foy & Thompson, 1986). Since Purkinje cells inhibit interpositus cells, decreased Purkinje cell activity would increase interpositus cell activity. The onset of decreased Purkinje cell activity can precede the onset of the behavioral CR by as much as 85 ms within a trial. So, Purkinje cell activity and interpositus neuron activity are the shortest onset latency neuronal responses within trials in the memory trace circuit that show learning-induced changes, and are also the first loci in the circuit from CS to CR where learning-induced changes are seen.

Prior to training, the corneal airpuff US elicits clear onset-evoked responses in neurons in the appropriate region of the dorsal accessory olive. (These neurons do not respond at all to the auditory CS, one of several lines of evidence excluding the inferior olive as the locus of the memory trace.) Similarly, prior to training, US onset evokes complex spikes (via climbing fibers from the inferior olive) in Purkinje cells that are activated by the US. In trained animals, complex spikes are seldom evoked in Purkinje cells by US onset, (a result that has very interesting implications to be explored in the last section of this chapter).

In summary, evidence to date argues that the memory traces for classical conditioning of discrete behavioral responses learned to deal with aversive USs—eyelid closure, limb flexion, head turn, and so forth—are stored in the cerebellum.

Learning Algorithms for Classical Conditioning

In order to develop computational models of learning networks, be they real circuits or "neural-like" networks, it is necessary to adopt a learning algorithm—some quantitative formulation of the process of plasticity that results in the establishment of a memory trace as a result of training, of repeated forward-paired presentations of CS and US. The most widely used algorithm has been Hebb's rule. Hebb (1949) assumed that if synaptic activation of a neuron by the CS pathway occurred conjointly with the development of an action potential in the neuron, then the CS-activated synaptic input to the neuron would be strengthened. This has been formalized and generalized to mean the pairwise cooccurrence of two events, similar to the earlier linear operator models of Hull (1943), Spence (1936), and Bush and Mosteller (1951).

Within the adaptive network literature, the Hebb rule is usually formulated as (see Donegan et al., 1989):

$$\Delta w_i = \beta a_i k \tag{1}$$

where a_i is the activity of input node i, β is a positive-valued learning rate parameter, k is the output activity of the adaptive node (or the activity of a special US training signal), and w_i is the associative strength of a CS (a continuous variable that can take on positive, zero, or negative values). The Hebb rule has many limitations (Rumelhart & McClelland, 1986). In the present context, it cannot account for many of the robust phenomena of classical conditioning, such as conditioned inhibition and blocking.

The most influential learning algorithm in classical conditioning was developed by Rescorla and Wagner (1972). The Rescorla–Wagner model proposes that phenomena of stimulus selection can be explained by assuming that the effectiveness of a US for producing associative learning depends on the relationship between the US and the expected outcome (Kamin, 1969; Rescorla, 1968; Wagner, 1969). In a critical experiment that well illustrates this principle, Kamin (1969) began by training animals with a light, the CS, which was paired with a shock US. In the second phase of training a compound stimulus, consisting of a light and a tone, was paired with the shock. Surprisingly, subjects learned very little about the tone–shock relationship, compared to control subjects who had received identical numbers of tone + light–shock trials, but no pretraining to the light. Kamin described this as a "blocking" effect because prior training of the light–shock association blocked learning of the tone–shock association during the second stage of training.

To account for blocking and related phenomena, the Rescorla–Wagner model assumes that changes in the strength of an association between a particular CS and its outcome is proportional to the degree to which the outcome is unexpected (or unpredicted), given all the stimulus elements present on that trial. To formally describe this assumption, Rescorla and Wagner used V_i to denote the strength of association between stimulus element CS_i and the US. If CS_i is followed by an unconditioned stimulus, then the change in association strength between CS_i and the US, ΔV_i, can be described by:

$$\Delta V_i = \alpha_i \beta_1 (\lambda_1 - \sum_{s \in S} V_s) \tag{2}$$

where α_i is a learning-rate parameter indexing the salience of CS_i, β_1 is a learning-rate parameter that reflects the salience of the US, λ_1 is equated with the maximum possible level of association strength supported by the US intensity, and $\Sigma_{s \in S} V_s$ is the sum of the associative strengths between all the stimulus elements present on that trial—including CS_i—and the US. If CS_i is presented and not reinforced, then the association between CS_i and the US decreases analogously by:

$$\Delta V_i = \alpha_i \beta_2 (\lambda_2 - \sum_{s \in S} V_s), \tag{3}$$

where λ_2 is generally taken to be 0, β_2 is a parameter that reflects the rate of learning on trials without US presentations, and the other parameters are as in Equation (2).

The model accounts for the blocking effect as follows: In Phase 1, CS_1 is paired with the US, and V_1 approaches the maximum associative value (e.g., $\lambda_I = 1$). In Phase 2, CS_1 is presented in compound with a neutral CS_2 (i.e., V_2 begins at zero) and the compound is followed by the US. In a comparison group, CS_1 and CS_2 enter the compound phase with no prior training (i.e., $V_1 = V_2 = 0$). By Equation (1), increments in associative strength accruing to the novel stimulus, CS_2, should be less in the blocking than in the comparison condition. By a similar logic, the model also accounts for such phenomena as contingency learning (Rescorla, 1968), multitrial overshadowing (Wagner, 1969), and conditioned inhibition (Wagner & Rescorla, 1972).

In a recent paper (Donegan et al., 1989), we noted, following a suggestion by Sutton and Barto (1981), that in Equations (2) and (3) of the Rescorla–Wagner model, the summation is taken over only those CSs which are present on a trial. With the introduction of a new variable, a_i which equals 1 if CS_i is present on a trial and 0 if CS_i is absent, and assuming that $\beta = \beta_i = \beta_2$, we can represent Equations (2) and (3) as:

$$\Delta w_i = a_i(\lambda \cdot \Sigma w_j a_j), \qquad (4)$$

where the summation is now over all possible CSs, and w_j, the weight from CS node $_j$ to the US, is equivalent to V_i in the Rescorla–Wagner formulation. The model now describes the associative changes prescribed for all CSs (present or absent). When the US is present $\lambda > 0$, otherwise $\lambda = 0$. Note that for CSs that are not present (i.e., when $a_i = 0$), no changes in associative strength are predicted.

As noted by Sutton and Barto (1981) this is the well-known least means squares (LMS) algorithm of adaptive network theory (Widrow & Hoff, 1960), also sometimes referred to as the *delta rule* (Rumelhart, Hinton, & Williams, 1986). These rules are called *error-correcting* rules because associative changes are driven by a discrepancy (error) between what the system expects and what actually occurs. For a given task, these rules can be shown to find, asymptotically, the optimal set of association weights that minimize the magnitude of the expected (squared) error. Adaptive network models generally presume that individual elements or nodes in the network have the computational power of altering the association weights according to the LMS rule of Equation (4). Within such a network, each adaptive element would have two types of inputs: multiple inputs corresponding to possible CSs, and a single special "teaching" input corresponding to the US (see, e.g., Sutton & Barto, 1981). The weights associated with the different CSs would change over trials according to the LMS rule so that the weighted sum of the incoming activation from the CS inputs, $\Sigma w_j a_j$, will come to predict, as closely as possible, the US input λ.

Equation (4) of the Rescorla–Wagner/LMS rule (henceforth, R–W/LMS) has a number of well-known computational advantages as a learning rule over the simpler Equation (1) of the Hebbian rule. When embedded within a network of associations, the R–W/LMS rule allows a system to learn complex discriminations that the Hebbian rule is unable to learn (for a more complete review of the differences between these algorithms, see Sutton & Barto, 1981). While the Hebbian rule has been used by adaptive network theorists to model associative memories (e.g., Amari, 1977; Anderson, Silverstein, Ritz, & Jones, 1977), most current "connectionist" models in cognitive psychology and artificial intelligence use the more powerful R–W/LMS rule for modeling associative changes. A further attraction of the R–W/LMS rule is that it has recently been generalized to train multilayer adaptive networks for solving complex learning tasks (LeCun, 1985, Parker, 1985, 1986; Rumelhart, Hinton, & Williams, 1986). In Donegan et al. (1989) we showed how the R–W/LMS rule can be expressed mathematically as two "Hebbian" terms: the standard conjoint activation term (Equation 1) and a reverse Hebbian rule. This formulation has interesting implications for current models of associative learning such as priming (Wagner, 1979) and SOP (Donegan & Wagner, 1987) (See Donegan et al., 1989, for details).

A Neural Circuit Example of the Error-Correcting Rule

A key aspect of the widely used R–W/LMS rule concerns the effect of degree of learning on the amount of additional associative strength added on a given CS–US trial, the error-correcting rule. At the beginning of training with a particular CS and US, $\Sigma V_s = 0$ [Equation (2)] and the amount of associative strength added by a CS–US trial is maximal ($\Delta V_i = \alpha_i \beta_1 \lambda_1$). As training continues and association strength between CS and US accrues, ΣV_s approaches λ_1, and the amount added by a CS–US pairing decreases. When learning is asymptotic and the accrued CS–US associative strength is maximum, then $\Sigma V_s = \lambda_1$ and $\Delta V_i = \alpha_i \beta_1 (\lambda_1 - \lambda_1) = 0$; no additional associative strength is added by additional CS–US pairings. A prosaic interpretation would hold that if there are n plastic synapses for a particular CS and US, then when learning is asymptotic, all n synapses are used up and no more are available to add more associative strength. Although possible, this interpretation seems unlikely given the dynamic and flexible aspects of learning and memory in real mammals.

In our current empirical work on the neural circuit essential for classical conditioning of discrete behavioral responses (see Figure 2.2), we have discovered a quite different mechanism that can account for the error-correcting aspect of the R–W/LMS formulation and related theories. It is, in fact, an emergent property of the organization of the neural circuit itself rather than a specialized synaptic process. As noted earlier, our evidence argues that the inferior olive-climbing fiber projection to the cerebellum is the essential US-reinforcing pathway. We hypothesize that conjoint activation of cerebellar Purkinje neurons by mossy-parallel fibers and climbing fibers (and analogously for interpositus neurons) is the critical associative event. Recall that as a result of learning, the CS comes to evoke a decrease in simple spike discharge frequency (parallel fiber activation) of Purkinje neurons in the CS period (the CS is hypothesized to activate the mossy fiber–granule cell–parallel fiber pathway). This learning-induced decrease in Purkinje cell simple spike responses is closely analogous to the phenomenon of long-term depression (LTD). In the decerebrate animal or in the cerebellar slice preparation, conjoint activation of climbing fibers and mossy or parallel fibers leads to a prolonged decrease in the excitability of Purkinje cells to parallel fiber activation (Ito, 1984). Ito termed this depression LTD and showed how it can account for plasticity of the vestibulo-ocular reflex. The mechanism of LTD is believed to involve an influx of calcium ions that act as a second messenger system. In this sense, at least, LTD is analogous to the "mirror image" process of long-term potentiation (LTP) in pyramidal neurons in the hippocampus.

One way to decrease the amount of associative strength added as a function of learning is to decrease the amount of reinforcement—in this instance the amount of climbing fiber activation of Purkinje and interpositus neurons as a function of learning. We assume that the amount of associative strength added on a given CS–US trial is proportional to the amount of climbing fiber activation of the cerebellum on that trial. We thus conceptualize reinforcement strength on a given trial as the proportion of effective climbing fibers activated by the US on that trial (by *effective* is meant all those activated by the US prior to training).

Our empirical evidence to date is consistent with this hypothesis. In recordings from Purkinje cells, as noted earlier, complex spikes evoked by climbing fibers are consistently evoked by the US (corneal airpuff) prior to training; in well-trained animals, the US typically does not evoke complex spikes (Donegan et al., 1985; Foy & Thompson, 1986). So there is a marked decrease in climbing fiber activation of Purkinje cells as a result of learning.

Two descending inhibitory systems from the interpositus nucleus have been described (see the dashed lines in Figure 2.2). One of these appears to originate in the red nucleus (which is activated by the interpositus). Thus, activation of the red nucleus can inhibit somatosensory activation of the inferior olive (Weiss, McCurdy, Houk, & Gibson, 1985). Weiss et al. (1985) suggest that this inhibition acts at the somatosensory nuclei that relay somatosensory information to the inferior olive. Recent anatomical and physiological evidence also indicates the existence of a powerful inhibitory pathway from the interpositus directly to the inferior olive (Andersson, Corwicz, & Hesslow, 1987; Nelson & Mugnioni, 1987).

Before learning, the tone CS does not result in any increase in the activity of interpositus neurons (see Figure 2.3). As training proceeds, neurons in the interpositus increase their patterns of discharge in the CS period such that they precede and predict the occurrence of the behavioral CR, as noted earlier (see Figure 2.3). In a well-trained animal, activation of neurons in the efferent CR pathway is massive in the CS period (Figure 2.3). We assume

that this activation of interpositus efferents includes the inhibitory efferents to the inferior olive and activation of the red nucleus inhibitory
efferents to the trigeminal nucleus. Thus, in addition to driving the behavioral CR, we argue that the efferent CR pathway also exerts a powerful
inhibitory influence on the essential US pathway. Then, as the CR is increasingly well learned, activation of the US pathway by the US is increasingly attenuated. In a well-trained animal, the US pathway might be
completely shut down. If so, then additional training will result in no
additional associative strength—the US has functionally ceased to occur at the
critical regions of memory trace formation in the cerebellum (see Figure 2.2).

In current studies we have obtained more direct evidence for this hypothesis by recording the activity of neurons in the dorsal accessory olive
activated by the corneal airpuff US onset (Steinmetz, Donegan, & Thompson, in preparation). US-alone presentations consistently evoke a phasic
increase in responses of these neurons (Us-onset-evoked). As the behavioral
CR (eyelid response) begins to develop, this US-onset-evoked response in
dorsal accessory olive neurons becomes markedly attentuated. Indeed, in a
well-trained animal, US-onset-evoked activity may be completely absent in
the dorsal accessory olive on trials where the animal gives a CR. But US-*alone*
presentations still evoke the same onset response that the US evoked prior
to training.

The learning-induced descending inhibitory systems of the essential
memory trace circuit provide a very straightforward explanation of blocking. Once learning has occurred to CS_1, presentations of CS_1 will shut down
the US-reinforcing pathway, so additional training of the compound CS_1,
CS_2–US will not result in associative strength accuring to CS_2.

The essence of this neural circuit is a negative feedback loop that inhibits
US activation of the memory trace formation network in proportion to the
degree of learning. It is striking that theoretical formulations of conditioning by Wagner and associates (priming—Wagner & Terry, 1975; SOP—
Donegan, 1980; Wagner, 1981) independently hypothesized precisely this
kind of negative feedback loop to inhibit activity in the US "node" in order
to account for certain behavioral phenomena of conditioning—namely,
latent inhibition, second-order conditioning, and conditioned diminution
of the UR. The relations among these theories and their applicability to
learning and memory circuits in the brain are explored in detail in Donegan et al., 1989.

Conclusion

The organization of the essential memory trace circuit for classical conditioning of discrete behavioral responses thus instantiates the error-
correcting rule aspect of the R–W/LMS algorithm by a negative feedback
loop, in close accord with the priming and SOP models of learning. We

would speculate that this may be a general principle that holds for all forms of basic associative learning: as learning develops, the effective US reinforcement system is actively inhibited by the associative memory system. This would limit the amount of associative strength that could accrue to any particular CS without "using up" the available plastic synapses.

More generally, work on the neurobiology of classical conditioning appears to be providing a powerful bridge to the broad field of associative network modeling, which forms an essential base for so many aspects of cognitive science today. Indeed, the empirical analysis of real neural circuits in the mammalian brain that serve to code and store memories may provide the key nexus that will make possible the development of parallel-distributed processing models of how the brain actually does process information, store and retrieve memories, and make intelligent decisions.

Acknowledgments

Supported in part by research grants from the National Science Foundation (BNS-8718300), the Office of Naval Research (N00014-88-K-0112), the Sloan Foundation, and the McKnight Foundation, and from funds provided by the University of Southern California. We depended heavily on previous publications in preparing this chapter, particularly Donegan, Gluck, and Thompson (1989) and Thompson, Donegan, and Lavond (1988). We express particular thanks to Nelson Donegan for his helpful suggestions.

References

Ackley, D. H., Hinton, G. E., & Sejnowski, T. J. (1985). A learning algorithm for Boltzmann machines. *Cognitive Science, 9*, 147–169.

Albus, J. S. (1971). A theory of cerebellar function. *Mathematical Bioscience, 10*, 25–61.

Amari, S. (1977). Neural theory of association and concept formation. *Biological Cybernetics, 26*, 175–185.

Anderson, J. A., Silverstein, J. W., Ritz, S. A., & Jones, R. S. (1977). Distinctive features, categorical perception, and probability learning: Some applications of a neural model. *Psychological Review, 84*, 413–451.

Andersson, G., Corwicz, M., & Hesslow, G. (1987). Effects of bicuculline on cerebellar inhibition of the inferior olive. *2nd World Congress of Neuroscience Abstract*, Budapest, 1888P (p. 5631).

Brown, J. S., Kalish, H. I., & Farber, I. E. (1951). Conditioned fear as revealed by magnitude of startle response to an auditory stimulus. *Journal of Experimental Psychology, 41*, 317–328.

Bush, R., & Mosteller, F. (1951). A model for stimulus generalization and discrimination. *Psychological Review, 58*, 413–423.

Chapman, P. F., Steinmetz, J. E., & Thompson, R. F. (1988). Classical conditioning does not occur when direct stimulation of the red nucleus or cerebellar nuclei is the unconditioned stimulus. *Brain Research, 441*, 97–104.

Cohen, D. H. (1980). The functional neuroanatomy of a conditioned response. In R. F.

Thompson, L. H. Hicks, & V. B. Shvyrkov (Eds.), *Neural mechanisms of goal-directed behavior and learning*. New York: Academic Press.

Davis, M. (1986). Pharmacological and anatomical analysis of fear conditioning using the fear-potentiated startle paradigm. *Behavioral Neuroscience, 100*(6), 814–824.

Davis, M., Gendelman, D. S., Tischler, M. D., & Gendelman, P. M. (1982). A primary acoustic startle circuit: Lesion and stimulation studies. *Journal of Neuroscience, 2*, 791–805.

Donegan, N. H. (1980). Priming produced facilitation or diminution of responding to a Pavlovian unconditioned. Unpublished doctoral dissertation, Yale University, New Haven, CT.

Donegan, N. H., Foy, M. R., & Thompson, R. F. (1985). Neuronal responses of the rabbit cerebellar cortex during performance of the classically conditioned eyelid response. *Neuroscience Abstract, 11*, 835 (Abstract No. 245.8).

Donegan, N. H., Gluck, M. A., & Thompson, R. F. (1989). Integrating behavioral and biological models of classical conditioning. In R. D. Hawkins & G. H. Bower (Eds.), *Computational models of learning in simple neural systems* (Vol. 22). New York: Academic Press.

Donegan, N. H., Lowry, R. W., and Thompson, R. F. (1983). Effects of lesioning cerebellar nuclei on conditioned leg-flexion responses. *Neuroscience Abstract, 9*, 331 (Abstract No. 100.7).

Donegan, N. H., & Wagner, A. R. (1987). Conditioned diminution and facilitation of the UCR: A sometimes-opponent-process interpretation. In I. Gormezano, W. Prokasy, & R. Thompson (Eds.), *Classical conditioning II: Behavioral, neurophysiological, and neurochemical studies in the rabbit*. Hillsdale, NJ: Erlbaum.

Eccles, J. C. (1977). An instruction-selection theory of learning in the cerebellar cortex. *Brain Research, 127*, 327–352.

Foy, M. R., & Thompson, R. F. (1986). Single unit analysis of Purkinje cell discharge in classically conditioned and untrained rabbits. *Neuroscience Abstract, 12*, 518.

Fuller, H. H., & Schlag, J. D. (1976). Determination of antidromic excitation by the collision test: Problems of interpretation. *Brain Research, 112*, 283–298.

Gluck, M. A., & Bower, G. H. (1988a). From conditioning to category learning: An adaptive network model. *Journal of Experimental Psychology: General, 117*(3), 225–244.

Gluck, M. A., & Bower, G. H. (1988b). Evaluating an adaptive network model of human learning. *Journal of Memory and Language, 27*, 166–195.

Gluck, M. A., & Thompson, R. F. (1987). Modeling the neural substates of associative learning and memory: A computational approach. *Psychological Review, 94*, 176–191.

Gormezano, I. (1972). Investigations of defense and reward conditioning in the rabbit. In A. H. Black & W. F. Prokasy (Eds.), *Classical conditioning: II. Current research and theory* (pp. 151–181). New York: Appleton-Century-Crofts.

Hebb, D. O. (1949). *The organization of behavior*. New York: Wiley.

Hinton, G. E., & Anderson, J. A. (1981). *Parallel models of associative memory*. Hillsdale, NJ: Erlbaum.

Hull, C. L. (1943). *Principles of behavior*. New York: Appleton-Century-Crofts.

Ito, M. (1974). Control mechanisms of cerebellar motor system. In F. O. Schmidt & R. G. Warden (Eds.), *The neurosciences, Third Study Program* (pp. 293–303). Cambridge, MA.: MIT Press.

Kamin, L. J. (1969). Predictability, surprise, attention and conditioning. In B. A. Campbell & R. M. Church (Eds.), *Punishment and aversive behavior* (pp. 279–296). New York: Appleton-Century-Crofts.

Kandel, E. R. (1976). *Cellular basis of behavior*. San Francisco: Freeman.

Kapp, B. S., Gallagher, M., Applegate, C. D., & Frysinger, R. C. (1982). The amygdala central nucleus: Contributions to conditioned cardiovascular responding during aversive Pavlovian conditioning in the rabbit. In C. D. Woody (Ed.), *Conditioning: Representation of involved neural functions* (pp. 581–599). New York: Plenum Press.

Kettner, R. E., & Thompson, R. F. (1985). Cochlear nucleus, inferior colliculus, and medial geniculate responses during the behavioral detection of threshold-level auditory stimuli in the rabbit. *Journal of the Acoustical Society of America*, 77(6), 2111–2127.

Lashley, K. S. (1929). *Brain mechanisms and intelligence. Chicago*: University of Chicago Press.

Lavond, D. G., Hembree, T. L., & Thompson, R. F. (1985). Effect of kainic acid lesions of the cerebellar interpositus nucleus on eyelid conditioning in the rabbit. *Brain Research*, *326*, 179–182.

LeCun, Y. (1985). Une procédure d'apprentissage pour réseau à seuil assymétrique. *Proceedings of Cognitiva*, *85*, Paris, 599–604.

Lincoln, J. S., McCormick, D. A., & Thompson, R. F. (1982). Ipsilateral cerebellar lesions prevent learning of the classically conditioned nictitating membrane/eyelid response. *Brain Research*, *242*, 190–193.

Lye, R. H., O'Boyle, D. J., Ramsden, R. T., & Schady, W. (1988). Effects of a unilateral cerebellar lesion on the acquisition of eyeblink conditioning in man. *Journal of Physiology*, *403*, 58.

Marr, D. (1969). A theory of cerebellar cortex. *Journal of Physiology*, *202*, 437–470.

Mauk, M. D., Steinmetz, J. E., & Thompson, R. F. (1986). Classical conditioning using stimulation of the inferior olive as the unconditioned stimulus. *Proceedings of the National Academy of Sciences*, *83*, 5349–5353.

Mauk, M. D., & Thompson, R. F. (1987). Retention of classically conditioned eyelid responses following acute decerebration. *Brain Research*, *403*, 89–95.

McCormick, D. A., Lavond, D. G., Clark, G. A., Kettner, R. E., Rising, C. E., & Thompson, R. F. (1981). The engram found?: Role of the cerebellum in classical conditioning of nictitating membrane and eyelid responses. *Bulletin of the Psychonomic Society*, *18*(3), 103–105.

McCormick, D. A., Steinmetz, J. E., & Thompson, R. F. (1985). Lesions of the inferior olivary complex cause extinction of the classically conditioned eyeblink response. *Brain Research*, *359*, 120–130.

McCormick, D. A., & Thompson, R. F. (1984a). Cerebellum: Essential involvement in the classically conditioned eyelid response. *Science*, *223*, 296–299.

McCormick, D. A., & Thompson, R. F. (1984b). Neuronal responses of the rabbit cerebellum during acquisition and performance of a classically conditioned nictitating membrane-eyelid response. *Journal of Neuroscience*, *4*(11), 2811.

Nelson, B., & Mugnioni, E. (1987). GABAergic innervation of the inferior olivary complex and experimental evidence for its origin. In *Symposium: The olivocerebellar system in motor control*. Turin, August 9–12, #9.

Norman, R. J., Buchwald, J. S., & Villablanca, J. R. (1977). Classical conditioning with auditory discrimination of the eyeblink in decerebrate cats. *Science*, *196*, 551–553.

Parker, D. (1985). *Learning logic* (Report No. 47). Cambridge, MA: Center for Computational Research in Economics and Management Science, MIT.

Parker, D. (1986). A comparison of algorithms for neuron-like cells. *Proceedings of the Neural Networks for Computing Conference*. Snowbird, Utah.

Pavlov, I. P. (1927). *Conditioned reflexes*. Oxford, England: Oxford University Press.

Rescorla, R. A. (1968). Probability of shock in the presence and absence of CS in fear conditioning. *Journal of Comparative Physiological Psychology*, *66*, 1–5.

Rescorla, R. A. (1988a). Behavioral studies of Pavlovian conditioning. *Annual Review of Neuroscience*, *11*, 329–352.

Rescorla, R. A. (1988b). Pavlovian conditioning: It's not what you think it is. *American Psychologist*, *43*(3), 151–160.

Rescorla, R. A., & Holland, P. C. (1982a). Behavioral studies of associative learning in animals. *Annual Review of Psychology*, *33*, 265–308.

Rescorla, R. A., & Holland, P. C. (1982b). Some behavioral approaches to the study of learning. In M. R. Rosenzweig & E. L. Bennett (Eds.), *Neural mechanisms of learning and memory*. Cambridge, MA: MIT Press.

Rescorla, R. A., & Wagner, A. R. (1972). A theory of Pavlovian conditioning: Variations in the effectiveness of reinforcement. In A. H. Black & W. F. Prokasy (Eds.), *Classical conditioning II: Current research and theory* (pp. 64–99). New York: Appleton-Century-Crofts.

Rosenberg, C. R., & Sejnowski, T. J. (1986). The spacing effect on Nettalk, a massively-parallel network. *Proceedings of the 8th Annual Conference of the Cognitive Science Society*, Amherst, MA.

Rumelhart, D. E., Hinton, G. E., & Williams, R. J. (1986). Learning internal representations by error propagation. In D. Rumelhart & J. McClelland (Eds.), *Parallel distributed processing: Explorations in the microstructure of cognition*. Vol. 1: *Foundations* (pp. 318–362). Cambridge, MA: MIT Press.

Rumelhart, D. E., & McClelland, J. L. (1986). *Parallel distributed processing: Explorations in the microstructure of cognition*: Vol. 1. *Foundations*. Cambridge, MA: MIT Press.

Schneiderman, N., McCabe, P. M., Haselton, J. R., Ellenberger, H. H., Jarrell, T. W., & Gentile, C. G. (1987). Neurobiological bases of conditioned bradycardia in rabbits. In I. Gormezano, W. F. Prokasy, & R. F. Thompson (Eds.), *Classical conditioning* (pp. 3, 35–36). London: Lawrence Erlbaum.

Shepherd, G. M., & Brayton, R. K. (1987). Logic operations are properties of computer-simulated interactions between excitable dendritic spines. *Neuroscience, 21*(1), 151–165.

Solomon, P. R., Lewis, J. L., LoTurco, J., Steinmetz, J. E., & Thompson, R. F. (1986). The role of the middle cerebellar peduncle in acquisition and retention of the rabbit's classically conditioned nictitating membrane response. *Bulletin of the Psychonomic Society, 24*(1), 75–78.

Spence, K. W. (1936). The nature of discrimination learning in animals. *Psychological Review, 43*, 427–449.

Steinmetz, J. E., Logan, C. G., Rosen, D. J., Thompson, J. K., Lavond, D. G., & Thompson, R. F. (1987). Initial localization of the acoustic conditioned stimulus projection system to the cerebellum essential for classical eyelid conditioning. *Proceedings of the National Academy of Sciences, 84*, 3531–3535.

Steinmetz, J. E., Rosen, D. J., Woodruff-Pak, D. S., Lavond, D. G., & Thompson, R. F. (1986). Rapid transfer of training occurs when direct mossy fiber stimulation is used as a conditioned stimulus for classical eyelid conditioning. *Neuroscience Research, 3*, 606–616.

Sutton, R. S., & Barto, A. G. (1981). Toward a modern theory of adaptive networks: Expectation and prediction. *Psychological Review, 88*, 135–170.

Thompson, R. F. (1986). The neurobiology of learning and memory. *Science, 233*, 941–947.

Thompson, R. F. (1988). The neural basis of basic associative learning of discrete behavioral responses. *Trends in Neurosciences, 11*(4), 152–155.

Thompson, R. F., Berger, T. W., Cegavski, C. F., Patterson, M. M., Roemer, R. A. D., Teyler, T. J., & Young, R. A. (1976). The search for the engram. *American Psychologist, 31*, 209–227.

Thompson, R. F., Donegan, N. H., & Lavond, D. G. (1988). The psychobiology of learning and memory. In R. C. Atkinson, R. J. Herrnstein, G. Lindzey, & R. D. Luce (Eds.), *Steven's handbook of experimental psychology* (Vol. 2, pp. 245–247). New York: Wiley.

Wagner, A. R. (1969). Stimulus selection and modified continuity theory. In G. Bower & J. Spence (Eds.), *The psychology of learning and motivation* (Vol. 3). New York: Academic Press.

Wagner, A. R. (1979). Habituation and memory. In A. Dickinson & R. A. Boakes (Eds.), *Mechanisms of learning and motivation*. Hillsdale, NJ: Lawrence Erlbaum.

Wagner, A. R. (1981). SOP: A model of automatic memory processing in animal behavior. In N. Spear & G. Miller (Eds.), *Information processing in animals: Memory mechanisms*. Hillsdale, NJ: Erlbaum.

Wagner, A. R., & Rescorla, R. A. (1972). Inhibition in Pavlovian conditioning: Applications of a theory. In R. A. Boakes & S. Halliday (Eds.), *Inhibition and learning* (pp. 301–336). New York: Academic Press.

Wagner, A. R., & Terry, W. S. (1975). Backward conditioning to a CS following an expected vs. a surprising UCS. *Animal Learning and Behavior, 3*, 370–374.

Weiss, C., McCurdy, M. L., Houk, J. C., & Gibson, A. R. (1985). Anatomy and physiology of dorsal column afferents to forelimb dorsal accessory olive. *Society of Neuroscience, Abstract, 11*, 182.

Widrow, G., & Hoff, M. E. (1960). Adaptive switching circuits. Institute of Radio Engineers, Western Electronic Show and Convention, *Convention Record, 4*, 96–194.

3

Drosophila's Role in Identifying the Building Blocks of Associative Learning and Memory

TIM TULLY

More than sixty years of genetic experiments, primarily on *E. coli* and *D. melanogaster*, put in place the conceptual groundwork to begin genetic dissections of more complex phenotypic traits, such as behavior (see Rubin, 1988). In its theoretical extreme, the goal of genetic dissection is to isolate many, if not all, of the genes involved with the expression of a given trait. By the early seventies, single-gene mutant analyses of chemotaxis in bacteria (Adler, 1966, 1969); phototaxis in fruit flies (Benzer, 1967); excitability in ciliates; and locomotion, chemotaxis, and responses to tactile cues in nematodes (Brenner, 1974) were underway. That single-gene mutations that disrupt these relatively simple behavioral responses could be generated quickly became apparent.

Here, I will describe what some may view as a more ambitious attempt to dissect genetically the complex biological phenomena of learning and memory. First, I will discuss the behavioral procedures used to study associative learning in *Drosophila* and to isolate mutant strains. Next, I will review what we know about underlying biochemical mechanisms. Finally, I will outline new steps taken to continue the genetic dissection of learning and memory in an expeditious manner.

Shock-Avoidance Learning in *Drosophila*

Quinn, Harris, and Benzer (1974) initiated a genetic dissection of learning in *Drosophila* by demonstrating shock-avoidance conditioning in wild-type (Canton-S) flies. They designed a discriminative conditioning procedure in which a group of about 40 flies were exposed sequentially to one odor (CS+) paired with electric shock (negative reinforcer) and a second odor (CS−) without shock. After three such training cycles, conditioned avoidance responses were tested by exposing the flies to the CS+ and then to the CS− in the absence of any reinforcement.

To begin conditioning, flies were introduced into the start tube (Figure 3.1) by holding the conditioning apparatus (which was a modification of

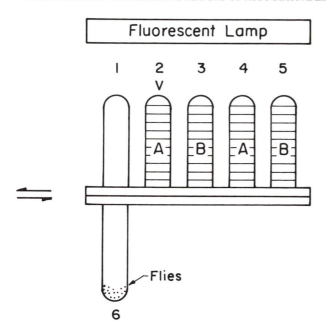

Figure 3.1   Shock-avoidance conditioning apparatus of Quinn, Harris, and Benzer (1974). Tube 1 is the rest tube, 2 and 3 are used during training, 4 and 5 are used during testing. Tube 6 is the start tube. Horizontal stripes in tubes indicate shock grids. (A) and (B) denote odorants 3-octanol and 4-methylcylcohexanol, respectively. (V) indicates voltage on the grid. See text for details.

Benzer's countercurrent apparatus for phototaxis) vertically and shaking the flies to the bottom. The start tube was shifted into register with the proper grid tube, and then the apparatus was laid horizontally in front of a fluorescent lamp. Induced by their positive phototactic responses, flies ran from the start tube into a grid tube, which was coated with either octanol (OCT; tube 2) or methylcyclohexanol (MCH; tube 3). For each training cycle, flies were exposed sequentially to the rest tube (60 s), tube 2 (15 s), rest tube (60 s), and tube 3 (15 s). Flies that ran into a grid tube containing one of the odors received 90 V (AC) of continuous electric shock. During the test cycle, which followed training by 60 s, flies were exposed to grid tubes 4 and 5, which were coated with either OCT or MCH but were not electrified. After 15 s, the number of flies avoiding the grid tubes was counted visually.

A learning index was calculated as the fraction of flies avoiding the CS+ minus the fraction of flies avoiding the CS− during the test cycle. OCT was CS+ and MCH was CS− for one group of flies; MCH was CS+ and OCT was CS− for a second group of flies. Finally, the learning indices from these two reciprocally trained groups were averaged to produce one learning index for a complete experiment (Λ). If flies failed to associate an odor with shock, then they would avoid the CS+ and CS− equally, and the learning index would be zero. On the other hand, if all flies learned to associate an odor with shock, they all of them would avoid the CS+, none would avoid the CS− and the learning index would be one.

Quinn, Harris, and Benzer, (1974) claimed that calculation of a learning index for a population in this manner necessarily yields an average measure of *associative* learning that is not biased by nonassociative learning. The CS− serves as an explicitly unpaired control (Rescorla, 1967), and changes in avoidance responses to the CS− can reflect the combined and interactive effects of habituation and sensitization. Moreover, by averaging indices from reciprocally trained groups, the overall learning index can be greater than zero only when more flies avoid the CS+ than the CS−.

Trained and tested this way, 67% of wild-type (Canton-S) flies avoided the CS+ and 33% avoided the CS−, yielding a learning index of 0.34 ± 0.02. Quinn et al. (1974) showed that this conditioned avoidance behavior could be extinguished when trained flies were exposed repeatedly to odor cues without reinforcement. Following extinction, the same flies could be re-trained to avoid the original CS−. If flies were left undisturbed after training, retention of conditioned avoidance responses lasted for at least one hour. When the usual training procedure was repeated four times at two-hour intervals, conditioned avoidance responses lasted at least 24 hours ($\Lambda = 0.12 \pm 0.02$).

Isolation of Single-Gene Mutants

As is obvious in the preceding description of the shock-avoidance condition-ing procedure, learning in individual flies cannot be determined—a limita-tion in experimental design viewed as a serious shortcoming by some (see Hirsch & Holliday, 1988). In fact, Quinn et al. " . . . sought to devise a paradigm suitable for mutant isolation, in which flies can be trained and tested en masse." In this manner, relatively accurate estimates of average associative learning levels for populations of flies could be obtained rapidly. Thus, one can argue that this procedural "limitation" actually represented conceptual progress, thereby permitting the isolation of single-gene mutations the average effects of which produced lower associative learning levels in mutant populations (see Tully, 1986).

Byers at Cal Tech and Quinn and co-workers at Princeton set out to isolate X-linked mutations that disrupt normal learning in Canton-S (wild-type) flies. Individual male Can-S flies were mutagenized chemically with ethyl methanesulfonate (EMS). Their progeny then were mated in specific genetic crosses with special "balancer" strains to produce many popula-tions, each derived from a single male, in which the X-chromosomes were identical within and among individuals. Flies from each of these potentially mutant strains then were assayed for shock-avoidance learning. A mutant strain was retained for further behavioral analyses if it yielded a learning index less than 0.05 (versus 0.33 for Can-S flies).

Based on such a selection criterion, four mutant strains were isolated— *dunce, rutabaga, turnip,* and *cabbage* (Aceves-Pina & Quinn 1979; Dudai, Jan, Byers, Quinn, & Benzer, 1976; Quinn, Sziber, & Booker, 1979). In addition, a fifth mutant strain was found by shocking flies ten times in the presence of one odor and then waiting 45 min after training before testing. Under these conditions, flies from the *amnesiac* strain learned normally but forgot four times faster than wild-type flies (Quinn et al, 1979).

Although the EMS concentration used during mutagenesis causes one mutation per chromosome on average (Lewis & Bacher, 1968), the possibility still existed that two or more mutations could be responsible for lower learning or memory in any particular mutant stock. Accordingly, genetic and chromosomal mapping experiments and complementation analyses have established that lower scores in the *dunce, rutabaga,* and *amnesiac* strains result from the effects of separate, single-gene mutations (Aceves-Pina & Quinn, 1979; Booker & Quinn, 1981; Byers, Davis, & Kiger, 1981; Livingstone, 1985; Quinn et al., 1979; Tully & Gergen, 1985). In contrast, recent data now suggest that low learning scores in *turnip* flies result from the effects of at least two mutations, one X-linked and one autosomally linked (Tully, 1988).

Naive flies from each mutant strain reacted normally to electric shock and showed normal olfactory acuity to the odors used in the learning assay. They also displayed normal phototactic behavior in the conditioning apparatus. Thus, these mutations did not appear to affect the behavioral responses required to perform during the shock-avoidance procedure. However, other phenotypic effects have been reported for two of the learning mutants. The *dunce* mutation affects both learning and female fertility (Kiger, 1977), while flies from the *turnip* strain have abnormal nerve and muscle structures in larvae and adults (Hall, 1982). These results are not too surprising, since we might expect single-gene mutations such as *dunce* to produce (abnormal) gene products that have pleiotropic effects on other components of higher order systems. In *turnip* flies, on the other hand, it is not clear whether the various phenotypic effects result from one gene or several (see above).

Since four mutant strains failed to show any conditioned responses in the shock-avoidance procedure, they immediately were christened "learning" mutants. The possibility existed, however, that these mutations affected retention, rather than acquisition, of conditioned responses. This became glaringly apparent when *amnesiac* was isolated. Perhaps memory decay was so fast in mutant flies that conditioned responses were forgotten between training cycles or between training and testing. Dudai (1979, 1983) modified the original shock-avoidance procedure in two ways. First, flies were exposed to only one CS during training (as was the case during the screen for *amnesiac*). Second, trained flies were transported within 30 s to the choice point of a T-maze (Dudai et al., 1976) with odorants (the CS and a novel odor) spread on copper grids inside the T-maze arms. The choice chamber was positioned so that its arms were perpendicular to a light source, thereby eliminating the potentially confounding effects of phototactic responses during the test trial. With these modifications, Dudai showed that *dunce*, *rutabaga*, and *cabbage* flies learn associatively, but their memories decay much faster that normal flies (*turnip* flies were not tested).

The low learning or memory scores produced by flies from these mutant strains also are not task-specific. Many behavioral assays have been developed that are thought to involve associative learning at least to some degree. Hungry flies, for instance, have been rewarded with sucrose in the presence of olfactory cues (Tempel, Bonini, Dawson, & Quinn, 1983), and visual (color) cues have been used as CSs in shock-avoidance procedures (Quinn et al., 1974; Spatz, Emmans, & Reichert, 1974; also see Dudai & Bicker, 1978; Folkers, 1982; Menne & Spatz, 1977). Headless flies even can be trained to extend, or to retract, their legs in response to negative reinforcement (Booker & Quinn, 1981). Interestingly, individual males can be conditioned to inhibit their courtship behaviors if they court fertilized females (Siegel & Hall, 1979). Learning in this context probably is mediated by olfactory and gustatory cues (Gailey, Jackson, Siegel, 1984; Tompkins, Siegel, Gailey, & Hall, 1983) and appears to be associative (Ackerman & Siegel, 1986). Importantly, the "learning" mutants perform more poorly in all of these behavioral assays. In several tasks, some learning is apparent initially, but memory decays rapidly in mutant strains, corroborating the idea that the extant mutations may affect memory rather than learning.

Classical Conditioning in Normal and Mutant Flies

The shock-avoidance procedure of Quinn et al. (1974) used light to induce flies to move into the odor tubes during training and testing. This feature of the experiment made the conditioning procedure instrumental; flies did not receive negative reinforcement in the presence of the CS (odor) unlessthey continued to move toward the light source and onto the shock grid. Moreover, flies with weaker phototactic responses may not have received as many acquisition "trials" during training. To correctly avoid odor tubes during the test trial, flies were required to elicit two conditioned responses: avoidance of a specific odor (CS+) and inhibition of phototaxis. These considerations suggested that stronger olfactory avoidance behavior might be detectable if phototactic behavior was not required during the conditioning procedure.

Tully and Quinn (1985) modified the T-maze chamber of Dudai et al. (1976) so that carefully controlled currents of air could be drawn through it (Figure 3.2). In this manner, the instrumental shock-avoidance procedure was adapted to a classical (Pavlovian) procedure, where flies always received negative reinforcement in the presence of the CS+. About 150 flies were sequestered in a closed chamber and were trained by exposing them se-

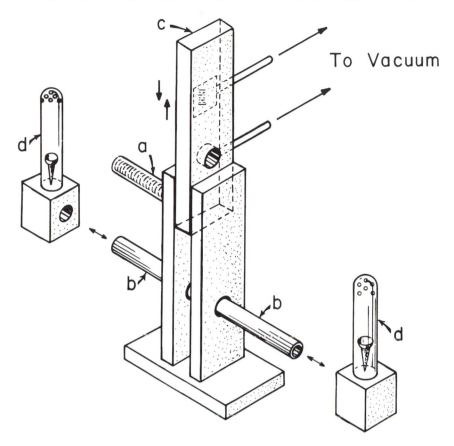

Figure 3.2   Classical conditioning apparatus of Tully and Quinn (1985). About 150 flies are sequestered in a closed grid tube (a) and trained by exposing them sequentially to two odors (3-octanol or 4-methylcyclohexanol), which are housed in odor tubes (d) and delivered in air currents. Flies receive twelve 1.25-s 60-V (DC) shock pulses during a 60-s presentation of the first odor (CS+) but not during a 60-s presentation of a second, control odor (CS−). To test for conditioned odor avoidance, flies are transported via a sliding center compartment (c) to a T-maze choice point between converging currents of the two odors. During this 120-s test trial, most (wild-type) flies avoid the T-maze arm (b) containing the CS+ by walking into the opposite arm containing CS−.

quentially to two odors (either OCT or MCH) delivered in air currents. Flies received twelve 1.25-s, 60-V (DC) electric shock pulses—one pulse every five seconds during a 60-s presentation of the first odor (CS+) but not during a 60-s presentation of the second odor (CS−). To test for conditioned avoidance, flies were tapped gently into a sliding compartment and were transported to the T-maze choice point, between converging currents of OCT and MCH. Typically, 93% of wild-type (Canton-S) flies avoided the CS+, 4% avoided the CS−, and 3% remained at the choice point, yielding a learning index of 0.89 (Figure 3.3A).

A

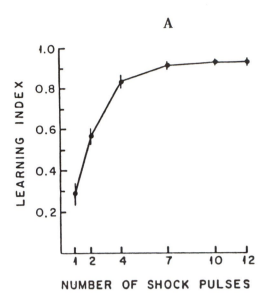

B

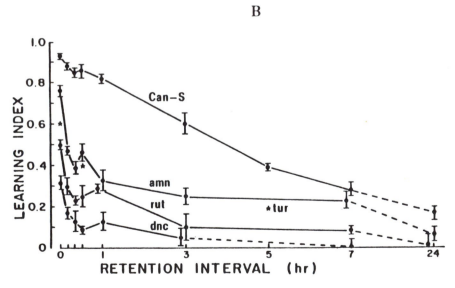

Figure 3.3 Acquisition and retention of classically conditioned avoidance responses in normal and mutant *Drosophila*. (A) Different groups of wild-type flies received from 1 to 12 shock pulses during training and then were tested for conditioned avoidance behavior immediately afterward. Acquisition is a function of the number of shock pulses received during training. (B) Different groups of flies received the usual 12 shock pulses during training and then were tested at various intervals afterward. Retention curves are drawn for wild-type (Can-S), *amnesiac* (amn), *rutabaga* (rut), and *dunce* (dnc) flies. Data were collected only at three retention intervals for *turnip* (tur) flies, so these values are represented as individual points (stars).

Tully and Quinn (1985) demonstrated empirically that the average learning index for a population of flies is not biased by nonassociative changes in avoidance behavior. Different groups of flies were trained with odors alone, shock alone, or odors and shock explicitly unpaired. Then, avoidance responses were quantified by presenting OCT and MCH simultaneously at the T-maze choice point. For each of these traditional nonassociative control groups, the average learning index was zero. Another set of experiments, designed to detect any changes in avoidance responses produced by the nonassociative control procedures, revealed avoidance response decrements in each of the three control groups. Thus, the classical conditioning procedure did produce some nonassociative changes in avoidance behavior, but these effects did not influence the *associative* learning index.

With the better experimental control of stimulus presentations in the classical conditioning procedure, Tully and Quinn (1985) also were able to show in wild-type flies several properties of classical conditioning that are similar to classically conditioned behaviors in other species (see Mackintosh, 1974; Carew & Sahley, 1986). Associative learning levels were a function of CS or US saliencies; conditioned responses were extinguished by presenting odor cues after training in the absence of reinforcement. Classically conditioned flies also showed two of the more important properties of associative learning: CS–US order dependence and temporal specificity. When the US preceded the CS+ during training (backward conditioning), the average learning index was zero. When the interval of time between the offset of odor presentation and the onset of shock increased (trace conditioning), the average learning index decreased.

Given such strong learning in wild-type flies, Tully and Quinn (1985) went on to compare learning and memory among wild-type and mutant flies (Figure 3.3B). From these data, three general conclusions were drawn. (1) The mutants *amnesiac, rutabaga, dunce,* and *turnip* all are capable of moderate initial levels of associative learning. (2) Mutant memory decay curves look similar. They differ quantitatively rather than qualitatively. (3) Mutant memory curves appear to be composed of (at least) two components. Their memories decay very rapidly during the first 30 minutes after training; decay rates slow considerably thereafter.

Earlier work on retrograde amnesia revealed that memory formation in *Drosophila*, as in most vertebrates, is composed of an anesthesia-sensitive, short-term phase and an anesthesia-resistant, long-term phase (Quinn & Dudai, 1976; Tempel et al., 1983). We have confirmed these observations using the classical conditioning procedure (Tully et al., 1990; Tully & Block, unpublished). Wild-type flies, trained in the usual way, were anesthetized at various times afterwards by placing them in a test tube and submerging the tube in 0°C water for two minutes. Conditioned avoidance then was tested three hours after training. As seen in Figure 3.4, cold-shock anesthesia administered shortly after training produces larger decrements in 3-hr memory than cold-shock at later intervals. These results indicate that an

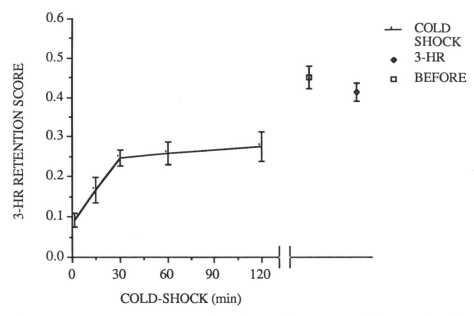

Figure 3.4 Retrograde amnesia in wild-type flies. Different groups of flies experienced cold-shock anesthesia 3.5, 15, 30, 60, or 120 min after training in the usual classical conditioning procedure. Compared to control groups that were not cold-shocked (normal 3-h retention) or that were cold-shocked 60 min before training, cold-shock within 30 min after training produces a loss of 3-h memory. These results indicate that an anesthesia-resistant phase of memory begins to form during training and reaches maximum levels 30 to 60 min later.

anesthesia-resistant phase of memory begins to form during training and reaches a maximum level 60 minutes later. Interestingly, this maximal level of memory after cold-shock in wild-type flies is equivalent to retention levels for *amnesiac* flies one or more hours after training (see Figure 3.3). Taken together, these results suggested to Tully and Quinn (1985) that the *amnesiac, rutabaga, dunce,* and *turnip* mutations might affect short-term memory, leaving long-term memory formation substantially intact. More recent experiments have confirmed the existence of an anesthesia-resistant phase of memory in *amnesiac* (Tully et al., 1990) and *rutabaga* flies (Dudai, Corfas, & Hazvi, 1988).

Biochemistry of Learning and Memory in *Drosophila*

From the first mutagenesis at Cal Tech, two mutant alleles of the *dunce* gene were isolated (see Byers et al., 1981). Significantly, females homozygous for the *dunce*[2] allele also were sterile. Genetic recombination experiments indicated that both of these phenotypic effects mapped to the distal X-chromosome between the morphological markers *yellow* and *chocolate*.

Concurrently, Kiger and associates were studying the genetics of cyclic nucleotide hydrolysis and had identified a region of the distal X-chromosome around chromomere 3D4, which affected cAMP hydrolysis in flies carrying deletions or duplications of that region (Kiger & Golanty, 1977, 1979). Kiger (1977) also noticed that a large fraction of females lacking chromomere 3D4 were sterile. Genetic complementation analyses with other X-linked mutations producing sterility in females identified two mutations (*M11* and *M14*) that mapped to the region near chromomere 3D4. Guided by these clues, Byers showed that the *M11* and *M14* mutations also failed to complement the female sterility associated with the *dunce²* mutation, suggesting that these three mutations (and *dunce¹*) are alleles of the same gene.

Subsequent genetic, biochemical, and behavioral experiments strongly suggested that *dunce* is the structural gene for a cAMP-specific phosphodiesterase (PDEII; Byers et al., 1981). A total of six alleles of the *dunce* gene were found. The dnc^{M11}, dnc^{M14}, and dnc^{ML} mutations are amorphic, producing no functional PDEII. The dnc^1, dnc^2, and dnc^{CK} mutations are hypomorphic, producing 15 to 75% reductions in PDEII activity (Saltz & Kiger, 1984). Flies homoallelic or heteroallelic for any of these six mutations are sterile (Kiger, Davis, Saltz, Fletcher, & Bowling, 1981) and have elevated levels of cAMP (Byers et al., 1981; Davis & Kiger, 1981). PDEII from dnc^1 flies is more thermolabile than normal, and PDEII from dnc^2 flies has a reduced affinity for its cAMP substrate (Kauvar, 1982). Furthermore, females (or males) carrying zero, one, two, or three doses of chromomere 3D4 have 0%, 50% (or 100%), 100% (or 167%), or 150% of normal PDEII activity (Shotwell, 1983). Finally, flies homozygous for dnc^1, dnc2, dnc^{M11}, and dnc^{M14}, or flies deficient for the 3D4 chromomere, show no associative learning in the shock-avoidance procedure (Byers, 1980; Byers et al., 1981).

More recently, the *dunce* gene has been cloned, providing direct proof that the *dunce* gene encodes a phosphodiesterase (Chen, Denome, & Davis, 1986). Substantial sequence homology exists between the predicted amino acid sequence of dnc^+ and those of bovine Ca^{2+}/calmodulin-dependent PDE and yeast cAMP-dependent PDE. Furthermore, the *dunce* gene product and the RII subunit of cAMP-dependent protein kinase share a short but perfect sequence homology, which may correspond to the cAMP binding domain. Interestingly, weak sequence homology exists between the *dunce* protein and the precursor of *Aplysia* egg-laying hormone. This homology is confined to its own exon, leading to the speculation that alternative RNA splicing may code for a *dunce* protein involved with egg-laying in *Drosophila*—providing a molecular model for the observed pleiotropic effects of *dunce* mutations.

The discovery that *dunce* mutations affect phosphodiesterase activity was immediately exciting and relevant, because the cAMP cell-signaling pathway already had been implicated with associative learning from work by Kandel and associates on the gill-withdrawal reflex in the marine mollusk *Aplysia*. In their model of heterosynaptic facilitation, which was derived from detailed behavioral, neurophysiological, and biochemical data, Kandel and Schwartz (1982) proposed that behavioral sensitization (a form of nonassociative learning) involves the monoamine-activated adenylate cyclase pathway. Strong electric shock to the tail stimulates facilitatory interneurons in the abdominal ganglion that release serotonin and other neurotransmitters onto the presynaptic terminal of a tactile sensory neuron innervating skin near the siphon and gill. Via the receptor/G-protein membrane complex, adenylate cyclase is activated and intracellular cAMP levels increase. This in turn activates a cAMP-dependent protein kinase, which then phosphorylates potassium channels in the presynaptic terminal membrane. Phosphorylated potassium channels impede the flow of K^+ ions out of the cell after the sensory neuron fires an action potential. As a consequence, more Ca^{2+} ions flow into the cell during membrane repolarization. Finally, higher Ca^{2+} levels in the presynaptic terminal promote vesicle binding and enhance neurotransmitter (acetylcholine) release from the sensory neuron onto the postsynaptic terminals of a motor neuron that drives gill withdrawal.

By conferring on adenylate cyclase the ability to be activated by increases in intracellular Ca^{2+} in the presynaptic terminal of an actively firing sensory neuron, Hawkins et al. (1983) suggest a model of activity-dependent neuromodulation to explain associative learning produced by classical conditioning of the gill-withdrawal reflex (weak tactile stimulus to siphon paired with strong tail shock). Thus, the *Aplysia* models for nonassociative and associative learning share much of the same biochemical machinery (Kandel et al., 1983). This mechanistic relation also is supported by behavioral data from *Drosophila*: All of the memory mutants show abnormal sensitization or habituation (Duerr & Quinn, 1982).

Given these observations, a search began for lesions of the cAMP pathway in the other extant memory mutants. Surprisingly, Livingstone, Sziber, and Quinn (1984) found that adenylate cyclase activity from homogenates of *rutabaga* abdomens was abnormally low, apparently due to a much lower V_{max} for the enzyme. More detailed biochemical experiments showed that octopamine, GppNHp (a GTP analogue) and NaF stimulation of adenylate cyclase activity was normal in *rutabaga* flies, suggesting that the neurotransmitter receptor and stimulatory G-protein subunit (G_s) components of the signal transducing membrane complex were not affected. GppNHp inhibition of manganese-stimulated adenylate cyclase activity also was normal in *rutabaga* flies, implying that the inhibitory G-protein subunit (G_i) functioned normally.

In contrast, activation of adenylate cyclase by ligands thought to interact directly with the catalytic subunit revealed differences between wild-type and *rutabaga* flies (Livingstone et al., 1984). In particular, adenylate cyclase activation by low concentrations of Ca^{2+} ($< 10^{-6}$ mol) was completely absent in *rutabaga* flies, whereas the inhibition of adenylate cyclase at higher calcium concentrations was similar in both wild-type and *rutabaga* flies. The absence of activation of adenylate cyclase in *rutabaga* flies did not result from an abnormal interaction of Ca^{2+} with calmodulin; adenylate cyclase from *rutabaga* homogenates that had been stripped of all endogenous calmodulin did not respond to externally applied bovine calmodulin (in the presence of Ca^{2+}). Dudai and Zvi (1985) also found that adenylate cyclase from *rutabaga* flies (a) did not respond normally to forskolin, (b) had a higher K_m for the Mg^{2+}–ATP substrate complex and (c) had a lower K_m for free Mg^{2+}.

Genetic experiments have localized the defects in Ca^{2+} and forskolin activation of adenylate cyclase, along with the learning defect, to chromomeres 12F5-7 on the X-chromosome (Dudai et al., 1985; Livingstone, 1985; Livingstone et al., 1984). Furthermore, Ca^{2+}-stimulated adenylate cyclase activity was proportional to the number of *rutabaga*[+] gene copies in the genome, providing genetic evidence that the *rutabaga* mutation affects the structural gene for adenylate cyclase (Livingstone, 1985).

A few tantalizing reports of biochemical lesions in other mutations that may affect learning have appeared. Flies homozygous for temperature-sensitive alleles of the *Ddc* (dopa decarboxylase-deficient) gene do not synthesize dopamine or serotonin (Livingstone & Tempel, 1983) and failed to learn several associative tasks (Tempel, Livingstone, & Quinn, 1984) at restrictive temperatures. These behavioral effects, however, have not been replicated (Tully, 1988). EGTA-extractable protein kinase C activity is almost completely absent from *turnip* head-membrane homogenates (Smith, Choi, Tully, & Quinn, 1986). Flies carrying a *turnip* X-chromosome and one containing a deletion of chromomeres 18A5 to 18D1-2 also show no PKC activity, but they show normal classical conditioning (Choi, Tully, & Quinn, unpublished). These results indicate that a mutation affecting PKC activity in *turnip* flies maps within the region of the deletion, but another mutation in the same mutant strain affects associative learning. Moreover, because flies with no EGTA-extractable PKC activity from head membranes can learn normally, it is doubtful that PKC is involved with olfactory learning in fruit flies (cf Akers, Lovinger, Colley, Linden, & Routtenberg, 1986; Farley & Auerbach, 1986; Hochner, Braha, Klein, & Kandel, 1986; Lynch & Baudry, 1984). On a more hopeful note, however, flies homozygous for *Shaker*[5] show learning and memory deficits in classical conditioning (Cowan & Siegel, 1986). The *Shaker* gene has been cloned and encodes a class of voltage-dependent potassium channels (Papazian, Schwarz, Tempel, Jan, & Jan, 1987; Tempel, Papazian, Schwarz, Jan & Jan, 1987). Although results from *Aplysia* implicate voltage-independent potassium channels with learning, the *Shaker* data at least are generally consistent with the *Aplysia* model of associative learning.

To date, this is about all the information that a genetic dissection of learning and memory in fruit flies has yielded in the fifteen years since Quinn et al. (1974) first demonstrated associative learning in *Drosophila*. Single-gene mutations affecting learning in the *turnip* and *cabbage* strains have yet to be mapped genetically. The *rutabaga* and *amnesiac* genes have been mapped cytologically to small regions of the X-chromosome. The *dunce* gene has been cloned, but progress toward cloning the *rutabaga* and *amnesiac* genes has been hampered by the lack of several chromosomal breakpoints nearby and by the fact that only one mutant allele exists. Both of these genetic shortcomings prevent the identification of these genes at the molecular level via DNA "landmarks" associated with mutant alleles.

Nevertheless, the fact that *dunce* and *rutabaga* mutations both affect components of the cAMP-cell signaling pathway demonstrates convincingly that genetic dissection in fruit flies can help to identify underlying components of learning and memory processes. These two mutations were generated randomly and were isolated independently because they disrupted associative learning. The fact that they both affect the cAMP pathway cannot be coincidental. But how can we expedite further genetic dissection?

Next Step with New Genetic Technology

We have begun another mutagenesis to isolate new genes involved with learning and memory (Tully et al., 1990). At the behavioral level, we are screening mutant strains for deficits in 3-hr retention after classical conditioning (see Figure 3.3B) Once a putative mutant strain is isolated, detailed behavioral analyses will determine whether deficits in (1) peripheral factors such as olfactory acuity, sensitivity to shock, or locomotor activity, (2) learning/short-term memory, or (3) long-term memory are responsible for the lower 3-hr retention scores. With this approach, we will identify mutations affecting learning and short-term memory as well as those that may affect long-term memory specifically. Significantly, behavioral tests used in past mutant screens were not designed to detect this latter class of mutation.

At the genetic level, we are using P–transposable elements to create new, single-gene mutations in autosomal loci (see Figure 3.5; Cooley, Berg, & Spradling, 1988, Robertson et al., 1988). This scheme takes advantage of molecular work, which has identified two functional domains in the DNA sequence of naturally occurring P-elements (Karess & Rubin, 1984). Long-terminal repeats (LTRs) at either end of a P-element are required for site recognition and integration into the host genome. Transposase, which is encoded by DNA located between the LTRs, greatly catalyzes the process of transposition.

More recently, a P-element (Δ2-3) has been genetically engineered that produces high levels of transposase but cannot itself transpose to other locations in the host genome (Robertson et al., 1988). Conversely, "mutator" P-elements have been engineered that do not produce any transposase but can transpose at high frequency in the presence of transposase. Of particular importance is the fact that these mutator elements contain DNA sequences encoding proteins involved with the expression of selectable phenotypic markers such as eye color, so that single transposition events can be identified phenotypically, thereby greatly improving the efficiency of muta-genesis. Furthermore, mutator P-elements also contain plasmid DNA se-quences to facilitate direct cloning of host DNA fragments flanking the mutator element into bacteria (Cooley et al., 1988).

Mutator inserts can be localized cytologically by hybridizing a DNA probe corresponding to sequences contained within the mutator element to polytene chromosomes in situ. This approach to mapping the gene requires only a few hours of work over two weeks, as opposed to hundreds of hours of behavioral testing over several months to map the gene with traditional recombinant or deletion mapping techniques. This technical improvement in mapping, along with the use of selectable markers during the muta-genesis, ultimately will save years of work on each new mutant strain.

Theoretical estimates suggest that there might be 70 genes in *Drosophila* that, when mutated, will affect our learning or memory phenotypes. This consideration, along with the efficiency of P-element mutagenesis, indicates that 0.5% of P-element inserts should disrupt learning or memory genes. Currently, we have screened more than 700 mutant strains and about 6% show low 3-hr retention scores. Due to the measurement error inherent in our behavioral screen, many of these putative mutants will not produce consistently lower memory scores across several retention intervals. Thus, behavioral characterization of putative mutant strains most likely will elimi-nate some from further consideration. False positives also may arise for genetic reasons. The possibility exists that we may inadvertently isolate naturally occurring "mutations," which exist in the parental strains at low frequency, during the mutagenesis. For this reason, reversion analysis of excision sublines for each putative mutant is necessary.

Significantly, reversion analysis of putative mutant strains is facilitated greatly by the presence of a selectable marker within the mutator element. Excision of the mutator element from the mutant strain can be achieved easily by reintroducing transposase via genetic crosses. If the mutator inser-tion actually disrupts the function of a gene involved with learning or memory, then some excision events will restore normal function of the host gene, thereby reverting mutant behavior to normal levels in the excision sublines. On the other hand, if the mutant phenotype is associated with a naturally occurring mutation rather than the mutator insertion, then all of the excision sublines still will show learning or memory deficits. Usually, about 100 excision events must be analyzed to distinguish between these two alternative outcomes. Here again, we expect to identify a few more false positives.

We currently can screen about 150 mutant strains per week. Compared to most other mutageneses, this is proceeding at glacial speed. Our rate-limiting step, however, is measurement of the behavioral phenotype. Here, we are constrained to be thorough to ensure the valid identification of mutations affecting learning or memory—a task made somewhat easier

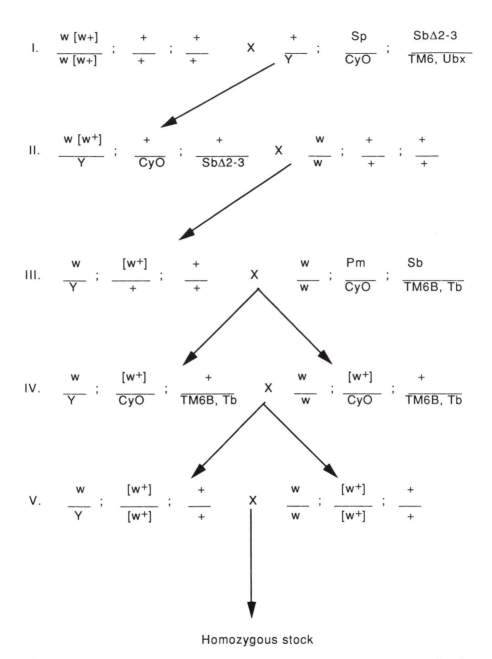

Figure 3.5 Mating scheme to produce P-element insertional mutants. In Cross I, females carrying a stable mutator P-element on the X chromosome are mated en masse to males carrying the transposase donor element on a third chromosome marked with the dominant morphological marker *Stubble-bristles (Sb)*. Mutator elements do not produce their own transposase and, therefore, cannot transpose in the absence of an "external" source of transposase. In this example, the mutator elements also contain DNA encoding the wild-type product of the *white-eye* gene. Accordingly, flies homozygous for a *white* mutation

with the development of a classical conditioning procedure that produces strong associative learning and long-lasting memory. The primary advantage of our mutagenesis is that, once identified, genes involved with learning and memory can be genetically characterized and cloned much more quickly than in the past.

In the next ten years, we expect to have cloned several of these genes. With luck, the biochemical functions of some of the corresponding gene products already will be known. We may identify other components of the cAMP second messenger system or other biochemical pathways. Yet the function of many gene products will be unknown. In any case, DNA probes from cloned genes, or antibodies raised against their gene products expressed in vitro, can be used to detect the presence of such components in other model systems of learning and memory. In this manner, the molecular components of learning and memory can be identified without making any a priori assumptions about the mechanisms involved. Indeed, cloned genes from *Drosophila* may come to represent the vanguard of our efforts to unravel the molecular biology of associative learning and memory.

Acknowledgments

I thank my colleagues S. C. Boynton, C. Brandes, J.-M. Dura, R. Mihalek, T. Preat, and A. Villella for their hard work on this project and their continuing commentary on these issues. This work was supported by NIH grants GM33205 and NS25621 and by the McKnight Foundation.

---

will be white-eyed if they do not carry at least one copy of the mutator somewhere in their genome, or they will be red-eyed if they do carry a copy of the mutator. Mutator and transposase donor elements both are carried in male progeny from Cross I, and trans position of the mutator to other (pseudorandom) sites in the genome occurs at a high frequency in both somatic and germ-line tissue of these males. In Cross II, these males are mated to females homozygous for a *white* mutation. In this manner, male progeny can be recovered that (1) do not carry the original mutator X chromosome; (2) carry at least one transposition to an autosome and, therefore, are red-eyed; and (3) no longer carry the transposase donor third chromosome (not *Sb*). Most of these males represent independent transposition events, so they are mated individually to white-eyed females in Cross III. The females used in Cross III carry "balancers" for the autosomes (*CyO* and *TM6*), which prevent viable recombinants between homologues and which carry dominant morphological markers to identify flies carrying the balancers. Use of the balancers in Cross III permits the isolation of sibling progeny, all of which are heterozygous for the same mutator insertion (and which are red-eyed). In Cross IV, these heterozygous siblings are mated together to obtain progeny that are homozygous (and red-eyed) for the same mutator insertion and which do not carry any balancer chromosomes. Finally, these homozygous flies are interbred in Cross V to produce pure-breeding mutant strains. Flies within each mutant strain are homozygous for the same mutator insertion. Flies from different strains carry mutator insertions into different regions of the autosomes. See the text for a description of how these strains are screened for mutations affecting learning or memory.

References

Aceves-Pina, E. O., & Quinn, W. G. (1979). Learning in normal and mutant *Drosphila* larvae. *Science, 206,* 93–96.

Ackerman, S. L., & Siegel, R. W. (1986). Chemically reinforced conditioned courtship in *Drosophila*: Responses of wild-type and the *dunce, amnesiac,* and *don giovanni* mutants. *Journal of Neurogenetics, 3,* 111–123.

Adler, J. (1966). Chemotaxis in bacteria. *Science, 153,* 708–716.

Adler, J. (1969). Chemorecptors in bacteria. *Science, 166,* 1588–1597.

Akers, R. F., Lovinger, D. M., Colley, P. A., Linden, D. J., & Routtenberg, A. (1986). Translocation of protein kinase C activity may mediate hippocampal long-term potentiation. *Science, 231,* 587–589.

Benzer, S. (1967). Behavioral mutants of *Drosophila* isolated by countercurrent distribution. *Proceedings of the National Academy of Sciences, U.S.A., 58,* 1112–1119.

Booker R., & Quinn, W. G. (1981). Conditioning of leg position in normal and mutant *Drosophila. Proceedings of the National Academy of Sciences, U.S.A., 78,* 3940–3944.

Brenner, S. (1974). The genetics of *Caenorhabditis elegans. Genetics, 77,* 71–104.

Byers, D. (1980). Studies on learning and cyclic AMP phosphodiesterase of the *dunce* mutant of *Drosophila melanogaster.* Unpublished doctoral dissertation, California Institute of Technology, Pasadena.

Byers, D., Davis. R. L., & Kiger, J. A. (1981). Defect in cyclic AMP phosphodiesterase due to the *dunce* mutation of learning in *Drosophila melanogaster. Nature, 289,* 79–81.

Carew, T. J., & Sahley, C. L. (1986). Invertebrate learning and memory: From behavior to molecules. *Annual Review of Neuroscience, 9,* 435–487.

Chen, C.-N., Denome, S., & Davis. R. L. (1986). Molecular analysis of cDNA clones and the corresponding genomic coding sequences of the *Drosophila dunce+* gene, the structural gene for cAMP-dependent phosphodiesterase. *Proceedings of the National Academy of Sciences (U.S.A.), 83,* 9313–9317.

Cooley, L., Berg, C., & Spradling, A. (1988). Controlling P element insertional mutagenesis. *Trends in Genetics, 4,* 254–258.

Cowan, T., & Siegel, R. W. (1986). *Drosophila* mutations that alter ionic conduction disrupt acquisition and retention of a conditioned odor avoidance response. *Journal of Neurogenetics, 3,* 187–201.

Davis, R. L., & Kiger, J. A. (1981). *Dunce* mutants of *Drosophila melanogaster*: Mutants defective in the cyclic AMP phosphodiesterase enzyme system. *Journal of Cell Biology, 90,* 101–107.

Dudai, Y. (1979). Behavioral plasticity in a *Drosophila* mutant, *dunce* [DB276]. *Journal of Comparative Physiology, 130,* 271–275.

Dudai, Y. (1983). Mutations affect storage and use of memory differentially in *Drosophila. Proceedings of the National Academy of Sciences, (U.S.A.), 80,* 5445–5448.

Dudai, Y., & Bicker, G. (1978). Comparison of visual and olfactory learning in *Drosophila. Naturwissenschaften, 65,* 495–496.

Dudai, Y., Corfas, G., & Hazvi, S. (1988). What is the possible contribution of Ca^{2+}-stimulated adenylate cyclase to acquisition, consolidation and retention of an associative olfactory memory in *Drosophila? Journal of Comparative Physiology, A 162,* 101–109.

Dudai, Y., Jan, Y.-N., Byers, D., Quinn, W. G., & Benzer, S. (1976). *Dunce,* a mutant of *Drosophila* deficient in learning. *Proceedings of the National Academy of Sciences (U.S.A.), 73,* 1684–1688.

Dudai, Y., Sher, B., Segal, D., & Yovell, Y. (1985). Defective responsiveness of adenylate cyclase to forskolin in the *Drosophila* memory mutant *rutabaga. Journal of Neurogenetics, 2,* 365–380.

Dudai, Y., & Zvi, S. (1985). Multiple defects in the activity of adenylate cyclase from the *Drosophila* memory mutant *rutabaga. Journal of Neurochemistry, 45,* 355–364.

Duerr, J. S., & Quinn, W. G. (1982). Three *Drosophila* mutations that block associative learning also affect habituation and sensitization. *Proceedings of the National Academy of Sciences (U.S.A.), 79*, 3646–3650.

Farley J., & Auerbach, S. (1986). Protein kinase C activation induces conductance changes in *Hermissenda* photoreceptors like those seen in associative learning. *Nature, 319*, 220–223.

Folkers, E. (1982). Visual learning and memory of *Drosophila melanogaster* wild-type CS and the mutants *dunce, amnesiac, turnip* and *rutabaga. Journal of Insect Physiology, 28*, 535–539.

Gailey, D. A., Jackson, F. R., & Siegel, R. W. (1984). Conditioning mutations in *Drosophila melanogaster* affect an experience-dependent behavioral modification in courting males. *Genetics, 106*, 613–623.

Hall, J. C. (1982). Genetics of the nervous system in *Drosophila. Quarterly Reviews of Biophysics, 15*, 223–479.

Hawkins, R. D., Abrams, T. W., Carew, T. J., & Kandel, E. R. (1983). A cellular mechanism of classical conditioning in *Aplysia*: Activity-dependent amplification of presynaptic facilitation. *Science, 219*, 400–405.

Hirsch, J., & Holliday, M. (1988). A fundamental distinction in the analysis and interpretation of behavior. *Journal of Comparative Psychology, 102*, 372–377.

Hochner, B., Braha, O., Klein, M., & Kandel, E. R. (1986). Distinct processes in presynaptic facilitation contribute to sensitization and dishabituation in *Aplysia*: Possible involvement of C kinase in dishabituation. *Society of Neuroscience, Abstract, 12*, 1340.

Kandel, E. R., Abrams, T., Bernier, L., Carew, T. J., Hawkins, R. D., & Schwartz, J. H. (1983). Classical conditioning and sensitization share aspects of the same molecular cascade in *Aplysia. Cold Spring Harbor Symposia on Quantitative Biology, 48*, 821–830.

Kandel, E. R., & Schwartz, J. H. (1982). Molecular biology of learning: Modulation of transmitter release. *Science, 218*, 433–443.

Karess, R. E., & Rubin, G. M. (1984). Analysis of P transposable element functions in *Drosophila. Cell, 38*, 135–146.

Kauvar, L. M. (1982). Defective cyclic adenosine $3'5'$ monophosphate phosphodiesterase in the *Drosophila* memory mutant *dunce. Journal of Neuroscience, 2*, 1347–1358.

Kiger, J. A. (1977). The consequences of nullosomy for a chromosomal region affecting cyclic AMP phosphodiesterase activity in *Drosophila. Genetics, 85*, 623–628.

Kiger, J. A., Davis, R. L., Saltz, H. K., Fletcher, T., & Bowling, M. (1981). Genetic analysis of cyclic nucleotide phosphodiesterase in *Drosophila melanogaster. Advances in Cyclic Nucleotide Research, 14*, 273–288.

Kiger, J. A., & Golanty, E. (1977). A cytogenetic analysis of cyclic nucleotide phosphodiesterase activities in *Drosophila. Genetics, 85*, 609–622.

Kiger, J. A., & Golanty, E. (1979). A genetically distinct form of cyclic AMP phosphodiesterase associated with chromosome 3D4 in *Drosophila melanogaster. Genetics, 91*, 521–535.

Lewis, E. B., & Bacher, F. (1968). Method of feeding ethyl methane sulfonate to *Drosophila* males. *Drosophila Information Service, 43*, 193.

Livingstone, M. S. (1985). Genetic dissection of *Drosophila* adenylate cyclase. *Proceedings of the National Academy of Sciences (U.S.A.), 82*, 5992–5996.

Livingstone, M. S., Sziber, P. P., & Quinn, W. G. (1984). Loss of calcium/calmodulin responsiveness in adenylate cyclase of *rutabaga*, a *Drosophila* learning mutant. *Cell, 37*, 205–215.

Livingstone, M. S., & Tempel, B. L. (1983). Genetic dissection of monoamine transmitter synthesis in *Drosophila. Nature, 303*, 67–70.

Lynch, G., & Baudry, M. (1984). The biochemistry of memory: A new and specific hypothesis. *Science, 224*, 1057–1063.

Mackintosh, N. J. (1974). *The psychology of animal learning.* New York: Academic Press.

Menne, D., & Spatz, H.-Ch. (1977). Colour vision in *Drosophila melanogaster. Journal of Comparative Physiology, 114*, 301–312.

Papazian, D. M., Schwarz, T. L., Tempel, B. L., Jan, Y. N., & Jan, L. Y. (1987). Cloning of genomic and complementary DNA from *Shaker*, a putative potassium channel gene from *Drosophila*. *Science, 237*, 749–753.

Quinn, W. G., & Dudai, Y. (1976). Memory phases in *Drosophila*. *Nature, 262*, 576–577.

Quinn, W. G., Harris, W. A., & Benzer, S. (1974). Conditioned behavior in *Drosophila melanogaster*. *Proceedings of the National Academy of Sciences (U.S.A.), 71*, 708–712.

Quinn, W. G., Sziber, P. P., & Booker, R. (1979). The *Drosophila* memory mutant *amnesiac*. *Nature, 277*, 212–214.

Rescorla, R. A. (1967). Pavlovian conditioning and its proper control procedures. *Psychological Review, 74*, 71–80.

Robertson, H. M., Preston, C. R., Phillis, R. W., Johnson-Schlitz, D. M., Benz, W. K., & Engels, W. R. (1988). A stable genomic source of P element transposase in *Drosophila melanogaster*. *Genetics, 118*, 461–470.

Rubin, G. M. (1988). *Drosophila melanogaster* as an experimental organism. *Science, 240*, 1453–1459.

Sacktor, T. C., O'Brian, C. A., Weinstein, I. B., & Schwartz, J. H. (1986) Translocation from cytosol to membrane of protein kinase C after stimulation of *Aplysia* neurons with serotonin. *Society of Neuroscience, Abstract, 12*, 1340.

Saltz, H. K., & Kiger, J. A. (1984). Genetic analysis of chromomere 3D4 in *Drosophila melanogaster*. II. Regulatory sites for the *dunce* gene. *Genetics, 108*, 377–392.

Shotwell, S. L. (1983). Cyclic adenosine 3':5'-monophosphate phosphodiesterase and its role in learning in *Drosophila*. *Journal of Neuroscience, 3*, 739–747.

Siegel, R. W., & Hall, J. C. (1979). Conditioned responses in courtship behavior of normal and mutant *Drosophila*. *Proceedings of the National Academy of Sciences (U.S.A.), 76*, 3430–3434.

Smith, R. F., Choi, K.-W., Tully, T., & Quinn, W. G. (1986). Deficient protein kinase C activity in *turnip*, a *Drosophila* learning mutant. *Society of Neuroscience, Abstract, 12*, 399.

Spatz, H.-Ch., Emmans, A., & Reichert, H. (1974). Associative learning of *Drosophila melanogaster*. *Nature, 248*, 359–361.

Tempel, B. L., Bonini, N., Dawson, D. R., & Quinn, W. G. (1983). Reward learning in normal and mutant *Drosophila*. *Proceedings of the National Academy of Sciences (U.S.A.), 80*, 1482–1486.

Tempel, B. L., Livingstone, M. S., & Quinn, W. G. (1984). Mutations in the dopa decarboxylase gene affect learning in *Drosophila*. *Proceedings of the National Academy of Sciences (U.S.A.), 81*, 3577–3581.

Tempel, B. L., Papazian, D. M., Schwarz, T. L., Jan, Y. N., & Jan, L. Y. (1987). Sequence of a probable potassium channel component encoded at *Shaker* locus of *Drosophila*. *Science, 237*, 770–775.

Tompkins, L., Siegel, R. W., Gailey, D. A., & Hall, J. C. (1983). Conditioned courtship in *Drosophila* and its mediation by association of chemical cues. *Behavior Genetics, 13*, 565–578.

Tully, T. (1986). Measuring learning in individual flies is not necessary to study the effects of single-gene mutations in *Drosophila*: A reply to Holliday and Hirsch. *Behavior Genetics, 16*, 449–455.

Tully, T. (1988). On the road to a better understanding of learning and memory in *Drosophila melanogaster*. In G. Hertting & H.-Ch. Spatz (Eds.), *Modulation of synaptic transmission and plasticity in nervous systems* (pp. 401–418). Berlin: Springer-Verlag.

Tully, T., & Gergen, J. P. (1986). Deletion mapping of the *Drosophila* memory mutant *amnesiac*. *Journal of Neurogenetics, 3*, 33–47.

Tully, T., & Quinn, W. G. (1985). Classical conditioning and retention in normal and mutant *Drosophila melanogaster*. *Journal of Comparative Physiology, 157*, 263–277.

Tully, T., Boynton, S., Brandes, C., Dura, J. M., Mihalek, R., Preat, T., & Villella, A. (in press). Genetic dissection of memory formation in *Drosophila melanogaster*. *Cold Spring Harbor Symposia on Quantitative Biology, 55*.

4

Simple Conditioning

N. J. MACKINTOSH

The experimental and theoretical analysis of Pavlovian and instrumental conditioning may seem simple, even boring or trivial, to cognitive psychologists studying human thinking, reasoning, or problem solving, but for neurobiologists interested in the structural basis of learning and memory, most modern psychological studies of conditioning are far too complex. The preferred subjects of study are rats, rabbits, and pigeons rather than simple invertebrates with more manageable numbers of neurons, and the techniques employed are enormously varied and no longer used to address familiar questions about acquisition functions or the course of extinction. Worse still, the theoretical analyses favored in the past 25 years have been marked by a sharp increase in complexity and even by appeal to apparently cognitive processes. Just when it seemed that neurobiologists were at last equipped to elucidate the physical substrate of simple conditioning, conditioning theorists moved the goalposts, claiming that conditioning is not so simple after all.

The traditional analysis of conditioning was in terms of stimulus-response theory, derived from classical reflexology: if innate behavior can be analyzed as a set of S-R reflexes, learned behavior is most economically described as a set of new S-R units. And this analysis certainly captures the obvious fact that, in order to infer that learning has taken place, all that we can ever observe in any conditioning experiment is a change in behavior (i.e., a change in a response to a stimulus). Moreover, if learning is to be of adaptive significance, it is not enough that it change an animal's knowledge of the world; it must also change the animal's behavior.

Nevertheless, there are good reasons for suggesting that conditioning should be conceptualized first as the acquisition of knowledge about the world, and only then as a potential change in behavior. Conditioning experiments arrange relationships or contingencies between events—in a Pavlovian experiment a relation between CS and US or reinforcer, in an instrumental experiment a relation between a response and a reinforcer. If we take this as our point of departure, then a natural and economical description of what is learned as a result of conditioning is an association between these paired events, CS and US or response and reinforcer, or at least between their central "representations."

The Laws of Association

One of the tasks of an associative theory of conditioning, then, is to elucidate the laws of association—that is, to specify the conditions under which these associations will be formed. Classical associationist theory and traditional conditioning theory concentrated on the role of temporal contiguity, and there can certainly be no question of the latter's importance. As can be seen in Figure 4.1, in a variety of Pavlovian preparations, conditioned responding is sensitive to variation in the temporal interval between CS and US, being best established when the CS precedes the US by a short (but not too short) interval, and declining rapidly as this interval increases. Quite why conditioning should be disrupted by very short intervals between CS and US remains something of a puzzle: Some of the more obvious and simple explanations would seem to be refuted by a feature of the results shown in Figure 4.1 that may not be immediately apparent. Although the general shape of the function relating conditioning to the CS–US interval is much the same across a variety of preparations, the absolute values of these intervals differ enormously from one preparation to another. The rabbit's eyelid CR is best conditioned when CS precedes US by less than a second, but this is far too short an interval for optimal conditioning of suppression of appetitive responding to a CS signaling shock, whereas food aversions are conditioned at intervals orders of magnitude longer than those that will sustain conditioning of either of these response systems. Although it is rash to compare across studies that differed in species, stimuli, and procedures, there is good evidence that these differences are real, and can be observed in a single experiment in the same subject, with the same CS and US (e.g., Schneiderman, 1972).

A final point needs emphasis: It is not the absolute temporal interval between CS and US that determines the level of conditioning, so much as the relative proximity of CS to US. Two examples will suffice. Food aversions can be conditioned, as Figure 4.1 makes clear, to a novel-flavored substance, say a weak vinegar solution, even though it is ingested an hour or more before the injection of lithium that makes the rat ill. But a rat that drinks some other novel flavor after drinking vinegar and shortly before becoming ill will condition an aversion to this second flavor rather than to the original vinegar (Kaye, Gambini, & Mackintosh, 1988; Revusky, 1971). That conditioning should occur to a stimulus that immediately precedes a US at the expense of less proximal stimuli may not seem particularly surprising. But in fact it contradicts traditional, simple theories of conditioning: The very same interval that produced good conditioning to the vinegar solution in the absence of other events results in little or no conditioning when another stimulus intervenes between vinegar and illness. In order to understand when conditioning will occur, one must consider the CS–US episode not in isolation, but in the context of other events in which it is embedded. But CSs are always presented in the context of

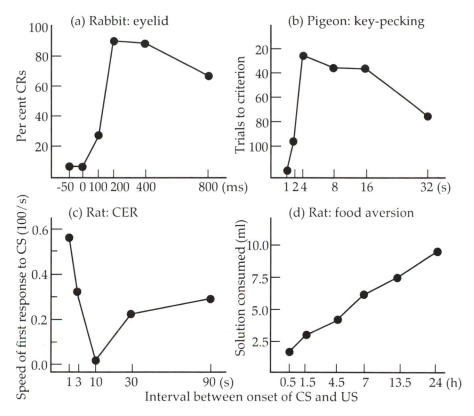

Figure 4.1 Conditioning as a function of the interval between onset of CS and onset of US; in (a), (b), and (c) the procedure is delay conditioning where the CS does not terminate until the US is presented and thus varies in duration across experimental conditions; in (d) the procedure is trace conditioning where the CS is of fixed duration (here 10 min) and thus terminates before the US onset. The results represent (a) the percentage of responses over 640 acquisition trials and (b) the number of trials required to attain a criterion of conditioning (here plotted so that successful conditioning is shown by high scores on the ordinate); (c) and (d) show the results of a test trial following a single conditioning trial (successful conditioning is shown by low scores on the ordinate). [(a) After Smith, Coleman, & Gormezano, 1969; (b) after Gibbon, Baldock, Locurto, Gold, & Terrace, 1977; (c) after Yeo, 1974; (d) after Andrews & Braveman, 1975.] [From Mackintosh, 1983.]

other events, if only that of a particular experimental apparatus. And there is evidence that conditioning to these contextual stimuli can also affect the level of conditioning that accrues to the experimenter's CS. For even in the absence of other potential CSs, it is still not the actual interval between a particular CS and US that determines conditioning to that CS, but the relationship between this interval and the interval between successive USs or trials (Gibbon, Baldock, Locurto, Gold, & Terrace, 1977). Informally, successful conditioning to the CS depends on its signaling a significant reduction in the time to wait before the next US occurs. If the US occurs at frequent intervals in this context anyway, the CS provides little additional information about its next occurrence.

Constant conjunction was a second popular law of traditional association-ist theory, translated in the terminology of conditioning experiments into the probability of reinforcement. Pavlovian conditioning occurs more rapidly, and often to a higher asymptote, the more consistently the CS predicts the occurrence of the US. But here, too, it is the relative, not the absolute, probability of US given CS that determines conditioning to the CS. In experiments on eyelid conditioning in rabbits and conditioned suppression in rats, Wagner, Logan, Haberlandt, and Price (1968) showed that animals would condition quite adequately to a light paired on only 50% of trials with the US, even though it was accompanied by one or the other of two tones, no better correlated with the occurrence of the US than the light. But if one tone consistently accompanied the light on reinforced trials, and the other on trials when no US occurred, then the first tone became established as a signal for the US, and essentially no conditioning accrued to the light at all.

Here also, conditioning to the experimental context in which CS and US are presented may act to interfere with conditioning to the CS itself. Successful conditioning requires that the CS predict an increase in the prevailing probability of reinforcement. Rescorla (1968) showed that condi-tioning was an increasing function of the difference between the probability of the US occurring in the presence of the CS and the probability of the US occurring in the absence of the CS. A given probability of reinforcement, just like a given degree of temporal proximity, may or may not result in successful conditioning, depending on what happens at other times.

Although all the experiments so far cited have been Pavlovian, it would have been equally possible to provide operant examples. Just as successful Pavlovian conditioning requires that the CS provide information about an impending US, so successful instrumental conditioning requires that the occurrence of reinforcement be predicted by the subject's actions. Temporal contiguity between response and reinforcement is an important determi-nant of conditioning, but once again it is relative rather than absolute proximity to reinforcement that counts (Williams, 1975). One still unre-solved discrepancy between Pavlovian and instrumental conditioning seems to be that instrumental responding can be maintained by very low proba-bilities of reinforcement, but even so it is the contingency between response and reinforcement, rather than the absolute probability of reinforcement, that matters, and if the occurrence of the reinforcer is better predicted by some other event, this will interfere with conditioning of the required instrumental response (Dickinson & Charnock, 1985; Mackintosh & Dickin-son, 1979).

An instrumental response that produces reinforcement in one situation may not do so in another. If the rat runs when placed in the start-box of a runway, he will get food; but running around in his home cage will only tire him out. If his lever pressing earns food only when a light is switched on, he will learn to confine lever pressing to occasions when the light is on. The light is a discriminative stimulus, which, in Skinner's words, "sets the occasion" for lever pressing (Skinner, 1938). Skinner insisted that there was an irreducible, three-term relationship between discriminative stimulus, response and reinforcer. In associative terms, this would amount to saying that the discriminative stimulus is not (or not only) associated with the response or with the reinforcer alone, but with the relationship between them: It retrieves a representation of the contingency relating the two.

The plausibility of this analysis has recently been increased by evidence of similar hierarchical association in Pavlovian experiments. If the experimenter arranges that a given CS will be paired with the US only in the presence of another stimulus X never in its absence (i.e., $X + CS \rightarrow US; CS \rightarrow 0$), then X may be said to set the occasion when the CS predicts the US. Under certain circumstances, this occasion-setting function of X may be quite independent of any direct associations it may form with the US, for it remains intact in spite of repeated presentation of X alone, sufficient to extinguish any such associations (Holland, 1989).

The associative rules suggested by this necessarily cursory account are notably more complex than those suggested by traditional accounts of temporal contiguity and constant conjunction. But they make a great deal of sense. A CS will be associated with a US, or a response with a reinforcer, only if it is the best available predictor of that US or reinforcer, and this association will be retrieved only when the animal is placed in the context where the association is held. If animals attributed the occurrence of a reinforcer to an event that only occasionally preceded it, and that by a long interval of time, when some other event regularly and immediately preceded that reinforcer, they would be affected by every chance conjunction of events, and would have failed to grasp the causal structure of their world. By conditioning selectively in this manner, animals will succeed in attributing reinforcers to their most probable causes.

Knowledge and Action

Knowledge of the causal structure of the world will be of selective advantage, however, only if it leads to adaptive action. And if the adaptive significance of conditioning depends on the behavior it generates, one might wonder why it should be necessary to suggest such an elaborate mechanism for producing that behavior. The virtue of S-R theory is that it places the emphasis directly where it belongs—on the change in behavior that provides the function of conditioning. But there are good reasons for insisting that the connections between CS and CR, or between discriminative stimulus and instrumental response, are much more indirect than S-R theory allows, and are better captured by the idea that the CS retrieves a representation of the US or that the disciminative stimulus retrieves a representation of the response–reinforcer relation.

In the first place, a change in the value put on the reinforcer by the subject will have immediate and appropriate effects on conditioned responding, effects that, although intuitively unsurprising, are quite unpredicted by S-R theory. According to S-R theory, a CR to a CS paired with food, or an instrumental response established by food reinforcement, is elicited by the CS or discriminative stimulus as a consequence of the association formed between them. The formation of these associations may have depended on stimulus and response having been followed by reinforcement, but the associations themselves do not include any representation of the food reinforcer. Thus the integrity of the association, and hence the performance of the response, will be unaffected by any later change in the value put on the reinforcer. If an aversion is subsequently conditioned to the food, for example by pairing it with an injection of lithium, animals should still, on initial test, continue to perform the original response. Neither in Pavlovian nor in instrumental experiments is this the normal outcome: Typically, even when tested in extinction and therefore without opportunity to sample the reinforcer again, animals rapidly desist from responding (Holland & Rescorla, 1975; Adams & Dickinson, 1981). Put informally, they must be credited with the knowledge that, say, lever pressing produces food, but that this particular food now makes them ill.

It is also quite clear, however, that the information conveyed by a CS may be independent of any overt response it elicits. If a small, localized visual stimulus signals the delivery of food, pigeons will approach and peck it. But neither a diffuse, overhead light nor an auditory signal paired with the delivery of food will elicit a pecking CR. This is not, however, because they are ineffective CSs: Experiments on second-order conditioning (see Table 4.1) establish quite clearly that they have been associated with food and are, in this respect, functionally equivalent to the localized CS. All three stimuli will equally act to reinforce second-order conditioning to another CS with which they are paired (Leyland, 1977; Nairne & Rescorla, 1981).

Exactly the same principle has been documented by Holland (1977) in work with rats: A visual CS paired with food elicits one CR, an auditory CS a quite different CR. But if either stimulus is now paired with a third (in the absence of food) they will be equally effective at establishing second-order

Table 4.1. Second-Order Conditioning

	Stage 1	Stage 2	Test
	CS1 → US	CS2 → CS1	CS2

Note: Once CS1 has been conditioned to a particular US, it will also have acquired the ability itself to reinforce further conditioning to a CS2 paired with it, even though CS2 has never been paired with the US.

conditioning to it, and the second-order CR now elicited by the third CS will be exactly the same regardless of which first-order CS it was paired with.

Such functional equivalence implies that, in spite of the quite different CRs different CSs may come to elicit, when paired with the same US they are associated with some similar consequence—presumably some central representation of that US. The more general implication is that there is a variety of ways of assessing what an animal has learned as a consequence of a conditioning episode, and the overt CR elicited by the CS should not be accorded special status. Another example is provided by blocking experiments (see Table 4.2). Kamin (1968) showed that if an animal received conditioning trials to a compound CS consisting of two elements, little conditioning would accrue to the second element if the animal had previously been conditioned to the first. The ability to block subsequent conditioning to the second element thus provides a nice measure of prior conditioning to the first. But this ability is quite independent of the nature of the CRs the various stimuli elicit. Diffuse visual and auditory stimuli can block the acquisition of pecking to a localized visual CS in pigeons (Blanchard & Honig, 1976; Leyland & Mackintosh, 1978), whereas a visual CS can block conditioning to an auditory CS in Holland's preparation and vice versa (Holland, 1977).

Although results such as these emphasize the importance of the distinction between learning and performance, between an animal's knowledge and the action that knowledge may lead to, they hardly help to explain just why conditioning should produce the changes in behavior that are actually observed. The adaptive significance of instrumental action is so obvious as hardly to need comment, but that does not necessarily shed light on the nature of the processes that generate action from knowledge of a response–reinforcer relation. Some theorists at least have argued that genuine instrumental action cannot be explained in simple associative terms at all, but must involve propositional knowledge and inference (Dickinson, 1985). But even if Pavlovian CRs are elicited by CSs as a consequence of a simple associative mechanism, there remains much uncertainty about the precise mechanism. Pavlov's and Konorski's stimulus-substitution theory, according to which the CS, by virtue of its association with the US, activates a representation of the US and therefore equally activates those responses normally elicited by presentation of the US itself, has been frequently dismissed, but

Table 4.2. Blocking

Groups	Stage 1	Stage 2	Test
Blocking	CS1 → US	CS1 + CS2 → US	CS2
Control	—	CS1 + CS2 → US	CS2

Note: In Stage 2, both groups receive conditioning trials to a compound of CS1 and CS2 paired with the US. Although this is sufficient to produce conditioning to CS2 in the control group, prior establishment of CS1 as a signal for the US will block conditioning to CS2.

probably remains the best starting point for an adequate theory (Mackintosh, 1983). The central feature of Pavlovian conditioning, after all, is that most CRs are recognizably related to the US whose occurrence they anticipate. And it is obvious enough why they should do so. It makes obvious adaptive sense to approach, make contact with, and engage in food-related activity in response to a stimulus associated with food, or to retreat from and respond defensively toward a threatening stimulus, and careful experimental analysis of particular CRs has documented the advantage their occurrence can confer (Hollis, 1984).

Simpler Conditioning?

The processes of conditioning suggested by the vertebrate data are not necessarily particularly simple, and although remarkable progress has been made in tracing some of the pathways involved in the establishment of one discrete skeletal CR in vertebrates, the rabbit eyelid CR (see Chapter 2), the more popular approach to the neurobiological basis of learning has been to study supposedly simpler invertebrates.

It is hardly likely that the mechanisms of conditioning that we now see in mammals and other vertebrates should have appeared suddenly, and intact, on the evolutionary scene. They presumably have evolved from earlier and simpler precursors, not necessarily just once, but many times. Although it is therefore clear that the invertebrates favored by neurobiologists for the study of learning, most notably molluscs, are by no stretch of the imagination to be regarded as ancestral to modern vertebrates (their last common ancestors were probably related to modern coelenterates), it is still worth asking whether such animals will reveal some of the simpler processes of conditioning seen in vertebrates—or at least the precursors to those processes. I am not concerned with the detailed biochemical or biophysical mechanisms underlying neural plasticity, as much as I am with the logic of the system. If fully fledged vertebrate conditioning evolved from simpler systems, both associative and nonassociative, what might these processes have looked like, and can we find evidence of them in supposedly simpler animals alive today?

The most systematic attempt to answer these questions is found in the work of Hawkins and Kandel (1984; Hawkins, 1989) on *Aplysia*. Briefly, and necessarily crudely, they have argued that one process that may produce associative learning in a Pavlovian conditioning experiment is time-dependent sensitization. Repeated presentation of a stimulus initially elicits a set of responses that will eventually habituate. But this habituation may be counteracted by a process of sensitization, whereby presentation of a second, strong stimulus may temporarily increase the vigor of the responses elicited by the first. Habituation and sensitization thus provide two opposed processes for modulating inborn reflexes. But sensitization may also come to provide a mechanism of associative learning if its action becomes narrowly constrained in time, such that only if the sensitizing stimulus (now called the US) occurs in close temporal proximity to another stimulus (a CS) will it augment responding (a CR) to that CS. In this simple mechanism of conditioning, therefore, the CR is not an indirect reflection of the association formed between the CS and a representation of the US, it is a response already elicited by the CS (although possibly below threshold), now augmented by the temporal conjunction of CS and US.

In spite of this difference in emphasis, Hawkins and Kandel's account provides a natural enough description of at least some conditioning preparations. Many CSs do initially elicit, in weak form, a response that, although habituating if the CS is repeatedly presented alone, can be strengthened by pairing the CS with a US. Rats are initially reluctant to drink a novel-flavored solution; this neophobia will habituate with repeated exposure to the solution, but pairing the solution with an injection of lithium will increase their tendency to reject it. But there are surely reasons to question whether this provides a plausible account of many of the phenomena of Pavlovian conditioning outlined earlier. As I have argued, at least in vertebrates, the CR actually elicited by a CS may be only an imperfect reflection of what an animal has learned about that CS: Different CRs to different CSs associated with the same US may conceal identical knowledge about the CS–US relation. This is one reason for supposing that CSs are associated with a central representation of the US with which they have been paired. Another is that the nature of the CR elicited by a CS is not just dependent on the nature of the CS, it also depends on the nature of the US. A CS paired with the delivery of food will elicit one set of CRs (approach, contact, salivation, consummatory responses, etc.); one associated with the delivery of water will elicit a partially overlapping set; whereas CSs associated with various aversive USs will elicit a quite different set—although again with substantial overlap between them. Even if all these responses are already connected (below threshold) to the CS, the function of a US can hardly be to augment them all. It is hard to see how a CS can come to elicit responses appropriate to (determined by) a particular impending US unless it can activate a representation of that US.

Arguments such as these are certainly not sufficient to rule out time-dependent sensitization as one mechanism of learning in Pavlovian experiments. But if the essential function of conditioning is to enable animals to predict the occurrence of USs and act appropriately in anticipation of that occurrence, then a rather more plausible precursor, and one for which there is good evidence, is the phemomenon of pseudoconditioning. The presentation of a US can sometimes cause another, neutral stimulus to elicit a response characteristically related to the US, even though there is no temporal association between the two stimuli. Thus a recently fed marine worm (*Nereis*) will be more likely to approach and explore a novel change in illumination, whereas one recently shocked will be more likely to retreat defensively when exposed to the same stimulus (Evans, 1966a, 1966b). Given a statistically regular world, it is not difficult to imagine the adaptive significance of such behavior. If the occurrence of food predicts the availability of more, it will pay to behave in a food-related way in response to any changes of stimulus once one has recently encountered food. If predators lurk around, it will pay to behave defensively for some time after one has encountered danger. But if it pays to behave in a manner appropriate to a particular US just because one has recently encountered it, it will be all the more appropriate to behave thus in response to a stimulus that predicts the occurrence of that US. As Wells (1975) has argued, it is not difficult to see how the processes of pseudo-conditioning could have evolved into something very like the process of Pavlovian conditioning sketched here.

References

Adams, C. D., & Dickinson, A. (1981). Instrumental responding following reinforcers' devaluation. *Quarterly Journal of Experimental Psychology, 33B,* 109–121.

Andrews, A. E., & Braveman, N. S. (1975). The combined effects of dosage level and interstimulus interval on the formation of one-trial poison-based aversions in rats. *Animal Learning and Behavior, 3,* 287–289.

Blanchard, R., & Honig, W. K. (1976). Surprise value of food determines its effectiveness as a reinforcer. *Journal of Experimental Psychology: Animal Behavior Processes, 2,* 67–74.

Dickinson, A. (1985). Action and habits: The development of behavioural autonomy. *Philosophical Transactions of the Royal Society, B308,* 67–78.

Dickinson, A., & Charnock, D. J. (1985). Contingency effects with maintained instrumental reinforcement. *Quarterly Journal of Experimental Psychology, 37B,* 397–416.

Evans, S. M. (1966a). Non-associative avoidance learning in nereid polychaetes. *Animal Behaviour, 14,* 102–106.

Evans, S. M. (1966b). Non-associative behavioural modifications in the polychaete *Nereis diversicolor. Animal Behaviour, 14,* 107–119.

Gibbon, J., Baldock, M. D., Locurto, C. M., Gold, L., & Terrace, H. S. (1977). Trial and intertrial durations in autoshaping. *Journal of Experimental Psychology: Animal Behavior Processes, 3,* 264–284.

Hawkins, R. D. (1989). A biologically realistic neural network model for higher-order features of classical conditioning. In R. G. M. Morris (Ed.), *Parallel distributed processing—Implications for psychology and neurobiology* (pp. 214–247). Oxford: Oxford University Press.

Hawkins, R. D., & Kandel, E. R. (1984). Is there a cell-biological alphabet for simple forms of learning? *Psychological Review, 91,* 375–391.

Holland, P. C. (1977). Conditioned stimulus as a determinant of the form of the Pavlovian conditioned response. *Journal of Experimental Psychology: Animal Behavior Processes, 3,* 77–104.

Holland, P. C. (1989). Occasion setting with simultaneous compounds in rats. *Journal of Experimental Psychology: Animal Behavior Processes, 15,* 183–193.

Holland, P. C., & Rescorla, R. A. (1975). Second-order conditioning with food unconditioned stimulus. *Journal of Comparative and Physiological Psychology, 88,* 459–467.

Hollis, K. L. (1984). The biological function of Pavlovian conditioning: The best defense is a good offense. *Journal of Experimental Psychology: Animal Behavior Processes, 10,* 413–425.

Kamin, L. J. (1968). "Attention-like" processes in classical conditioning. In M. R. Jones (Ed.), *Miami symposium on the prediction of behavior: Aversive stimulation* (pp. 9–33). Miami, FL: University of Miami Press.

Kaye, H., Gambini, B., & Mackintosh, N. J. (1988). A dissociation between one-trial overshadowing and the effect of a distractor on habituation. *Quarterly Journal of Experimental Psychology, 40B*, 31–47.

Leyland, C. M. (1977). Higher order autoshaping. *Quarterly Journal of Experimental Psychology, 29*, 607–619.

Leyland, C. M., & Mackintosh, N. J. (1978). Blocking of first- and second-order autoshaping in pigeons. *Animal Learning and Behavior, 6*, 391–394.

Mackintosh, N. J. (1983). *Conditioning and associative learning*. Oxford, England: Oxford University Press.

Mackintosh, N. J., & Dickinson, A. (1979). Instrumental (Type II) conditioning. In A. Dickinson & R. A. Boakes (Eds.). *Mechanisms of learning and motivation* (p. 143). Hillsdale, NJ: Lawrence Erlbaum.

Nairne, J. S., & Rescorla, R. A. (1981). Second-order conditioning with diffuse auditory reinforcers in the pigeon. *Learning and Motivation, 12*, 65.

Rescorla, R. A. (1968). Probability of shock in the presence and absence of CS in fear conditioning. *Journal of Comparative and Physiological Psychology, 66*, 1–5.

Revusky, S. (1971). The role of interference in association over a delay. In W. K. Honig & P. H. R. James (Eds.), *Animal memory* (pp. 155–213). New York: Academic Press.

Schneiderman, N. (1972). Response system divergencies in aversive classical conditioning. In A. H. Black & W. F. Prokasy (Eds.), *Classical conditioning II: Current research and theory* (p. 341). New York: Appleton-Century-Crofts.

Skinner, B. F. (1938). *The behavior of organisms*. New York: Appleton-Century-Crofts.

Wagner, A. R., Logan, R. A., Haberlandt, K., & Price, T. (1968). Stimulus selection in animal discrimination learning. *Journal of Experimental Psychology, 76*, 171–180.

Wells, M. J. (1975). Evolution and associative learning. In P. N. R. Usherwood & D. R. Newth (Eds.), *"Simple" nervous systems* (p. 445). London: Arnold.

Williams, B. A. (1975). The blocking of reinforcement control. *Journal of the Experimental Analysis of Behavior, 24*, 215–225.

5

Morphological Basis of Short- and Long-Term Memory in Aplysia

CRAIG H. BAILEY

One of the last frontiers in neuroscience is an understanding of the mechanisms by which learning is acquired and memory is retained. To achieve this goal may ultimately require that the biochemical, biophysical, and morphological processes that underlie learning and memory be specified in appropriate cellular detail.

The tractable central nervous systems of several higher invertebrates have proven particularly accessible for the cellular analysis of behavioral problems and have begun to reveal aspects of the mechanisms that underlie various elementary forms of learning and memory (Fentress, 1976; Hoyle, 1977; Kandel, 1976). Here I will focus on one such preparation, the marine mollusc *Aplysia californica*, to consider the morphological basis for the changes in synaptic effectiveness that underlie simple forms of learning and memory. I will review two aspects of this research designed to address the question of how the structure of the synapse may be related to its plastic capabilities. First, I will describe what we now know about the functional architecture of the synapse in *Aplysia* and how this relates to our current understanding of synaptic organization in the vertebrate CNS. Second, in an attempt to illustrate the utility that structural approaches hold for elucidating the mechanisms that underlie various forms of behaviorally relevant synaptic plasticity, I will focus on a specific set of identified synapses that we have used as a model system for exploring the morphological relationships between short- and long-term memory.

Functional Architecture of the Synapse in *Aplysia*

Chemical synapses in the mature *Aplysia* CNS are preferentially located at small, irregularly shaped varicose expansions that occur along or at the end of fine neurites (Bailey & Chen, 1983; Bailey, Thompson, Castellucci, & Kandel, 1979). A synaptic contact in *Aplysia* typically involves a single presynaptic element—a varicosity that contains varying numbers of synaptic vesicles, mitochondria, and elements of the cytoskeleton, and a single postsynaptic element, which is most often a small diameter spine but can also be a larger neurite, intervaricose segment, or varicosity, or a neuronal or glial cell body (Bailey, Hawkins, Chen, & Kandel, 1981; Bailey & Thompson, 1979; Bailey et al., 1979; Coggeshall, 1967; Graubard, 1978; Gillette & Pomeranz, 1975; Jordan & Nicaise, 1971; Shkolnik & Schwartz, 1980; Tremblay, Colonnier, & McLennan, 1979). Only a restricted portion of the total membrane surface area of the pre- and postsynaptic elements is structurally modified for synaptic transmission (Bailey, Kandel, & Chen, 1981; Bailey et al., 1979). When appropriately visualized, these specialized sites appear similar to the active zones (Couteaux & Pecot-Dechavassine, 1970; Heuser & Reese, 1981; Heuser et al., 1979) described at central synapses in other higher invertebrates and vertebrates (Gray, 1959; King, 1976; Muller & McMahan, 1976; Palay, 1958; Pfenninger, Sandri, Akert, & Eugster, 1969; Vrensen, & Cardozo, 1981; Wood, Pfenninger, & Cohen, 1977) and consist of a presynaptic component that is essentially a directional apparatus modified for loading vesicles along the membrane for subsequent fusion and release, and a postsynaptic component that is modified for receiving the chemical information and triggering subsequent electrical and in some cases biochemical events.

In conventionally fixed tissue, a typical presynaptic terminal contains a population of synaptic vesicles converging on a small patch of specialized membrane. This focal modified region—the active zone—at *Aplysia* synapses is characterized by rigidly parallel pre- and postsynaptic membranes bounded by a widened synaptic cleft that contains some form of electron-dense material (Figure 5.1). Small lucent vesicles are preferentially found clustered along these sites and electron-dense material is attached to the cytoplasmic leaflet of both pre- and postsynaptic membranes. The presynaptic specialization often appears as small, truncated, discrete dense projections. The postsynaptic specialization, by comparison, is much less pronounced and gives the appearance of a very thin sheet.

These paramembranous components of the active zone can be isolated and studied in detail by applying selective cytochemical stains (Bailey, Kandel, & Chen, 1981; Bloom & Aghajanian, 1968; Pfenninger et al., 1969). When viewed in transversely sectioned junctions (Figure 5.2a), the cytochemically stained active zone consists of three components: a series of evenly spaced presynaptic dense projections, a uniform synaptic cleft density, and a thin, sheetlike postsynaptic specialization 8 to 10 nm thick. The three-dimensional organization of dense projections within the presynaptic area is revealed in en face views (Figure 5.2b), and can exist in a precise hexagonal arrangement similar to the coherent lattice that has been described in vertebrates (Pfenninger et al., 1969).

The intramembranous architecture of the active zone, as revealed by freeze-fracture, corresponds nicely with the cytochemical views (Bailey & Chen, 1981). Presynaptic specializations appear as shallow depressions with associated clusters of intramembranous particles in P-face views similar in

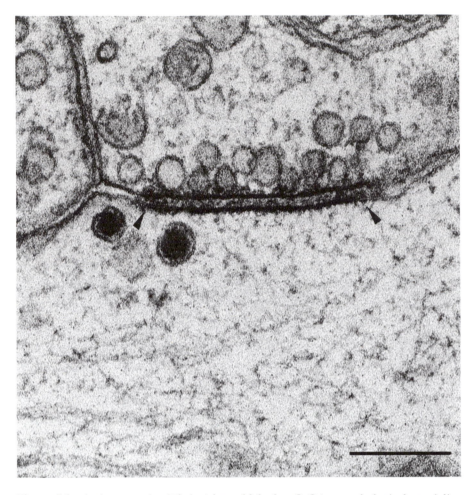

Figure 5.1 Active zone in *Aplysia* (glutaraldehyde + OsO₄): morphological specializations common to *Aplysia* active zones are illustrated between the arrowheads. See text for details. Scale = 0.25 μm. (From Bailey, Kandel, & Chen, 1981.)

size and shape to E-PTA-stained en face plaques. Postsynaptic specializations occur as loose accumulations of large intramembranous particles, which often appear at the tip of an invading spine. This array of particles is coextensive with the clustering of vesicles in the presynaptic cytoplasm. All of these features are similar to the intramembranous organization of excitatory vertebrate spine synapses found in cortical areas such as the cerebellum, hippocampus, and dendate gyrus (Landis & Reese, 1974; for review see Peters, Palay, & deF. Webster, 1976).

The results of these cytochemical and freeze-fracture studies demonstrate that *Aplysia* synapses have discrete active zones whose para- and intramembranous organization can be similar to release sites described in other

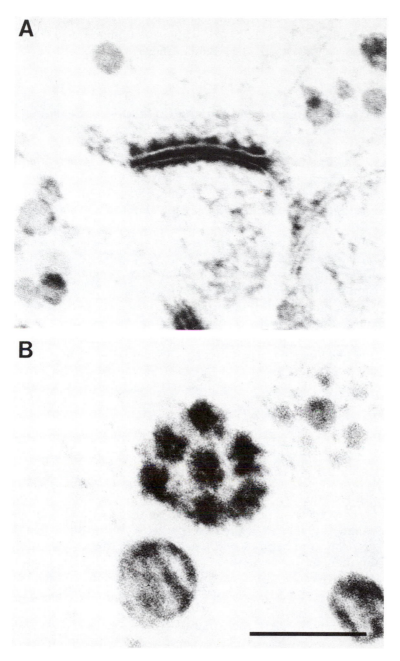

Figure 5.2 Active zone as visualized in 0.5 μm thick sections (E-PTA). (A) A row of seven regularly spaced dense projections is clearly illustrated in this transversely sectioned active zone. (B) En face view of a synaptic disk revealing a precise hexagonal array of dense projections. This geometry may represent the elemental unit of active zone organization in *Aplysia*. Scale = 0.25 μm. (From Bailey, Kandel, & Chen, 1981.)

higher invertebrates as well as vertebrate central synapses. We next asked whether these specialized sites were plastic rather than immutable components of the synapse, and in particular, whether they could be modified by learning and memory.

A Simple Behavioral System for Morphological Studies of Learning and Memory: Structural Plasticity at Identified Synapses

Toward this end we have exploited the cellular specificity of the gill- and siphon-withdrawal reflex in *Aplysia californica*. This simple behavioral system undergoes two elementary forms of nonassociative learning—habituation and sensitization. *Habituation* is probably the simplest and most common form of learning found in the animal kingdom. It is a process by which an animal learns through repeated exposure that the consequences of a weak stimulus are neither noxious nor rewarding. *Sensitization*, on the other hand, is the mirror-image process of habituation. Through it an animal learns to strengthen its defensive reflexes and to respond vigorously to previously neutral or indifferent stimuli after it has been exposed to a novel or noxious stimulus. Both habituation and sensitization can exist in a short-term form lasting minutes to hours (Carew, Castellucci, & Kandel, 1971; Pinkser, Kupfermann, Castellucci, & Kandel, 1970) and in a long-term form that persists for several weeks (Carew, Pinsker, & Kandel, 1972; Frost, Castellucci, Hawkins, & Kandel, 1985; Pinkser, Hening, Carew, & Kandel, 1973).

Through the efforts of Kandel and his colleagues, several aspects of the biophysical and biochemical mechanisms that underlie short-term habituation and sensitization have been understood. These involve changes in synaptic effectiveness produced by modulation of the Ca^{2+} current at a common locus—the presynaptic terminals of identified mechanoreceptor sensory neurons. Habituation results from a homosynaptic depression in the number of quanta released per impulse from sensory neuron terminals (Castellucci & Kandel, 1974) and is due to a reduced Ca^{++} influx (Klein & Kandel, 1980). Sensitization is produced heterosynaptically by facilitating neurons that are thought to release their transmitter onto the presynaptic terminals of sensory neurons, thereby activating an adenylate cyclase that increases synthesis of cyclic AMP. Serotonin can mimic the physiological effects of sensitization and is one of the candidates for the facilitatory transmitter. The increased cAMP, via a cAMP-dependent protein kinase, acts to increase Ca^{++} influx by reducing an opposing K^+ current, thereby prolonging the action potential and enhancing transmitter release (Castellucci et al., 1980; Klein & Kandel, 1978, 1980). Less well characterized are the morphological mechanisms that underlie habituation and sensitization—in particular the role that structural alterations at sensory neuron synapses may play in mediating the transition of their short-term form to one of longer duration.

To address these issues, we have begun to examine the nature and extent of morphological changes at sensory neuron synapses that may accompany both short- and long-term memory (Bailey & Chen, 1983, 1987, 1988a, 1988b, 1988c). For both types of studies, we have identified the presynaptic varicosities of sensory neurons by the intrasomatic injection of horseradish peroxidase (HRP) as originally described by Muller and McMahan (1976). We have increased the methodological power of the HRP technique by combining it with the analysis of serial sections. This approach has made it possible to quantitatively study complete reconstructions of unequivocally labeled identified synapses in both control and behaviorally modified animals.

Morphological Basis of Long-Term Memory

Long-Term Memory Alters the Morphology of Sensory Neuron Active Zones

Our initial ultrastructural studies of long-term memory (Bailey & Chen, 1983) utilized three groups of animals: untrained (control) animals, and animals trained for long-term habituation or sensitization following the protocols of Carew et al. (1972) and Pinsker et al. (1973). To examine in detail the fine structure of the synaptic terminals of injected cells, a total of 14 sensory neurons from 6 animals was analyzed through a blind procedure. We completely reconstructed over 300 sensory neuron synapses, a task that required the collection of nearly 5,000 thin sections and the quantitative analysis of over 12,000 HRP-labeled profiles.

We first compared the frequency of active zones at sensory neuron varicosities in control and long-term habituated and sensitized animals (Figure 5.3). Two surprising features emerged from this initial survey. The first was that only 41% of the sensory neuron varicosities in naive animals have an active zone. The second was that long-term behavioral training produces a dramatic modulation of active zone number decreasing them to 12% in habituated animals and increasing it to 65% in sensitized animals.

We next examined the total surface membrane area of reconstructed sensory neuron active zones and the vesicle complement associated with each release site in the three behavioral groups, and found a trend parallel to that of active zone number. Compared to the valves for control animals, these values were smaller in habituated animals and larger in sensitized animals.

These results demonstrated that clear structural changes could accompany behavioral modification in *Aplysia* and indicated, for the first time, that these changes could be detected at the level of identified synapses that were critically involved in the behavior. Moreover, our initial ultrastructural studies demonstrated that active zones were modifiable components of the synapse and suggested that even elementary learning experiences such as the acquisition of habituation and sensitization could alter synaptic morphology to modulate the functional expression of neural connections.

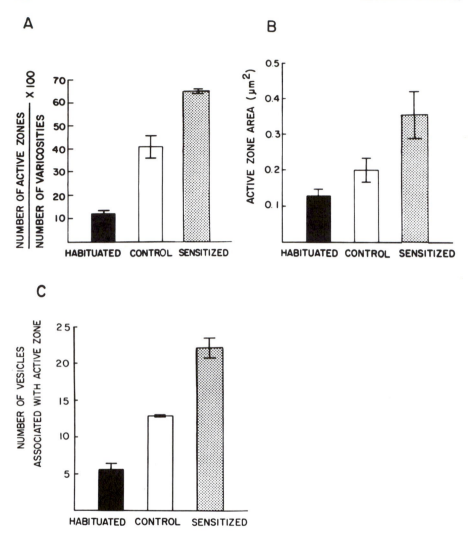

Figure 5.3 Effects of long-term memory on active zone morphology at identified sensory neuron synapses. (A) Incidence of active zones. (B) Active zone area. (C) Vesicle complement. Each bar represents the mean score of two animals ± SEM in each behavioral group. (Data taken from Bailey & Chen, 1983; figure from Bailey & Kandel, 1985.)

Long-Term Memory Modulates the Total Number of Sensory Neuron Synapses

An additional possibility, which we could not rule out at the time, was that long-term memory might also affect the total number of presynaptic varicosities per sensory neuron. We have addressed this question directly by quantitatively analyzing the total axonal arbor of single HRP-filled sensory neurons from control and behaviorally modified animals (Bailey & Chen, 1988a). To accomplish this, the neuropil arbor of each neuron was com-

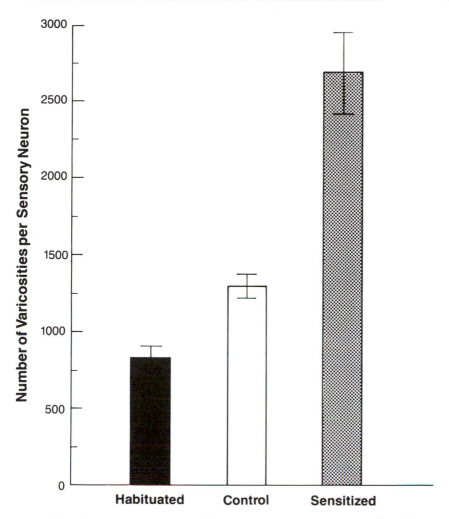

Figure 5.4 Total number of varicosities per sensory neuron in control and long-term behaviorally modified animals. Each bar represents the mean±SEM. Control groups from both experiments were not significantly different from one another and have been combined (1306±82 varicosities, mean±SEM, $n = 16$). The total number of varicosities per sensory neuron from each group was as follows: for long-term habituation animals, 836±75 varicosities (mean±SEM, $n = 10$, and for their controls, 1291±138 varicosities (mean±SEM, $n = 8$); and for long-term sensitized animals, 2697±277 varicosities (mean± SEM, $n = 10$), and for their controls, 1320±99 varicosities (mean±SEM, $n = 8$). The mean number of varicosities for long-term habituated animals is significantly less than the mean for their controls ($t = 3.05$, $P < 0.01$), and the mean for long-term sensitized animals is significantly larger than the mean for their controls ($t = 4.25$, $P < 0.01$). (From Bailey & Chen, 1988a.)

pletely reconstructed using serial 20-μm-slab-thick sections of EPON-embedded material and the total number of HRP-labeled varicosities for each cell was counted through a blind procedure. The neuropil arbors from 36 sensory neurons, taken from 36 animals, were completely reconstructed

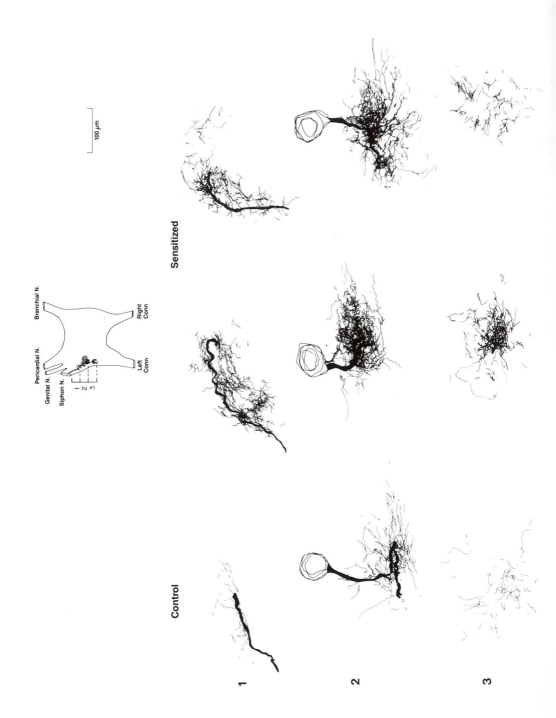

Control   Sensitized

100 μm

Pericardial N.   Branchial N.
Genital N.
Siphon N.
    1
    2
    3
                    Left   Right
                    Conn   Conn

1

2

3

84

in this fashion. The results are summarized in Figure 5.4 and reveal a trend parallel to our initial observations on active zone morphology, although represented here in an even more dramatic form because it portrays the effect of learning on each neuron's entire synaptic field. Sensory neurons from control animals had on average a total of 1,300 varicosities; this was reduced by 35% to about 850 varicosities per sensory neuron in long-term habituated animals and dramatically increased (that is, doubled to 2,700) in long-term sensitized animals. Graphic representations of completely reconstructed cells indicate that in addition to an increase in varicosity number, sensory neurons from long-term sensitized animals also display enlarged neuropil arbors (Figure 5.5).

We have recently begun to examine which class of structural changes at sensory neuron synapses might be necessary for the maintenance of long-term sensitization by exploring the time course over which they occur and in particular, their duration relative to the persistence of the memory assessed behaviorally (Bailey & Chen, 1989). By examining sensory neuron synapses at different intervals following the completion of training, we have found that the duration of changes in varicosity number as well as the incidence of active zones both parallel the behavioral time course for the retention of long-term sensitization (approximately 3 weeks). By contrast, increases in active zone size and vesicle complement are not present after 48 hours following training. These results suggest that alterations in the *number* of sensory neuron synapses are the most likely of the structural candidates to contribute to the maintenace of long-term sensitization.

Morphological Basis of Short-Term Memory

In order to explore the role that morphological alterations at sensory neuron synapses might play in the development of memory storage, we have also begun to examine the structural events that underlie a short-term memory trace (Bailey & Chen, 1988b).

Figure 5.5 Serial reconstruction of sensory neurons from long-term sensitized and control animals. Total extent of the neuropil arbors of sensory neurons from one control and two long-term sensitized animals are shown. In each case the rostral (row 3) to caudal (row 1) extent of the arbor is divided roughly into thirds. Each panel was produced by the superimposition of camera-lucida tracings of all HRP-labeled processes present in 17 consecutive slab-thick sections and represents a linear segment through the ganglion of roughly 340 μm. For each composite, ventral is up, dorsal is down, lateral is to the left, and medial is to the right. By examining images across each row (rows 1, 2, and 3), the viewer is comparing similar regions of each sensory neuron. In all cases, the arbor of long-term sensitized cells is markedly increased compared to control. (Siphon N.), (Genital N.), and so on, are various peripheral nerves of the abdominal ganglion, (Left Conn) and (Right Conn) are left and right connectives, fiber tracts connecting the abdominal ganglion with other ganglia. (From Bailey & Chen, 1988a.)

For our studies of short-term memory, we have used an in vitro sensory neuron preparation to produce the synaptic depression that underlies short-term habituation in individual cells while simultaneously labeling their synaptic terminals with intrasomatic pulses of small amounts of HRP. With this approach it is possible to monitor the sensory-to-follower cell EPSP during experimental sessions of more than one hour in duration. This connection with follower cells remains intact and decrements with kinetics that are similar to the production of synaptic depression without the introduction of label. Parallel groups of unstimulated cells treated in exactly the same way but without induced synaptic depression were used as controls.

The quantitative analysis of synaptic structure followed the same methodological approaches as those used in our study of long-term memory. To examine in detail the fine structure of synaptic terminals from control and experimental cells, we completely reconstructed 422 sensory neuron varicosities, a task that required 4,685 serial thin sections and the quanitative analysis of 17,354 HRP-labeled profiles.

We found that short-term memory is not accompanied by any obvious changes in either the incidence or the size of sensory neuron active zones. This contrasts sharply with the reduction in the number and surface area of active zones that occurs during long-term habituation. Taken together, these observations suggest that modulation of active-zone morphology may represent one of the initial phases in a family of structural and functional changes at sensory neuron synapses that underlie the transition of a short-term memory trace to one of longer duration.

A potential morphological substrate of short-term habituation was revealed by examining the relationship between the active zone and neighboring vesicle populations. Toward that end, the volume of the presynaptic terminal lying immediately above each active zone was divided into different sectors. The total vesicle population was defined as all vesicles within a vertical distance of 240 nm over the entire active zone area. A readily releasable pool was defined as all vesicles that came within 30 nm (the approximate height of individual dense projections at *Aplysia* synapses) of the presynaptic active zone membrane.

Habituation leads to a depletion of synaptic vesicles immediately adjacent to the active zone—that is, there is a dramatic reduction (by almost half) in the size of the releasable pool of vesicles at habituated terminals compared to control. Dividing this releasable pool by the total number of vesicles associated with the active zone yields a morphological expression that may be correlated with the ease with which vesicles can be brought down and loaded at the presynaptic membrane. Approximately 30% of the overlying vesicles are positioned next to the presynaptic membrane in control synapses, but only 12% are found adjacent to the membrane at depressed terminals. Since the total vesicle population is about the same for both experimental and control groups, this observation suggested that a smaller percentage of the potentially available overlying vesicles are being brought down and loaded at depressed active zones and provided the first direct morphological evidence that an impairment in vesicle mobilization may account, in part, for the homosynaptic depression that underlies short-term habituation.

This finding supports an earlier proposal by Kandel (1976) and is consistent with the model of vesicle depletion and altered mobilization predicted for synaptic depression by the theoretical studies of Gingrich and Byrne (1985). Moreover, it provides a possible morphological explanation for the results of electrophysiological studies on dishabituation by Hochner, Klein, Schacher, and Kandel (1986), where it was found that following homosynaptic depression at sensory neuron synapses, prolonging the duration of the action potential has little effect on transmitter release. Nonetheless, serotonin and cyclic AMP are still capable of enhancing release, perhaps by increasing the availability of releasable transmitter. These studies suggest that habituation leads to vesicle depletion from the active zone and a failure to mobilize overlying vesicles. To overcome habituation, therefore, a dishabituating stimulus would first have to mobilize vesicles into sensory neuron active zones.

An Overall View

The short- and long-term forms of habituation and sensitization of the gill-withdrawal reflex in *Aplysia* share a critical locus of plasticity—the monosynaptic connections between identified sensory neurons and their central target cells (Kandel & Schwartz, 1982). This reduced behavioral system therefore offers an opportunity to explore the anatomical basis of both forms of learning and both forms of memory with a realistic hope of comparing them at the mechanistic level (Bailey & Kandel, 1985). We have begun to address these issues by examining the nature, extent, and time course of morphological changes at sensory neuron synapses that accompany short- and long-term memory.

Our studies have revealed clear differences in the family of morphological events at the synapse that underlie memories of differing durations. The transient nature of short-term memories probably involves covalent modifications of preexisting proteins (Kandel, Klein, Castellucci, Schacher, Goelet, 1986) and appears to be restricted to modest structural rearrangements reflected primarily by shifts in the proximity of vesicles near the active zone. In sharp contrast, the prolonged duration of long-term memories may be dependent upon protein synthesis (Agranoff, Davies, & Brink, 1967; Davis & Squire, 1984; Montarolo et al., 1986) and is accompanied by more pronounced and potentially more enduring structural alterations that are reflected by changes in both the number pf presynaptic terminals as well as their active zone morphology (Figure 5.6).

The nature and extent of these morphological changes at sensory neuron

LONG - TERM HABITUATION

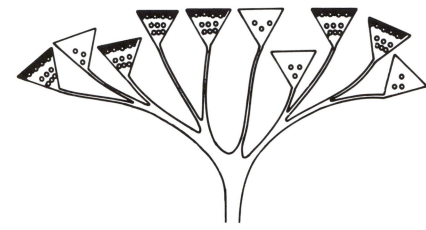

LONG - TERM SENSITIZATION

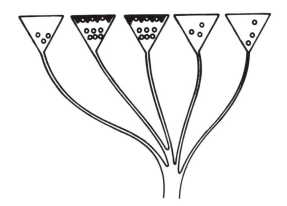

CONTROL

synapses during long-term memory are consistent with the altered behavioral efficacy of long-term habituation and sensitization in *Aplysia*
(Carew, et al., 1972; Pinkser et al., 1973) and with electrophysiological
studies that indicate a prolonged alteration in the strength of sensory
neuron connections following long-term training (Castellucci, Carew, &
Kandel 1978; Frost et al., 1985). Moreover, the changes in active zone
structure that we have observed during learning and memory in *Aplysia* are
consistent with an emerging trend based upon studies in both invertebrates
(Atwood & Marin, 1983; Rheuben, 1985; Atwood & Wojtowicz, 1986; Chiang
& Govind, 1986) as well as some reports in vertebrates (Herrera, Grinnell, &
Wolowske, 1985; Walrond & Reese, 1985) that suggest that certain aspects of
active zone morphology can be correlated with differences in synaptic
effectiveness (for review see Atwood & Lnenicka, 1986).

The increase in synapse number during long-term sensitization in *Aplysia*
is strikingly similar to results from studies in the vertebrate CNS that
indicate increases in the number of synapses following various forms of
environmental manipulations and learning tasks (for review see Greenough, 1984; Greenough & Chang, 1985). This growing body of morphological evidence, originating initially from studies on mammalian development
and more recently embodying results from research on nonmammalian
vertebrates as well as higher invertebrates, suggests that alterations in the
number of synapses may contribute to memory formation and its maintenance (Greenough & Bailey, 1988). These similarities clearly indicate a role
for structural plasticity during behavioral modification and provide evidence for a fundamental notion—that the timekeeping steps for a long-term
memory trace may be specified by morphological change at the synapse.
Although the specific molecular mechanisms that underlie these structural
changes remain unknown—altered gene regulation and protein synthesis
are likely possibilities—the emerging parallels from the invertebrate and
vertebrate studies suggest that these mechanisms are highly conserved by
evolution and may share common features throughout phylogeny.

Acknowledgments

This work was supported by Grant MH 37134 from the National Institute of
Mental Health, GM 32099 from the National Institute of Health, and the
McKnight Endowment Fund for Neuroscience.

Figure 5.6 Morphological model of long-term habituation and sensitization in *Aplysia*.
Long-term memory is accompanied by alterations in both the number of sensory neuron
varicosities and their active zones (solid triangles). These changes are differentially
expressed depending upon the type of training. Habituation leads to a decrease in these
morphological features, whereas sensitization leads to an increase. Open circles represent synaptic vesicles. (From Bailey & Chen, 1988a.)

References

Agranoff, B. W., Davies, R. E., & Brink, J. J. (1967). Actinomycin D blocks formation of memory of shock avoidance in goldfish. *Science, 158,* 1600–1601.

Atwood, H. L., & Lnenicka, G. A. (1986). Structure and function in synapses: Emerging correlations. *Trends in Neurosciences, 9,* 248–250.

Atwood, H. L., & Marin, L. (1983). Ultrastructure of synapses with different transmitter-releasing characteristics on motor axon terminals of a crab, *Hyas areneas. Cell and Tissue Research, 231,* 103–115.

Atwood, H. L., & Wojtowicz, J. M. (1986). Short-term and long-term plasticity and physiological differentiation of crustacean motor systems. *International Review of Neurobiology 28,* 275–362.

Bailey, C. H., & Chen, M. (1981). The active zone at *Aplysia* synapses: Intramembranous organization. *Society of Neuroscience, Abstract, 7,* 114.

Bailey, C. H., & Chen, M. (1983). Morphological basis of long-term habituation and sensitization in *Aplysia. Science, 220,* 91–93.

Bailey, C. H., & Chen, M. (1988a). Long-term memory in *Aplysia* modulates the total number of varicosities of single identified sensory neurons. *Proceedings of the National Academy of Sciences (U.S.A.), 85,* 2373–2377.

Bailey, C. H., & Chen, M. (1988b). Morphological basis of short-term habituation in *Aplysia. Journal of Neuroscience* (in press).

Bailey, C. H., & Chen, M. (1989). Time course of structural changes at identified sensory neuron synapses during long-term sensitization in *Aplysia. Journal of Neuroscience, 9,* 1774–1780.

Bailey, C. H., Hawkins, R. D., Chen, M., & Kandel, E. R. (1981). Interneurons involved in mediation and modulation of gill-withdrawal reflex in *Aplysia.* IV. Morphological basis of presynaptic facilitation. *Journal of Neurophysiology, 45,* 340–360.

Bailey, C. H., & Kandel, E. R. (1985). Molecular approaches to the study of short- and long-term memory. In C. W. Coen (Ed.), *Functions of the brain* (pp. 98–129). New York: Clarendon Press.

Bailey, C. H., Kandel, P., & Chen, M. (1981). The active zone at *Aplysia* synapses: Organization of presynaptic dense projection. *Journal of Neurophysiology, 46,* 356–368.

Bailey, C. H., & Thompson, E. B. (1979). Indented synapses in *Aplysia. Brain Research, 173,* 13–20.

Bailey, C. H., Thompson, E. B., Castellucci, V. F., & Kandel, E. R. (1979). Ultrastructure of the synapses of sensory neurons that mediate the gill-withdrawal reflex in *Aplysia. Journal of Neurocytology, 8,* 415–444.

Bloom, F. E., & Aghajanian, G. K. (1968). Fine structural cytochemical analysis of the staining of synaptic junctions with phosphotungstic acid. *Journal of Ultrastructure Research, 22,* 361–375.

Carew, T. J., Castellucci, V. F., & Kandel, E. R. (1971). An analysis of dishabituation and sensitization of the gill-withdrawal reflex in *Aplysia. International Journal of Neuroscience, 2,* 79–98.

Carew, T. J., Pinsker, H. M., & Kandel, E. R. (1972). Long-term habituation of a defensive withdrawal reflex in *Aplysia. Science, 175,* 451–454.

Castellucci, V. F., Carew, T. J., & Kandel, E. R. (1978). Cellular analysis of long-term habituation of the gill-withdrawal reflex of *Aplysia californica. Science, 202,* 1306–1308.

Castellucci, V. F., & Kandel, E. R. (1974). A quantal analysis of the synaptic depression underlying habituation of the gill-withdrawal reflex in *Aplysia. Proceedings of the National Academy of Sciences (U.S.A.), 71,* 5004–5008.

Castellucci, V. F., Kandel, E. R., Schwartz, J. H., Wilson, F. D., Nairn, A. C., & Greengard, P. (1980). Intracellular injection of the catalytic subunit of cyclic AMP-dependent protein kinase simulates facilitation of transmitter release underlying behavioral sensitization in *Aplysia*. *Proceedings of the National Academy of Sciences (U.S.A.)*, *77*, 7492–7496.

Chiang, R. G., & Govind, C. K. (1986). Reorganization of synaptic ultrastructure at facilitated lobster neuromuscular terminals. *Journal of Neurocytology*, *15*, 63–74.

Coggeshall, R. E. (1967). A light and electron microscope study of the abdominal ganglion of *Aplysia californica*. *Journal of Neurophysiology*, *30*, 1263–1287.

Couteaux, R., & Pecot-Dechavassine, M. (1970). Vesicules synaptiques et poches au niveau des zones actives de la jonction neuromusculaire. *Comptes Rendus de l'Académie des Sciences*, *271*, 2346–2349.

Davis, H. P., & Squire, L. R. (1984). Protein synthesis and memory: A review. *Psychological Bulletin*, *96*, 518–559.

Fentress, J. C. (1976). *Simpler networks and behavior*. Sunderland, MA: Sinauer Press.

Frost, W. N., Castellucci, V. F., Hawkins, R. D., & Kandel, E. R. (1985). Monosynaptic connections made by the sensory neurons of the gill- and siphon-withdrawal reflex in *Aplysia* participate in the storage of long-term memory for sensitization. *Proceedings of the National Academy of Sciences (U.S.A.)*, *82*, 8266–8269.

Gillette, R., & Pomeranz, B. (1975). Ultrastructural correlates of interneuronal function in the abdominal ganglion of *Aplysia californica*. *Journal of Neurobiology*, *6*, 463–474.

Gingrich, K. J., & Byrne, J. H. (1985). Simulation of synaptic depression, posttetanic potentiation and presynaptic facilitation of synaptic potentials from sensory neurons mediating gill-withdrawal reflex in *Aplysia*. *Journal of Neurophysiology*, *53*, 652–669.

Graubard, K. (1978). Serial synapses in *Aplysia*. *Journal of Neurobiology*, *9*, 325–328.

Gray, E. G. (1959). Axosomatic and axodendritic synapses of the cerebral cortex: An electron microscopic study. *Journal of Anatomy*, *83*, 420–433.

Greenough, W. T. (1984). Structural correlates of information storage in the mammalian brain: A review and hypothesis. *Trends in Neurosciences*, *7*, 229–233.

Greenough, T. T., & Bailey, C. H. (1988). The anatomy of memory: Convergence of results across a diversity of tests. *Trends in Neurosciences*, *11*, 142–147.

Greenough, W. T., & Chang, F.-L. F. (1985). Synaptic structural correlates of information storage in mammalian nervous systems. In C. W. Cotman (Ed.), *Synaptic plasticity* (pp. 335–372). New York: Guilford Press.

Herrera, A. A., Grinnell, A. P., & Wolowske, B. (1985). Ultrastructural correlates of naturally occurring differences in transmitter release efficacy in frog motor nerve terminals. *Journal of Neurocytology*, *14*, 193–202.

Heuser, J. E., & Reese, T. S. (1981). Structural changes after transmitter release at the frog neuromuscular junction. *Journal of Cell Biology*, *88*, 564–580.

Heuser, J. E., Reese, T. S., Dennis, M. J., Jan, Y., Jan, L., & Evans, L. (1979). Synaptic vesicle exocytosis captured by quick freezing and correlated with quantal transmitter release. *Journal of Cell Biology*, *81*, 275–300.

Hochner, B., Klein, M., Schacher, S., & Kandel, E. R. (1986). Additional components in the cellular mechanism of presynaptic facilitation contributes to behavioral dishabituation in *Aplysia*. *Proceedings of the National Academy of Sciences (U.S.A.)*, *83*, 8794–8798.

Hoyle, G. (1977). *Identified neurons and behavior of arthropods*. New York: Plenum Press.

Jourdan, F., & Nicaise, G. (1971). L'ultrastructure des synapses dans le ganglion pleural de l'*Aplysie*. *Journal de Microscopie (Paris)*, *11*, 69–70.

Kandel, E. R. (1976). *Cellular basis of behavior: An introduction to behavioral neurobiology*. San Francisco: W. H. Freeman.

Kandel, E. R., Klein, M., Castellucci, V., Schacher, S., & Goelet, P. (1986). Some principles emerging from the study of short- and long-term memory. *Neuroscience Research*, *3*, 198 520.

Kandel, E. R., & Schwartz, J. H. (1982). Molecular biology of an elementary form of learning: Modulation of transmitter release by cyclic AMP. *Science*, *218*, 433–443.

King, D. G. (1976). Organization of crustacean neuropil. I. Patterns of synaptic connections in lobster stomatogastric ganglion. *Journal of Neurocytology*, 5, 207–237.

Klein, M., & Kandel, E. R. (1978). Presynaptic modulation of voltage-dependent Ca^{++}
current: Mechanism for behavioral sensitization in *Aplysia californica*. *Proceedings of
the National Academy of Sciences (U.S.A.)*, 75(7), 3512–3516.

Klein, M., & Kandel, E. R. (1980). Mechanism of calcium current modulation underlying
presynaptic facilitation and behavioral sensitization in *Aplysia*. *Proceedings of the
National Academy of Sciences (U.S.A.)*, 77, 6912–6916.

Landis, D. M. D., & Reese, T. S. (1974). Differences in membrane structure between
excitatory and inhibitory synapses in the cerebellar cortex. *Journal of Comparative
Neurology*, 155, 93–126.

Montarolo, P. G., Goelet, P., Castellucci, V. F., Morgan, J., Kandel, E. R., & Schacher, S.
(1986). A critical period of macromolecular synthesis in long-term heterosynaptic
facilitation in *Aplysia*. *Science*, 234, 1249–1254.

Muller, K. J., & McMahan, U. J. (1976). The shapes of sensory and motor neurons and the
distribution of their synapses in ganglia of the leech: A study using intracellular
injection of horseradish peroxidase. *Proceedings of the Royal Society of London*, B194,
481–499.

Palay, S. L. (1958). The morphology of synapses in the central nervous system. *Experimental
Cell Research*, 5, 275–293.

Peters, A., Palay, S. L., & Webster, H. deF. (1976). *The fine structure of the nervous system. The
neurons and supporting cells*. Philadelphia: Saunders.

Pinsker, H. M., Hening, W. A., Carew, T. J., & Kandel, E. R. (1973). Long-term sensitization
of a defensive withdrawal reflex in *Aplysia*. *Science*, 182, 1039–1042.

Pinsker, H. M., Kupfermann, I., Castellucci, V. F., & Kandel, E. R. (1970). Habituation and
dishabituation of the gill-withdrawal reflex in *Aplysia*. *Science*, 167, 1740–1742.

Pfenninger, K., Sandri, C., Akert, K., & Eugster, C. H. (1969). Contribution to the problem
of structural organization of the presynaptic area. *Brain Research*, 12, 10–18.

Rheuben, M. D. (1985). Quantitative comparison of the structural features of slow and fast
neuromuscular junctions in *Manduca*. *Journal of Neuroscience*, 5, 1704–1716.

Shkolnik, L. J., & Schwartz, J. H. (1980). Genesis and maturation of serotonergic vesicles
in the identified giant cerebral neuron of *Aplysia*. *Journal of Neurophysiology*, 43,
945–967.

Tremblay, J. P., Colonnier, M., & McLennan, H. (1979). An electron microscopic study of
synaptic contacts in the abdominal ganglion of *Aplysia californica*. *Journal of Comparative Neurology*, 188, 367–396.

Vrensen, G., & Cardozo, J. Nuñes (1981). Changes in size and shape of synaptic connections after visual training: An ultrastructural approach of synaptic plasticity. *Brain
Research*, 218, 79–97.

Walrond, J. P., & Reese, T. S. (1985). Structure of axon terminals and active zones at
synapses on lizard twitch and tonic muscle fibers. *Journal of Neuroscience*, 5, 1118–
1131.

Wood, M. R., Pfenninger, K. H., & Cohen, M. J. (1977). Two types of presynaptic configurations in insect central synapses: An ultrastructural analysis. *Brain Research*, 130,
25–45.

6

Connectionist and Neural Modeling: Converging in the Hippocampus

BRUCE L. MCNAUGHTON / PAUL SMOLENSKY

This chapter arose out of an ongoing dialogue between the authors that began with the usual neurobiologist's complaint that current connectionist models (1) were grossly oversimplified and (2) made numerous assumptions that were not remotely biologically plausible, and with the usual cognitive scientist's retort that (1) not enough is known about the brain to provide either useful insights or constraints for the study of cognitive computation and (2) these insights might not be relevant in any case, because the enterprise of cognitive science is not to understand how the brain works, but to arrive at a general and practical understanding of the nature of cognition.

Our objective here is to argue (somewhat loosely) that we are on the threshold of a major change in the relation between connectionist cognitive modeling and neuroscience. In the first part, we argue that connectionist models, even (or perhaps especially) grossly oversimplified ones, can provide a badly needed conceptual framework for neuroscience, and can form the basis of biologically responsible neural models. In the second part, we go on to argue that, although connectionist modeling and empirical neuroscience are quite different enterprises—connectionist modeling having absorbed relatively little of the dramatic advances in neuroscience since the era of Hebb and McCulloch and Pitts—this picture is beginning to change. Using examples from our own area of collaboration, we hope to illustrate how recent advances in understanding the neural basis of spatial representation and memory now provide one of several emerging opportunities to bring together in a detailed and mutually productive way the enterprises of computational neuroscience and connectionist cognitive modeling.

From Connectionist to Neural Modeling in the Hippocampus

Even the most fundamental, grossly oversimplified connectionist model accounts, in an intuitive way, for many of the features of associative learning and memory. What is surprising is that even such a simple connectionist scheme can also be translated into a "neuronal" model that appears to account for a surprisingly broad range of physiological and anatomical data obtained from the mammalian hippocampal formation, a structure that is crucial for the establishment of memories of the sort that have been variously characterized as cognitive, declarative, episodic, or contextual (to list a few), and in lower mammals is best illustrated by memory for spatial relationships.

The connectionist scheme we shall consider is based on the ideas of the earliest connectionist modelers such as Steinbuch, Marr, Willshaw, Kohonen, and Anderson. These pioneers were among the first to suggest explicit theoretical formulations for the manner in which information, represented as distributed patterns of activity over a set of input lines, might be learned by a network through the adaptive modification of connections among its units. The essence of the definition of learning in these models was the ability to recall a complete, or nearly complete, representation of a previous input using only a reduced and/or corrupted subset of the original pattern. This definition encompasses both heteroassociative (i.e., paired-associate) learning and the more general case of autoassociative learning, in which the subset of the original experience to be used as a recall cue is not explicitly defined. In both cases, the goal is to reconstruct (reactivate) the missing components (units) of the original experience on the basis of increased associative strength between these components and the ones used to elicit recall. The remarkable feature of such models is that, within limits, the same network (indeed, the same set of connections) can be used to store multiple patterns without interference.

We present in the following what is perhaps the simplest possible version of such a distributed associative memory, expressed in a form that, we believe, brings out the relevance of such theoretical ideas to known neurobiological phenomena. Our version consists simply of a single set or layer of "units" with two sets of inputs (see Figure 6.1). One set, which can be thought of as a "forcing," "teaching," or "detonator" pathway, is connected to the layer in a one-to-one fashion, and with a fixed connection strength sufficient to guarantee output from its target cell. The other input set is connected to the layer exhaustively (i.e., each input unit contacts every unit of the layer), with connections that initially have zero (or very low) strength but can be changed to a strength of one according to the simplest variant of Hebb's rule: Whenever one of these inputs and a unit in the layer are simultaneously generating output, the connection weight goes *irreversibly* from zero to one.

The "activation" function of this scheme is also of the simplest variety. If the activation of a unit is one or greater, its output is one on the next (discrete-time) cycle. We also need some way of informing each unit in the layer about how many of its inputs on the modifiable pathway are currently active. The reason for this is very simple. The objective is to create a

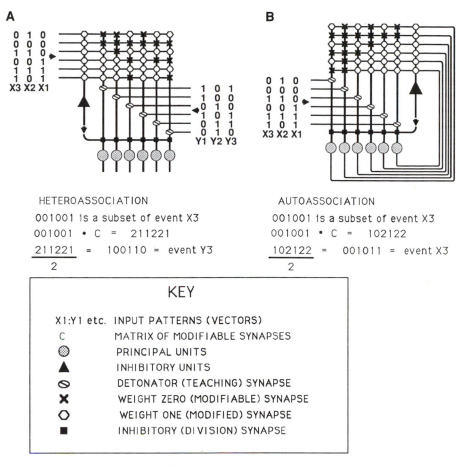

A

```
0 1 0
0 0 0
1 1 0
0 0 1
1 1 1
1 0 1
X3 X2 X1
```

```
1 0 1
1 0 0
0 1 0
1 0 1
0 1 1
0 1 0
Y1 Y2 Y3
```

B

```
0 1 0
0 0 0
1 1 0
0 0 1
1 1 1
1 0 1
X3 X2 X1
```

HETEROASSOCIATION

001001 is a subset of event X3

001001 • C = 211221

$$\frac{211221}{2} = 100110 = \text{event Y3}$$

AUTOASSOCIATION

001001 is a subset of event X3

001001 • C = 102122

$$\frac{102122}{2} = 001011 = \text{event X3}$$

KEY

X1:Y1 etc.	INPUT PATTERNS (VECTORS)
C	MATRIX OF MODIFIABLE SYNAPSES
⊛	PRINCIPAL UNITS
▲	INHIBITORY UNITS
⊘	DETONATOR (TEACHING) SYNAPSE
✕	WEIGHT ZERO (MODIFIABLE) SYNAPSE
◇	WEIGHT ONE (MODIFIED) SYNAPSE
■	INHIBITORY (DIVISION) SYNAPSE

Figure 6.1 Two versions of the Hebb–Marr connectionist theory for associative memory in neural circuits. Heteroassociation is the ability to recall one pattern (e.g., Y3) that was previously paired with a different pattern (e.g., X3), given only one member of the pair (or only part of one member). The more general form is autoassociation, which is the ability to recall a complete pattern from fragmentary input. The operation of these nets is outlined in the text. (Adapted from McNaughton & Morris, 1987.)

situation in which paired-associate learning can take place for multiple pairs. We "train" the system on pairs of patterns on the two input pathways, and later on we probe the system with an input on the modifiable pathway only. The system should respond with the corresponding event of the paired associate. In order to discover whether the current input is one to which a given unit has been trained to respond, each unit in the layer must determine whether *all* of its currently active inputs have been involved in a previous pairing. If so, then it should produce output. If not, then the current input pattern must not belong to the set of inputs for which the unit has been tuned to respond, and the unit should remain silent. This operation has sometimes been referred to as *thresholding*. Notice that, to the extent to which previous inputs were similar (i.e., shared some common elements), *some* of a unit's inputs will likely have been modified by other patterns. The simplest solution (which, as is often the case, turns out to be the most biologically plausible one) is to add up the total number of elements of the input that are currently active, and *divide* the activation on each unit in the layer by this amount. If the unit should respond, the result will be one. If not, the result will be less than one. A more conventional connectionist solution could have used a unit output threshold of anything greater than zero, and a learning rule that changed the weights from minus one to some very small positive value, to accomplish the same result using addition and subtraction only.

Such a simple binary-state network can perform a variety of interesting operations that have many (but clearly not all) of the hallmarks of mammalian associative learning. For example, if the bit string (or vector) on the teaching input represents the auditory response to a spoken name whereas the visual representation of the person's face is presented at the modifiable input, the system will subsequently recall the name when shown only the face. A number of such pairings can be stored in the same network, even if the names and faces are somewhat similar to one another. Moreover, within limits, the correct name will be recalled even if some of the features of the corresponding face have been masked.

As shown so elegantly by Marr, Kohonen, and others, replacing the modifiable *input* to the layer with modifiable *feedback* connections can enable the more general form of learning known as autoassociation, or the ability to recall a given configuration from some subset of it. It was Marr who made the first proposals about how such computations might be implemented using physiologically plausible mechanisms and anatomically plausible connections. Two networks based on his idea that the normalization (division) operation alluded to earlier is carried out by inhibitory interneurons are presented in Figure 6.1. We wish to consider briefly the extent to which inferences drawn from these nets appear to be verified by neurophysiological study, and can at least serve as a useful conceptual framework both for the design of new experiments and for the interpretation of what might otherwise be an embarrassment of disjointed experimental observations.

The first area of correspondence concerns four fundamental logical characteristics of the weight-change mechanism in the modifiable pathway. First, the weight should change only if the *receiving* cell has been activated beyond its output threshold. Second, for a given input line, only those weights where the necessary conditions on receiving cells are in place should change (this is really a corollary of the first property). Third, for a given cell in the layer, only those of its inputs that were involved in a conjunction of inputs on the modifiable and "teaching" inputs should be enhanced. Finally, according to this simple formulation, the weight change should be binary in nature. It is now quite well established that a number of pathways in the hippocampus can undergo weight changes whose physiological mechanisms conform to at least the first three of these conditions. This phenomenon, known either as long-term enhancement or as long-term potentiation, is induced by the coincidence of presynaptic activity and postsynaptic depolarization to a level beyond that possible from the activity of a single synapse. Hence more than one input must cooperate in its induction. Several groups have confirmed the specificity with respect to the sending and receiving units that is implicit in the formulation of the simple scheme under consideration. Although the binary character of the enhancement process has yet to be confirmed, experiments motivated by consideration of this or similar schemes are underway in several laboratories.

According to the conceptual model, the principal neurons of a system incorporating these principles should contain mostly weak but modifiable inputs, but also a very small proportion of considerably stronger ones. Of course, exact one-to-one input on the teaching pathway would be simplest, but this is not a necessary condition for the general principles to work. As it turns out, the molecular layer inputs to the granule cells of the fascia dentata of the hippocampus contain two populations of synapses, one of which constitutes only 1 to 2% of the other but is 10 to 20 times more powerful.

The model makes several predictions with regard to the consequence of excessive use of its modifiable inputs. As the number of these storage synapses actually used approaches 50%, recall performance starts to degrade. This phenomenon can be mimicked by artificially enhanced perforant path synapses to the fascia dentata using electrical activation. This treatment results in the predicted corruption of existing information and impaired acquisition of new information. The other side of this coin is that blockade of the modification process should cause anterograde acquisition impairment only, leaving existing information intact. Evidence in support of this inference has also been found.

Apart from some variant of Hebb's rule to control weight changes, the model also depends heavily on certain characteristics of inhibitory interneurons. The first of these is that inhibition should be equivalent to a division operation (at least approximately). It is now well established that the equilibrium potential for the major type of recurrent and feed-forward inhibition in the hippocampus is at or near the cell resting potential. Hence, most of the effect of inhibition is through a change in the membrane conductance. Ohm's law ($v = i/g$) thus makes inhibition mathematically equivalent to division.

In the model, there need only be a single inhibitory neuron, provided that it can take a good statistical sample of the input density and feed-forward inhibition to all the principal cells. In the hippocampus, the main inhibitory neurons number less than 1% of the principal cells, but do have excitatory (feed-forward) as well as feedback connections that are widely distributed. The astute reader may have noticed that, using a fixed, discrete-time cycle, our models will actually fail—because the inhibitory division operation appropriate for a particular input event must already be in place at the principal cells by the time the excitation from the same event reaches the integration site where output is initiated (i.e., the soma or initial segment of the axon). The hippocampus appears to solve this problem by making the inhibitory cells reach their discharge threshold substantially faster than the principal cells do, thus ensuring the onset of the IPSP before the principal cells can fire a spike. Moreover, this IPSP latency depends on the magnitude of the input event. If the rise of the IPSP conductance is approximately linear in time, this latency variation ensures that the IPSP magnitude at any time during this rising phase is proportional to the input density, thus concurring with another implication of the current scheme.

The essence of a distributed connectionist model is that information is contained in the population "vector" of unit activities (whether the elements are discrete or continuous as in some models). Each unit must therefore take part in numerous representations. A number of laboratories have now independently confirmed the original observations of O'Keefe, Ranck, and others, that each hippocampal pyramidal cell takes part in the representation of different parts of numerous different environments. On the other hand, arguments concerning economic and efficient use of storage predict that the overall proportion of cells that are active in the representation of a given event must be low. This is more clearly apparent in the hippocampus than perhaps in any other part of the nervous system. Some cells can be completely silent for tens of minutes, and yet fire vigorously and robustly when the animal enters the appropriate location.

The model makes the opposite prediction for the inhibitory cells. According to theory, these cells should not care *which* of their inputs are currently active, only *how many*. Accordingly, they should be active whenever there is input to the system (for example, during exploratory activity), but should convey no information within the information domain of the principal cells. Hippocampal inhibitory interneurons are most active when the animal is engaged in exploration, and exhibit almost no detectable spatial biases.

Most important, according to the model, there should be no apparent "learning" period during which the response properties of the cells are gradually shaped. The full response is determined by the fixed-weight ("hardwired") detonator synapses. However, after several exposures to the adequate stimulus configuration, the cells should perform "pattern completion," in that subsets of the initially necessary stimuli can be removed with no apparent effect. As nearly as can be measured, hippocampal cells exhibit their full place-specific responses upon first exposure to their appropriate place fields, yet, in familiar environments, most of the controlling stimuli can be removed without disruption of the fields. For example, place cells are not affected by turning out the lights, even when it can be shown that the fields are based on visual features of the remote spatial stimuli.

Finally, the feedback network of Figure 6.1b works only under the constraint that information is presented to a silent network, otherwise the current input interferes with the previous one. Because these networks reverberate (to use Hebb's term), they must be explicitly shut down before new information arrives. What better interpretation can be placed on the well-known theta rhythm of the hippocampal EEG, a 7-Hz periodic and global silencing of the entire hippocampal system?

Having illustrated how even the most trivial of connectionist schemes can provide a badly needed framework for thinking about neurobiological data, let us now attempt to show how neurobiological data might inspire connectionist computational models, again within the realm of the problem of encoding and storing "cognitive" representations of spatial relationships.

The *connectionally plausible* model we discuss derives from the attempt to explain a neurophysiological observation, first made by O'Keefe, and since extended by several others. The essence of the observation is as follows. An animal is released from a known starting location (A) in a familiar environment. It makes a left turn into another part of the environment (B) and the appropriate spatially selective cells commence firing as they always do at B. However, on this occasion, the experimenter has removed *all* of the visual features of the environment that were previously shown to determine the spatially selective firing at point B. The cells seem to fire on the basis of the knowledge that a left turn made from location A leads to location B. A little introspection at this point on how we navigate in a familiar environment makes this observation not too surprising (although the suggestion that a rat may do the same thing may surprise some connectionists). A little further neurophysiological investigation turns up the fact that a substantial part of the rat's parietal cortex (which feeds into the hippocampal formation) appears to be devoted to a robust and highly reliable representation of the rat's state of motion—whether, for example, it is turning left, turning right, or moving straight ahead. Pursuing a developing hunch a little further, we learn that all spatially selective firing in the rat's hippocampus is abolished when the animal is restrained from moving, and has learned to sit motionless without struggling. What does this have to do with spatial cognition or connectionist modeling?

Suppose that a major purpose of exploratory activity is to learn a set of conditional relations between representations of locations, and representations of the movements that link them. In our example, we can symbolize an instance of this as $AL \rightarrow B$, where AL is the compound or "configural" representation of a left turn at A, and B is the representation of location B. Suppose the memory also contains the representation $BL \rightarrow C$ and others. It is now conceivable that the organism might be able to activate a representation of C from A even without executing the represented movements. A connectionist/neural model of how this might work is suggested in Figure 6.2. One prediction of this scheme is that the time required to activate the representation of C from A will be linearly related to the length of the previously experienced path linking them. While we have no data as yet on

A Sequence of States During Initial Exploration

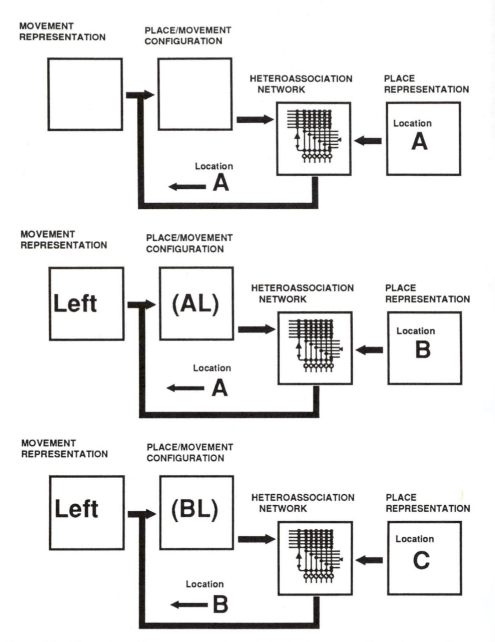

Figure 6.2 Illustration of how representations of place/movement configurations could be constructed and used to recall sequences of location representations using only information about the starting location and the sequence of movements. See text for details. It is of interest to note that if one places the "place representation" box in the

B Recall of Place Representations Given Start Location Only

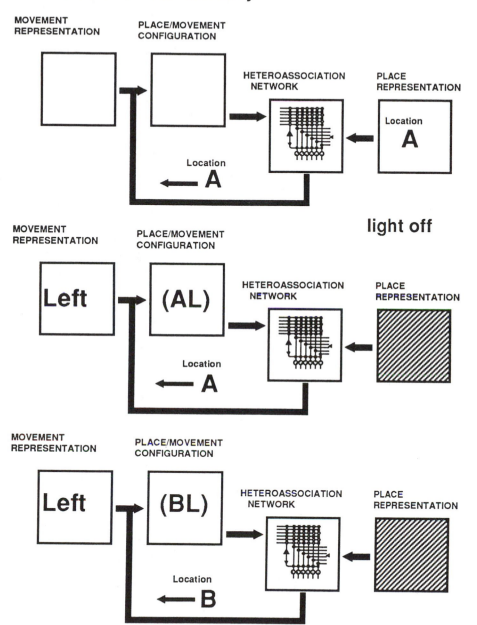

inferotemporal cortex and the "place/movement configuration" box in the posterior parietal lobe, one can develop at least a partial explanation for why these two streams of visual processing appear to be maintained separate in the primate cortex.

101

this point, the elegant experiments on "mental rotation" performed by Shepard and his colleagues suggest that there is a linear relation between the degree of rotation physically required to bring two identical objects into congruence and the reaction time before the objects are indeed identified as identical (as opposed to mirror images).

From Neural to Connectionist Modeling of Spatial Cognition

Connectionist Models versus Neural Models

Let's turn now from the relation between connectionist models and the hippocampus, and consider the general issue of how connectionist models relate to neural models. In this context, it is important to realize that connectionist modeling has several motivations in addition to neuroscience. Here we shall consider three.

The motivation farthest from neuroscience is that of designing a new class of analog computers: new architectures based on continuous variables interacting through differential equations as opposed to digital devices interacting by the discrete changing of bits. It's interesting from the perspective both of optical computers and of electronic computers to try to understand what kinds of computations these machines can do, and in particular, what kinds of intelligent behavior they can exhibit.

A second motivation comes from the theory of computation. It is extremely interesting to try to develop a new set of mathematical tools for understanding computation (and thereby cognitive processes) that derive basically from the kinds of processes that go on in physical systems: tools that derive more from the body side of the mind–body problem than the traditional ones, which stem mostly from the mind side, and involve understanding computation purely in terms of notions from discrete mathematics.

From the perspective of psychology, the most interesting motivation is the third one. The enterprise of formally modeling cognition has always been a difficult one because of a particular dilemma that faces us whenever we try to characterize intelligent behavior with any kind of precision. We are drawn in two contradictory directions. On the one hand, many intelligent processes are characterized in terms of the hard rules that define proper behavior or competence in the domain. Language and reasoning are the central cases: Sets of rules seem to characterize what it means to do the task properly. On the other hand, when we look in detail at human behavior, we see all sorts of effects that lead us to talk about statistical sources rather than rule-based sources of behavior. On the one hand we are driven to think that the mind is basically a device for manipulating abstract symbolic structures according to certain rules. On the other hand we are driven to viewing the mind as a machine that is very sensitive to the statistical environment in which it operates. The second view leads us down the road towards numerical computation, and the first view leads us in the opposite direction, to symbolic computation.

Connectionist modeling offers a subtle, interesting, and potentially very powerful way out of this dilemma. Connectionist systems are basically statistical inference and statistical learning systems, so they directly afford a modeling paradigm for the statistical side of the dilemma. On the other hand, the higher level emergent properties of connectionist systems offer the potential to provide symbolic representations and processes. Carrying out this research program will take many years, but hopeful beginnings are in place.

So connectionist models have a number of motivations that have nothing directly to do with neuroscience; it is therefore not surprising that these models are often driven by constraints that do not come from our knowledge of the structure of the nervous system. But the neuroscience motivation for these models is a central one, too, and it can be stated as follows: What are the consequences of taking a *minimal set* of the mechanisms that seem to characterize computation in the nervous system and trying to model higher level cognitive phenomena within these constraints? It is worthwhile to explicitly consider the extent to which connectionist models can properly be viewed as neural models.

Another way to ask this question is, "Is the level at which connectionist modeling is done the neural level?" It is quite helpful to distinguish what might be called the *syntactic* and the *semantic* aspects of this question; the former concerns the *processes* in the model, while the latter concerns the data or *representations* assumed. The syntactic question is, "Are the processes found in connectionist architectures really faithful to what we know about the neural architecture?" This question pertains to *internal mechanisms*; the other question pertains to the correspondence between the variables being manipulated in the system and those external things they *represent*. In neuroscience, the variables are cell activities and the like, and what they represent are their behavioral correlates or the environmental stimuli corresponding to their activity: the sorts of things one measures in behaving rats. The semantic question, then, is, "Do the things represented in connectionist models by the activity of individual units correspond well to the things represented in the nervous system by activity of individual neurons?"

These are the two questions we need to think about in regard to whether connectionist models should be viewed as neural models. Our answers to these questions are "Yes and no." Regarding the syntactic question, there is little doubt that *relative to symbolic models*—which are the only other game in town for doing cognitive modeling—the connectionist architecture is much closer to the neural architecture. That is something connectionists often find it necessary to bring up when talking to neuroscientists. At this point it is a relative comparison that has to be kept in mind. When neuroscientists talk about how absurd it is to think of these networks as being like the nervous system, we suggest that they go read about production systems and language acquisition devices.

The semantic question is not quite so simple. In connectionist representations, there tend to be two kinds of units: those whose meanings are hand-wired by the modeler, and those whose meanings are the result of self-organization during learning. Hard-wired connectionist units tend to represent information in ways that are not impressively different from symbolic models, whereas internally generated representations tend to be very different from symbolic representations—in fact, at present these internal representations are by and large incomprehensible. The neural plausibility of these representations is not easy to assess. Certainly it is not difficult to identify neural representations that are as incomprehensible as connectionist internal representations, but, with a few notable exceptions, it remains to be demonstrated that the connectionist models can provide insight into these neural representations. It is also possible to identify neural representations in higher association cortex that appear closely tied to what might be considered symbolic representations. Thus, high symbolic content is not necessarily a strike against the neural plausibility of some of the hand-wired connectionist representations—but again, it remains undemonstrated that these connectionist representations provide more neural insight than do their close symbolic counterparts. So, although it seems clear that on the syntactic question, connectionist models constitute a major step from traditional cognitive modeling toward modeling at the neural level, on the semantic question the issue remains ambiguous.

To say that the connectionist architecture more closely characterizes the nervous system than does the symbolic architecture is of course not to say that the connectionist architecture is as faithful as it could be to the neural architecture. If you were trying to do a mathematical model of what is known about the architecture of the nervous system, or if you were trying to do modeling based on what is known about the representations in the firings of individual cells, you would not build the models that connectionists tend to build (as we will soon elaborate). The conclusion is that, although these connectionist models are much closer to operating at the neural level than are symbolic models and therefore are much more helpful in understanding how neural processes can realize mental computation, they should not be confused with serious models of the nervous system (neural models). Another way of saying this is that the level at which most connectionist modeling is done is intermediate between the symbolic and neurophysiological levels.

Table 6.1 provides a more specific view of how the prototypical connectionist architecture compares with neural architecture. It shows the first dozen or so different dimensions on which it seems sensible to compare the neural architecture with the kinds of dynamic systems that connectionists use in their models. The table is intended to go more or less from the most

Table 6.1. Relations Between the Neural and Subsymbolic Architectures[a]

Cerebral cortex		Connectionist dynamical systems
State defined by continuous numerical variables (potentials, synaptic areas, . . .)	+	State defined by continuous numerical variables (activations, connection strengths)
State variables change continuously in time	+	State variables change continuously in time
Interneuron interaction parameters changeable; seat of knowledge	+	Interunit interaction parameters changeable; seat of knowledge
Huge number of state variables	+	Large number of state variables
High interactional complexity (highly non-homogeneous interactions)	+	High interactional complexity (highly non-homogeneous interactions)
Neurons located in 2 + 1-d space	−	Units have no spatial location
have dense connectivity to nearby neurons:	−	uniformly dense
have geometrically mapped connectivity to distant neurons	−	connections
Synapses located in 3-d space; locations strongly affect signal interactions	−	Connections have no spatial location
Distal projections between areas have intricate topology	−	Distal projections between node pools have simple topology
Distal interactions mediated by discrete signals	−	All interactions nondiscrete
Intricate signal intergration at single neuron	−	Signal integration is linear
Numerous signal types	−	Single signal type

[a]The intent of this table is to raise questions and provoke discussion of general issues we consider important rather than to provide definitive answers. The various assertions should therefore be taken as rough approximations rather than precise descriptions. (Reprinted from Smolensky, 1988, with permission of *The Behavioral and Brain Sciences*.)

general properties of the system to finer and finer levels of characterization. At the most general level, in the cerebral cortex the computational state of the system is presumably defined by a large number of continuous variables such as potentials and synaptic efficacies. We have a similar situation in the dynamic systems that connectionists use: A large number of numerical variables define the state of this system, and these variables change continuously in time. The interaction between neurons seems to be the seat of knowledge in the nervous system, and the corresponding interactions are the seat of knowledge in connectionist models. The cortex has very high complexity—the interactions vary largely from cell to cell: There is no spatial homogeneity of the interaction such as that characteristic of the physical systems, such as gases and magnets, that are typically studied in statistical mechanics. The same thing is true of connectionist models. At about this point, however, the correspondence between connectionist and neural architecture starts to break down. In connectionist models, units tend to have no spatial location: Whether a unit is connected to another unit is not a function of the units' two- or three-d locations because they don't have any. In many parts of the nervous system that tends not to be the case. Anatomical maps of the hippocampus and other complex structures in the nervous system tend to show many little boxes and arrows indicating complex topologies of directed connections between different modules. This is unlike most connectionist models, which tend to have extremely simple connectivity topologies. In cortex, the discrete character of signals—action potentials—is often (although not always) of paramount importance to the computation performed. Prototypical connectionist models, on the other hand, are based on interactions involving continuous signals. More-over, whereas connectionist units are pointlike, a major component of neural integration involves spatial interactions over the surface of the dendritic tree of the neuron. Finally, the nervous system is characterized by a tremendous richness of signal type—different chemical transmitters with different spatial and temporal domains of action. This richness is absent from most connectionist schemes.

So about halfway down Table 6.1, where the +s become −s, it stops being appropriate to view connectionist models as neural models. If you ask how much of what is going on in the brain can be explained with these models, you are not likely to get extremely far. Yet it is fairly clear that progress in connectionist modeling would be impeded by too much insistence on neural faithfulness. Thus, for the time being, it is best to view connectionist modeling and neural modeling as interacting but separate enterprises. This situation, however, is in the process of changing. In the remainder of this chapter we will argue that the time is ripe to bring connectionist and neural modeling into intimate contact.

Connectionist Modeling of Spatial Cognition and the Hippocampus

Where are the big opportunities for connectionist modeling to derive needed insights from neuroscience?

Some connectionists believe that if we just had the right equation to capture the magic learning rule that the brain actually uses, all the problems of cognitive science would be solved. We do not hold this view, so we are not looking for the magic learning trick that the brain uses to solve all the problems of cognition. (But we wouldn't mind if we found it.)

Connectionism constitutes a major departure from previous cognitive modeling in using numerical activity vectors as the medium of information processing. At present, our ideas of how to process information in this form are quite limited, and recent advances in neuroscience would seem to be a logical place to turn for new inspiration.

One of the most important problems for connectionism is that of representation. Connectionist cognitive models nearly always rely on assumptions about how information in the cognitive domain is represented in terms of activity patterns. This tends to be an ad hoc business, not one guided systematically by empirical constraints. But now the data (such as those alluded to earlier) concerning how spatial information is actually represented in firing patterns in the hippocampus are beginning to hold the promise of real empirical constraints on representations that connectionist models may be able to exploit.

Another important aspect of the modeling enterprise about which connectionists tend to have few clever ideas is modularization: There has been little insight into how processing being done by connectionist mechanisms should be carved up into pieces. Of course the brain has done a lot of that, and if we can understand how the brain has done it then we might start to see the computational principles involved.

So far we have identified some key places where insights from neuroscience might help connectionists in their current cognitive modeling endeavors. In addition, there are some routes through which insights from neuroscience could conceivably substantially alter the nature of connectionist models.

Is the fact that the brain is laid out geometrically a constraint that we should incorporate into connectionist cognitive models? Until now, the answer to this question has in general been "No," with the exception of certain intrinsically geometrical processes such as vision. If we were to discover that the brain is exploiting the geometry in a general and powerful way, connectionist models might well take a radical new direction in which computation is performed geometrically.

Is the fact that in many cases neural information is communicated through discrete rather than continuous signals computationally important? What is the computational significance of the fact that in the nervous system, synaptic modification takes place over a range of time scales, rather than merely slowly as in standard connectionist models? How does the brain use innate modular structure to enhance learning and processing? For each of these questions, new insights into neural computation would bear on fundamental technical assumptions of current connectionist models and could offer important new directions.

Why do we believe that the hippocampus is a good place to look for answers to these kind of questions? To do a good job of building a neural model of a cognitive process we need at least seven things.

First, we need a neural structure that is implicated in doing major processing in some cognitive task. The hippocampus appears to play a major role in learning, memory, high-level associations, spatial maps, and spatial navigation. Now, if these are not cognitive processes, what are they? Admittedly we are not talking about language or reasoning, but some day when we get rats to do that we will reappraise our options.

Next, if we want to have some view of the modularity implicit in the processing, we need some understanding of the topography of connectivity. For the hippocampus, modern neuroanatomical investigations are rapidly providing the basis for such understanding.

Third, we need to know about the dynamic interaction of the units. In the hippocampus, the physiology of how the cells communicate and their intrinsic dynamics are beginning to be well understood, perhaps better than for any other neural structure. Fortunately, unlike the cerebral cortex, the hippocampus appears to contain relatively few distinct classes of neurons.

Fourth, we need to know the details of the learning process. We know quite a bit about the synaptic enhancement processes that go on in hippocampus, yet this may only be scratching the surface of the number of different mechanisms for changing connection weights that the brain may use for different purposes.

Fifth, we need good data on how the relevant information is encoded. From single-unit studies of animals freely behaving in spatially extended environments, such as those pioneered by O'Keefe, we have quite a lot of data about behavioral and sensory correlates of hippocampal cell activity, and how these relate to spatial representation. We can thus make a serious start at addressing the general issue of the information content of individual units.

Sixth, we need good behavioral data on the cognitive task. Beginning with the era of Tolman, Lashley, and Hebb, a rich literature has developed on the spatial behavior of rodents and the nature of the spatial information that is encoded.

Finally, we need a good understanding of the inputs and outputs to the structure. Unfortunately, we are only beginning to understand this at the anatomical level, and it will likely be some time before the nature of the information exchanged is well understood. Especially to a computer scientist, knowing neither the inputs nor the outputs of a computational system is a bit of a disadvantage. But six out of seven is pretty good in this game.

Conclusion

So let us conclude. We are trying to understand the mechanisms underlying cognitive processes—an extremely difficult problem, for which we need lots of constraints. We can think of at least three different kinds of constraints: behavioral constraints, computational constraints, and neural constraints. If we can put them all together, maybe we can crack this problem.

Now, what would it really take to make this work?

First, we need a computational framework that is capable of addressing both behavioral and neural data. Connectionism is uniquely well situated to perform that role: It is located halfway between psychology and neuroscience, and is capable of making contact with both kinds of data if we know just how to push it.

Second, we need a behavioral realm that is accessible both to connectionist modeling and to neural investigation. If we tried to study conscious reasoning, we would run into two rather serious problems: It is very difficult to do any connectionist modeling at that level of cognition, and it is very difficult to get adequately detailed neural data. So we need a behavioral realm that is accessible both to connectionist modeling and to neuroscience, but is still cognitive. Spatial cognition fits the bill.

Finally, we need a neural system that is well studied and amenable to modeling in a connectionist style: The computational properties of the system must be fairly well understood and not so far from connectionist models that we cannot bridge that gap, and there must be well-documented relations between brain and behavior. What we know about the hippocampus seems to fit this bill, too.

For these reasons, we are optimistic that it is now possible to achieve some convergence between connectionist cognitive modeling and biologically responsible neural modeling via spatial cognition in the hippocampus.

Both connectionist modelers and neural modelers are ultimately after the same goal: unearthing the internal mechanisms that give rise to the richness of human behavior; bridging the gap between mental activity and physical activity; in short, solving the mind-body problem. Given the tremendous complexity of this task, strategies of simplification are necessary: connectionist modelers have chosen to simplify the physical mechanisms and to deal with as much as possible of the complexity of interesting behaviors, whereas neural modelers have chosen to simplify the physical mechanisms less and compromise as needed on the complexity of the behaviors studied. So, although the ultimate goals of connectionist and neural modeling may be the same, the methodological strategies are quite different and have led to rather different modeling enterprises. Our hope is that in the current age of interaction between connectionists and neuroscientists, it will be possible to restore to each side some of what has so far been strategically put aside.

References

Our intent in this chapter has been to try to come to grips with general principles and to avoid getting bogged down in too much detail. Accordingly, we have not included reference to explicit research findings of the many people who have contributed far in excess of our own small contributions to bringing our two disciplines to the point where we can have this dialogue. Our references, therefore, are restricted to two published position papers, one by each of us, in which interested readers can find entry to the relevant literature.

McNaughton, B. L., & Morris, R. G. M. (1987). Hippocampal synaptic enhancement and information storage within a distributed memory system. *Trends in Neurosciences*, *10*, 408–425.

Smolensky, P. (1988). On the proper treatment of connectionism. *The Behavioral and Brain Sciences*, *11*, 1–74.

7

Long-Term Potentiation and Memory Operations in Cortical Networks

GARY LYNCH / JOHN LARSON / URSULA STAUBLI / JOSÉ AMBROS-INGERSON /
RICHARD GRANGER

The phenomenology of memory is sufficiently complex, varied in nature, and enmeshed in larger psychological structures that it provides few agreed-upon benchmarks for theory development. Thus, it is difficult to decide a priori on behavioral grounds alone what is fundamental to many aspects of memory as opposed to what is likely to be a derived, secondary feature, or even to pick examples of simple versus complex memorial operations. Psychological studies have been successful in identifying what appear to be qualitatively different forms of memory, an undoubtedly important step in the development of general hypotheses. However, there is no obvious way to decide if the list of such systems is very long or to establish criteria by which specific instances of behavior should be included (or not) into an extant category.

A complementary approach to behavioral analyses of the memory system would be to seek fundamental elements and organizing principles in the neuronal machinery that encodes and processes information. That is, rather than searching for brain correlates of a particular behavioral feature, it may be possible to form de novo ideas about memory from the analysis of specific neurobiological properties. This route has the disadvantage that it would not necessarily lead, at least initially, to aspects of memory that have been studied in behavioral experiments; it could instead emphasize features that have received little experimental attention and that intuitively do not seem fundamental. The advantages of a biologically oriented strategy would be the same as those associated with any attempt to account for higher order phenomena in terms of more molecular substrates (e.g., the known properties of the substrates provide a natural scheme for organizing and interpreting the more macroscopic effects that emerge from them; hypotheses are likely to be more explicit (formal) and generalizable, etc.)

The major roadblock to efforts of this type has been the absence of information about the cellular events that are responsible for the formation of memories. However, this situation may have changed with the advent of discoveries regarding the long-term potentiation (LTP) effect. While the

110

data are by no means conclusive, there is now significant evidence implicating LTP in the encoding process. If we assume this to be the case, then the issue becomes one of how to convert information about LTP into meaningful predictions about behavior. One obvious approach is to describe the potentiation effect in terms of a set of spatiotemporal activity "rules" that dictate when and where it will appear and then incorporate these rules, along with the characteristic features of the expressed potentiation, into a computer simulation of a particular brain network. One could then ask if the network level effects that emerge are actually seen in behavior. The present review describes work that was prompted by this line of argument and begins with a summary of experimental work intended to define rules that govern the induction of LTP. Later sections will then describe some of the effects that appear when these rules are incorporated into simulations of a cortical region.

Long-Term Potentiation (LTP) and Memory

The long-term potentiation effect has been most thoroughly analyzed in the Schaffer-commissural (S-C) fibers that originate in one of the two great subfields of hippocampal pyramidal cells (CA3) and terminate in the other (CA1). When a single pulse of electrical stimulation is delivered to a subpopulation of S-C axons, it generates EPSPs with reasonably constant peak amplitudes in the CA1 cells; bursts of high-frequency stimulation result in a potentiation of the responses elicited by subsequent single pulses of stimulation (see Figure 7.1). Several characteristics of this potentiation effect have been identified, some of which seem particularly appropriate for a memory mechanism: (1) only those synapses that received the high-frequency activation experience an increase in their strength (Lynch, Dunwiddie, & Gribkoff, 1977; Andersen, Sundberg, Sveen, & Wigstrom, 1977)—LTP is selective; (2) the effect saturates at about a 50 to 100% increment in synaptic efficacy; (3) fractional amounts of stable LTP can be induced; (4) when induced with optimal stimulation patterns (see the next section), the potentiation does not detectably decay over periods of weeks (Staubli & Lynch, 1987). These points are illustrated in Figure 7.2.

Two further points should be made about the characteristics of LTP. First, there is evidence that the effect can be reversed by several seconds of low frequency stimulation (Barrionuevo, Schottler, & Lynch, 1980) and does not reappear even after a delay of 24 hours (Staubli & Lynch, 1990). Neural network research has illustrated the computational utility of a correlational rule for synaptic plasticity (i.e., a modified "Hebb synapse") in which inactive synapses on active neurons are weakened while active contacts are strengthened. The LTP rule seems to be somewhat different: Inactive synapses are not necessarily affected by induction of LTP in neighboring inputs whereas synaptic weakening (i.e., LTP reversal) involves a different

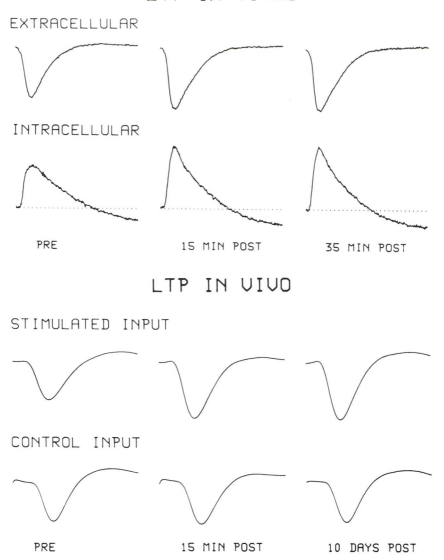

Figure 7.1 Manifestations of LTP in the hippocampal slice and the intact hippocampus of the freely moving rat. (Top): Intracellular and extracellular records of EPSPs evoked by S-C stimulation in vitro before and after theta burst stimulation (bursts of 4 pulses at 100 Hz repeated 10 times at 5 Hz). Each record is an average of 4 consecutive responses evoked at 10-s intervals. Both the single-cell and population EPSPs exhibited a potentiation that did not noticeably decay between 15 and 35 min after the burst stimulation. (Bottom) Population EPSPs evoked in the stratum radiatum of field CA1 in response to ipsilateral and contralateral CA3 stimulation in the freely behaving rat. Theta burst stimulation (2 episodes of 5 bursts given at 5 Hz) of the ipsilateral (stimulated) input resulted in LTP that was stable for 10 days; the control input (contralateral projection) exhibited stable responses that were unaffected by stimulation of the ipsilateral projection. Calibration bars for population EPSPs: 1 mV, 5 ms; for intracellular EPSPs: 5 mV, 5 ms.

112

rhythmic activity pattern than that used to strengthen transmission. Second, stable LTP is not the only synaptic facilitation to follow high-frequency stimulation—and in particular a transient type of potentiation lasting for many minutes or even hours has also been observed (Racine, Milgram, & Hafner, 1983; Larson, Wong, & Lynch, 1986).

The stability and synapse specificity of LTP are properties that would be expected of an encoding process as they would be needed to account for the stability and capacity of the memory system. While there has not been a great deal of work done on the role of LTP in memory, three lines of evidence do point to a relationship. First, two very different types of drugs (one that blocks a subclass of glutamate receptors and a second that inhibits proteolytic enzymes) that suppress the induction of LTP also interfere with the learning of spatial and olfactory cues (Collingridge, Kehl, & McLennan, 1983; Morris, Anderson, Lynch, & Baudry, 1986; Staubli, Baudry, & Lynch, 1984, 1985; Staubli, Larson, Baudry, Thibault, & Lynch, 1988; Staubli, Thibault, DiLorenzo, & Lynch, 1989). These compounds do not affect avoidance conditioning tasks. Conversely, protein synthesis inhibitors disrupt avoidance conditioning but have little effect on the learning problems that are susceptible to inhibitors of LTP (Staubli, Faraday, & Lynch, 1985). This pharmacological "double dissociation" points to the conclusion that different chemistries are involved in certain simple stimulus-response forms of learning versus the rapid acquisition of recognition memories for large numbers of environmental cues. Second, experiments using electrical stimulation of the lateral olfactory tract as a substitute for odors indicate that LTP develops in synapses driven by the stimulation as the animal learns to recognize it (the stimulation) as a disciminative cue (Roman, Staubli, & Lynch, 1987). Thus, induction of LTP correlates with the development of stable memory. Third, patterns of stimulation of the S-C pathways that mimic hippocampal rhythms occurring during learning prove to be optimal for inducing LTP, suggesting a relationship between the potentiation effect and the physiological activity that appears when hippocampus is presumably encoding information. This last point will be developed in the next section.

Spatiotemporal Activity Patterns and the Induction of LTP

Brain activity patterns are usually thought of in terms of psychological state and information transfer, but recent studies have shown that they are also linked to the induction of LTP. These observations provide the first steps toward the development of biologically based synaptic learning rules that can be incorporated into neural network models.

The hippocampus of rats (and of a number of other species) exhibits a sinusoidal 5-Hz EEG rhythm known as *theta* when the animals are exploring their environments and learning (see Bland (1986) for review). Theta waves

INCREMENTAL POTENTIATION

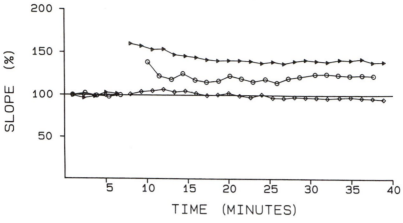

SATURATION

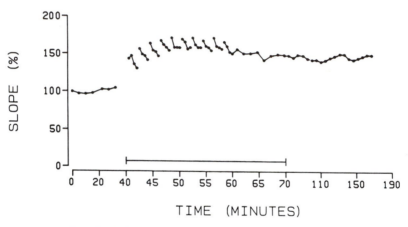

STABILITY

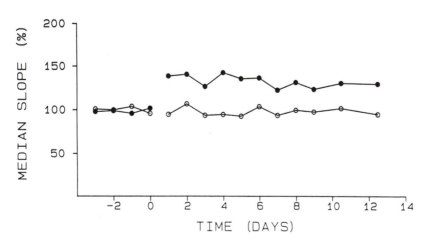

114

in hippocampus are synchronized with inhalation cycles (sniffing) during the acquisition of odor cues (Macrides, Eichenbaum, & Forbes, 1982), and therefore appear to be part of a sampling pattern. Not surprisingly, the firing patterns of individual cells in hippocampus reflect the global theta rhythm; in particular, neurons have been observed to emit 3 to 5 action potentials in <50-ms bursts with the bursts separated by 200 ms when rats are learning spatial (Hill, 1978) or olfactory (Eichenbaum, Kuperstein, Fagan, & Nagode, 1987) problems. These parameters prove to be optimal for inducing LTP—thus, brief bursts (4 pulses in 30 ms) elicit robust LTP in the S-C projections when the bursts are separated by the period of the theta wave (200 ms); shorter or longer intervals between bursts are considerably less effective (Larson, Wong, & Lynch, 1986). The theta pattern gains its efficacy in producing synaptic changes because of the refractory nature of inhibitory potentials (IPSPs) that typically accompany excitatory responses in hippocampus and other forebrain structures. That is, the S-C inputs generate EPSPs on target CA1 cells, but also activate GABAergic inter-neurons that impose a potent IPSP on the pyramidal neurons that effec-tively truncates the EPSP. Having been activated, the IPSP (lasting from 50 to 100 ms) enters a refractory period such that subsequent stimulation evokes prolonged EPSPs with little inhibition; this effect is maximal 200 ms after the first stimulation (Larson & Lynch, 1986; McCarren & Alger, 1985). The absence of the IPSP allows the longer EPSPs within a burst to summate together and thus produce a very profound level of depolarization (Larson & Lynch, 1986; 1988; 1989). These effects are illustrated in Figure 7.3.

The link between the enhanced depolarization and LTP induction is straightforward. Studies from a number of laboratories have shown that synapses in hippocampus (and many other forebrain structures) contain two classes of glutamic acid transmitter receptors, one of which (the NMDA class) is associated with an ionophore that is blocked in voltage-dependent

Figure 7.2 Characteristics of LTP in the S-C projections to field CA1 of hippocampus. (Top) Incremental nature of the LTP effect. The graph shows the initial slope of dendritic population EPSPs (plotted as percentage of baseline) before and after different burst stimulation paradigms in two in vitro experiments. (Triangles) Maximal potentiation following 10 bursts given at 5 Hz. (Diamonds) lack of change in responses evoked in an independent input to the same cells in the same experiment. (Circles) smaller but stable potentiation induced by one "primed" burst (i.e., a single burst to the fibers 200 ms after a priming burst to an independent input in a different experiment. (Middle) Saturation of LTP by repeated burst stimulation. Each gap in the graph indicates the application of a single primed burst to S-C fibers in a slice. Note that the EPSP potentiation saturates after about 5 bursts and that the final LTP remains stable for at least 2 h. (Bottom) Stability of LTP in chronic recordings from behaving rats. Stimulation electrodes were implanted bilaterally to activate S-C fibers; one input in each rat was given theta burst stimulation, the other served as a control. The graph shows the median for 8 rats of the initial slope of the population EPSP for the stimulated input (filled circles) and control input (open circles) for 4 days of baseline recording and 12–13 days after theta burst stimulation. Note that the LTP induced on the stimulated input remains stable over many days.

A

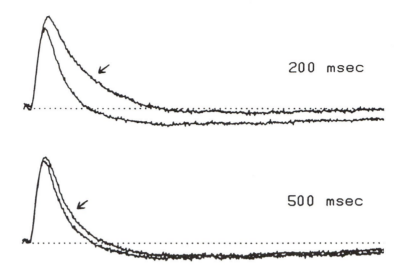

200 msec

500 msec

B

CONTROL

PRIMED

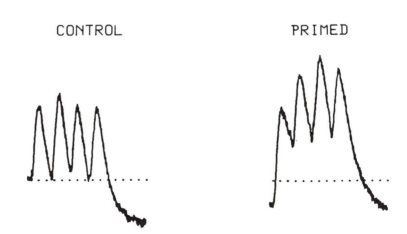

Figure 7.3 Effects of priming stimulation on intracellular single-pulse and burst responses. (A) Postsynaptic responses recorded under control conditions and 200 or 500 ms after a priming pulse to a separate input to the same target cell (arrows). Priming causes a prolongation of the EPSP that appears to result from a suppression of the IPSP; the effect is maximal 200 ms after the priming stimulation. Calibration bar: 10mV, 10 ms. (B) Postsynaptic responses to a high-frequency burst (4 pulses at 100 Hz) with (primed) or without (control) a priming pulse to a separate input 200 ms earlier. The EPSP prolongation produced by priming considerably enhances the depolarization produced by the burst. Calibration bar: 5mV, 20 ms.

116

fashion by magnesium ions (Mayer, Westbrook, & Guthrie, 1984; Nowak, Bregestovski, Ascher, Herbet, & Prochiantz, 1984). Activation of the receptor under conditions in which the ion channel is unblocked is an essential step in the induction of LTP—apparently because the ionophore passes calcium (MacDermott, Mayer, Westbrook, Smith, & Barker, 1986), the experimentally established trigger for LTP (Lynch, Larson, Kelso, Barrionuevo, & Schottler, 1983; Malenka, Kauer, Zucker, & Nicoll, 1988), into the target cell. Studies on the CA1 system have shown that the enhanced depolarization generated by theta burst stimultion does indeed elicit a sizable NMDA receptor-mediated response component that is absent under normal circumstances (Larson & Lynch, 1988).

It is important to recognize that the theta burst paradigm does not require that sequential bursts be delivered to the same population of axons. Thus, one input to a given collection of CA1 target cells will cause the IPSPs generated by surrounding interneurons to enter their refractory period; a second, completely separate set of afferents activated 200 ms later will then generate EPSPs that are largely free of inhibitory responses and that therefore produce the enhanced depolarization needed for activation of the NMDA receptor system. In essence, one input can "prime" a target cell for the arrival of a second; when input 1 and input 2 are paired in this way, LTP develops in input 2 with no change occurring in input 1 (Larson & Lynch, 1986). Given that theta is likely to be a sampling rhythm, this result suggests that a cue detected on one sample will pave the way for the encoding of a second cue detected on the following sample.

These results provide a rule for the interactions that occur between successive peaks of the theta rhythm; further study revealed that a very different set of principles governs interactions within a single peak. It can be assumed that the cells that discharge on a single segment of the theta wave will not be in complete synchrony and that bursts from different cells will arrive at a given CA1 target neuron sequentially. These conditions were reproduced in slices of hippocampus using three separate electrodes that activated different subpopulations of S-C axons. Bursts were then delivered to each electrode, with the burst to the first electrode beginning 20 ms before that to the second electrode and 40 ms before that to the third electrode; these bursts were applied 200 ms after priming pulses had been used to suppress IPSPs. This complex stimulation pattern produced the following result: Induction of LTP in the earliest arriving input was facilitated by the later arrivals whereas induction of LTP in later arrivals was suppressed by the first inputs (Larson & Lynch, 1989).

Expression of LTP

There has been considerable controversy regarding the locus and nature of the cellular changes responsible for the expression of stable LTP. Pharmacological studies have revealed a surprising aspect of LTP that imposes

strong constraints on ideas about the substrates of the effect and in particular points to a postsynaptic origin. A potentially important computational
rule also emerges from these results.

As noted earlier, synapses in hippocampus contain at least two classes of
postsynaptic transmitter receptors, one of which (the NMDA receptor)
operates through an ionic channel that is partially blocked under most
conditions. The second type of receptor (AMPA receptor) is more typical: It
opens a cation channel that is not voltage gated and mediates synaptic
transmission to single pulses of stimulation. A composite postsynaptic
response to which both the AMPA and the NMDA receptors contribute can
be generated by blocking IPSPs (e.g., with priming stimulation) and reducing extracellular magnesium concentrations (Muller, Joly, & Lynch, 1989;
Muller & Lynch, 1988). It would be assumed that increased release of
transmitter would have an equal or greater effect on the NMDA receptor
component of the response than on the AMPA component, and this has
been confirmed experimentally using manipulations that enhance release
(Muller & Lynch, 1988). LTP, however, selectively augmented the AMPA
receptor-generated potential and had little if any effect on the size of
NMDA receptor-mediated responses. Remarkably enough, LTP appears to
be induced by one class of receptors but expressed by a second (Muller, Joly,
& Lynch, 1988).

These results strongly suggest that LTP is due to one of three chages: (1)
modifications of AMPA receptors, (2) alterations in their coupling with
ionic channels, or (3) a biophysical modification that enhances AMPA
currents while having little impact on NMDA receptor-generated currents.
Electron microscopic studies have shown that LTP in field CA1 is accompanied by structural changes including a rounding of spines and an increase in the incidence of synapses attached to short spines (Chang &
Greenough, 1984; Lee, Oliver, Schottler, & Lynch, 1981; Lee, Schottler,
Oliver, & Lynch, 1980). It is not clear whether these effects represent the
formation of new contacts or a modification of existing ones, although
preliminary data from a new study point to the latter conclusion. This is
particularly interesting in the present context because modeling studies
suggest that appropriate changes in spine morphology will enhance fast
synaptic currents but have little effect on more slowly rising currents
(Wilson, 1984). Since the AMPA current is in fact much faster than the
NMDA current, we might well imagine that the LTP-related changes in
spine morphology would produce the pattern of pharmacological results
described earlier. However, the possibility also exists that shape changes
alter the surface chemistry of the spine in such a way as to allow for a larger
population of AMPA receptors to maintain themselves in the membrane or
to associate with ionophores. This argument is less satisfying because it
provides no a priori explanation for why such effects should be selective to
the AMPA receptors.

The finding that LTP is induced via one class of synaptic receptors and
expressed through another means that the potentiation effect cannot be

incorporated into models as a simple change in synaptic strength. Recall that under normal circumstances synaptic responses to single stimulation pulses are mediated by AMPA receptors whereas responses to primed bursts contain both AMPA *and* NMDA mediated components. Since LTP selectively increases the potentials generated by the AMPA subclass, we might expect it to have a larger effect on responses to single spikes than to bursts since the latter (but not the former) contain a nonpotentiated component. Physiological experiments have in fact shown that the difference between burst responses generated by "naive" versus potentiated synapses is somewhat less than the difference obtained with single-pulse responses (Lynch, Granger, Larson, & Baudry, 1989). Part of this effect can be traced to the absence of potentiation of the NMDA receptors, but part of it also is likely to be due to nonlinear summation by dendrites.

If we think of LTP as a synaptic learning rule, the preceding observations point to the curious idea that the effects of prior experience are more pronounced when the system is in performance mode (i.e., single pulses or nonrhythmic repetitive firing) than when it is in learning mode (i.e., theta bursting).

Implementing LTP in a Simple Model of a Cortical Network

Exploration of the functional properties that emerge from the characteristics of LTP requires analysis of a network of neurons and thus computer simulations of particular brain circuitries. We have selected the initial stages of the olfactory system, and in particular the superficial layers of the piriform cortex, for this purpose. Several factors led to this choice. First, layers I and II of the cortex are unusually simple and well described, features that are essential for any modeling effort. Second, the superficial portions of the piriform contain high concentrations of NMDA receptors (Monaghan & Cotman, 1985), and physiological studies have shown that the two major pathways in these layers exhibit the same pharmacological profile as those described earlier for hippocampus (Jung, Larson, & Lynch, 1990b). Third, elegant work by other investigators indicates that the excitatory-inhibitory machinery present in hippocampus is also present in the piriform (cf Tseng & Haberly, 1988, and references therein). These two points suggest that the LTP rules developed for field CA1 should be applicable in piriform cortex, and this has been partially confirmed in experiments using in vitro slices (Jung, Larson, & Lynch, 1990a). Fourth, the major extrinsic input to the superficial layers of the piriform cortex (the projections from the olfactory bulb carried by the lateral olfactory tract) has a known and stereotyped operating rhythm that corresponds to the theta rhythm used as a starting point in the studies on hippocampus. As described earlier, rats and other small mammals (rabbits, cats, small primates) sniff odors in a 4- to 8-Hz pattern and this is reflected in the physiological activity of the bulb–piriform/entorhinal cortex–hippocampus sequence. De-

ciding upon appropriate input patterns is one of the more difficult challenges facing efforts to build simulations of cortical networks and the
existence of a sampling pattern with a dominant rhythm is one of the most
attractive features of the olfactory system for modeling work.

The use of piriform cortex as a model may also somewhat shorten the
distance from results obtained with simulations to predictions about behavior. The olfactory cortex, of which the piriform constitutes the largest
single component, receives the output of the main olfactory bulb, which
itself is the recipient of the olfactory nerve; the olfactory system diverges
after, not before, the cortex. Moreover, the cortex is only two synapses
removed from the receptors in the nasal epithelium (i.e., receptor cells
project to mitral cells in the bulb, which project to the superficial layers of
olfactory cortex). The proximity of cortex to the physical stimulus of the
olfactory system and the absence of parallel pathways makes it likely that
cortical operations will be manifested in even the most fundamental of
olfactory behaviors. Accordingly, phenomena that emerge from simulations
of the superficial layers of the cortex should have recognizable counterparts
in behavior to the extent that the simulations are valid representations.

Design of the Simulation

The outer two layers of piriform cortex are composed of a dendritic zone
(layer I) and a densely packed cell body layer (layer II) from which the
dendrites emerge. The LOT travels at the top of the dendritic field and
emits fine collaterals that terminate in the outer half of layer I (layer Ia)
(Haberly & Price, 1977). The inner portions of the dendrites are innervated
by an associational system generated by the cell bodies in layer II (Haberly
and Price, 1978). It is important to note that both sets of projections are very
loosely organized and that in general the olfactory cortex lacks the topographic mapping that characterizes thalamocortical organization in the
other sensory modalities. A second feature of the system that deserves
attention is the rostrocaudal variation in the size of the associational versus
LOT projections. The bulbar projections (LOT) become progressively less
dense with increasing distance from the bulb whereas the associational
projections from layer II have a caudal orientation and hence become more
dense with distance from the bulb (Schwob & Price, 1978). (Rostral piriform
does receive associational projections, but these originate from layer III
cells and from areas of olfactory cortex outside the piriform.) Finally, layer
II neurons have extensive basal dendritic trees (in layer III) that are innervated by collateral axons originating from neighboring layer II cells. The
piriform cortex also contains feed-forward and feedback interneurons that
generate dense but spatially restricted axonal fields.

The simulation of the cortex incorporates these details, although in
greatly simplified form. Thus, the LOT inputs become progressively less
dense whereas associational projections become more frequent in the ros-

trocaudal dimension. Interneurons are much less common than layer II primary cells (by a ratio of 1 to 20), with each of them contacting ten neighboring cells; this divides the network into a set of "patches," each of which is dominated by a feed-forward and a feedback interneuron. These patches tend to act in a "winner-take-all" fashion such that one or two intensely active layer II cells will activate local feedback interneurons to a degree that suppresses firing by other layer II cells within the patch.

The network also includes after-hyperpolarizing potentials (AHPs) that develop when a neuron has fired intensely. These last for several hundred milliseconds and tend to prevent a neuron from firing again for that interval.

Finally, the model uses the incremental, saturable, and stable features of LTP described in earlier sections; the differential effects of LTP on single-pulse responses versus bursts are also included. The detailed spatiotemporal time rules have not yet been incorporated.

Results from the Simulation

The cortical layer receives its inputs from the bulb at a 5-Hz theta rhythm. This sampling rhythm allows for induction of LTP when the LOT inputs are bursting (i.e., when the network is in "learning mode") and interacts with the inhibitory processes embedded in the anatomy of the network during performance mode. That is, excitatory responses (EPSPs) are much briefer than are the intervals between sampling periods, but two forms of inhibition present in the model are not; as a result, different groups of cells fire to a fixed cue on different cycles or samples of the cue. More specifically, the local patch circuitry along with cell-specific hyperpolarizing after potentials in strongly firing cells blocks those cells that respond most strongly to the first sample of a cue from firing on subsequent cycles, causing the network to recruit different neurons on the second and subsequent cycles. In essence, the network generates different representations of the same cue over different sampling periods (Lynch & Granger, 1989; Lynch, Granger, & Larson, 1989). Analysis of these sequences revealed that they provide different levels of representation such that early responses are nearly identical across a range of similar, separately learned cues whereas later patterns of response are increasingly unique representations of individual cues. For example, if the network is taught ten cues that share a common large component (say 50% of the total cue), then on subsequent testing the network's first response will be the same to any of the cues whereas its later response will be unique to the cue now present, even if that cue closely resembles another member of the group (e.g., the two signals overlap by 90%). Results from a simulation of this type are shown in Figure 7.4.

In summary, the combination of anatomical design, physiological features, and multiple sampling that characterizes the piriform cortex, when coupled with basic characteristics of LTP, yields a system that builds percep-

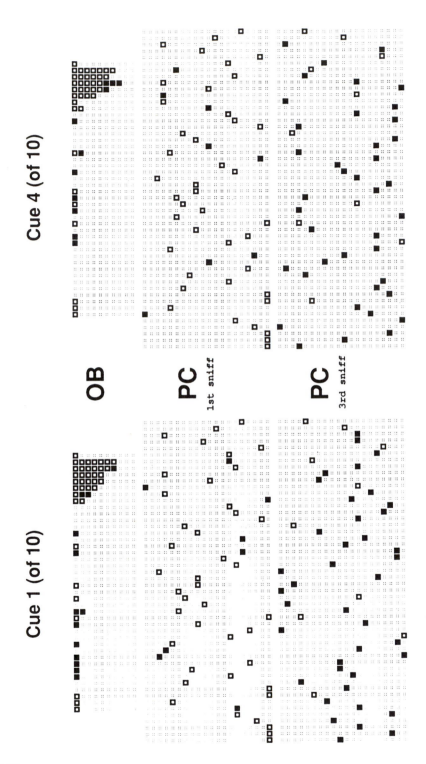

Cue 4 (of 10)

Cue 1 (of 10)

OB

PC
1st sniff

PC
3rd sniff

tual hierarchies—that is, a system that both groups (categorizes) stimuli with some degree of similarity and individuates stimuli that are very similar (Granger, Ambros-Ingerson, & Lynch, 1989). Readout of these seemingly opposing types of representation of the same cue is made possible by repetitive sampling. Perceptual hierarchies are a common feature of human perception (e.g., successive recognition as in *animal, bird, robin*) and indeed have such evident advantages that they seem a necessary part of recognition memory. The results from the simulation suggest that rather than being a feature of complex interactions between brain networks, hierarchical organization will be an implicit part of memory operations of even very simple cortical networks and therefore should be present in all aspects of animal behavior.

Interconnecting the Olfactory Bulb with the Superficial Layers of the Piriform Cortex

Behavioral properties will emerge from interactions between as well as within networks, and it can be assumed that many design features are present to insure that these interactions proceed smoothly. The outer layers of piriform cortex not only receive input from the bulb, they also direct a large projection back to the bulb both monosynaptically and via a relay in the anterior olfactory nucleus. Thus bulb and cortex form a loop. We have begun exploring the significance of these arrangements, and in particular how the induction of LTP in the cortex affects their interactions, using a very simplified version of the bulb.

The olfactory bulb is composed of a layer of mitral/tufted cells that are innervated by the olfactory nerve and that project to the superficial layers of the cortex. Two layers of inhibitory cells are present: a group of superficial periglomerular neurons that interact with the distal tips of the mitral cell dendrites at the sites where contact with the olfactory nerve is achieved, and a deeper layer of granule cells that form dendrodendritic contacts with the more proximal portions of the mitral cell dendrites. This latter arrangement has a number of unusual features, two of which are incorporated into the bulb model. First, the primary branch of the mitral cell dendrites emits a

---

Figure 7.4   Representation of the response of a simulation of the piriform cortex (PC) to two similar cues, with a cue being defined as a set of active olfactory bulb (OB) cells. The network had previously been trained on a collection of 10 cues (including the two used in the illustration) that share a common set of active OB cells. The open squares denote cells that are responsive to both cues; filled squares indicate those that are responsive to one but not the other. The PC model repetitively samples ("sniffs") its inputs, and the first and third such episodes are illustrated in the middle and bottom panels of the figure. Note that the first response of the network to the two cues is quite similar in that 84% of the active cells are triggered by either cue. However, the output of the network to the third sample is very different for the two cues (i.e., 80% of the cells activated by one cue are not driven by the other).

number of oblique processes that travel for great distances across the bulb. Second, the mitral–granule cell dendrodendritic contacts are bidirectional—with the mitral cell exciting the granule cell, which in turn inhibits the mitral cell. The granule cell dendrites are more restricted in their distribution than are those of the mitral neurons. These arrangements would seem to produce a system in which local excitation of a group of mitral cells by a subpopulation of olfactory nerve fibers will excite granule cell dendrites over a large area of the bulb, producing an inhibition of other mitral cells in that area. The possible functional significance of this can be appreciated when it is realized that single odors excite defined patches of the bulb (Lancet, Greer, Kauer, & Shepherd, 1982); thus, presentation of a particular simple odorant should produce a focus of active mitral cells surrounded by a broad area of inhibition. When complex signals are presented, then several patches would be excited, but this effect would be moderated by local inhibition from granule cells triggered by oblique dendrites of mitral cells in other, even quite distant, excited patches. For purposes of a simple model, it was assumed that the granule cell inhibitory system serves to normalize the output of the bulb, such that a relatively constant amount of output (number of cells, rate of firing) occurs across substantial variations in the number and intensity of odor components.

How does the feedback from cortex fit into this organization? In principle, the activation of a local group of granule cells could impose inhibition over a patch of mitral cells that react to a particular odor component and thereby modify the signal being sent on to the cortex. The chief difficulty with this idea is that there is little evidence for any topographical relationship in the feedback from cortex to bulb (just as there is little evidence for such relationships in bulb to cortex).

The general question at issue here is how a very loosely organized loop of connections can exploit the topography present in the initial stages of olfactory processing. One possibility was suggested by a simulation involving a "developmental" period during which the network was exposed to hundreds of simple cues and the connections between cortex and granule cell layer of bulb allowed to vary in strength using the modified "Hebb rule"; that is, active contacts on active granule cells were strengthened whereas inactive contacts on these cells were weakened. At the end of "development," LTP was instituted in the cortex, plasticity in the corticobulbar system was terminated, and the combined simulation allowed to run as before. Under these conditions, learning of a cue by the cortical network led to a robust suppression of the patch of bulbar cells that responded to the cue and that generated the initial input to the cortex. A particularly interesting case emerged when a complex cue activated two or more bulb patches to varying degrees; when this signal was learned, then feedback from cortex suppressed each of the active foci scattered across the bulb. Thus, a learned response pattern in cortex consisting of cells with no particular spatial relationship produced a coherent pattern of inhibition in the bulb.

The potential utility of this system became apparent when the combined simulation was presented with two previously learned complex "odors" (e.g., with components *ABC* versus components *XYZ*) where one odor (*ABC*) was much stronger than the other. Under these circumstances the previously formed response pattern to the stronger odor appeared in early samples but was then replaced by the representation of the weaker signal (see Figure 7.5). This result occurred because the learned cortical response to the dominant odor suppressed most of the representation in cortex of the weaker signal; as a result, the feedback was directed primarily at the bulb patches activated by the stronger input. Renormalization following this inhibition enhanced activity in those bulb zones responsive to the secondary odor to the point at which they would trigger the cortical representation of that odor. The bulb–cortex model was thus able to use repetitive sampling to dissect a very complicated signal into two previously learned, complex signals.

It need not be emphasized that using Hebbian learning during a "developmental" period involves the introduction of purely hypothetical events into the simulation. The results suggest one way in which the many elements present in the bulbar–cortical system interact and provides an unexpected solution to a difficult, though presumably commonplace problem. Whether the system actually develops in this way can only be addressed through physiological and behavioral experimentation.

Discussion

Computer simulations have begun to provide ideas about the network-level properties that emerge from some of the more salient aspects of long-term potentiation, but the functional consequences of the complex timing rules that govern the induction of the effect remain to be explored. Some insight into this may come when the bulbar–piriform model is confronted with more realistic representations of olfactory cues. We assume that odors, and in particular complex odors, will generate responses with distinct rise and fall times in different patches of the bulb. The output of the bulb to cortex will thus consist of different collections of cells from different areas firing in a sequence that is a characteristic for a particular odor. The LTP sequence rule (i.e., the first input suppresses induction of potentiation in later inputs; later inputs facilitate LTP induction in early inputs) operates over the time frame of one sample cycle and thus should have a major impact on which cells will gain strengthened synapses. Also, the strength of inhalation can vary in a stereotyped fashion across successive peaks in the theta pattern of sampling (Youngentob, Mozell, Sheehe, & Hornung, 1987),

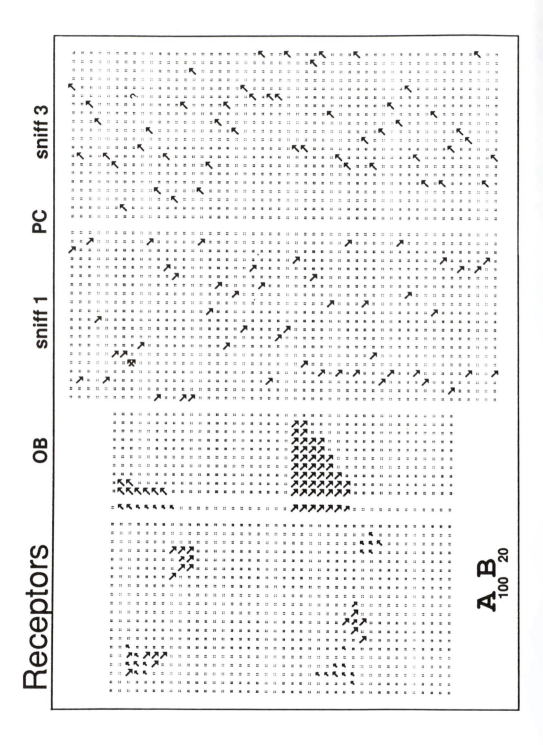

a feature that could result in the addition of components to the input signal. The "between-peaks" LTP rule (i.e., first input primes target cells for a second input arriving 200 ms later) would then operate upon the successively arriving signals, again determining which neurons would develop potentiated contacts. These events may permit encoding in some form of the temporal as well as the spatial topography of the input signal, a possibility that can be tested using simulations that implement neurons with more realistic biophysical properties than those used in the bulbar–piriform models described above.

A more generalized formulation of the LTP rules should also promote efforts to understand their functional significance. Hitherto, descriptions have been limited to statements relating specific spatial and temporal aspects of afferent activity to the occurrence or nonoccurrence of LTP. As a deeper appreciation of the cellular events responsible for those relationships (IPSP refractory period, temporal properties of the NMDA receptor–ionophore complex, etc.) is achieved, it will be possible to define single-neuron equations that handle a very wide range of input patterns in terms of their efficacy in inducing LTP.

The simulation work to date, limited though it is, has already resulted in several observations about the types of memory operations that should emerge from networks using LTP. Perhaps the most striking of these is that some type of organization of memory seems to emerge as many representations are encoded. The incremental, saturable, and stable characteristics of LTP all contribute this. Thus, shared components of cues produce maximally potentiated contacts that drive a subset of cells that defines a categorical response that dominates the output of the network to early samples of a cue. The stability of potentiation ensures that these representations are not disturbed by subsequent learning. The differential effects of the potentiation effect during learning and performance modes (i.e., the reduced difference between potentiated and naive contacts during theta bursting) ensures that cells with maximally potentiated synapses do not become "attractors" upon which new inputs preferentially strengthen their synapses. In a sense, this feature of LTP prevents categories from becoming excessively broad (see Granger, Ambros-Ingerson, & Lynch, 1989, for a discussion of other network properties that define category width).

Figure 7.5   First sniff and third sniff simulated responses to a cue that models a mixture of two cues: 100 parts of (A; represented by southeast directional arrows) plus 20 parts of (B; represented with northeast directional arrows). The olfactory bulb responses to each component have been collected into two separate groups of contiguous cells in order to illustrate the relative intensities of the components; in actuality, these cells would be found in several scattered patches of active neurons. The network has been trained on (A) and (B) separately, and the PC (piriform cortex) layer can be seen to produce the pattern elicited by the presentation of (A) alone on the first sniff and the pattern of (B) alone on the third sniff, thus in effect segmenting the input cue into the learned components. No such effect is obtained before the network is trained on the mixture's components.

The simulations also emphasize the point that multiple sampling is needed both to form potentiated contacts and to retrieve the information they encode. The networks use LTP and the theta rhythm to carry out categorizing and individuating operations. Sequential sampling also plays a pivotol role in detecting and recognizing a complex signal masked by an equally complex but much stronger input.

Finally, there remain the related questions of the validity and utility of the network models. Do the cortical circuits used as a model actually work in the manner described by the simulations, and can we use the results so far obtained to develop hypotheses about the nature of memory? As discussed, one of the primary reasons for using piriform cortex as a basis for simulation work is that network-level phenomena present in that region can reasonably be expected to manifest themselves in a recognizable fashion in behavior. So we can now ask if rats actually do form perceptual hierarchies for odors and use multiple sniffing to uncover complex, masked signals.

Beyond this, the networks exhibit a variety of unexpected memory effects that can be tested in animals. For example, the model has difficulty in distinguishing two signals *Abc* versus *Abd* when the unit *Ab* is dominant and has been learned previously. Without prior learning the same cues are easily distinguished and learned. When the network does eventually acquire the discrimination, it does so by learning the *c* and *d* components of the signals. In essence, prior experience "blinds" the network to an obvious distinction and forces it to use a strategy other than that normally employed. Behavioral work has already established that rats readily learn problems of the *Abc* versus *Abd* variety and do not on subsequent trials recognize the components *c* and *d* (Staubli, Fraser, Faraday, & Lynch, 1987). Studies are now in progress to test the predictions of the simulation regarding the effects of prior learning in this task.

In summary, the modeling enterprise does lead to ideas about fundamental features of memory that have not been stressed by more traditional approaches. Tests of these ideas and the evolution of the simulations toward greater biological validity may therefore produce novel theories about how experience is encoded, organized, and utilized by cortical circuitries.

References

Andersen, P., Sundberg, S. H., Sveen, O., & Wigstrom, H. (1977). Specific long-lasting potentiation of synaptic transmission in hippocampal slices. *Nature (London), 266,* 736–737.

Barrionuevo, G., Schottler, F., & Lynch, G. (1980). The effects of repetitive low frequency stimulation on control and "potentiated" synaptic responses in the hippocampus. *Life Sciences, 27,* 2385–2391.

Bland, B. H. (1986). The physiology and pharmacology of hippocampal formation theta rhythms. *Progress in Neurobiology, 26,* 1–54.

Chang, F. L.-F., & Greenough, W. T. (1984). Transient and enduring morphological correlates of synaptic activity and efficacy change in the rat hippocampal slice. *Brain Research, 309*, 35–46.

Collingridge, G. L., Kehl, S. J., & McLennan, H. (1983). Excitatory amino acids in synaptic transmission in the Schaffer collateral-commissural pathway of the rat hippocampus. *Journal of Physiology (London), 334*, 33–46.

Eichenbaum, H., Kuperstein, M., Fagan, A., & Nagode, J. (1987). Cue-sampling and goal-approach correlates of hippocampal unit activity in rats performing an odor-discrimination task. *Journal of Neuroscience, 7*, 716–732.

Granger, R., Ambros-Ingerson, J., & Lynch, G. (1989). Derivation of encoding characteristics of layer II cerebral cortex. *Journal of Cognitive Neuroscience, 1*, 52–78.

Haberly, L. B., & Price, J. L. (1977). The axonal projection of the mitral and tufted cells of the olfactory bulb in the rat. *Brain Research, 129*, 152–157.

Haberly, L. B., & Price, J. L. (1978). Association and commissural fiber systems of the olfactory cortex of the rat. I. Systems originating in the piriform cortex and adjacent areas. *Journal of Comparative Neurology, 178*, 711–740.

Hill, A. J. (1978). First occurrence of hippocampal spatial firing in a new environment. *Experimental Neurology, 62*, 282–297.

Jung, M. W., Larson, J., & Lynch, G. (1990a). Long-term potentiation of monosynaptic EPSPs in rat piriform cortex in vitro. *Synapse, 6*, 279–283.

Jung, M. W., Larson, J., & Lynch, G. (1990b). Role of NMDA and non-NMDA receptors in synaptic transmission in rat piriform cortex. *Experimental Brain Research, 82*, 451–455.

Lancet, D., Greer, C. A., Kauer, J. S., & Shepherd, G. M. (1982). Mapping of odor-related neuronal activity in the olfactory bulb by high-resolution 2-deoxyglucose autoradiography. *Proceedings of the National Academy of Sciences (U.S.A.), 79*, 670–674.

Larson, J, & Lynch, G. (1986). Induction of synaptic potentiation in hippocampus by patterned stimulation involves two events. *Science, 232*, 985–988.

Larson, J., & Lynch, G. (1988). Role of N-methyl-D-aspartate receptors in the induction of synaptic potentiation by burst stimulation patterned after the hippocampal theta rhythm. *Brain Research, 441*, 111–118.

Larson, J., & Lynch, G. (1989). Theta pattern stimulation and the induction of LTP: The sequence in which synapses are stimulated determines the degree to which they potentiate. *Brain Research, 489*, 49–58.

Larson, J., Wong, D., & Lynch, G. (1986). Patterned stimulation at the theta frequency is optimal for induction of hippocampal long-term potentiation. *Brain Research, 368*, 347–350.

Lee, K., Oliver, M., Schottler, F., & Lynch, G. (1981). Electron microscopic studies of brain slices: The effects of high frequency stimulation on dendritic ultrastructure. In G. Kerkut & H. V. Wheal (Eds.), *Electrical Activity in Isolated Mammalian C.N.S. Preparations* (pp. 189–212). New York: Academic Press.

Lee, K., Schottler, F., Oliver, M., & Lynch, G. (1980). Brief bursts of high frequency stimulation produce two types of structural changes in rat hippocampus. *Journal of Neurophysiology, 44*, 247–258.

Lynch, G. S., Dunwiddie, T., & Gribkoff, V. (1977). Heterosynaptic depression: A postsynaptic correlate of long term potentiation. *Nature (London), 266*, 737–739.

Lynch, G., & Granger, R. (1989). Simulation and analysis of a simple cortical network. *Psychology of Learning and Motivation, 23*, 205–241.

Lynch, G., Granger, R., & Larson, J. (1989). Some possible functions of simple cortical networks suggested by computer modeling. In J. H. Byrne & W. O. Berry (Eds.), *Neural models of plasticity: Experimental and theoretical approaches* (pp. 329–362). San Diego: Academic Press.

Lynch, G., Granger, R., Larson, J. & Baudry, M. (1989). Cortical encoding of memory: Hypotheses derived from analysis and simulation of physiological learning rules in anatomical structures. In L. Nadel, L. A. Cooper, P. Culicover, & R. M. Harnish

(Eds.), *Neural connections, mental computation* (pp. 180–224). Cambridge, MA: MIT Press.

Lynch, G., Larson, J., Kelso, S., Barrionuevo, G., & Schottler, F. (1983). Intracellular injections of EGTA block induction of hippocampal long-term potentiation. *Nature (London), 305,* 719–721.

MacDermott, A. B., Mayer, M. L., Westbrook, G. L., Smith, S. J., & Barker, J. L. (1986). NMDA-receptor activation increases cytoplasmic calcium concentration in cultured spinal cord neurones. *Nature (London), 321,* 519–522.

Macrides, F., Eichenbaum, H. B., & Forbes, W. B. (1982). Temporal relationship between sniffing and the limbic theta rhythm during odor discrimination reversal learning. *Journal of Neuroscience, 2,* 1705–1717.

Malenka, R. C., Kauer, J. A., Zucker, R. S., & Nicoll, R. A. (1988). Postsynaptic calcium is sufficient for potentiation of hippocampal synaptic transmission. *Science, 242,* 81–84.

Mayer, M. L., Westbrook, G. L., & Guthrie, P. B. (1984). Voltage-dependent block by Mg^{2+} of NMDA responses in spinal cord neurons. *Nature (London), 309,* 261–263.

McCarren, M., & Alger, B. E. (1985). Use-dependent depression of IPSPs in rat hippocampal pyramidal cells *in vitro. Journal of Neurophysiology, 53,* 557–571.

Monaghan, D. T., & Cotman, C. W. (1985). Distribution of N-methyl-D-aspartate-sensitive L-[^{3}H]glutamate-binding sites in rat brain. *Journal of Neuroscience, 5,* 2909–2919.

Morris, R. G. M., Anderson, E., Lynch, G. S., & Baudry, M. (1986). Selective impairment of learning and blockade of long-term potentiation by an N-methyl-D-aspartate receptor antagonist, AP5. *Nature (London), 319,* 774–776.

Muller, D., Joly, M., & Lynch, G. (1988). Contributions of quisqualate and NMDA receptors to the induction and expression of LTP. *Science, 242,* 1694–1697.

Muller, D., & Lynch, G. (1988). Long-term potentiation differentially affects two components of synaptic responses in hippocampus. *Proceedings of the National Academy of Sciences (U.S.A.), 85,* 9346–9350.

Nowak, L., Bregestovski, P., Ascher, P., Herbet, A., & Prochiantz, A. (1984). Magnesium gates glutamate-activated channels in mouse central neurones. *Nature (London), 307,* 462–465.

Racine, R. J., Milgram, N. W., & Hafner, S. (1983). Long-term potentiation phenomena in the rat limbic forebrain. *Brain Research, 260,* 217–231.

Roman, F., Staubli, U., & Lynch, G. (1987). Evidence for synaptic potentiation in a cortical network during learning. *Brain Research, 418,* 221–226.

Schwob, J. E., & Price, J. L. (1978). The cortical projection of the olfactory bulb: Development in fetal and neonatal rats correlated with quantitative variation in adult rats. *Brain Research, 151,* 369–374.

Staubli, U., Baudry, M., & Lynch, G. (1984). Leupeptin, a thiol-proteinase inhibitor, causes a selective impairment of maze performance in rats. *Behavioral and Neural Biology, 40,* 58–69.

Staubli, U., Baudry, M., & Lynch, G. (1985). Olfactory discrimination learning is blocked by leupeptin, a thiol-protease inhibitor. *Brain Research, 337,* 333–336.

Staubli, U., Faraday, R., & Lynch, G. (1985). Pharmacological dissociation of memory: Anisomycin, a protein synthesis inhibitor, and leupeptin, a protease inhibitor, block different learning tasks. *Behavioral and Neural Biology, 43,* 287–297.

Staubli, U., Fraser, D., Faraday, R., & Lynch, G. (1987). Olfaction and the "data" memory system in rats. *Behavioral Neuroscience, 101,* 757–765.

Staubli, U., Larson, J., Baudry, M., Thibault, O., & Lynch, G. (1988). Chronic administration of a thiol-proteinase inhibitor blocks long-term potentiation of synaptic responses. *Brain Research, 444,* 153–158.

Staubli, U., & Lynch, G. (1987). Stable hippocampal long-term potentiation elicited by "theta" pattern stimulation. *Brain Research, 435,* 227–234.

Staubli, U., & Lynch, G. (1990). Stable depression of potentiated synaptic responses in the hippocampus with 1–5 Hz stimulation. *Brain Research, 513*, 113–118.

Staubli, U., Thibault, O., DiLorenzo, M., & Lynch, G. (1989). Antagonism of NMDA receptors impairs acquisition but not retention of olfactory memory. *Behavioral Neuroscience, 103*, 54–60.

Tseng, G.-F., & Haberly, L. B. (1988). Characterization of synaptically mediated fast and slow inhibitory processes in piriform cortex in an *in vitro* slice preparation. *Journal of Neurophysiology, 59*, 1352–1376.

Wilson, C. J. (1984). Passive cable properties of dendritic spines and spiny neurons. *Journal of Neuroscience, 4*, 281–297.

Youngentob, S. L., Mozell, M. M., Sheebe, P. R., & Hornung, D. E. (1987). A quantitative analysis of sniffing strategies in rats performing odor detection tasks. *Physiology and Behavior, 41*, 59–69.

8

An Experimental Psychologist's View of Cognitive Science

MICHAEL J. WATKINS

Experimental psychology has become a member of the family of disciplines known collectively as cognitive science. Some would say that this is the reason it has grown so much over the course of the past two or three decades, and maybe they are right. But I want to accentuate the negative, to voice concern about what experimental psychology has given up in becoming part of cognitive science.

The very name *cognitive science* hints at the reasons for my qualms. The choice of *cognitive* to describe the new field was natural enough, and one to which psychologists of all people should have no objections. As Skinner has remarked of the term, "Is there any field of psychology today in which something does not seem to be gained by adding that charming adjective to the occasional noun?" (1987). The noun in the present case, namely *science*, also has desirable connotations, but as a rule it is used to name only what might be considered to be borderline sciences—culinary science, library science, computer science, social science, political science, and Christian science, for example. Without wishing to pass judgment on these usages, I suspect that they fail to capture what most people would consider to be the core meaning of the term, which surely has to do with the activity of people earnestly at work uncovering the most basic secrets of Mother Nature. Certainly, when asked to think of a science, most people are less likely to think of culinary science or library science than of physics or chemistry. Of course, we should not make too much of how a field chooses to name itself, but I believe that it is science in exactly this core sense that experimental psychology has given up as the price of joining the cognitive science club.

The cognitive science approach to psychology emerged in the 1950s and 1960s. Psychology had, at least in the United States, fallen under the heavy yoke of behaviorism, and in doing so had—as some wag once put it—lost its mind. Research and theorizing were largely limited by the possibilities inherent in associationism, the principles of which were being sought by watching rats earn their keep running through mazes or pressing levers. When World War II and the ensuing Cold War presented psychology with new and urgent problems (e.g., the shooting down of moving objects), the principles of associationism were simply not up to the job. New thinking was needed, and the response to this need was sweeping. Behaviorism was largely brushed aside, and mainstream experimental psychology became caught up with contemporaneous developments in other fields, developments that coalesced into what became cognitive science.

The concern of cognitive science is with intelligent systems considered as computational mechanisms. Thus, by assuming the perspective of cognitive science, psychologists have come to think of people, not as associators of stimuli, but as processors of information. The machine metaphor was not new—it had, for example, formed the basis of La Mettrie's (1748/1912) psychology—but the information-processing developments in some of the other branches of cognitive science gave it tremendous impetus. It has spawned a vast amount of theorizing and an array of research methods that would have been hard to imagine under the austere regimen of behaviorism. To single out one of the more important examples, it led naturally to mental chronometry and to the reintroduction of reaction time as a pertinent measure of performance.

As exciting as these developments might seem, in my opinion the influence of cognitive science on cognitive psychology has been at best a mixed blessing. Various aspects of my misgivings have been expressed in previous publications (Watkins, 1978, 1980, 1981, 1984, 1989, 1990; Wright & Watkins, 1987), and what follows is a partial summary and an extension of these views.

Cognitive Science and Mind

One of my main misgivings about the cognitive science approach to experimental psychology concerns its treatment of conscious mind. Most of the cognitive psychologists whom I know are entirely untroubled in this regard, and are quick to point out that they are infinitely more respectful of the mind than were the behaviorists they succeeded. This much must be granted, for the behaviorists eschewed the mind as a matter of policy. On the other hand, behaviorism never did achieve complete domination of the field, not even in the United States. Indeed, the behaviorist era saw the development of such nonbehaviorist psychologies as the social psychology of Heider (1944), humanistic clinical psychology (Maslow, 1962; Rogers, 1951), and the more mainstream experimental psychologies of Bartlett (1932) and those advocating the "new look" at perception (e.g., Postman, Bruner, & McGinnies, 1948), all of which were much more serious about the mind than was not only behaviorism, but also the vast majority of today's experimental psychology.

There are no doubt those who would argue that the entire driving force in today's cognitive psychology is a quest to provide an information-processing account of performance and that such an account constitutes a theory of mind. This is a pernicious argument. If we are to concede that the focus of interest in today's theorizing is on mind, then it is on a surrogate mind, a mind replete with stores and processes of the sort found in computers but not necessarily in conscious experience. To define a mind so broadly is to reduce it to a metaphor and thereby to discourage consideration of mind in the literal sense. I do not mean to imply, of course, that the cognitive science era has produced no research at all on the nature of conscious mind. The experiments of Brooks (1968) suggesting that information we hold in mind can take distinct pictorial and verbal forms and those of Shepard and his colleagues (e.g., Shepard & Metzler, 1971) that appear to capture an ability to rotate visual images are just two examples of first-class explorations of real mind. But such research is very much the exception. Whether some component process of one of today's theories is identified with a phenomenon of conscious mind is usually a matter of either secondary importance or no importance at all. And even when such an identification is made, there is typically little if any attempt to verify it.

Of the countless examples that would illustrate this latter point, I will choose one that we are currently investigating in our laboratory. It concerns what is perhaps the most widely held theory of how the frequencies of various sets of events are estimated. According to this theory, which we may call the recall-estimate theory, frequency estimation for a given set of events involves covert recall of the events, and the estimation is based on the number of events so recalled. This theory does a better job than the vast majority of today's theories in identifying the processes it postulates with conscious phenomena. But, characteristically, it lacks compelling support. Its main empirical support consists of demonstrations that variables that affect recall have a similar effect on frequency estimation. For example, the number of exemplars recalled and the number estimated to have been presented have both been found to increase when their familiarity (Tversky & Kahneman, 1973) and exposure time in the presentation list (Lewandowsky & Smith, 1983) are increased, and when the exemplars are blocked in the presentation list so that those of a given category occur one after the other (Greene, 1989b). But such correlations hardly constitute a strong evidential basis for the theory. Other support no doubt comes from the compelling experience of individual instances coming to mind when an estimate is made. My own introspections indicate that although typically they do occur, even when unbidden and unwanted, such recollections do not as a rule form the basis of my frequency estimates. I do not doubt that there are exceptions in certain situations, although, interestingly enough, I do not believe that such exceptions include the main experimental procedure by which frequency estimation has been studied.

In this procedure, the critical events are presentations of names of exemplars of a given conceptual category. Typically, words from two or more categories are presented in the same list, and the subjects' task is to estimate the number of presented words from each category. Denny C. LeCompte and I have recently conducted an experiment involving this procedure. The study list comprised 300 words exemplifying 20 conceptual categories, with 6, 12, 18, or 24 words per category. List presentation was followed by a test in which the category names were presented at 10-second intervals. For each category, half the subjects estimated the number of exemplars that had been presented, and the other half recalled as many of the exemplars as they could. All responses were oral and were recorded on audiotape. They were subsequently analyzed to determine how actual category size affected the responses in these tasks. According to recall-estimate theory, recall performance should have carried at least as much information about actual category size as did the category-size estimations. The data, however, indicated just the opposite. The degree to which the number of words from a category that were presented was "predicted" by the number recalled increased monotonically over the course of the 10-second recall interval, but it was never as high as the degree to which the number of presented words was predicted by the frequency estimates, even though the frequency estimates were given after an average of less than 4 seconds. In other words, the recall-estimate theory of frequency judgment did not adequately account for the level of performance in frequency estimation. Subsequent experiments have attested to the robustness of this conclusion. As limited as this example may be, I believe it illustrates well a characteristically cavalier disregard (shown in so much of today's theorizing) for the validity of whatever appeal is made to conscious processes.

The cognitive science perspective not only fails to encourage verification of whatever conscious experience or experiences a theory might appeal to but also seriously distorts the way in which we think about such experiences. A particularly common distortion concerns how the flow of conscious events is controlled, and specifically the degree to which this control is vested in the information-processing being itself as opposed to its stimulus environment. Apparently, there is challenge enough in devising mechanisms that operate in the most inert of stimulus environments, for subject control is exaggerated and stimulus control slighted. I have discussed this point in rather general terms elsewhere (Watkins, 1989), and will illustrate it here with specific examples. There are examples aplenty of both the exaggeration of subject control and the slighting of stimulus control, and I will settle for one of each, again choosing from research topics in which we in our laboratory happen to have an active interest.

The exaggeration of subject control is especially well illustrated by the role experimental psychologists attribute to rehearsal. I suspect that lay people would consider rehearsal reasonably neutral with respect to the willful-versus-stimulus control issue. We may, for example, rehearse our part in a play under the control of the director or of the script, or we may run through it mentally with no particular external control. Perhaps more typically, rehearsal may be somewhere between these two possibilities. Thus, we speak our lines partly on the basis of the script or the prompter and partly from memory, with the object of the rehearsal being to reduce stimulus control to the lines and actions of other players and whatever other cues will be available during actual performances. But cognitive psychologists almost always restrict usage of the term to rehearsal that is memory-based and therefore more reasonably characterized as being under the control of the rehearser. In fact, in what to my knowledge is the most comprehensive review of the rehearsal research, Johnson (1980, p. 265) explicitly restricts the term *rehearsal* to memory-based activity. On the other hand, experimental psychologists see no problem in applying the term to certain activities that do not include what anyone else would surely regard as the essence of rehearsal, namely repetition. Johnson illustrates this bias, too, in declaring that "a definition equating rehearsal with repetition is overly restrictive" (p. 265). Further consideration of how experimental psychologists have redefined rehearsal can be found in Watkins and Peynir- cioglu (1982), but the important point for the present purpose is that they have done so in a way consistent with a prominent bias of the cognitive science approach, namely that of exaggerating willful control.

Definitions aside, there is a rare level of agreement among experimental psychologists that it is largely through memory-based rehearsal that certain items of information are maintained in conscious mind and, perhaps more important, committed to memory proper. This idea has served as one of the cornerstones of the modal memory model (e.g., Atkinson & Shiffrin, 1968; Waugh & Norman, 1965), a model that has done so much to promote the cognitive science perspective within experimental psychology and one that remains very much a part of contemporary theorizing. And yet, as with the recall-estimate theory of frequency judgments, the idea that rememberer- controlled rehearsal plays such an important role in remembering is re- markably lacking in empirical support. Perhaps the most influential evi- dence for rememberer-controlled rehearsal playing a major role in memory is a robust correlation between the likelihood of recalling a list item and the number of times the item is rehearsed during list presentation, as indicated in an overt rehearsal procedure (e.g., Rundus, 1971). In this procedure the list items are presented slowly, and the subjects are instructed to rehearse aloud by calling out the items during presentation as they come into mind. One of the problems with this procedure is that it forces rehearsal where none might otherwise occur. Moreover, what the experimenter chooses to characterize as "rehearsals" given during a succession of "rehearsal inter- vals" could equally well be called recall responses given during a succession of miniature recall tests. Looked at this way, the evidence that the overt rehearsal findings provide for the role of rehearsal in memorization re- duces to nothing more than evidence of a correlation between recall in a series of small unofficial tests given during the course of list presentation and recall in an official test given after list presentation. For this and other reasons that have been discussed in some detail in Watkins & Peynircioglu (1982), I am convinced that the role played by memory-based rehearsal in the memory process has been greatly exaggerated.

Let us turn now from the exaggeration of subject control to the slighting of stimulus control. The steadfast refusal, which I have already noted, of experimental psychologists to use the term *rehearsal* with reference to stimulus-driven activities could be said to illustrate the slighting of stimulus control. But the example I would like to focus on concerns the relatively high-fidelity retention of something just heard, a phenomenon known as echoic memory. Among other ways, echoic memory has been implicated by the modality effect and the suffix effect. Both of these effects refer to the immediate reproduction of a short sequence of items, usually random arrangements of digits or other verbal items. More specifically, they refer to the recall of items from toward the end of such sequences, and especially the last item. By modality effect is meant the greater likelihood of such items being recalled when the sequence is presented auditorily rather than visually; by suffix effect is meant the virtual elimination of this auditory recency advantage when a nominally irrelevant auditory item is presented immediately following presentation of the sequence. Both effects strongly imply a transient sensory memory for auditory events, and hence stimulus control. Because information-processing systems excel in the efficient transformation of information from one form to another, contemporary theorizing tends to downplay the significance of the sensory form of the information at the time of its "registration" by the system. Thus, throughout the cognitive science era the most popular account of short-term memory has featured a short-term store that is amodal—that is, ahistoric with respect to whether the information is presented auditorily or visually (Atkinson & Shiffrin, 1968; Waugh & Norman, 1965). How, then, have the modality and suffix effects been explained?

By far the best known account of both effects remains that given by Crowder and Morton (1969). Their explanation concedes a store specifically for auditory information, but severely limits its significance. The amodal short-term store is not only retained, but continues to carry the main burden of short-term memory performance, leaving the auditory store only an auxiliary role. Specifically, this store is assigned to the periphery of the information-processing system, upstream of the main short-term store and beyond the reach of conscious control. The information it holds is assumed to be in a raw precategorical form (i.e., in a form prior to analysis into linguistic units), and is commonly assumed to be limited to a single verbal item (Balota & Engle, 1981; Morton, Marcus, & Ottley, 1981) and to have a useful life of no more than about 2 seconds (Darwin, Turvey, & Crowder, 1972). The modality effect results from the use of this information to supplement information held in the main short-term store, and the suffix effect results from the displacement of this information by information that pertains to the suffix item and does not benefit recall.

Even though Crowder and Morton's (1969) precategorical acoustic storage account of the modality and suffix effects has not gone unchallenged (see Penney, 1989), so obviously does it embrace the spirit of the information-processing metaphor and so thoroughly does it trivialize stimulus control that it has proven virtually immune to contrary evidence. Thus, it remains the best known account of the modality and suffix effects despite evidence that the memory underlying these effects (a) extends over several positions rather than over just the last one (Greene, 1989a; Mayes, 1988), (b) survives for at least an order of magnitude longer than 2 seconds (Engle & Roberts, 1982; Watkins & Todres, 1980), (c) is in a postcategorical rather than a precategorical form (Watkins & Watkins, 1973), and (d) is far from peripheral (Ayers, Jonides, Reitman, Egan, & Howard, 1979).

These, then, are two examples of a bias toward attributing control over the cognitive processes to an internal as opposed to an external locus, to an information-processing system as opposed to the environment. In the case of rehearsal, the internal control is overplayed; in the case of echoic memory external control is underplayed. As I have argued more fully elsewhere (Watkins, 1989), the bias is pervasive. Information-processing devices extract from the environment only that which is necessary for performing the task at hand. The stimulus that brought the information, with its abundance of dimensions and lushness of detail, would convey reality to the human mind and yet is reduced to an abstract symbol at the mercy of the "system." Demonstrations by Gibson (1950, 1966) and others of how experience is shaped by the stimulus environment have done little to moderate this view.

So it is that the cognitive science approach to psychology fails to foster serious study of the mind and even distorts our views on the way the mind is controlled. But as central as the mind is—or ought to be—to psychology, the cognitive science approach has other, no less disturbing consequences. To a large extent they arise, directly or indirectly, from a commitment to mechanism.

Cognitive Science and Mechanism

Experimental psychology conducted from a cognitive science perspective is mechanistic, its purpose being to provide a mechanistic account of performance. Its theories, whether they take the form of computer programs or flowcharts or, as is more commonly the case, more or less general information-processing descriptions, are invariably blueprints for artifacts that would mimic aspects of behavior. In other words, experimental psychology has become a branch of artificial intelligence.

A major incentive for adopting this mode of psychologizing seems to be its compatibility with other fields of study. Of particular interest here is its compatibility with cognitive neuropsychology. As a branch of cognitive science, cognitive neuropsychology also seeks to provide a mechanistic account of behavior. Its distinguishing feature is its commitment to an account in terms of the functioning of the brain. Many cognitive psychologists freely incorporate neuropsychological constructs into their own theorizing. I believe this is a fundamental mistake. The relation between cognitive neuropsychology and psychology is asymmetric: Cognitive neuropsychology needs psychology, but psychology does not need cognitive neuropsychology. Cognitive neuropsychologists seek the neurological substrate of cognition; they may or may not steep themselves in the latest findings of cognitive psychologists, but they must of necessity use psychological concepts. Cognitive psychologists deal with cognition per se, the physiological underpinnings of which are an entirely separate issue. The special techniques of cognitive neuropsychology are relevant only to the physical substrate of psychology, and not to psychology itself. The secrets of psychology are no more likely to yield to, say, today's radioactive tracings of regional cerebral blood flow than they did to yesterday's electromyographic tracings. This is not to say, of course, that psychology is not a biological science. The biological and ecological significance of our various cognitive abilities is a legitimate realm of psychological inquiry, as is a comparison of these abilities across species and at different stages of the life cycle. Furthermore, the cognitive deficits of the neuropsychologist's patients are relevant to the psychological mission. But while the onus on cognitive neuropsychologists is to seek the physiological bases for those distinctions that have proven important in cognitive psychology, cognitive psychologists should pursue their studies unfettered by the physiological evidence of cognitive neuropsychologists.

The problem with incorporating physiological concepts into psychological theories is the problem with mechanism, and hence cognitive science, generally. A mechanistic account of how our state of mind or behavior at a given point in time is affected by an event that occurred at an earlier time—whether milliseconds or decades earlier—must link the effect to the event through a causal chain unbroken in either time or space. Except where this chain is fully expressed in conscious experience and therefore describable in purely psychological terms, such theorizing is beyond the realm of psychology. Appeal to such physiological constructs as memory traces may be entirely appropriate for those concerned with the physiological substrate of memory, and appeal to such constructs as memory stores may be entirely appropriate for those concerned with simulating performance in a memory task, but neither is nor ever could be appropriate for explaining the psychological phenomenon of memory itself.

Of course, psychologists do not use such terms as memory trace in quite the same way as do those whose chief concern is with the physiological bases of psychological phenomena, and they do not use such terms as *memory store* in quite the same way as do those whose chief concern is the actual construction of intelligent machines. Most obviously, psychologists use these terms in a tentative or hypothetical way; most important, they use these terms with no accountability. Those seeking physiological substrates are constrained by Mother Nature, for they are scientists; those constructing intelligent machines are constrained by the laws of physics and economics, for they are engineers; but those who use these terms to account for psychological phenomena are under no constraint at all. As I have argued elsewhere (Watkins, 1990), the cognitive science perspective allows any number of interpretations of any given psychological phenomenon. Such interpretations enjoy a series of safeguards against falsification, and on the rare occasion when an interpretation is falsified, several new theories arise to take its place. The result is that the cognitive science approach to psychology is distinguished by rampant theorizing—so much so, in fact, that just about all experimental psychologists have their very own theory, or at least their own variation of a theory.

Is there nothing, amid all this theorizing, that is of value and at the same time uniquely the product of the cognitive science perspective? Proving the null hypothesis of no benefit is infeasible, but it would be an instructive exercise to consider what would appear to be likely examples of a valuable contribution. For present purposes, I will mention the example that my cognitive science colleagues have most often suggested to me—namely, information theory.

Information theory was developed in the field of telecommunications engineering. Its core idea is one of information transmission through a more or less noisy communication channel of limited capacity. An important paper on the subject (Shannon, 1948) caught the attention of George Miller, who saw how information theory could be applied to humans, and promptly introduced it to psychologists (Miller & Frick, 1949). Its thoroughly quantitative nature and its considerable generality proved irresistible to many other bright and mostly young experimental psychologists, who proceeded to adopt its conceptual framework as their own. It was not long, however, before the quantitative heart of the theory was largely abandoned. Seven years after the Miller and Frick paper, there appeared a paper in which the author complained that for seven years he had been persecuted by an integer. Seven, plus or minus two, was the integer, and Miller (1956) the persecuted author. The problem was that in at least some tasks performance was, contrary to the theory, substantially invariant of the informational content of the items involved. To be fair, information theory is properly credited with alerting experimental psychologists to the importance not just of what the stimulus actually is, but of what it might have been, though even on this point it is worth noting that at least one psychologist (Gibson, 1966, p. 286; see also Garner, 1988, p. 33, for a comment on this reference) seems to have arrived at this insight independently. In any event, there is now general agreement that the enduring influence of information theory on psychology has been one of general theoretical orientation, and specifically the assumption of the idea of a person as an information-processing system. This brings us full circle: A cognitive science approach to experimental psychology is highly compatible with information theory, but information theory's principal effect on experimental psychology has derived, not from its particulars, but rather from its promotion of the cognitive science approach.

Whatever the merits of my claim that the cognitive science approach to experimental psychology has done little to further our understanding of mind and behavior, there can be no denying that it has achieved an unprecedented dominance of the field. Indeed, few active researchers are even aware that other approaches are possible. Research reports that adopt another approach are usually dismissed as inadequate, and nonmechanistic explanations are usually dismissed as no explanations at all. Such, alas, is the tyranny of cognitive science.

Summary and Concluding Comments

Experimental psychologists have assumed a cognitive science perspective, and in doing so they have radically changed their terminology, their theories, and their procedures. These changes have, apparently, helped spur the field's enormous growth over the last 30 years or so, but they have also changed its objectives: Today's experimental psychology is characterized by a thoroughgoing commitment to the artificial intelligence agenda of simulating performance.

Underlying this commitment is a renewed allegiance to mechanism as a form of explanation. To be fair, cognitive scientists are not alone in their allegiance to mechanism. Notwithstanding the success that, for example, nineteenth-century physicists had with field theory, there is a reluctance among scientists of many different sorts to concede the possibility that the behavior of a body at one point in space can be accounted for in terms of another body at a different point in space without there being a direct physical connection between the two. In just the same way, experimental psychologists resist the idea that an event witnessed at one point in time can have an effect on the witness's experience or behavior at a later point in time in the absence of some sort of representation of the event to bridge the temporal gap between the two points in time.

The rejection of action at a distance in favor of mechanism as an explanatory concept may be apt for most sciences. For psychology, however, it is a fundamental mistake, for unless an event has remained continuously in conscious mind, any representation of the event is necessarily beyond the reach of psychological enquiry. Our task as psychologists should be to chart the effects of such events, including their variation with conditions. Such labors will reveal regularities that, enshrined as more or less general principles or laws, will constitute truly psychological explanations.

The conceptual reorientation wrought by the cognitive science revolution finds particularly stark expression in the attribution of control over mind and behavior: Under behaviorism, stimulus control was exaggerated and willful control largely ignored, whereas nowadays willful control is exaggerated and stimulus control largely ignored. Their parsimony notwithstanding, such one-sided perspectives are, I believe, fundamentally misguided. By all means let us explore each source of control as fully as we can, but our theorizing and overall research objectives should reflect the interplay of the two sources, for this interplay is essential for mind, for behavior, even for life itself.

Aside from the issues of mechanism and control, the cognitive perspective has failed to inspire experimental psychologists with much zeal for exploring the nature of mind. I suspect that the ghosts of William James and Wilhelm Wundt would be sadly disappointed on being told how little their own studies of the mind had been advanced, and they would not be fooled by the enthusiasm of the telling. When the goal is to design artificial intelligence, validity becomes secondary. Thus, there may be little reason to choose among adequate alternative simulations, and even if such choices are made, they are made in accordance with an engineering or economic criterion rather than a criterion based on psychological validity.

Another, though not unrelated, casualty of the cognitive science orientation of experimental psychology is intuition. Intuition is an indispensable ingredient of science in the core sense, and yet it gets little respect in experimental psychology. Because just about any conceivable finding can be simulated in any number of ways, data viewed from a cognitive science perspective are not of interest in their own right, not even when they violate intuition. In fact, any relation to intuition is suppressed as a matter of policy. There is an unwritten rule in experimental psychology, enthusiastically enforced by journal editors, that any violation of intuition be covered up with a concatenation of psychologically empty hypothetical constructs. Significant progress will be difficult until the folly of this policy is appreciated.

As I write this chapter, the spacecraft *Voyager II* is sending back vast quantities of information about the planet Neptune. By all reports, this information is astonishing the scientists who are receiving it, and yet at the same time it is exciting them. To be sure, explanations are feverishly being sought, but the information has a legitimacy of its own, and is regarded as more secure, more real, than any theory that might be proffered to explain it. This state of affairs is as it should be. Unexpected findings in any science should be treasured as significant messages from Mother Nature, as things to be shared with fellow researchers for their serious thought, as challenges to correct the thinking that made them unexpected. And such a correction should not be confused with the comparatively trivial exercise of simulating the finding.

Cognitive science is without question a legitimate discipline, and the design of artifacts that simulate performance on well-chosen tasks is an important part of this discipline. But by taking on the goal of simulation as their own, experimental psychologists are eviscerating their field. Most, no doubt, are beyond reform, and perhaps the best that can be hoped for in the foreseeable future is that they can be persuaded to exercise more tolerance for those who favor another way, to cease insisting that researchers wrap their findings in simulations, and to allow that mind and behavior are still so poorly understood that they may yet yield treasures to those who adopt a more exploratory approach.

Acknowledgment

The writing of this chapter was supported by National Institute of Mental Health Grant MH35873.

References

Atkinson, R. C., & Shiffrin, R. M. (1968). Human memory: A proposed system and its control processes. In K. W. Spence & J. T. Spence (Eds.), *The psychology of learning and motivation: Advances in research and theory* (Vol. 2). New York: Academic Press.

Ayers, T. J., Jonides, J., Reitman, J. S., Egan, J. C., & Howard, D. A. (1979). Differing suffix effects for the same physical suffix. *Journal of Experimental Psychology, 5*, 315–321.

Balota, D. A., & Engle, R. W. (1981). Structural and strategic factors in the stimulus suffix effect. *Journal of Verbal Learning and Verbal Behavior, 20*, 346–357.

Bartlett, F. C. (1932). *Remembering: An experimental and social study*. Cambridge, England: Cambridge University Press.

Brooks, L. R. (1968). Spatial and verbal components of the act of recall. *Canadian Journal of Psychology, 22*, 349–368.

Crowder, R. G., & Morton, J. (1969). Precategorical acoustic storage (PAS). *Perception & Psychophysics, 5*, 365–373.

Darwin, C. J., Turvey, M. T., & Crowder, R. G. (1972). An auditory analogue of the Sperling partial report procedure. *Cognitive Psychology, 3*, 255–267.

Engle, R. W., & Roberts, J. S. (1982). How long does the modality effect persist? *Bulletin of the Psychonomic Society, 19*, 343–346.

Garner, W. R. (1988). The contribution of information theory to psychology. *The making of cognitive science: Essays in honor of George A. Miller*. Cambridge, England: Cambridge University Press.

Gibson, J. J. (1950). *The perception of the visual world*. Boston: Houghton Mifflin.

Gibson, J. J. (1966). *The senses considered as perceptual systems*. Boston: Houghton Mifflin.

Greene, R. L. (1989a). Immediate serial recall of mixed-modality lists. *Journal of Experimental Psychology: Learning, Memory, and Cognition, 15*, 266–274.

Greene, R. L. (1989b). On the relationship between categorical frequency estimation and cued recall. *Memory & Cognition, 17*, 235–239.

Heider, F. (1944). Social perception and phenomenal causality. *Psychological Review, 51*, 358–374.

Johnson, R. E. (1980). Memory-based rehearsal. In G. Bower (Ed.), *The psychology of learning and motivation* (Vol.14). New York: Academic Press.

La Mettrie, J. O. de (1912). *Man a machine*. (G. Bussey and M. Calkins, Trans.). LaSalle, IL: Open Court. (Original work published 1748)

Lewandowsky, S., & Smith, P. W. (1983). The effect of increasing the memorability of category instances on estimates of category size. *Memory & Cognition, 11*, 347–350.

Maslow, A. H. (1962). *Toward a psychology of being*. Princeton, NJ: Van Nostrand.

Mayes, J. T. (1988). On the nature of echoic persistence: Experiments with running memory. *Journal of Experimental Psychology: Learning, Memory, and Cognition, 14*, 278–288.

Miller, G. A. (1956). The magical number seven, plus or minus two: Some limits on our capacity to process information. *Psychological Review, 63*, 81–97.

Miller, G. A., & Frick, F. C. (1949). Statistical behavioristics and sequence of responses. *Psychological Reviews, 56*, 311–324.

Morton, J., Marcus, S. M., & Ottley, P. (1981). The acoustic correlate of "speechlike": A use of the suffix effect. *Journal of Experimental Psychology: General, 110*, 568–593.

Penney, C. G. (1989). Modality effects and the structure of short-term verbal memory. *Memory & Cognition, 17*, 398–422.

Postman, L., Bruner, J. S., & McGinnies, E. (1948). Personal values as selective factors in perception. *Journal of Abnormal Social Psychology, 43*, 142–154.

Rogers, C. R. (1951). *Client-centered therapy*, Boston: Houghton Mifflin.

Rundus, D. (1971). Analysis of rehearsal processes in free recall. *Journal of Experimental Psychology, 89*, 63–77.

Shannon, C. E. (1948). A mathematical theory of communication. *Bell Systems Technical Journal, 27*, 379–423.

Shepard, R. N., & Metzler, J. (1971). Mental rotation of three-dimensional objects. *Science, 171*, 701–703.

Skinner, B. F. (1987). Whatever happened to psychology as the science of behavior? *American Psychologist, 42*, 780–786.

Tversky, A., & Kahneman, D. (1973). Availability: A heuristic for judging frequency and probability. *Cognitive Psychology, 5*, 207–232.

Watkins, M. J. (1978). Theoretical issues. In M. M. Gruneberg & P. E. Morris (Eds.), *Aspects of memory* (pp. 40–60). London: Methuen.

Watkins, M. J. (1980). Theories of memory. Review of L. G. Nilsson (Ed.), *Perspectives on memory research. Science, 207*, 755–756.

Watkins, M. J. (1981). Human memory and the information processing metaphor. *Cognition, 10*, 331–336.

Watkins, M. J. (1984). Models as toothbrushes. Commentary on "The Maltese Cross: A new simplistic model for memory" by D. E. Broadbent. *The Behavioral and Brain Sciences, 7*, 55–94.

Watkins, M. J. (1989). Willful and nonwillful determinants of memory. In H. L. Roediger, III, & F. I. M. Craik (Eds.), *Varieties of memory and consciousness: Essays in honor of Endel Tulving* (pp. 59–71). Hillsdale, NJ: Lawrence Erlbaum.

Watkins, M. J. (1990). Mediationism and the obfuscation of memory. *American Psychologist, 45*, 328–335.

Watkins, M. J., & Peynircioglu, Z. F. (1982). A perspective on rehearsal. In G. H. Bower (Ed.), *The psychology of learning and motivation* (Vol. 16, pp. 153–190). New York: Academic Press.

Watkins, M. J., & Todres, A. K. (1980). Suffix effects manifest and concealed: Further evidence for a 20-second echo. *Journal of Verbal Learning and Verbal Behavior, 19*, 46–53.

Watkins, M. J., & Watkins, O. C. (1973). The postcategorical status of the modality effect in serial recall. *Journal of Experimental Psychology, 99*, 226–230.

Waugh, N. C., & Norman, D. A. (1965). Primary memory. *Psychological Review, 72*, 89–104.

Wright, A. A., & Watkins, M. J. (1987). Animal learning and memory and their relation to human learning and memory. *Learning and Motivation, 18*, 131–146.

Commentary on Part I

The opening chapter by Churchland and Sejnowski provides an excellent introduction to cognitive neuroscience. They note that approaches to cognition and brain function can be loosely divided into two forms. The first, the *top-down* approach, is one that starts with an analysis of perception, consciousness, and learning in purely behavioral terms, ignoring biology completely. The second, the *bottom-up* approach, begins with an analysis of the most basic biological systems of the brain and argues that an understanding of these systems will eventually lead to explanations of behavior. Contemporary cognitive neuroscience involves an integration of these two approaches.

Churchland and Sejnowski discuss three different notions of levels: levels of analysis, levels of organization, and processing levels. In Marr's (1982) discussion of levels of analysis, in which computational, algorithmic, and implementational levels were considered, it was argued that analysis at a higher level was largely independent of the levels below it. Churchland and Sejnowski propose that analysis at the implementational level can be invaluable in devising likely algorithms of how a task is accomplished by the brain.

In considering different structural levels they note that although molecules, synapses, neurons, networks, layers, maps, and systems can be separated conceptually, they are not detachable physically. The level that is selected is really defined by the techniques available for studying the phenomenon of interest. Furthermore, the lower (more molecular) levels are more easily defined and studied than are the higher ones. However, to quote Marr (1982): "Trying to understand perception by studying only neurons is like trying to understand bird flight by studying only feathers: it just cannot be done."

Churchland and Sejnowski proceed to illustrate how the interaction between psychology and physiology has been fruitful with examples from color vision. They then consider some of the techniques used by cognitive neuroscientists, some of which are considered in much more detail by other contributors in this volume. Finally, they discuss cognitive models. They consider them valuable but point out that an almost infinite number of models can be generated if the brain is considered to be a black box. A distinction is made between *realistic* and *simplifying* models. A realistic model of some set of cognitive processes must take into account the biological structure and function of the brain. A simplifying model need not take into account all aspects of brain function, only some chosen subset. Churchland and Sejnowski suggest that these two approaches may be integrated in the development of accounts of information processing.

The second chapter, by Thompson and Gluck, provides an ideal model for studying the neurobiology of behavior. They choose to study a behavior that has been well characterized based on a long history of research dating back to Pavlov's studies of classical conditioning. They then apply a number of different neurobiological techniques (e.g., lesioning and microstimulation) to elucidate the underlying systems that are necessary for such associative learning to take place. A detailed analysis is provided of conditioned fear and of eyelid conditioning from input to output, both behaviorally and biologically. Based on this analysis, Thompson and Gluck develop a functioning computational model of the identified neural networks and circuits and, in so doing, attempt to create a realistic model of associative learning as described by Churchland and Sejnowski.

It is clear from their introduction that they are committed to using an understanding of a comparatively simple behavioral process as a vehicle for comprehending higher mental functions in humans and that they believe that cognitive psychology, neuroscience, and computer science can be usefully integrated. They see the next stage in the evolution of cognitive neuroscience as a process of building on our understanding of simple cognitive systems.

Tully and Mackintosh are also concerned with understanding the determinants of the classical conditioned response, but they approach the problem from two very different directions. Tully is interested in the molecular basis of associative learning. Therefore, he chooses to study a relatively simple organism, the fruit fly. Fruit flies are capable of a well-characterized form of associative learning; it is also fairly easy to produce mutations and many generations in short periods of time. Tully discusses a number of strains that show impaired ability to either learn or remember a simple conditioned response. In this way he uses a "pathological" behavior as a vehicle for understanding normal learning. This strategy has been frequently used to study human behavior with great success and is discussed in greater detail in Parts II and IV. The biological methods available allow a detailed description of behavioral differences at a molecular (genetic) level.

Like Thompson and Gluck, Tully uses behavior as an assay, but he also presents quite a detailed analysis at the behavioral level. For example, he systematically investigates whether the deficits in a conditioned response occur in acquisition, retention, or retrieval. He also describes studies that use differences in CS and compares the effects of positive and negative reinforcement. Evidence presented reveals that, at one level, the impairments are specific—namely, that a response is learned normally but rapidly forgotten. At another level, however, the impairments may be considered general in that they occur in many experiments using different CS and US conditions. Tully's attention to the subtleties of behavior considerably enhances the biological data. Without such an analysis the significance of the biological findings would have been lost. The research is programmatic, and there can be little doubt that in the next few years it will lead to a greater understanding of the molecular mechanisms involved in associative learning.

Tully's research logically raises the issue of whether genetic studies in humans can add to our understanding of cognition. There is an extensive literature based on behavioral genetic studies, which typically use a battery of cognitive tests aimed at assessing broad functions such as intelligence (see, e.g., Scarr & Carter-Saltzman, 1982). These tests were designed not to answer questions about cognitive mechanisms but to assess performance in educational settings, in the workplace, and in clinical populations. The data are remarkably consistent in showing that cognitive performance shows a relatively high degree of heritability. However, it should be appreciated that any heritability estimate is specific to the population and environment studied and therefore carries no weight for generalizations. Perhaps of greater value will be pedigree studies using techniques like those that have been recently used to investigate the genetics of various psychiatric disorders, such as schizophrenia and affective disorder (Martin, 1987; Lander, 1988).

Mackintosh, unlike Tully in Chapter 3 and Thompson and Gluck in Chapter 2, is interested in defining and accounting for determinants of Pavlovian conditioning in behavioral rather than neurobiological terms. He begins by redefining what is implied by Pavlovian conditioning and notes that many cognitivists have dismissed this area as being trivial because they consider this unit of cognition too simple for the development of models that may be used to characterize higher mental functions. Perhaps if the term *associative learning* were used instead of *conditioning*, cognitivists would show a greater interest in the area. At the same time, however, Mackintosh points out that conditioning has many subtleties that are ignored by neurobiologists working on simple systems. For that reason he questions the generality of their findings. He argues that the traditional analysis of stimulus-response theory is inadequate by referring to experiments on blocking and second-order conditioning. He also emphasizes the importance of distinguishing between learning and performance. Investigators interested in the biology of learning and memory often tend to ignore this distinction. Clearly Mackintosh demonstrates the value of a purely behavioral analysis of associative learning.

In Chapter 5, Bailey provides a description of the morphological changes that accompany short- and long-term memory in a very simple invertebrate system. The terms *short-* and *long-term memory* are very familiar to experimental psychologists and many cognitive models have been built on this distinction. Yet it should also be clear that in Bailey's analysis of *Aplysia* these terms need not correspond to the same terms as they are used extensively in human memory research. In choosing this rather simple system, Bailey shows that it is possible to relate memory changes to the functional and structural architecture of single neurons and neuronal networks. Long-term memory is characterized by changes in the number of active zones (synapses) that increase following sensitization and decrease following habituation. In contrast, no changes in the number of active zones are observed during short-term memory formation. Instead, a depletion in synaptic vesicles (which store the chemical transmitter) accompanies short-term habituation. Parallels are emerging between invertebrate and vertebrate studies that examine changes in the number of synapses during long-term memory. This morphological correspondence indicates that learning may resemble a form of neuronal growth and cellular differentiation across a broad segment of the animal kingdom and suggests that synapse formation may be a highly conserved mechanism accompanying long-term behavioral modifications.

Bailey's research may also have implications for treating cognitive dysfunctions. It is interesting to speculate on whether the simple system that Bailey discusses might be a useful model for studying the effects of agents that might improve cognitive function. This issue is considered in more detail in Part IV.

Chapter 6 by McNaughton and Smolensky addresses the issue of whether a dialogue between a neurobiologist and a computationalist provides a useful approach to advancing our understanding of cognition. The goals and background of the computationalist are rather different from those of the neurobiologist. In discussing visual information processing, Marr (1982) considers three questions: what, why, and how. The computationalist is generally concerned with *what* an information-processing device does and *why* the computation is appropriate. The neurobiologist places a greater emphasis on *how* representations and algorithms are realized physically.

There are various computationalist positions. As noted by Churchland and Sejnowski, some modelers attempt to include as much of the known biology of the system that is being modeled as possible. In so doing, they produce a model system that resembles the way a biological organism processes information. These *realistic* brain models, however, may incorporate so much complexity that they become as poorly understood as the brain itself. On the other hand they may omit some crucial feature, thereby invalidating the results (see Sejnowski et al, 1988). In other, simplified models, certain important biological principles are incorporated, but some features are deliberately omitted. The "neurons" featured in many of these models are not, therefore, biological neurons, even though they share some of their properties. The properties of the units in many of the parallel-distributed processing (PDP) models, which have been the subject of much recent work, were inspired by basic properties of neural hardware, however, their physiological plausibility and neural bases were not their primary appeal. Instead, it was the hope that they would offer "computationally sufficient and psychologically accurate mechanistic accounts of the phenomena of human cognition" (McClelland, Rumelhart, & Hinton, 1986). Yet some computationalists would argue that there is no need to include any knowledge of neurobiology in their models. The McNaughton and Smolensky position is that a computational model of spatial processing can be achieved using some clues and constraints derived from the knowledge of hippocampal function. The end product clearly has to work in terms of input–output relationships, but is likely to differ from real biological systems.

Lynch and colleagues point out that much of cognitive psychology, particularly in the area of memory, has been concerned with identifying different forms of memory (see Part II). This provides one possible starting point for exploring the neurobiology of memory processes. Lynch et al. approach the problem in a different way, by analyzing the neuronal hardware, and believe that an understanding of the basic neuronal machinery may help in identifying new ideas and concepts about memory. To do this successfully, one needs a lead linking some cellular events to behavior. This lead has come from the study of long-term potentiation (LTP)—the increase in synaptic efficacy that can persist for days following a brief period of stimulation. It has been studied most extensively in the hippocampus, but can also be observed in other brain areas.

To define the neuronal hardware, Lynch et al. begin with a detailed description of the neuroanatomy and neurophysiology of the olfactory bulb and pyriform cortex. To a nonneuroscientist this involves a knowledge base that is unfamiliar. In their analysis Lynch et al. pay little attention to behavior. The logic of their approach, however, is similar to that used by McNaughton and Smolensky. The issue is to what extent can one design a computer system with rules and processes that fit the biological properties of central nervous system circuitry. Some of the events that must be simulated are timing rules for the integration of information, the contingent steps for first suppressing and then facilitating LTP, and the organizational properties of such a system in encoding input. Can one simulate how a network of neurons might begin by responding in an undifferentiated fashion to different stimuli, but show differentiated responses with repeated exposure?

From the analysis of the anatomy and structure of the olfactory bulb and pyriform cortex, Lynch et al. generate a computer simulation of olfactory learning. From that simulation come certain predictions about behavior. For example, they describe how an organism can discriminate between different odors given various types of prior experience. Most important, they make predictions based on their model. Thus they predict that an organism that has prior experience of a stimulus *Ab* will be impaired in its ability to discriminate the compound stimuli *Abc* from *Abd*. This can clearly be tested in the laboratory.

In Chapter 8 Watkins questions the need to consider the brain in discussing cognition, posing a challenge to cognitive neuroscientists. He begins with a brief history of cognitive science viewed from the vantage point of experimental psychology. The tendency of cognitive scientists to ignore issues of consciousness and mind is criticized. Watkins argues that, even when cognitive scientists do consider conscious experiences, the information-processing approach distorts the way they think about them. He illustrates this with examples from the study of rehearsal and of echoic memory.

The second half of the chapter discusses mechanistic accounts of cognitive phenomena. Any mechanism proposed to account for the effect of some event on subsequent behavior must involve a causal chain of events unbroken in either time or space. Watkins asserts that unless this chain can be fully expressed in conscious experience and thus be described in purely psychological terms (i.e., at the very highest level in Churchland and Sejnowski's framework), such a mechanism should be considered outside the realm of psychology. In other words, a conscious experience at time 1 and a related subsequent experience or behavior at time 2 are the proper concern of psychological theory. Anything that occurs between time 1 and time 2 that is not part of consciousness and cannot be studied using the methods of experimental psychology is not.

He points out that neurobiological issues are not uninteresting but are irrelevant for discussing psychological phenomena. The reader may be surprised to find such an opinion expressed in a volume on cognitive neuroscience, and yet Watkins provides a valuable cautionary perspective for those who have been seduced by the impressive biological studies of the other contributors. Other disciplines have often appeared capable of making major contributions to the study of psychology, but Watkins suggests that they have not contributed to our understanding of cognition. He compares the electromyographic tracings of yesterday with the imaging techniques of today. Moreover, are the neural networks of today going to contribute any more to psychology than the McCulloch–Pitts neuron of the 1940s?

Several other points emerge from Watkins' discussion of mechanism. One concerns the proliferation of psychological models among experimental psychologists. He notes that most psychologists have their very own theory. While these have provided frameworks within which various aspects of cognitive behavior can be discussed, it is unclear whether they have advanced our understanding of cognition. A second point concerns the relationship between experimental psychology and neuropsychology. Watkins criticizes experimental psychologists for incorporating neuropsychological constructs into their own theories. He argues that neuropsychologists need psychology but the reverse is not true. This is because cognitive neuropsychologists are interested in structure–function relationships (i.e., are trying to answer Marr's *how* question). Experimental psychologists, on the other hand, are interested in cognitive functions independent of the biological mechanisms of their implementation (i.e., are trying to answer Marr's *what* question). However, a neuropsychological study cast purely in behavioral terms may provide useful insights to the experimental psychologist (see Chapter 21 in Part IV).

Thompson and Gluck, Tully, and Bailey all successfully exploit paradigms that are well characterized behaviorally. They then use these paradigms as assays for discovering the behavioral effects of biological manipulations. It is therefore not surprising that the behaviors they choose are relatively simple. It is logical to try to understand simple systems before investigating the psychobiology of more complex cognitive behaviors. Where does this leave the study of complex behaviors? If studying the properties of relatively small populations of neurons casts light on the mechanisms of associative learning, might the study of the properties of multiple neuronal networks help us understand more complex cognitive systems? Do new properties emerge when several networks become integrated? This question might be considered from the perspectives of evolutionary biology and of computational science. Mackintosh raises the question of how associative learning evolved in discussing the value of research into simple systems. The discussion can be carried further, and has formed the basis of a history of research in comparative psychology. The interested reader is referred to Hinde (1970) and MacPhail (1982).

In computational science there are currently technological limits on the kinds of simulations that can be implemented. Computational brain models are generally simulated on digital computers and must perform the many parallel operations that the brain accomplishes. Some investigators are trying to circumvent this problem by developing hardware devices having components that mimic the brain's circuits directly. This technology may eventually lead to an understanding of some higher mental functions, built on the properties of these simpler systems (see Sejnowski et al., 1988).

It is clear from the contributions in this section that we are able to understand and describe a number of simple behavioral systems in considerable detail. This work does not merely provide us with an understanding of habituation and sensitization in the sea snail, or of the classical conditioning of the rabbit's eyelid, but potentially allows us to understand some cognitive phenomena in humans. Thompson and Gluck provide preliminary evidence of the generalization of some of their data to a single human patient. There is evidence too that further characterization of classical conditioning in humans using carefully selected patient populations and the currently available imaging techniques in both normals and patients will lead to new insights (e.g., see Chapter 23 by Gur and Gur). Classical conditioning also appears to be a powerful tool in examining age-related changes in learning and memory (Solomon et al., 1988; Woodruff-Pak, 1988). In Part IV, Siegel uses a detailed analysis of classical conditioning to suggest methods of treating drug dependence.

We conclude this commentary by asking two questions that relate directly to the top-down and bottom-up approaches discussed by Churchland and Sejnowski. First, has the analysis of behavior helped our ability to describe cognitive phenomena at lower levels of analysis? Second, have biological studies enhanced our understanding of cognition at higher levels of analysis? We believe that the answer to the first question is a resounding "Yes."

The biological studies by Thompson and Gluck and by Tully both required a detailed behavioral understanding of associative learning. Their use of the knowledge of behavior derived from experimental psychology makes their findings more compelling to the neurobiologist. For example, if Tully merely demonstrated that he could breed fruit flies that differed in their ability to perform one particular task, then these findings would be of limited value. What makes the program of research exciting is the ability to probe the specificity of the biological manipulations—that is, What is meant by a smart or dull fruit fly? Similarly, without the behavioral documentation of phenomena such as blocking and second-order conditioning, Thompson and Gluck would be limited in their ability to test their computational models.

The answer to the second question is less clear. Many of the contributors to the section have a research agenda that is based on the belief that the answer to the question will be "Yes." McNaughton and Smolensky, Thompson and Gluck and Lynch et al. all provide pictures of associative learning that would look rather different and far less complete had they presented only the computational analysis and not attended to biology. However, it is less clear how these contributions have helped our understanding at the purely behavioral level. The analysis of associative learning provided by Mackintosh was developed exclusively from behavioral studies. It has not been enriched by the detailed neurobiological description of events occurring between information input and behavioral output. This is consistent with Watkins's position that neurobiological studies have not contributed to behavioral explanations of cognitive phenomena. Furthermore, Watkins believes that this situation will not change in the future. We are more optimistic. The current computational models show considerable promise. Besides being able to account for a wide range of classical conditioning phenomena, such as second-order conditioning, blocking, overshadowing, and compound conditioning, they may also make predictions about the results of experiments not yet performed. Lynch et al. illustrate this in their discussion of olfactory associative learning. If cognitivists are to pay attention to what neuroscience can offer, neuroscientists must be able to answer questions cast in solely behavioral terms that could not be answered merely on the basis of behavioral data. We believe that our current knowledge and theory will allow us to do just that.

References

Hinde, R. A. (1970). *Animal behaviour. A synthesis of ethology and comparative psychology.* New York: McGraw-Hill.

Lander, E. S. (1988). Splitting schizophrenia. *Nature, 336,* 105–106.

MacPhail, E. M. (1982). *Brain and intelligence in vertebrates.* Oxford, England: Clarendon Press.

Marr, D. (1982). *Vision.* New York: W. H. Freeman.

Martin, J. B. (1987). Molecular genetics: Applications to the clinical neurosciences. *Science*, *238*, 765–772.

McClelland, J. L., Rumelhart, D. E., & Hinton, G. E. (1986). The appeal of parallel distributed processing. In *Parallel distributed processing* (Vol. 1, pp. 3–44). Cambridge MA: MIT Press.

Scarr, S., & Carter-Saltzman, L. (1982). Genetics and intelligence. In R. J. Sternberg (Ed.), *Handbook of human intelligence* (pp. 792–896). Cambridge, England: Cambridge University Press.

Sejnowski, T. J., Koch, C., & Churchland, P. (1988). Computational neuroscience. *Science*, *241*, 1299–1306.

Solomon, P. R., Beal, M. F., & Pendlebury, W. A. (1988). Age-related disruption of classical conditioning: A model systems approach to memory disorders. *Neurobiology of Aging*, *9*, 535–546.

Woodruff-Pak, D. S. (1988). Aging and classical conditioning: Parallel studies in rabbits and humans. *Neurobiology of Aging*, *9*, 511–522.

II

Dissociations and Models

All the contributors to Part II of the volume have an interest in understanding some broad aspect of cognition—in particular, memory. For some, this understanding comes from neurobiological studies of memory; for others, from studying memory failures or normal human memory. They all conclude that memory is not a unitary phenomenon but appears to exist in a number of different forms. The contributors are concerned with uncovering the distinctive features and characteristics of these various types of memory.

The notion that memory is not unitary is certainly not new. A major area of interest in nineteenth-century neurological science was uncovering specific components of cognitive functioning. Since then, many classic distinctions have been uncovered such as those between long- and short-term memory and between pattern and language processing. Much recent work has been based on a dissociation between implicit and explicit memory systems (Schacter, 1987; Tulving & Schacter, 1990).

In the commentary that follows we consider the value of distinguishing between different cognitive functions. Do the distinctions imply differences in mechanisms, or do they merely provide a heuristic for describing cognitive phenomena? What methods can be used to uncover distinct cognitive processes? Is it possible to provide a unified theory of cognition based on a consideration of all the distinctions?

Part II opens with a chapter by Weiskrantz that discusses the difficulties of determining whether two cognitive processes are distinct. Schacter and Nadel focus on a particular area of cognitive functioning and examine the evidence for different forms of spatial learning. Olton et al. focus on comparative studies of hippocampal function in memory. Such studies are most important if we are to use animal data to argue for dissociations in human cognition. The chapters by Johnson and Hirst, by Kesner, and by Moscovitch and Umilta all consider cognition in much broader terms. Rather than considering any specific domain such as spatial learning, they each provide a framework for discussing all forms of learning and memory. They discuss

evidence for dissociations in their chapters and incorporate these dissociations into their global models of learning and memory.

References

Schacter, D. L. (1987). Implicit memory: History and current status. *Journal of Experimental Psychology: Learning, Memory and Cognition, 13*, 501–522.
Tulving, E., & Schacter, D. L. (1990). Priming and human memory systems. *Science, 247*, 301–306.

9

Dissociations and Associates in Neuropsychology

L. WEISKRANTZ

In the narrow sense, neuropsychology is the study of dysfunction following brain damage. It has a dual justification for its existence and popularity: to provide clinical profiles for practical purposes, and to enable inferences to be drawn about the organization and functioning of the brain in its normal state. I want to suggest that the narrow sense is far too narrow for the second of these purposes, and that a restricted focus on the first justification can impede the advancing of the second. The logic by which inferences are drawn, the argument will run, depends crucially on dissociations rather than associations, but the field also requires a broader interaction with associates in other disciplines.

The clinical profile is a compendium of associations. Brain dysfunction X is characterized by behavioral changes A *and* B *and* C. . . . and so on. From the practical point of view the profile is an absolute prerequisite for diagnosis and care. Much effort is expended currently, for example, on the profiles of the various putative species of dementia and amnesia, each of which will have an impressively long list of symptoms, joined together importantly by the conjunction *and*. To see how this conjunction can affect inferences about mechanism, perhaps it is instructive to take some simpler examples of dysfunctions from outside the complexities of cognitive neuropsychology as such, where the mechanism is actually known.

For example, *tinnitus* is a buzzing in the ear, related to abnormal physiological activity in the inner ear. Any sufferer of tinnitus will readily testify to its widely disturbing qualities. In a formal sense, not only will tinnitus impair auditory discrimination, it will doubtless also impair auditory short-term memory, auditory recognition, and speech perception, and will have effects outside the auditory mode as such—cross-modal transfer, sleep loss, headache, faulty attention, fatigue, irritability and social aggression, and so on. Unquestionably, on a battery of neuropsychological tests, there could be a range of both verbal and nonverbal deficits. As it happens, the sufferer can tell us of the buzzing. But if he or she could not, if it were an animal or an aphasic human, how would we try to arrive at a reasonable guess about mechanism? (And, of course, for many serious disorders in the cognitive domain as well as for normal functioning, one *is* unable to give a clear verbal account.)

Tinnitus is an example of a disorder that disrupts because it interferes, literally by injecting noise, and because it produces a widely ranging number of consequential effects of the interference. It is not hard to think of neuropsychological counterparts in the cognitive domain, "cognitive buzzing" as it were. For example, reversible retrograde amnesia extending backward in time over a long interval (years) could be considered to be like a noisy memory retrieval system (Weiskrantz, 1966, 1985), and proactive and retroactive interference phenomena in memory provide other examples of disruption of this general type.

Another disorder from outside of neuropsychology that provides a useful example of a known mechanism is myopia, or nearsightedness. This, too, has a wide range of consequential deficits, including some that overlap with those resulting from tinnitus and perhaps also extending (if the myopia is not corrected) to such effects as bruised shins and reading difficulties. But myopia differs from tinnitus in showing a clear threshold effect (i.e., the deficit is apparent only beyond a certain distance). Again, neuropsychological analogies come to mind, as in "crowding" effects in reading disorders, or in the problems that "backward readers" have with infrequent but not with frequent words. Indeed, neuropsychology is replete with threshold types of deficits in which task difficulty is a critical variable (i.e., where subjects have no deficit until the task has a certain level of difficulty).

One final example is a deficit due to a specific absence (e.g., color blindness or loss of detection of high auditory frequencies). Even in this case there could be a range of measurable deficits that impinge consequentially in everyday life beyond the strictly limited range of color or auditory discrimination as such. And if the category of color blindness is rod monochromatopsia, in which the absence of retinal cones leaves a pure rod retina, there will be a complex pattern of changes in acuity, spectral sensitivity, and other threshold effects heavily dependent upon whether the eye is or is not dark-adapted, and so forth, with their own consequential effects in turn.

Those three examples provide analogues of three common species of neuropsychological deficits: modulatory or interference deficits, threshold deficits in a particular mode or category, and loss of specific capacities (or specific facilitation of a specific capacity, as may be the case with changes in motivation following hypothalmic lesions, for example). But the classification of those examples is more secure than in neuropsychology because we started with a disruption to known mechanisms. Without such knowledge, we perforce must work backward from the clusters of associated clinical deficits, and what is clear is that from the study of such associative clusters we do not easily or directly proceed to mechanism or interpretation.

If we did not know of the actual causes, all three examples would provide broad spectra of difficulties for the sufferer, from bruised shins to memory problems to social difficulties. We could not proceed to interpretation until we began to dissect and disentangle the clusters. It is clear that the clusters are formed of two main types of constituents. Some of the deficits are, so to speak, "optional." It is not obligatory to be socially irritable with tinnitus—some sufferers manage not to be. But it *is* obligatory to have difficulties in auditory discrimination. Optional but associated deficits in neuropsychology are especially common because of the anatomical proximity and even intermingling of putatively independent pathways in the brain and because of the uncooperative character of brain lesions, which are rarely restricted to just the region or pathway one would like for the understanding of mechanism.

The other type of constituent is of consequentiality. Consider a simple example. In all three examples given here, there will be a difficulty in some types of cross-modal transfer, not because cross-modal transfer as such is impaired, but because of the intramodal impairment. To take a neuropsychological analogue, bearing on a classical issue in the interpretation of frontal lobe dysfunction, if there is a strong tendency to perseverate (i.e., to repeat recent responses), this will consequently certainly impair short-term memory performance. On the other hand, a short-term memory impairment will not necessarily yield perseverative responding. Nor, obviously, will an impairment in cross-modal transfer necessarily always be associated with tinnitus.

Nonobligatory associates, if pursued, take us on one route, leading ultimately to double dissociations and inferences about independent systems or independent subsystems. Consequential associates take us on another route, ultimately leading to single, asymmetrical dissociations and inferences about hierarchies within a system. More of this shortly. But it is clear that, while the description of clusters is of great practical value for clinical management and diagnosis, the interpretation of mechanism cannot begin until we separate out those constituents that occur secondarily as a result of other constituents from those that occur optionally and are not essential and invariant. All three categories of deficits in our examples (interference, threshold, and category loss) will be superficially alike, in form if not in actual detail, until such a preliminary dissection takes place.

Inferences that a specific capacity has been lost or damaged or facilitated stem from and require double dissociations of deficits. The reasoning behind this has often been rehearsed and is familiar to neuropsychologists (cf. Teuber, 1955; Weiskrantz, 1968). To consider the example of color blindness: The demonstration that a person has a difficulty in discriminating between or identifying colors is not sufficient to demonstrate specificity, because color might be a more difficult or fragile dimension than other dimensions and hence be more sensitively disturbed even though the disturbance is more general. That is, the deficit might be of a threshold type. If another treatment affects a noncolor dimension and leaves color intact, task sensitivity as such can be discounted. If discrimination on that other dimension and color discrimination can both be set to threshold values, so much the better. As it happens, double dissociation between color and achromatic discrimination has not been demonstrated in primates. But achromatopsia has been reported as a relatively specific if rare disorder in man (Heywood, Wilson, & Cowey, 1987; Meadows, 1974; Sacks & Wasserman, 1987) and dissociable from various aspects of form and orientation discrimination (Warrington, 1985). Given that in rats achromatic discrimination and orientation discrimination have been doubly dissociated (by lesions in the posterior thalamus versus the striate cortex, Legg & Cowey, 1977), it seems likely that an achromatic/chromatic discrimination double dissociation will also eventually be found in the primate (with, perhaps, posterior thalamic versus V4 cortical lesions).

But leaving that particular example aside, "optional" deficits in a profile are precisely the clues to the possibility of independent systems, because the comparison of different profiles may uncover double dissociations. For example, if some but not all amnesic patients have short-term memory deficits, but all have long-term memory deficits, and if in some other category of patients all have short-term problems but not all have long-term memory problems, putative independence of these two types of memory processes can be entertained. In fact, just such a double dissociation has been found (Warrington, 1982), and other multiple memory systems have been proposed from such an approach (Warrington, 1979; Weiskrantz, 1987). Probably the most important single impetus to the development of concepts in neuropsychology historically came from the uncovering of double dissociations. In neurology, too, its importance goes back to the classical period—e.g., in separating the "motor cortex" from the "sensory cortex" (cf. Ferrier, 1888).

Consequential but asymmetrical types of relationships between deficits lead to hypotheses about hierarchical arrangements, as in logical tree structures or in diverging or converging sequential pathways. To take an example of the former, a subject who has trouble discriminating forms will also, pari passu, have difficulty in classifying them according to their meaning. Good form discrimination is a prerequisite for the other. An example of the latter is found in comparing damage to striate cortex with damage to cortical stages to which striate cortex subsequently projects: All categories of visual discrimination are affected by striate lesions, but only subcategories of vision by lesions at later stages. If deficit B always is present when deficit A is present, but not vice versa, we can infer that A is either superordinate to B or prior in a processing sequence. We can also infer more complex arrangements (e.g., independent subcomponents related hierarchically to a superordinate structure). The relationship of various "modular" components of visual association cortex to striate cortex processing is such an example (Cowey, 1979; Zeki, 1978, 1982).

But there is a problem: The inferences that flow from double dissociation or from asymmetrical single dissociations are not themselves obligatory. Put in another way, double dissociation is necessary for inferring independence, but it is not sufficient. Similarly, single dissociations are necessary for inferring hierarchical structure but are not sufficient. The reasoning behind these conclusions is different for each of the two categories, but they lead to a similar conclusion: namely, that while single and double dissociations are necessary for dissecting and disentangling associative conglomerates of deficits in a clinical profile in order to make interpretative inroads, the inferences that flow from the dissociations are heuristic and pragmatic rather than logically inescapable. The arguments have been analyzed in more detail elsewhere (Shallice, 1988; Weiskrantz, 1968), but can be reviewed briefly here.

Double dissociations can emerge even when there is no independence—for example, when the relationship between severity of lesion and performance is not monotonic. Given optimal levels for performance assumed, for example, by the Yerkes–Dodson Law, or any other inverted U-shape function between an independent variable and performance, different amounts of change in that variable can lead to nonlinear and nonmonotonic changes in performance. As regards single dissociations, in order to demonstrate that the dissociation is genuinely asymmetrical logically, it would be necessary to have control groups set for the appropriate level for each task to determine whether the other task does or does not change as a consequence. But even if this is possible (and in practice it is very difficult to achieve), and even when two treatments yield asymmetrical single dissociations, it is still an open possibility (although there would be strong impetus for entertaining a hierarchical hypothesis) that some further treatment will doubly dissociate the two tasks. This is tantamount to proving the universal negative, an impossibility. That this is not just an academic point can be seen from two instructive and counterintuitive examples, one in the field of agnosia, the other in visual perception. In the study of agnosia, it used to be commonly assumed, at least since Goldstein's contributions (1939), that words of abstract and concrete meaning were hierarchically arranged—i.e., that any neurological patient who had difficulty with the meaning of concrete words would inevitably also have trouble with abstract ones, but not vice versa. Recently, strong evidence has been produced that, in fact, concrete meanings can be severely impaired without any defect in the comprehension of abstract concepts (Warrington & Shallice, 1984), and therefore may well reflect quite independent modes of processing.

Perceptual deficits caused by occipital damage not uncommonly yield dense field defects for stationary stimuli, but the patient can nevertheless see moving stimuli within the field defect (Riddoch, 1917). As moving stimuli are more salient and attention-capturing, it is natural to assume that this is just a threshold effect, an example of a single asymmetrical dissociation—that is, it would be assumed that anyone who could not see movement also could not see stationary stimuli. A recent, well-documented study has demonstrated a patient who has a specific loss of movement perception but normal perception of stationary stimuli (Zihl et al, 1983). We shall return to this surprising example because it is possible to help understand it by considering relevant current physiological studies.

To sum up so far: Brain dysfunction typically yields a conglomeration of psychological deficits that, although important for diagnosis and care, do not help to disentangle the underlying mechanisms. For this it is essential to analyze the dissociations and consider which deficits may be obligatory, which are optional and hence independent, and which are consequential on an associated deficit, so as to uncover putative independence (double dissociations) or hierarchical dependences (single dissociations). But we have seen that although double dissociations are a necessary condition for demonstrating independence, they are not a sufficient condition. Similarly, although single, asymmetrical dissociations are a necessary condition for demonstrating a hierarchical tree or sequence, they are not sufficient. For both categories of dissociations, double or single, exceptions are logically possible. Hence to demonstrate independence or hierarchical structure from the study of dissociations in isolation is tantamount to demonstrating the universal negative; it is impossible.

A similar constraint applies to those deficits that are assumed or appear to be associated in an obligatory way (i.e., invariably coexist). It is always possible that a treatment will be found that will dissociate them. The unexpected revelations emerging from the study of implicit processing provide examples of the disproof of the universal negative. For example, it is natural to assume that if a subject cannot "see" a stimulus, then discrimination of that stimulus from another will be impossible, but "blindsight" is an example that this can be possible (Weiskrantz, 1986). Similar disjunctions between explicit and implicit processing are emerging in several areas of neuropsychology, including memory, perception, reading, and speech (cf Schacter, McAndrews, & Moscovitch, 1988). One of the reasons they are surprising is that they reveal dissociations between components where usually none were assumed to be possible.

Given the impossibility of drawing inescapable and firm inferences from the study of dissociations (or their lack), where do we go from here? Clearly, dissociations yield important information, but in the form of possible hypotheses, of relatively isolated and threadbare candidates rather than elected winners. However, the possible hypotheses can be given substance and support, and may even have been initiated, by appeal to associated disciplines that also bear on mechanism. These disciplines include neuroanatomy, neurophysiology, neurochemistry, and neuropathology (i.e., the whole gamut of disciplines in the neurosciences as well as the substance of theoretical and empirical experimental psychology, extending into neuronal network modeling). At the more complex end of the gamut of cognitive dysfunctions, it may be difficult to achieve an appeal to this complete range of supporting associated disciplines. The study of dyslexia, for example, has been deeply enriched by information-processing types of theories (and vice versa), but it is too soon to achieve convergence with developments in the neurosciences. The double dissociations among various subtypes of concrete categories and abstract categories in the study of agnosias impels one to think of how different networks might be arranged to allow such a differentiation to occur, but here, too, it is too soon for a direct tie-up with the other neurosciences. However, with PET scan and other advances, one can anticipate bridging developments. The study of memory disorders has started to see this fruitful convergence from several routes: neuroanatomy, neurochemistry, theoretical experimental psychology, and neuronal networking.

The example cited earlier of a specific loss of movement perception provides a nice illustration of how physiological and anatomical information can help elucidate a neuropsychological phenomenon. The bilateral lesion in that case lies putatively in area MT, a region that in the monkey is especially sensitive to moving stimuli. Other regions are shown to be electrophysiologically sensitive to different properties of visual stimuli (Zeki, 1978, 1982) such as color, and so the movement disorder can be placed within the broader context of a general modular organization for visual perception, and hence also can be related to a broad range of other perceptual disorders. MT, as it happens, has an anatomical input from striate cortex as well as an independent input from the superior colliculus, and this fact helps understand why even complete striate cortex removal may preserve a sensitivity for transient stimuli as one of the features in "blindsight" (Rodman, Gross, & Albright, 1986).

And so forth. There are countless examples. Associations among deficits are conglomerates that must be dissected by dissociations, but even dissociations do not suffice. Nothing is gained by preaching to the converted. Many neuropsychologists accept that neuropsychology advances when convergence occurs among several avenues of inquiry, and that the proper definition of neuropsychology should be not merely the study of psychological aspects of brain lesions, but brain dysfunction in relation to anatomy, physiology, neurochemistry, and the psychological investigation and theoretical enquiry of the normal—in fact what traditionally has been called *physiological psychology*. In this broader definition, neuropsychology not only emerges as a practical clinical application, but is uniquely placed at the crossroads between neurobiology, the neurological clinic, and theories of information processing and empirical aspects of psychological capacities. It both bolsters and is bolstered by this one species of association, from which dissociation profitably is resisted.

References

Cowey, A. (1979). Cortical maps and visual perception. The Grindley Memorial Lecture. *Quarterly Journal of Experimental Psychology* 31, 1–17.

Ferrier, D. (1888). Cerebral localization in its practical relations. Paper read before the Neurological Society, Great Britain, December 20, 1888.

Goldstein, K. (1939). *The organism.* New York: American Book.

Heywood, C. A., Wilson, B., & Cowey, A. (1987). A case study of cortical colour "blind-ness" with relatively intact achromatic discrimination. *Journal of Neurology, Neurosurgery, and Psychiatry, 50,* 22.

Legg, C. R., & Cowey, A. (1977). Effects of subcortical lesions on visual intensity discriminations in the rat. *Physiology and Behavior, 19,* 635–646.

Meadows, J. C. (1974). Disturbed perception of colours associated with localized cerebral lesions. *Brain, 97,* 615–632.

Riddoch, G. (1917). Dissociation of visual perceptions due to occipital injuries, with especial reference to appreciation of movement. *Brain, 40,* 15–57.

Rodman, H. R., Gross, C. G., & Albright, T. D. (1986). Removal of striate cortex does not abolish responsiveness of neurons in visual area MT of the macaque. *Neuroscience Abstracts, 12,* 1369.

Sacks, O., & Wasserman, R. (1987). The case of the colorblind painter. *New York Review of Books, 34,* 25–34.

Schacter, D., McAndrews, M. P., & Moscovitch, M. (1988). Access to consciousness: Dissociations between implicit and explicit knowledge in neuropsychological syndromes. In L. Weiskrantz (Ed.), *Thought without language* (pp. 242–278). Oxford, England: Oxford University Press.

Shallice, T. (1988). *From neuropsychology to mental structure.* Cambridge, England: Cambridge University Press.

Teuber, H.-L. (1955). Physiological psychology. *Annual Review of Psychology, 6,* 267–296.

Warrington, E. K. (1979). Neuropsychological evidence for multiple memory systems. In *Brain and mind* (pp. 153–166). Ciba Foundation Series 69. North-Holland/London: Elsevier.

Warrington, E. K. (1982). The double dissociation of short- and long-term memory deficits. In L. S. Cermak (Ed.), *Human memory and amnesia.* Hillsdale, NJ: Lawrence Erlbaum.

Warrington, E. K. (1985). Agnosia: The impairment of object recognition. In J. A. M. Frederiks (Ed.), *Handbook of clinical neurology* (Vol. 1, pp. 333–349). London: Elsevier.

Warrington, E. K., & Shallice, T. (1984). Category specific semantic impairments. *Brain, 107,* 829–854.

Weiskrantz, L. (1966). Experimental studies of amnesia. In C. W. M. Whitty & O. L. Zangwill (Eds.), *Amnesia* (pp. 1–35). London: Butterworths.

Weiskrantz, L. (1968). Some traps and pontifications. In L. Weiskrantz (Ed.), *Analysis of behavioral change* (pp. 415–429). New York: Harper & Row.

Weiskrantz, L. (1985). On issues and theories of the human amnesiac syndrome. In N. M. Weinberger, J. L. McGaugh, & G. Lynch (Eds.), *Memory systems of the brain: Animal and human cognitive processes* (pp. 380–415). New York: Guilford Press.

Weiskrantz, L. (1986). *Blindsight: A case study and implications.* Oxford, England: Oxford University Press.

Weiskrantz, L. (1987). Neuroanatomy of memory and amnesia: A case for multiple memory systems. *Human Neurobiology, 6,* 93–105.

Zeki, S. M. (1978). Functional specialization in the visual cortex of the rhesus monkey. *Nature, 274,* 423–428.

Zeki, S. M. (1982). The mapping of visual functions in the cerebral cortex. In Y. Katsuki, R. Norgren, & M. Sato (Eds.), *Brain mechanisms of sensation* (pp. 105–128). New York: Wiley.

Zihl, J., von Cramon, D., & Mai, N. (1983). Selective disturbance of movement vision after bilateral brain damage. *Brain, 106,* 313–340.

10

Varieties of Spatial Memory: A Problem for Cognitive Neuroscience

DANIEL L. SCHACTER / LYNN NADEL

One of the significant virtues of cognitive neuroscience is that it encourages and even demands adherence to the general logic of *converging operations* (e.g., Garner, Hake, & Eriksen, 1956): the evaluation of theoretical ideas on the basis of whether or not they are supported by independent lines of evidence. This general logic can be applied within a specific discipline by comparing patterns of results generated by different experimental procedures. Within the context of cognitive neuroscience, however, the most revealing insights may come from applying the logic of converging operations *across* disciplinary boundaries (cf. Nadel & O'Keefe, 1974). If a theoretical idea is supported by evidence from more than one of the disciplines that contribute to cognitive neuroscience, then we can express a higher level of confidence concerning the validity of that idea than we could about a theory that is supported only by data from a single discipline.

In this chapter we apply the logic of converging operations, both within and across disciplines, to an issue that is of considerable importance in cognitive neuroscience: the nature of spatial memory. We consider data from cognitive psychology, neuropsychology, and psychobiology, and argue that they converge on a common theme: several *varieties* or *forms* of spatial memory can be distinguished, and hence, spatial memory should not be viewed as a unitary or monolithic entity. There are several reasons why we have chosen to focus on spatial memory in general and on the theme of nonunitary spatial memory in particular. First, spatial memory has been investigated actively in cognitive psychology, neuropsychology, and psychobiology, and there has been at least some cross-disciplinary communication concerning the issue (e.g., Ellen & Thinus-Blanc, 1987; O'Keefe & Nadel, 1978). Spatial memory thus appears particularly well suited to an interdisciplinary analysis. Second, there are good ecological and evolutionary reasons for believing that spatial memory capacities are critical to an organism's survival and, because of their pervasive and fundamental importance, may be the basis of memory for various kinds of nonspatial information (e.g., Neisser, 1987; O'Keefe & Nadel, 1978; Sherry & Schacter, 1987). Third, the resolution of various fundamental issues in the study of memory and cognition turns on an adequate conceptualization of spatial memory, including the role of the hippocampus in memory, the structure and function of cognitive maps, and the nature of mental representation. Fourth, the general theme that memory is a nonunitary entity has occupied a prominent role in recent theoretical discussions (e.g., Cohen, 1984; Johnson, 1983; Mishkin, Malamut, & Bachevalier, 1984; Nadel & Zola-Morgan, 1984; O'Keefe & Nadel, 1978; Olton, Becker, & Handelmann, 1979; Schacter, 1985, 1987a; Sherry & Schacter, 1987; Squire, 1986, 1987; Tulving, 1983, 1985; Warrington & Weiskrantz, 1982). We think that it is important to consider the extent to which, and sense in which, the nonunitary idea holds within the domain of spatial memory.

In view of these considerations, we now turn our attention to studies of spatial memory, examine whether different varieties of spatial memory can be delineated, and assess whether there is any interdisciplinary convergence concerning the issue.

Spatial Memory in Cognitive Psychology

Cognitive psychologists have investigated spatial memory with a variety of paradigms and procedures, including recall and recognition of the location of objects or words that were presented during a single study trial (e.g., Mandler, Seegmiller, & Day, 1977; Pezdek, Roman, & Sobolok, 1986), multi-trial learning of various kinds of routes (e.g., Presson & Halzerigg, 1984), and retrieval of knowledge concerning familiar and unfamiliar maps (e.g., Maki, 1981; McNamara, Ratcliff, & McKoon, 1984; Thorndyke, 1981). However, the fact that different paradigms and procedures have been used to assess different aspects of spatial knowledge need not imply that varieties of spatial memory can be distinguished at the level of *mechanism*; performance in the various paradigms might well be subserved by a common mechanism. To argue for multiple spatial memory mechanisms, it is necessary to cite evidence for experimental dissociations between different kinds of spatial memory tasks. We will consider three domains in which some such evidence exists: (1) primary versus secondary spatial learning, (2) memory for landmarks, routes, and configurations, and (3) encoding of spatial memories.

Primary versus Secondary Spatial Learning

People can learn and remember a great deal of information about the spatial characteristics of their environment—how to travel from home to work, where a favorite restaurant is located, the distance from their home town to the nearest coast, and so forth. There are at least two ways in which such information can be acquired and utilized: during actual experience navigating a spatial layout, or through symbolic representational processes, such as studying or drawing a map of a particular environment. We shall adopt the terminology of Presson and Halzerigg (1984), and refer to spatial memories that are based on direct experience of an environment as *primary*, whereas spatial memories that are acquired symbolically will be referred to as *secondary* (for similar, though not identical, distinctions, see Byrne, 1982; Thorndyke & Hayes-Roth, 1982). The fundamental question for the present purpose is whether primary and secondary spatial memory can be dissociated experimentally.

Evidence supporting such a distinction was reported in a study by Evans and Pezdek (1980). They required subjects to make judgments about the accuracy of spatial relations concerning local buildings (information that was acquired through direct experience) or distant states (information that was acquired through maps). For example, in the state condition, the names of three states would be shown at three different points on a two-dimensional display, in a manner that either did or did not preserve the relative east-west and north-south spatial relations that actually exist between them. An analogous procedure was used in the building condition. The critical experimental manipulation involved the degree to which the three state or building names were rotated from their Cartesian map coordinates: Name triads were presented at varying degrees of angular displacement, ranging from 0 to 180. Evans and Pezdek found that latency to judge the accuracy of the depicted spatial relations was a linear function of degree of rotation for states, but not for buildings. In a follow-up experiment, Evans and Pezdek studied subjects who were unfamiliar with the local buildings used in the previous experiment. These subjects learned the buildings from a map, instead of through direct experience in the spatial environment. Their latencies to judge the accuracy of depicted spatial relations among the buildings were a linear function of the degree of rotation of the judged triad, just as was observed in the state condition of the earlier experiment. These data suggest a fundamental difference in the representation and/or utilization of primary and secondary spatial memories.

Data pointing toward a similar conclusion were reported by Thorndyke and Hayes-Roth (1982). They examined subjects' memory for the spatial layout of a floor at the Rand Corporation. One group consisted of people who had been working on the floor for varying amounts of time, and had thus acquired spatial information about it from direct navigational experience. The other group consisted of people who had never been on the floor, and were taught the layout from a map. Spatial memory was tested by requiring subjects to make various judgments about 42 pairs of locations on the floor. Thorndyke and Hayes-Roth found that navigational learners were more accurate when they made judgments about the *route distance* separating two points (i.e., the distance one would have to walk to get from one place to another) than when they made judgments about the *Euclidean distance* separating two points (i.e., the straight-line distance). Map learners, however, were equally accurate in these two conditions. Map learners were more accurate than navigational learners on a task in which subjects were shown the location of one pair member and required to indicate the location of the second. In contrast, navigational learners were more accurate than map learners on orientation tasks in which subjects either actually stood at a particular location on the floor and had to point to another designated point or imagined themselves standing at a specified location and pointed to another. These differences between navigational and map learners provide further support for the idea that different mechanisms underlying primary and secondary spatial memory.

Presson & Halzerigg (1984) observed that the foregoing results are consistent with the primary/secondary distinction, but also noted that an alternative interpretation of the results is possible. In the primary or navigational conditions, subjects acquired spatial knowledge through *successive* exposures to different parts of the environment, whereas in the secondary or map learning conditions, subjects acquired spatial knowledge through a *simultaneous* exposure to the entire spatial layout. Thus, the observed differences might be attributable to differences between simultaneous and successive learning, rather than to any fundamental difference between primary and secondary spatial memory. To distinguish between these two interpretations, Presson and Halzerigg examined performance in three experimental conditions. Subjects learned a route by (1) navigating (walking) it blindfolded, (2) viewing a map of it, or (3) viewing, but not actually navigating, the route itself. The first two conditions were similar to those examined in previous studies insofar as the navigational task involved both *primary* and *successive* learning, and the map task involved both *secondary* and *simultaneous* learning.

The third condition, however, involved both *primary* and *simultaneous* learning. According to the distinction between primary and secondary spatial memory, performance in this task should resemble performance in the primary/successive condition. Alternatively, if the simultaneous versus successive nature of spatial learning is the critical variable, then performance in the third condition should resemble performance in the map learning condition, because both involve simultaneous learning. Spatial memory was tested by placing blindfolded subjects at various points in the route, either in the *same* orientation as during the learning or in a *different* orientation, and asking them to point to other route locations. Consistent with the results of previous studies, Presson and Halzerigg found that test orientation had little or no effect on the accuracy of spatial memory in the navigational condition, and had a large effect on accuracy in the map learning condition. More important, they found that test orientation had no effect on accuracy in the condition in which subjects viewed the route, thereby indicating that the effect of orientation depended on the primary/ secondary nature of spatial learning.

Two more recent studies have provided further evidence that supports the primary/secondary distinction. Scholl (1987) required college students to point to local places on campus, learned through primary navigational experience, and remote places in various parts of the Northeast, learned through maps. She found that subjects were faster to point to the remote places when they were aligned with Cartesian map coordinates (i.e., facing north) than when they were not (i.e., facing west); north versus west orientation did not influence latency to point to local places. In contrast, subjects were faster to point to local places that were in front of them than to local places that were behind them; this variable had no effect on performance with remote places. Sharps and Gollin (1987) examined the ability of young and old people to remember locations of items on maps and in an actual room. Although the young remembered more map locations than did the old, memory performance of the two groups did not differ for room locations.

The foregoing studies indicate clearly that memory for spatial locations that were learned through real-world navigational experience differs from memory for locations that were acquired from maps. The best documented difference between the two is that primary spatial memory is largely independent of various orientation manipulations at the time of retrieval that have a significant effect on secondary spatial memory.

Landmarks, Routes, and Configurations

A number of cognitive psychologists have distinguished among different kinds of knowledge that are acquired when people engage in primary learning of the spatial layout of a real-world environment. Three main types of spatial knowledge have been discussed: landmarks or reference points, route knowledge, and configural knowledge (e.g., Acredolo, 1981; Allen, 1987; Allen, Siegel & Rosinski, 1978; Hart & Berzok, 1982; Piaget, Inhelder, & Szeminska, 1960; Sadalla, Burroughs, & Staplin, 1980; Siegel, 1981; Siegel & White, 1975). *Landmarks* or *reference points* refer to salient, distinctive features of a spatial layout that are used as a basis for making judgments about various aspects of the layout; *route knowledge* refers to the spatiotemporal relations between specific environmental features or "simple linear-order knowledge structures involved in navigating from a starting point to an unseen destination by means of coordinating motor behavior with a sequence of perceptual events" (Allen, 1987, p. 274); and *configural knowledge* refers to multiply interconnected, maplike structures that "include information about the interrelationships among locations within multidimensional space" (Allen, 1987, p. 274). Although cognitive research has not yet demonstrated conclusively that distinct underlying processes are involved in the acquisition and retrieval of these different kinds of spatial knowledge, several studies consistent with this idea have been reported.

Some developmental evidence suggests that encoding and retrieval of landmark knowledge may precede acquisition or other sorts of spatial knowledge. For example, Acredolo, Pick, and Olsen (1975) studied children's ability to remember where a set of keys had been dropped in a hallway. When the keys were dropped near a conspicuous landmark, both preschoolers and older children remembered the drop location accurately; when no local landmark was present, however, preschoolers' performance was extremely poor, whereas the performance of the older children was still quite accurate. Allen (1981) showed second graders, fifth graders, and college students a sequence of slides that depicted a walk through an environment that could be subdivided on the basis of landmark-defined boundaries. After viewing the slide sequence twice, subjects were shown a *reference scene* and made memory-based distance judgments regarding which of two *comparison scenes* was closer to it. When the nearer comparison scene came from within the same landmark-defined subdivision as did the reference scene, and the other comparison scene did not, all subjects performed with a high degree of accuracy, and there was little developmental trend. By contrast, when both comparison scenes were drawn from the same subdivision as the reference scene, a strong developmental trend was observed: second graders performed at chance, fifth graders showed somewhat more accuracy, and college students' performance was significantly higher than either of the younger groups. One reasonable interpretation of these data is that even the youngest subjects were able to acquire and retrieve landmark information for making distance judgments. However, when landmark information was not useful for task performance, as in the experimental conditions in which comparison scenes were drawn from the same subdivision as the reference scene, the younger subjects were unable to retrieve other kinds of spatial information (perhaps involving route knowledge) that were available to college students. Allen's (1981) data, together with the results of Acredolo et al. (1975), suggest that processes involved in landmark utilization operate quite well in young children, whereas other kinds of spatial learning processes develop later (for similar evidence and further discussion, see Acredolo, 1981).

Consistent with the foregoing idea, Siegel (1981) has described evidence revealing developmental improvements in the acquisition and utilization of route knowledge. In addition, results reported by Shemyakin (1962; cited in Byrne, 1982) suggest that children acquire route knowledge before configural spatial knowledge. He found that when they were asked to draw familiar environments from memory, young children traced out well-known individual paths, whereas older children represented the overall layout of the space (but see Hart & Berzok, 1982). Further evidence indicating that configural spatial knowledge is acquired later than route knowledge has been discussed by Siegel (1981; Siegel & White, 1975), again suggesting that different processes are involved in these two kinds of spatial memory.

Several studies of adult populations have focused on the role of landmarks and reference points, and they indicate that performance on various memory-based spatial tests differs depending on whether or not landmarks and reference points are involved in task performance. Allen et al. (1978) showed subjects slide sequences depicting routes that had high or low landmark value. Memory-based distance judgments were more accurate for high than for low landmark sequences. Sadalla et al. (1980) asked college students to imagine themselves at a particular spot on campus, then showed them a picture of another campus location and had them judge the distance between the two places. Some of the displayed locations were rated by a separate group of subjects to be campus landmarks or reference points; others were not. Sadalla et al. found that subjects were faster to judge the proximity of reference points than that of nonreference points. They also found that subjects were faster to judge the proximity of a depicted location from an imagined reference point than from an imagined nonreference point. In a study of verification judgments regarding the relative east-west location of cities, Maki (1981) found that when two target cities came from the same state, verification latencies decreased as between-city distance increased. When target cities came from different states, however, there was no effect of intercity distance on verification latency. These results suggest that state boundaries function as spatial reference points, and that within-state distance judgments are based on different processes than are between-state judgments.

Encoding of Spatial Memories

A recurring issue in the cognitive literature concerns whether encoding of spatial information is achieved through *automatic* or *effortful* processes (cf. Hasher & Zacks, 1979). Evidence for automatic encoding of spatial information has been provided by studies in which specific instructions to prepare for a test of location memory and general instructions to prepare for an unspecified "memory test" yielded similar levels of memory for spatial location (e.g., Rothkopf, 1971; von Wright, Gebhard, & Karttunen, 1975). However, these experiments have been criticized on the grounds that subjects who were preparing for an unspecified memory test may have used strategic, effortful processes that enhanced spatial memory (e.g., Naveh-Benjamin, 1987; Schacter, 1987b). A more appropriate test of the automatic/effortful idea is to compare the performance of subjects who do not expect *any* memory test with that of subjects who expect a spatial memory test. Under these conditions, Mandler, Seegmiller, and Day (1977) found suggestive, though not conclusive evidence for automatic encoding, whereas Naveh-Benjamin (1987) reported clear and consistent evidence that spatial encoding is not entirely automatic and draws upon effortful processes.

The relevance of the automatic/effortful distinction to the present concerns is that some evidence suggests that whether or not one obtains evidence for automaticity depends on the kind of spatial memory that is assessed. In contrast to Naveh-Benjamin's (1987) findings on memory for spatial location, Frederickson and Bartlett (1987) reported that memory for the spatial (i.e., left/right) orientation of pictures was unaffected by specific instructions to prepare for an orientation test. Some years earlier, Mandler and Parker (1976) reported results that also indicate that encoding manipulations have different effects on memory for spatial location and orientation. They showed subjects organized and disorganized pictures of common scenes, and found that memory for the spatial locations of specific items was more accurate for organized than disorganized pictures. In contrast, memory for the spatial orientation of individual items was not influenced by the organization manipulation.

A related series of experiments that highlights a similar theme has been reported recently by Pezdek et al. (1986). They presented several lines of evidence that indicated that spatial location is encoded—perhaps automatically—as an integral aspect of the memory representation of objects, but not words. For example, in a developmental study they found that memory for spatial location of objects, but not words, increased with age. They also found that increasing the length of the retention interval led to poorer recall of objects and words, whereas relocation accuracy decreased only for objects. These data point to fundamental differences in the processes underlying memory for the locations of objects and words, and are thus consistent with the general idea that varieties of spatial memory can be distinguished.

In summary, cognitive research has produced several kinds of evidence that favor a nonunitary view of spatial memory, including differential effects of experimental variables on primary and secondary spatial memory, developmental differences in the acquisition of landmark, route, and configural knowledge, and dissociation between spatial memory for objects and words.

Spatial Memory in Neuropsychology

During the past 100 years, neuropsychologists have described cases of brain-damaged patients with a variety of spatial memory and learning deficits (for review, see De Renzi, 1982). These deficits include problems in remembering the location of experimentally displayed materials (DeRenzi, Faglioni, & Previdi, 1977; Gutbrod, Cohen, Maier, & Meier, 1987; Schacter, Moscovitch, Tulving, McLachlan, & Freedman, 1986; Smith & Milner, 1981); learning and remembering routes in laboratory mazes (DeRenzi, Faglioni, & Villa, 1977; Milner, 1965) and in the real world (Hecaen, Tzortzis, & Rondot, 1980; Paterson & Zangwill, 1945; Whitely & Warrington, 1978; Whitty & Newcombe, 1973); recognizing once-familiar topographical landmarks (DeRenzi, et al., 1977; Habib & Sirigu, 1987; Hecaen et. al., 1980; Landis, Cummings, Benson, & Palmer, 1986; Pallis, 1955; Paterson & Zangwill, 1945; Whitely & Warrington, 1978); and retrieving information from mental maps (Morrow, Ratcliff, & Johnston, 1985; van der Linden & Seron, 1987). Such deficits have generally been observed in conjunction with damage to the right hippocampus (e.g., Smith & Milner, 1981; Smith, 1987), the right parahippocampal gyrus (Habib & Sirigu, 1987), or the right parietal lobe and other posterior regions of the right hemisphere (e.g., DeRenzi, Faglioni, & Previdi, 1977; Gutbrod et al., 1987; Hecaen et al., 1980; Landis et al., 1986; Morrow et al., 1985). Neuropsychological research has established that spatial memory and learning deficits are selective in the sense that they can occur in the context of normal memory for nonspatial information, as measured by the Wechsler Memory Scale (DeRenzi, Faglioni, & Villa, 1977) and tests of item recall and recognition (Smith & Milner, 1981; Whitely & Warrington, 1978). The critical question for the present purposes concerns whether there is evidence for selectivity *within* the domain of spatial memory and learning—that is, whether specific spatial memory deficits can be dissociated from one another.

Although neuropsychological research has not yet provided an unequivocal answer to the foregoing question, it has produced some highly suggestive evidence. Consider first studies of short-term and long-term memory for the spatial location of experimentally presented materials. Smith and Milner (1981) used a task in which patients studied a two-dimensional array of objects and were required to recall the location of each object, either immediately following presentation or after a four-minute delay. On the delayed test, patients with right-temporal-lobe and large hippocampal lesions were seriously impaired relative to normal control subjects, patients with left-temporal-lobe lesions, and patients with left- and right-frontal-lobe lesions. On the immediate test, however, the right-temporal/hippocampal patients were indistinguishable from all the other subject groups. DeRenzi, Faglioni, and Previdi (1977) assessed short-term spatial memory with a test devised by Corsi (cited in Milner, 1971) in which patients are required to tap designated blocks that are displayed on a board among other blocks, and are then probed, either immediately or after brief delays of 9 and 18 seconds, concerning the location of the tapped blocks. DeRenzi et al. found that patients with right-hemisphere lesions performed more poorly on this task than did patients with left-hemispere lesions and control subjects. These patients also showed significant deficits on tests of long-term memory for spatial location. A contrasting pattern of results was produced in an earlier study by DeRenzi and Nichelli (1975). They reported that two patients with posterior right-hemisphere damage who were significantly impaired on the Corsi short-term spatial memory test showed normal learning of visual mazes that required long-term spatial memory. These results, together with those of Smith and Milner, point toward a double dissociation between short-term and long-term spatial memory, just as has been observed within the domain of verbal memory (e.g., Warrington & Shallice, 1969).

Further evidence suggestive of a dissociation between forms of spatial memory is provided by studies of patients with *topographical amnesia* (for review, see DeRenzi, 1982). Such patients are often characterized by generalized difficulties in recognizing previously familiar topographical landmarks, remembering frequently traveled routes and learning new ones, and recollecting or using spatial information from maps (e.g., Hecaen, Tzortzis, & Rondot, 1980; Paterson & Zangwill, 1945; Whitty & Newcombe, 1973). In several cases, however, different aspects of spatial memory appear to have been affected selectively. Pallis (1955) described a patient who failed to recognize buildings and other landmarks in his immediate environment, but could accurately describe familiar routes and could draw accurate maps of the country in which he lived. In contrast, a patient discussed by DeRenzi, Faglioni, and Villa (1977) was unable to recollect previously familiar routes, but performed well on a test of memory for the spatial location of experimentally presented geometric designs.

Van der Linden and Seron (1987) invoked Byrne's (1982) distinction between network maps (i.e., primary spatial representations that evolve from perceptual/motor activity in the real world) and vector maps (i.e., secondary spatial representations that code symbolic information depicting the world as viewed from above) to account for the pattern of performance observed in a patient whose spatial memory problems were produced by a right frontoparietal stroke. This patient showed severe impairments on tasks that presumably required vector representations—such as locating distant, but previously known, places on maps; detecting errors on falsified maps; and estimating distances between remote locations. In contrast, the patient performed relatively well on tasks that were designed to tap network representations, such as recalling frequently traveled local routes and answering various questions about such routes.

In contrast to the foregoing, Whitely and Warrington (1978) described evidence suggestive of an opposite dissociation—preserved vector or secondary spatial knowledge together with impaired network or primary spatial knowledge. Their patient, whose topographical amnesia resulted from a closed head injury, failed to recognize familiar buildings in his local environment and performed at near-chance levels on a recognition memory test for pictures of new, unfamiliar buildings that had been presented experimentally. This patient could, however, draw and use effectively maps of various routes. Several patients described by Landis et al. (1986) also appear to possess relatively preserved secondary spatial knowledge despite severe deficits in acquiring and retrieving primary spatial knowledge. For example, one of the patients lost his ability to recognize and navigate familiar environments; he had difficulty finding his way home and even "became lost in the corridor of his own apartment" (Landis et al., 1986, p. 133). Nevertheless, this patient "correctly located cities on a map of the United States and readily located his home on a road map" (p. 133). Another patient who had similar route-finding difficulties, and whose apartment of 20 years seemed totally unfamiliar, was nonetheless able to produce from memory an accurate map of her native country. These kinds of observations suggest that primary and secondary spatial knowledge can be dissociated in brain-damaged patients.

Much of the foregoing evidence is based on relatively uncontrolled clinical observation, and therefore must be treated with interpretive caution.

However, the neuropsychological evidence is consistent with, and provides a modest level of support for, the general thesis that different varieties of spatial memory exist.

Spatial Learning and Memory in Psychobiology

The study of spatial behavior has played a prominent role in the history of experimental psychobiology; research in complex spatial mazes constituted a major battleground on which proponents of Hullian S-R learning met (and temporarily vanquished) proponents of Tolmanian "cognitive map" learning (cf Tolman, Ritchie, and Kalish, 1946; Restle, 1957). At stake were some of the central issues of psychological thought: What is learned? How many kinds of learning are there? Is there anything special about spatial learning/behavior? A major resurgence of interest in spatial behavior in animals has occurred in the past few decades, spurred by interest in cognitive functions in general, and in the neural bases of spatial cognition in particular. A good deal of this literature is concerned with distinctions between forms of spatial knowledge—the central issue for Hull and Tolman—and the question of whether there are separate neural systems concerned with them.

According to Hull, *knowing* consists of being disposed to behave in particular ways; in this view there is but one variety of spatial learning and knowing, and it is like any other kind of learning. The advantage of this approach is that no special mechanisms need to be postulated to account for how organisms move effectively through space—their knowledge of space simply is couched in the form of movements. Tolman (1932, 1948) saw learning and the acquisition of knowledge in a very different way, assuming that animals acquired "representations" specifying what might be expected in particular situations, rather than (or perhaps in addition to) merely which behaviors the animal should execute. He defined the issue of spatial knowledge in terms of the possible existence of *cognitive maps*, internal representations of what we have referred to as *configural knowledge*. He asserted that animals form these maps as a result of exploring an environment, in the course of which they notice the location of, and relations among, various objects. The discovery of possible neural correlates of such a mapping process (O'Keefe & Dostrovsky, 1971) and the view that the *hippocampal formation* was the core of the brain's cognitive mapping capacity (O'Keefe & Nadel, 1978) gave new credence to Tolman's views. In the years since that early work, new details about the mechanisms involved in the mapping process, and information about the characteristics of different forms of spatial learning, has mounted (McNaughton, 1988; O'Keefe & Speakman, 1987, O'Keefe, 1988).

Behavioral research has also shown that rats use spatial cues in a configural, and not simply a routelike or associative fashion. Suzuki, Augerinos, and Black (1980) used a radial maze in which rats were allowed to obtain the rewards from four of the eight baited arms, then were briefly prevented from further responding while the cues around the room were either transposed, rotated, or left alone. When permitted to continue the trial, the rat's remaining choices should reflect whether it was using cues in an associative or configural fashion. If the rat associates each arm with cues nearby, then after the delay it will choose arms near the cues that, before the delay, were near unvisited arms. This should happen after both rotation and transposition manipulations to much the same extent. If the rat forms a map based on relations among the cues, rotation will shift all the rat's responses accordingly, but they will otherwise be "correct." Transposition, on the other hand, would create havoc, by destroying the animal's sense of where it is. Choices should be random, and they are.

Discovery of a relational (configural) spatial system in rats has been accompanied by similar findings in a large variety of species, including chimpanzees (Menzel, 1973), dogs (Chapuis, Thinus-Blanc & Poucet, 1983; Chapuis & Varlet, 1987), cats (Poucet, 1985), wolves (Peters, 1973), foxes (Fabrigoule & Maurel, 1982), and others. Menzel (1973), for example, studied the ability of his chimpanzees to solve the "traveling salesman" problem. Menzel placed food in a large number of locations around a field, carrying a chimpanzee along with him in a cage. When this animal was subsequently let free in the field, along with a number of his mates, he (and he alone) found the food in a highly efficient fashion. Menzel suggested that the best way to account for this seemingly optimal solution was to assume that the chimpanzee had a cognitive map of the field, and stored within it the various locations of the food. Fabrigoule and Maurel (1982) used radio tracking to determine the movements of foxes in their home forest. Though they did have certain preferred pathways, much of their movement occurred off this network, leading the authors to the conclusion that Tolman's view of spatial learning was more likely correct than was Hull's. The foxes' behavior was simply too variable to be accounted for by a mere set of Hullian "locomotor habits."

Multiple Systems

Much as Tolman (1948) asserted, animals have a number of different systems for solving spatial tasks. Evidence from psychobiological studies confirms this view in a variety of ways. For example, O'Keefe et al. (1975) assessed landmark learning and cognitive mapping after brain lesions disconnecting the hippocampal formation from the septal region. Rats were trained to obtain water from one of eight wells sunk into a circular maze. In one condition the correct well was marked by an obvious landmark—a light shining on the well—and in another condition the correct well was marked only by its location in the room. In this latter case only a configural strategy, as used in the Suzuki et al. (1980) study, will suffice for task solution. Rats with lesions were incapable of solving this spatial-configural, or *place learn-*

ing, task, but were if anything better than normal on the landmark task. Comparable dissociations have been reported between behavioral strategies based on the various forms of spatial knowledge discussed earlier: routes, landmarks, and maps. Lesions of the caudate nucleus disrupt route, but not landmark, learning (Abraham, Potegal, & Miller, 1983); cholinergic receptor blockade impairs map, but not landmark or route, strategies (Whishaw, 1985); interruption with dopaminergic function disrupts both landmark and map learning (Whishaw & Dunnett, 1985). In the Abraham et al. (1983) study rats were tested on two types of spatial tasks, one requiring the monitoring of passive transport, the other the use of olfactory guides. In the passive transport task, rats had to return to a starting point after being transported away from that point in an opaque, wheeled cart. No other salient cues were available to help the animal unambiguously locate the start/goal point. In the olfactory guide task, rats had to choose and follow a path marked by a scented string from among a group of paths and strings, the remainder of which were unscented. Lesions in the caudate nucleus disrupted the former task but had no effect on the latter. Lesions in the hippocampus affected neither task. Whishaw (1985) studied the impact of atropine sulfate, a cholinergic muscarinic blocker, on various spatial strategies in a water-maze task. Rats so treated were unimpaired in using fixed routes or landmarks to find a platform, but were impaired when required to use (place) strategies based on an internal spatial map to find a submerged platform. He concluded that map and route strategies depend to varying degrees on cholinergic systems. Whishaw and Dunnett (1985) also used the water maze, and showed that manipulation of a different neurotransmitter system—dopamine—has quite different effects upon spatial strategy use than does manipulation of the cholinergic system. In the case of dopamine disruption, both place learning and the use of landmarks were affected.

It is also now clear that the three kinds of spatial knowledge noted here—route, landmark, and map—have rather different maturational time-courses. We know that rats can learn to form associations more or less at birth, and that they can perform well in spatial mazes solvable by "route" strategies within the first week of life. The development of landmark and map learning in the water tank has been explored in a number of laboratories (Schenk, 1985; Rudy, Stadler-Morris, & Albert, 1987; Nadel & Wilner, 1989). In all cases it has been shown that landmark usage precedes "configural" usage—that is, landmark learning precedes map learning. The differences are on the order of 3 to 6 days, which is quite a lot in the rat (weaned at 21 days). Rudy et al. found that 17-day-old rats can use landmarks to help them locate a platform, but cannot use distal cues in a relational way to do so. The first signs of any ability to use cues in a relational/configural fashion appeared only when the rats were 20 days old.

This evidence from a variety of psychobiological approaches strongly supports the claim that there are several distinct varieties of spatial knowledge. These varieties have at least partially separate underlying neural

substrates, distinct developmental trajectories, and distinguishable behavioral characteristics. Available psychobiological evidence is thus broadly
consistent with the data from cognitive psychology and neuropsychology
discussed previously. In the remainder of this paper we draw together
convergent data from the three domains of interest to come to some conclusions about the underlying nature of spatial knowledge.

Concluding Comments

The major purpose of the present chapter was to discuss evidence from
cognitive psychology, neuropsychology, and psychobiology that bears on the
hypothesis that distinct varieties or forms of spatial memory can be distinguished. At a rather general level, empirical observations from these
different sectors of cognitive neuroscience lend convergent support to this
hypothesis: Research in cognitive psychology has produced experimental
dissociations between tasks that tap different aspects of spatial memory,
and has also documented developmental differences in the utilization of
various kinds of spatial knowledge; neuropsychological observations have
revealed that some kinds of spatial memory abilities can be severely impaired while others are relatively spared; and psychobiological research has
demonstrated that lesions to specific brain structures affect different spatial memory and learning abilities selectively.

   In view of the clear convergence from cognitive psychology, neuropsychology, and psychobiology on the general theme that varieties of spatial
memory are dissociable, it is crucial to consider whether similar convergence exists concerning the exact nature of and relations among the
various kinds of spatial memory. Here we find a rather more limited
amount of agreement across the three disciplines. Consider, for example,
the distinction between primary and secondary spatial memory. As discussed earlier, this distinction has received empirical support from both
cognitive studies (e.g., Evans & Pezdek, 1980; Presson & Halzerigg, 1984;
Scholl, 1987; Thorndyke & Hayes-Roth, 1982) and neuropsychological investigations (Landis et al., 1986; van der Linden & Seron, 1987; Whitely &
Warrington, 1978). However, we know of no psychobiological evidence that
supports the primary/secondary distinction. More important, we think it
unlikely that the concept of secondary spatial memory (i.e., storage and
retrieval of information acquired through symbolic representational processes rather than through direct navigational experience) can be meaningfully applied to nonhuman species. It is difficult to imagine any process
by which an animal could acquire spatial knowledge about its environment
other than through actual navigation of it. Accordingly, it seems reasonable
to argue that secondary spatial memory and knowledge represents a
uniquely human phenomenon. In view of the foregoing considerations, we

hypothesize that the acquisition and utilization of secondary spatial knowledge is mediated by symbolic processes that are not specifically dedicated to spatial functions, but can be used in the performance of certain spatial tasks (e.g., remembering information from maps). By this view, the processes involved in secondary spatial memory may be much the same as those that are involved in memory for various kinds of nonspatial information. In contrast, we would hypothesize that primary spatial memory is supported by processes that are dedicated specifically to spatial functions and are distinct from the processes that support nonspatial memory. Stated in evolutionary terms, primary spatial memory abilities can be thought of as *adaptations* that have been selected specifically to deal with spatial problems, whereas secondary spatial memory abilities can be thought of as *exaptations*—processes that can support the solution of certain kinds of spatial problems, but were selected initially by evolution to perform some other, nonspatial function (see Sherry & Schacter, 1987, for further discussion). One implication of this hypothesis is that it should be possible to observe neuropsychological cases in which *primary* spatial memory is impaired selectively while nonspatial memory continues to function normally (e.g., Smith & Milner, 1981). However, it should not be possible to observe cases in which *secondary* spatial memory is impaired and nonspatial memory is entirely spared, because secondary spatial memory presumably depends on some of the same processes that support nonspatial memory peformance. It would be desirable to examine these ideas systematically in future research.

Within the domain of primary spatial memory, there is some, albeit limited, cross-disciplinary convergence on the nature of the various forms of spatial memory. Cognitive studies of both children and adults indicate that landmarks and reference points play a special role in spatial memory (Acredolo et al., 1975; Allen, 1981; Allen et al., 1978; Sadalla et al., 1980), and there is suggestive neuropsychological evidence that memory for landmarks and routes may be dissociable (e.g., Pallis, 1955). Psychobiological studies (e.g., Abraham et al., 1983) provide support for this distinction as well. The psychobiological evidence also rather strongly supports a distinction between memory for route and configural knowledge (e.g., Whishaw, 1985; O'Keefe & Nadel, 1978), and studies of cognitive development provide at least some converging evidence for this distinction (Siegel, 1981). However, there have been few if any attempts to dissociate memory for route and configural knowledge in adult populations, although some suggestive observations have been reported (e.g., van der Linden & Seron, 1987). Various other dissociations among forms of spatial memory have as yet been documented only in individual sectors of cognitive neuroscience, including the dissociation between short- and long-term spatial memory (DeRenzi & Nichelli, 1975; DeRenzi et al., 1977; Smith & Milner, 1981), and differences between automatic and effortful encoding as a function of type of to-be-remembered material (e.g., Frederickson & Bartlett, 1987; Mandler et al., 1977; Pezdek et al., 1987).

The implications of our cross-disciplinary analysis of spatial memory are reasonably clear. Dissociations among forms of spatial memory have been documented in several sectors of cognitive neuroscience, but a great deal of further work will have to be done before any interdisciplinary consensus can be established concerning the nature of and relations among different types of spatial memory. In addition to exploring further the nature of the various kinds of spatial memory discussed in this chapter, one additional direction for future research that we think would be worth pursuing is suggested by the distinction between explicit and implicit forms of memory (Graf & Schacter, 1985; Schacter & Graf, 1986). Explicit memory refers to intentional recollection of recent experiences, as expressed on standard recall and recognition tests. Implicit memory refers to unintentional retrieval of previous experiences, often expressed by priming effects, on tasks that do not make explicit reference to any particular experience, such as identification of words from brief perceptual exposures (e.g., Jacoby & Dallas, 1981) or completion of word stems (e.g., Graf, Squire, & Mandler, and fragments (e.g., Tulving, Schacter, & Stark, 1982). Evidence from studies of normal subjects and amnesic patients has revealed a variety of striking dissociations between explicit and implicit memory (for review and discussion, see Richardson-Klavehn & Bjork, 1988; Schacter, 1987a).

The relevance of the explicit/implicit distinction to the present concerns is that standard tests of spatial memory can be characterized as *explicit* tests: They require intentional, deliberate retrieval of spatial information that was acquired during some prior episode or episodes. Could spatial memory also be assessed with implicit tests—that is, tests that are influenced by recently acquired spatial information but do not require explicit recollection of it? In view of the explicit/implicit dissociations observed in studies of nonspatial memory, it is quite possible that implicit tests of spatial memory would reveal a different picture than would standard explicit tests. For example, patients who have difficulty remembering spatial information on standard explicit tests might be able to demonstrate retention of such information on an appropriate implicit test. A necessary first step in exploring this possibility is to devise implicit tests of spatial memory, and we are currently doing so.

The next steps would be to produce experimental dissociations between implicit and explicit spatial memory in normal subjects, and then use the appropriate paradigm to study brain-damaged patients with spatial memory problems. A related step would involve exploring the issue in animal studies, although this would involve couching the distinction between implicit and explicit memory in terms that do not entail such constructs as intentional retrieval and conscious recollection. Investigation of spatial memory with implicit tests could represent a significant new direction for each of the disciplines that contribute to cognitive neuroscience.

Acknowledgments

Preparation of this chapter was supported by a Biomedical Research Support Grant from the University of Arizona to D. L. Schacter and Grant No. 87-2-13 from the Alfred P. Sloan Foundation to L. Nadel.

References

Abraham, L., Potegal, M. & Miller, S. (1983). Evidence for caudate nucleus involvement in an egocentric spatial task: Return from passive transport. *Physiological Psychology*, *11*, 11–17.

Acredolo, L. P. (1981). Small- and large-scale spatial concepts in infancy and childhood. In L. S. Liben, A. H. Patterson, & N. Newcombe (Eds.), *Spatial representation and behavior across the life span* (pp. 63–81). New York: Academic Press.

Acredolo, L. P., Pick, H. L., & Olsen, M. C. (1975). Environmental differentiation and familiarity as determinants of children's memory for spatial location. *Developmental Psychology*, *11*, 495–501.

Allen, G. L. (1981). A developmental perspective on the effects of "subdividing" macrospatial experience. *Journal of Experimental Psychology: Human Learning & Memory*, *7*, 120–132.

Allen, G. L. (1987). Cognitive influences on the acquisition of route knowledge in children and adults. In P. Ellen & C. Thinus-Blanc (Eds.), *Cognitive processes and spatial orientation in animal and man*. Boston: Martinus Nijhoff.

Allen, G. L., Siegel, A. W., & Rosinski, R. R. (1978). The role of perceptual context in structuring spatial knowledge. *Journal of Experimental Psychology: Human Learning & Memory*, *4*, 617–630.

Byrne, R. W. (1982). Geographical knowledge and orientation. In A. W. Ellis (Ed.), *Normality and pathology in cognitive functions*. London: Academic Press.

Chapuis, N., Thinus-Blanc, C., & Poucet, B. (1983). Dissociation of mechanisms involved in dogs' oriented displacements. *Quarterly Journal of Experimental Psychology*, *35B*, 213–219.

Chapuis, N., & Varlet, C. (1987). Short cuts by dogs in natural surroundings. *Quarterly Journal of Experimental Psychology*, *39B*, 49–64.

Cohen, N. J. (1984). Preserved learning capacity in amnesia: Evidence for multiple memory systems. In L. R. Squire & N. Butters (Eds.), *Neuropsychology of memory*. New York: Guilford Press.

De Renzi, E. (1982). *Disorders of space exploration and cognition*. New York: Wiley.

De Renzi, E., Faglioni, P., & Previdi, P. (1977). Spatial memory and hemispheric locus of lesion. *Cortex*, *13*, 424–433.

De Renzi, E., Faglioni, P., & Villa, P. (1977). Topographical amnesia. *Journal of Neurology, Neurosurgery, and Psychiatry*, *40*, 498–505.

De Renzi, E., & Nichelli, P. (1975). Verbal and non-verbal short-term memory impairment following hemispheric damage. *Cortex*, *11*, 341–354.

Ellen, P., & Thinus-Blanc, C. (Eds.) (1987). *Cognitive processes and spatial orientation in animals and man* (Vol II). Boston: Martinus Nijhoff.

Evans, G. W., & Pezdek, K. (1980). Cognitive mapping: Knowledge of real-world distance and location information. *Journal of Experimental Psychology: Human Learning and Memory*, *6*, 13–24.

Fabrigoule, C., & Maurel, D. (1982). Radio-tracking study of foxes' movements related to their home range. A cognitive map hypothesis. *Quarterly Journal of Experimental Psychology*, *34B*, 195–208.

Frederickson, R. E., & Bartlett, J. C. (1987). Cognitive impenetrability of memory for orientation. *Journal of Experimental Psychology: Learning, Memory, and Cognition, 13,* 269–277.

Garner, W., Hake, H. W., & Eriksen, C. W. (1956). Operationism and the concept of perception. *Psychological Review, 63,* 149–154.

Graf, P., & Schacter, D. L. (1985). Implicit and explicit memory for new associations in normal and amnesic subjects. *Journal of Experimental Psychology: Leanring, Memory, and Cognition, 11,* 501–518.

Graf, P., Squire, L. R., & Mandler, G. (1984). The information that amnesic patients do not forget. *Journal of Experimental Psychology: Learning, Memory, and Cognition, 10,* 164–178.

Gutbrod, K., Cohen, R., Maier, T., & Meier, E. (1987). Memory for spatial and temporal order in aphasics and right hemisphere damaged patients. *Cortex, 23,* 463–474.

Habib, M., & Sirigu, A. (1987). Pure topographical disorientation: A definition and anatomical basis. *Cortex, 23,* 73–85.

Hart, R., & Berzok, M. (1982). Children's strategies for mapping the geographic-scale environment. In M. Potegal (Ed.), *Spatial abilities: Development and physiological foundations.* New York: Academic Press.

Hasher, L., & Zacks, R. T. (1979). Automatic and effortful processes in memory. *Journal of Experimental Psychology: General, 108,* 356–388.

Hecaen, H., Tzortzis, C., & Rondot, P. (1980). Loss of topographical memory with learning deficits. *Cortex, 16,* 525–542.

Jacoby, L. L., & Dallas, M. (1981). On the relationship between autobiographical memory and perceptual learning. *Journal of Experimental Psychology: General, 110,* 306–340.

Johnson, M. (1983). A multiple-entry, modular memory system. In G. H. Bower (Ed.), *The psychology of learning and motivation* (Vol 17). New York: Academic Press.

Landis, T., Cummings, J. L., Benson, F., & Palmer, P. (1986). Loss of topographic familiarity: An environmental agnosia. *Archives of Neurology, 43,* 132–136.

Maki, R. H. (1981). Categorization and distance effects with spatial linear orders. *Journal of Experimental Psychology: Human Learning and Memory, 7,* 15–32.

Mandler, J. M., & Parker, R. E. (1976). Memory for descriptive and spatial information in complex pictures. *Journal of Experimental Psychology: Human Learning and Memory, 2,* 38–48.

Mandler, J. M., Seegmiller, D., & Day, J. (1977). On the coding of spatial information. *Memory & Cognition, 5,* 10–16.

McNamara, T. P., Ratcliff, R., & McKoon, G. (1984). The mental representation of knowledge acquired from maps. *Journal of Experimental Psychology: Learning, Memory, and Cognition, 10,* 723–732.

McNaughton, B. (1988). Neuronal mechanisms for spatial computation and information storage. In L. Nadel, L. A. Cooper, R. M. Harnish, & P. Culicover (Eds.), *Neural connections, mental computation.* Cambridge, MA: Bradford Books/MIT Press.

Menzel, E. W. (1973). Chimpanzee spatial memory organization. *Science, 182,* 943–945.

Milner, B. (1971). Interhemispheric differences in the localization of psychological processes in man. *British Medical Bulletin, 27,* 272–277.

Mishkin, M., Malamut, B., & Bachevalier, J. (1984). Memories and habits: Two neural systems. In J. L. McGaugh, G. Lynch, & N. M. Weinberger (Eds.), *Neurobiology of learning and memory.* New York: Guilford Press.

Morrow, L., Ratcliff, G., & Johnston, S. (1985). Spatial knowledge in patients with right hemisphere lesions. *Cognitive Neuropsychology, 2,* 211–228.

Nadel, L., & O'Keefe, J. (1974). The hippocampus in pieces and patches. In E. G. Gray & R. Bellairs (Eds.), *Essays on the nervous system* (pp. 367–390). Oxford, England: Clarendon Press.

Nadel, L., & Wilner, J. (1989). Some implications of postnatal maturation in the hippocampal formation. In V. Chan-Palay & C. Kohler (Eds.), *The hippocampus: New vistas*. New York: A. R. Liss.

Naveh-Benjamin, M. (1987). Coding of spatial location information: An automatic process? *Journal of Experimental Psychology: Learning, Memory, and Cognition, 13*, 595–605.

Neisser, U. (1987). A sense of where you are: Functions of the spatial module. In P. Ellen & C. Thinus-Blanc (Eds.), *Spatial orientation in animals and man* (Vol. II). Boston: Martinus Nijhoff.

O'Keefe, J. (1988). Computations the hippocampus might perform. In L. Nadel, L. A. Copper, R. M. Harnish, & P. Culicover (Eds.), *Neural connections, mental computation* (pp. 225–284). Cambridge, MA: Bradford Books/MIT Press.

O'Keefe, J., & Dostrovsky, J. (1971). The hippocampus as a spatial map. Preliminary evidence from unit activity in the freely-moving rat. *Brain Research, 34*, 171–175.

O'Keefe, J., & Nadel, L. (1978). *The hippocampus as a cognitive map*. Oxford: Clarendon Press.

O'Keefe, J., Nadel, L., Keightley, S., & Kill, D. (1975). Fornix lesions selectively abolish place learning in the rat. *Experimental Neurology, 48*, 152–166.

O'Keefe, J., & Speakman, A. (1987). Single unit activity in the rat hippocampus during a spatial memory task. *Experimental Brain Research, 68*, 1–27.

Olton, D. L., Becker, J. T., & Handelmann, G. E. (1979). Hippocampus, space, and memory. *Behavioral and Brain Sciences, 2*, 313–365.

Pallis, C. A. (1955). Impaired identification of locus and places with agnosia for colours. *Journal of Neurology, Neurosurgery and Psychiatry, 18*, 218–224.

Paterson, A., & Zangwill, O. L. (1945). A case of topographical disorientation associated with a unilateral cerebral lesion. *Brain, 68*, 188–211.

Peters, R. (1973). Cognitive maps in wolves and men. In W. F. E. Preiser (Ed.), *Environmental design research* (Vol 2, pp. 247–253). New York: Plenum.

Pezdek, K., Roman, Z., & Sobolok, K. G. (1986). Spatial memory for objects and words. *Journal of Experimental Psychology: Learning, Memory, and Cognition, 12*, 530–577.

Piaget, J., Inhelder, B., & Szeminska, A. (1980). *A child's conception of geometry*. New York: Basic Books.

Poucet, B. (1985). Spatial behaviour of cats in cue-controlled environments. *Quarterly Journal of Experimental Psychology, 37B*, 155–179.

Presson, C. C., & Halzerigg, M. D. (1984). Building spatial representations through primary and secondary learning. *Journal of Experimental Psychology: Learning, Memory, and Cognition, 10*, 716–722.

Restle, F. (1957). Discrimination of cues in mazes: A resolution of the "place-vs-response" question. *Psychological Review, 64*, 217–228.

Richardson-Klaven, A., & Bjork, R. A. (1988). Measures of memory. *Annual Review of Psychology, 39*, 475–543.

Rothkopf, E. Z. (1971). Incidental memory for location of information. *Journal of Verbal Learning and Verbal Behavior, 10*, 608–613.

Rudy, J. W., Stadler-Morris, S., & Albert, P. (1987). Ontogeny of spatial navigation behaviors in the rat: Dissociation of "proximal"- and "distal"-cue based behaviors. *Behavioral Neuroscience, 101*, 62–73.

Sadalla, E. K., Burroughs, W. J., & Staplin, L. J. (1980). Reference points in spatial cognition. *Journal of Experimental Psychology: Human Learning & Memory, 6*, 516–528.

Schacter, D. L. (1985). Multiple forms of memory in humans and animals. In N. Weinberger, J. McGaugh, & G. Lynch (Eds.), *Memory systems of the brain: Animal and human cognitive processes* (pp. 351–379). New York: Guilford Press.

Schacter, D. L. (1987a). Implicit memory: History and current status. *Journal of Experimental Psychology: Learning, Memory, and Cognition, 13*, 501–518.

Schacter, D. L. (1987b). Memory, amnesia, and frontal lobe dysfunction. *Psychobiology, 15*, 21–36.

Schacter, D. L., & Graf, P. (1986). Effects of elaborative processing on implicit and explicit memory for new associations. *Journal of Experimental Psychology: Learning, Memory, and Cognition, 12,* 432–444.

Schacter, D. L., Moscovitch, M., Tulving, E., McLachlan, D. R., & Freedman, M. (1986). Mnemonic precedence in amnesic patients: An analogue of the $\overline{AB}$ error in infants? *Child Development, 57,* 816–823.

Schenk, F. (1985). Development of place navigation in rats from weaning to puberty. *Behavioral and Neural Biology, 43,* 69–85.

Scholl, M. J. (1987). Cognitive maps as orienting schema. *Journal of Experimental Psychology: Learning, Memory, and Cognition, 13,* 615–628.

Sharps, M. J., & Gollin, E. S. (1987). Memory for object locations in young and elderly adults. *Journal of Gerontology, 42,* 336–341.

Sherry, D. F., & Schacter, D. L. (1987). The evolution of multiple memory systems. *Psychological Review, 94,* 439–454.

Siegel, A. W. (1981). The externalization of cognitive maps by children and adults: In search of ways to ask better questions. In L. S. Liben, A. H. Patterson, & N. Newcombe (Eds.), *Spatial representation and behavior across the life span* (pp. 167–194). New York: Academic Press.

Siegel, A. W., & White, S. H. (1975). The development of representations of large-scale environments. In H. Reese (Ed.), *Advances in child development and behavior* (Vol 10). New York: Academic Press.

Smith, M. L. (1987). The encoding and recall of spatial location after right hippocampal lesions in man. In P. Ellen & C. Thinus-Blanc (Eds.), *Cognitive processes and spatial orientation in animal and man.* Boston: Martinus Nijhoff.

Smith, M. L., & Milner, B. (1981). The role of the right hippocampus in the recall of spatial location. *Neuropsychologia, 19,* 781–793.

Squire, L. R. (1986). Mechanisms of memory. *Science, 232,* 1612–1619.

Squire, L. R. (1987). *Memory and brain.* New York: Oxford University Press.

Suzuki, S., Augerinos, G., & Black, A. H. (1980). Stimulus control of spatial behavior on the eight-arm maze in rats. *Learning and Motivation, 11,* 1–18.

Thorndyke, P. W. (1981). Distance estimation from cognitive maps. *Cognitive Psychology, 13,* 526–550.

Thorndyke, P. W., & Hayes-Roth, B. (1982). Differences in spatial knowledge acquired from maps and navigation. *Cognitive Psychology, 14,* 560–589.

Tolman, E. C. (1932). *Purposive behavior in animals and men.* New York: Century.

Tolman, E. C. (1948). Cognitive maps in rats and men. *Psychological Review, 55,* 189–208.

Tolman, E. C., Ritchie, B. F., & Kalish, D. (1946). Studies in spatial learning. I. Orientation and the short-cut. *Journal of Experimental Psychology, 36,* 13–24.

Tulving, E. (1983). *Elements of episodic memory.* Oxford, England: Oxford University Press.

Tulving, E. (1985). How many memory systems are there? *American Psychologist, 40,* 385–398.

Tulving, E., Schacter, D. L., & Stark, H. A. (1982). Priming effects in word-fragment completion are independent of recognition memory. *Journal of Experimental Psychology: Learning, Memory, and Cognition, 8,* 336–342.

Van der Linden, M., & Seron, X. (1987). A case of dissociation in topographical disorders: The selective breakdown of vector-map representation. In P. Ellen & C. Thinus-Blanc (Eds.), *Cognitive processes and spatial orientation in animal and man.* Boston: Martinus Nijhoff.

Von Wright, J. M., Gebhard, P., & Karttunen, M. (1975). A developmental study of the recall of spatial location. *Journal of Experimental Child Psychology, 20,* 181–190.

Warrington, E. K., & Shallice, T. (1969). The selective impairment of auditory verbal short-term memory. *Brain, 92,* 885–896.

Warrington, E. K., & Weiskrantz, L. (1982). Amnesia: A disconnection syndrome? *Neuropsychologia, 20,* 233–248.

Whishaw, I. Q. (1985). Cholinergic receptor blockade in the rat impairs locale but not taxon strategies for place navigation in a swimming pool. *Behavioral Neuroscience*, *99*, 979–1005.

Whishaw, I. Q., & Dunnett, S. B. (1985). Dopamine depletion, stimulation or blockade in the rat disrupts spatial navigation and locomotion dependent upon beacon or distal cues. *Behavioural Brain Research*, *18*, 11–29.

Whitely, A. M., & Warrington, E. K. (1978). Selective impairment of topographical memory: A single case study. *Journal of Neurology, Neurosurgery and Psychiatry*, *41*, 575–578.

Whitty, C. W. M., & Newcombe, F. (1973). R. C. Oldfield's study of visual and topographic disturbances in a right occipito-parietal lesion of 30 years duration. *Neuropsychologia*, *11*, 471–475.

11

A Comparative Analysis of the Role of the Hippocampal System in Memory

DAVID S. OLTON / CYNTHIA G. WIBLE / ALICJA L. MARKOWSKA

This chapter examines the role of the hippocampus and related neuroanatomical structures in memory as assessed by the analysis of the behavioral effects of lesions. Studies with human amnesics have shown that memory is a modular function; certain types of memory can be profoundly impaired after brain damage, while other functions, such as intelligence and language, can be completely intact. The use of rats and monkeys in lesion experiments has furthered the study of amnesia and of the relationship between memory and the brain.

This chapter begins with a brief description of the behavioral effects following lesions of the hippocampus and neuroanatomically related structures in humans, monkeys, and rats, and discusses some reasons for the apparent discrepancies between the results from monkeys and those from rats. It then presents a series of experiments designed to resolve these discrepancies.

This resolution is important to clarify some general issues in cognitive neuroscience and to answer some specific questions about the brain mechanisms involved in memory, topics that are discussed at the end of the chapter.

The Hippocampus and Amnesia

Medial temporal resection in humans produces a severe amnesia that is not specific to a particular modality or stimulus dimension (Corkin, 1984; Milner, 1959). H. M. is a well-studied case who underwent bilateral removal of the hippocampus for intractable epilepsy. The damage was not limited to the hippocampus, but included part of the hippocampal gyrus and amygdala (Scoville & Milner, 1957). H. M.'s short-term memory is intact, but he has a profound anterograde amnesia and a temporally limited retrograde amnesia. The amnesia is a circumscribed deficit, and other intellectual abilities are intact. Some types of memory are spared, including some cognitive and motor skills (Squire & Cohen, 1984).

Although significant breakthroughs have been made in the study of the neural basis of memory with amnesic humans, the lesions producing amnesia in humans inevitably involve partial destruction of several brain areas rather than complete destruction of only one (or a specified combination). Consequently, the exact role of different neural systems in amnesia remains to be determined. Experiments with animals are particularly important in resolving this difficulty because the extent and location of the lesion can be controlled. This control makes it possible to examine each brain area separately to determine its psychological function; if a lesion does impair performance, this impairment can be attributed to a particular brain structure.

Both rats and monkeys have been used to study the neural bases of amnesia. However, the data from these two species have not always led to a straightforward resolution of the experimental questions. For example, consider the results of lesions in the hippocampus, fornix, and amygdala. (1) In monkeys, lesions of the hippocampus impair performance in many nonspatial tasks (see review in Squire & Zola-Morgan, 1983). In rats, however, these lesions may differentially affect performance in spatial tasks, sparing performance in some types of nonspatial tasks (see review in Barnes, 1988). (2) In monkeys and humans, fornix lesions give mixed results; some memory impairments do occur, but these may be mild and relatively circumscribed (see review in Squire, 1987). In rats, fornix lesions usually produce severe impairments in a variety of memory tasks (see review in Olton, Becker, & Handelmann, 1979). (3) In monkeys, amygdala lesions by themselves produce a slight impairment; when combined with hippocampal lesions, they produce a marked increase in the magnitude of the impairment produced by the hippocampal lesions alone (Mishkin, 1978; Squire & Zola-Morgan, 1983). In rats, amygdala lesions do not impair memory by themselves (Raffaele & Olton, 1988), and they do not enhance the memory impairment produced by fimbria-fornix lesions (Eichenbaum, Fagan, & Cohen, 1986).

The apparent conflicts between the results from rats and those from monkeys might occur for at least three reasons. (1) Different tasks are usually given to rats and monkeys. Rats are often given spatial discriminations, and the nonspatial discriminations have rarely used three-dimensional objects like those given to monkeys. In contrast, monkeys are rarely given spatial discriminations and are usually tested with three-dimensional objects or two-dimensional figures, types of stimuli rarely used with rats. (2) Different lesions are often used in the two species. Fornix lesions are often used in rats, but not in monkeys.

Because animal models with both rats and monkeys are providing so much information about the neural systems involved in memory, choosing among these (and other) alternative explanations is important. If rats and monkeys have a fundamentally different organization of the neural systems involved in memory, then data from the two models cannot be used in an integrated fashion. If, on the other hand, variations in experimental parameters (types of behavioral tests, sites of lesions) are responsible for the different results, then information from the two models can be integrated.

The strategy used to resolve the issues just discussed is twofold. First, in order to determine the extent to which different tasks have produced discrepant results, similar tasks are designed for both species. Second, in order to determine the extent to which different lesion sites have produced discrepant results, equivalent lesions sites are used in both species.

Spatial Recent Memory: The Role of the Fornix

The fornix is a major extrinsic fiber connection of the hippocampus. In the context of a disconnection analysis, transection of the fornix should reproduce many of the behavioral effects that are produced by lesions of the hippocampus proper, and these behaviors should reflect failure of memory (Olton, Walker, & Gage, 1978; Olton, Walker, & Wolf, 1982).

In rats, destruction of the hippocampus proper produces a substantial deficit in spatial tasks that require recent memory. A *spatial* task is one in which the discriminative stimuli are the usual types of stimuli in a room that define a particular location in it. These may include many objects: doors, windows, lights, and so on. A *recent memory* task is one in which new information must be maintained for a relatively short period of time (seconds to hours) in order to perform correctly. Typically, recent memory is assessed with some version of a *delayed conditional discrimination* (DCD). At the beginning of a trial in a DCD, a stimulus is presented and then removed. A delay follows. At the end of the delay, the rat is given a choice between two or more different responses. The response that is correct at the end of the delay is conditional on the stimulus that was presented at the beginning of the delay. Consequently, in order to choose correctly at the end of the delay, the animal must remember the information that was presented at the beginning of the delay.

A typical spatial DCD might be conducted as follows. The apparatus is a maze in the shape of a T. The base of the T is the start area. The end of each arm is the goal. At the beginning of the trial, the rat is placed in the start and forced to go to the goal in one arm. The rat is then removed from the goal for a delay interval. At the end of the delay, the rat is replaced in the start and can choose between the two arms. Reinforcement is in only the arm not entered at the beginning of the trial. Consequently, the correct response at the end of the delay is to go to the arm that was not entered at the beginning of the delay (a spatial delayed non-match-to-sample).

Destruction of the hippocampus in rats consistently produces a substantial and enduring deficit in this type of spatial DCD. Complete lesions reduce choice accuracy to the level expected by chance, and no recovery of function takes place, even with extended testing (Olton, Becker, & Handelmann, 1979).

In rats, fornix transection produces an impairment very similar to that produced by hippocampal lesions (Olton, Walker, & Wolf, 1982). Comparable experiments with monkeys were unavailable. Although the behavioral effects of fornix transection in monkeys had been examined in numerous tasks (Bachevalier, Saunders, & Mishkin, 1985; Gaffan, 1974, 1977; Gaffan, Gaffan, & Harrison, 1984; Gaffan & Saunders, 1985; Mahut, 1972; Mahut & Zola, 1973; Zola-Morgan & Mahut, 1973), none of them required spatial recent memory as tested in the T-maze DCD experiment described earlier. Furthermore, the effects of fornix transection on performance in tasks affected by hippocampal lesions was not consistent. Thus, no information from monkeys was available for a direct comparison to that available from rats.

To rectify this situation, experiments for both rats and monkeys used a T-maze to investigate the effects of fornix transection on spatial recent memory. Monkeys were tested in the T-maze task because information from this type of task was unavailable. Rats were also tested because none of the previous studies had used the procedures similar to the ones that had to be used for the monkeys. Consequently, in order to provide the most comparable data in the two experiments, rats were tested with experimental parameters that were as similar as possible to those used for the monkeys. These experiments will be described in order, beginning with the monkeys (Murray, Davidson, Gaffan, Olton, & Suomi, 1989).

For the monkeys, the maze was made of a wood frame and a wire mesh enclosure, 0.6 m in cross section. The stem and each arm were each 1 m long. Five doors were placed as follows: One *start door,* located between the start box and the stem; two *choice point* doors, one located between the choice point and the entrance to each arm; two *goal box* doors, one located at the end of each arm in front of the goal box. Primate cages were used interchangeably as the start box and the two goal boxes, and were used to move the monkey around the room. Each cage had a small bowl that could hold the food reward. Many room cues were available, and the monkey could see these through the wire mesh enclosure.

Each trial had a sample run and a choice run. For the *sample run,* the choice point door and goal box door in one arm were raised. The cage with the monkey was placed at the start location, the start door was raised, the monkey entered the available arm, went to the goal box at the end of that arm, and obtained the reward there. The cage with the monkey was then interchanged with the start cage. For the *choice run,* the choice point door and the goal door in both arms were raised, giving the monkey access to both goal boxes. However, because food was located only in the goal box at the end of the arm not entered during the sample run, food was obtainable only if the monkey entered that arm. The start box door was raised. The monkey chose between the two arms and entered the goal box at the end of the chosen arm. If the correct arm was entered, the monkey ate the food there. If the incorrect arm was entered, no food was available.

Because of practical constraints in terms of moving the monkeys around the test apparatus, the following parameters were used. Four trials were given per day. The intertrial interval was one hour. The *interrun delay* (the time between the sample run and the choice run) was one minute. Each monkey was trained to a criterion of 36 correct responses in 40 consecutive trials with these parameters. Then the interrun delay was extended to 5 minutes and 15 minutes.

The fornix was transected during surgery. Histological examination showed complete transection of the fornix with minimal damage to the surrounding brain areas. Postoperatively, each monkey was tested with the one-minute interrun delay, 5 trials per day, using the same test procedure, to the same criterion. After criterion, each monkey was tested with the 5-minute and 15-minute interrun delay.

Postoperatively, control monkeys reached criterion in a mean of less than 50 trials. Fornix transection produced a substantial impairment; the mean number of trials to reach criterion was greater than 200.

A similar dissociation appeared during testing with longer interrun delays. Control monkeys had a slight drop in choice accuracy to a mean of 83 and 77%, respectively, at the 5-minute and 15-minute interrun delays. The mean value for monkeys with fornix transections at the same delays was 62 and 52%, respectively.

The experiment for the rats was as similar to that for the monkeys as possible (Markowska, Olton, Murray, & Gaffan, 1989). The T maze had the same general design, except that it was smaller. The stem and each arm were 29 cm long and 11 cm wide. The maze had a wooden frame with wire mesh, and had doors at the same five places as in the maze for the monkeys. Special enclosed boxes were similar to the transport cages used for monkeys. They served interchangeably as the two goal boxes and the start box. The maze was placed in a room with many stimuli, and the rats were able to see these through the wire mesh.

The same testing parameters were used for the number of trials per day, the intertrial interval, the interrun delay, and the criterion for testing at each delay. These parameters differed considerably from those used in previous experiments, which usually had more trials per day, a shorter interrun delay, and no enclosed goal box at the end of the maze.

Postoperatively, control rats reached criterion quickly, taking a mean of 6 trials to do so. Fornix transection produced a severe impairment. Only three of ten rats ever reached criterion, and they took a mean of 55 trials to do so.

A similar dissociation occurred with the longer interrun delays. Control rats had a mean choice accuracy of 85% and 83% during the 5-minute and 15-minute delays, respectively. Fornix-transected rats had a severe impairment, with 67% and 64% at the two delays, respectively.

These data are important because they provide the first direct comparison of the effects of similar lesions in similar spatial DCD tasks in both rats and monkeys. The results of this experiment demonstrate that transection of the fornix in rats and in monkeys had the same general effect on behavior, although the magnitude of that effect was different in the two species, being greater in rats than in monkeys. The presence of a behavioral deficit in both rats and monkeys following fornix transection demonstrates that the fornix must have a similar behavioral function in both species, at least in this type of task. The relatively smaller magnitude of the impairment in the monkeys may indicate that monkeys have at least one additional brain mechanism contributing to the type of memory assessed by this task, or that some "fornix" fibers are ectopic, and travel to the hippocampus outside of this compact bundle.

Object Discriminations

Many experiments with monkeys have used object discriminations to examine the role of hippocampus and memory. The objects were usually presented on a tray in front of the monkey, which made a choice by displacing one of the objects. If the correct object was displaced, reward was available from a cup beneath it. If the incorrect object was displaced, no reward was available.

Although the hippocampus in rats clearly plays a role in spatial cognition, its spatial functions are not exclusive, and it is also important for some types of nonspatial discriminations. However, the experiments using nonspatial stimuli have produced mixed results. In some tasks, lesions of the hippocampus or closely related structures produced behavioral impairments, whereas in other tasks, these lesions had little or no effect (Aggleton, Hunt, & Rawlins, 1986; Barnes, 1988; Becker & Olton, 1980; Eichenbaum et al., 1986; Raffaele & Olton, 1988). A direct comparison with the results obtained from experiments with monkeys is virtually impossible because the experiments with rats did not use procedures directly comparable to those used for monkeys. Of particular importance are the differences between the discriminative stimuli used for the two species. Whereas the discriminative stimuli for the monkeys were three-dimensional objects located on an otherwise homogeneous surface, the nonspatial stimuli for rats were typically of some other kind: different visual or tactual patterns, small compartments containing many different stimuli, or odors (Rothblatt & Hayes, 1987).

To help resolve some of the apparent discrepancies between the results of experiments with monkeys and those with rats, rodent versions of two different monkey experiments were designed with three-dimensional objects. The first, which is completed, demonstrates that the fornix in both species is not necessary for conditional place-object discriminations (Markowska et al., 1989; Murray et al., 1989). The second, which is in progress, is designed to determine the role of the hippocampus, fornix, and amygdala in concurrent object discriminations.

Conditional Place-Object Discriminations: The Role of the Fornix

The conditional spatial-object discriminations gave the animal a choice between two objects. The object that was correct depended on (was conditional on) some spatial characteristic of the task. In the first experiment, both the animal and the objects to be discriminated were in two different locations in the testing room. In one location, one object was correct. In the other location, the second object was correct. In the second conditional spatial-object discrimination, the objects were placed in the same location, but the animal's position varied. When the animal was on one side of the table facing the two objects, one object was correct. When the animal was on the other side of the table facing the two objects, the other object was correct. In the third experiment, the positions of the objects relative to the animal determined which one was correct. When the objects were on the left side of the animal, one object was correct. When the objects were on the right side of the animal, the other object was correct. This last experiment was arranged so that the absolute position of both the animal and the objects was irrelevant for correct performance; only the relative position of the objects with respect to the animal identified the correct object.

As in the experiment with the T maze described earlier, all aspects of the testing procedure were made as similar as possible for both rats and monkeys, with the obvious exception of the physical aspects of the task. Some other procedural differences (the use of correction trials for rats but not for monkeys and the details of the response for the two species) also varied, but these were kept as minimal as possible.

The results from all three conditional experiments were the same in both species (Markowska et al., 1989; Murray et al., 1989). Fornix lesions did not interfere with these conditional discriminations. These results obviously have implications for descriptions of fornix function in memory (Gaffan & Harrison, 1989). For the current discussion, however, their major importance is the similarity of behavioral results following the lesions in both species. Thus, equivalent testing for rats and monkeys can lead to the absence of a behavioral impairment (these conditional experiments) as well as the presence of one (the T-maze experiments).

Concurrent Discriminations: Hippocampus, Fornix, and Amygdala

As reviewed earlier (see page 187), a comparison of the results from experiments with monkeys and with rats produces three apparent discrepancies. The first concerns the role of the hippocampus in nonspatial discriminations. In monkeys, hippocampal lesions impaired performance in some types of nonspatial discriminations: 8-pair concurrent object description (Mahut, Zola-Morgan, & Moss, 1982) and delayed-non-match-to-sample as the delay between the sample and the choice increased (Mishkin, 1978; Zola-Morgan & Squire, 1986). In rats, hippocampal lesions had mixed results in nonspatial tasks (Aggleton, Hunt, & Rawlins, 1986; Barnes, 1988; Becker & Olton, 1980; O'Keefe & Nadel, 1978; Raffaele & Olton, 1988).

The second concerns the effects of fornix lesions in memory tasks. In rats, these lesions produced a severe and enduring impairment in many types of tasks (Olton, Becker, & Handelmann, 1979). In monkeys the effects were variable and often not very large (Gaffan & Harrison; 1989; Mahut & Moss, 1986; Squire, 1987).

This experiment was designed to address both of these issues. It adapted for rats the 8-pair concurrent object description previously used for monkeys (Mahut et al., 1982) and included the following groups with lesions: hippocampus, fornix, and amygdala.

The 8-pair concurrent object discrimination is being tested with four groups of rats. Three have lesions: hippocampus, fornix, or amygdala. The fourth is a control group. The experiment is identical in procedure to the one done with monkeys (Moss, Mahut, & Zola-Morgan, 1981). In that experiment, hippocampal lesions impaired performance of the 8-pair concurrent discrimination; fornix lesions did not.

The apparatus has a start area separated from a choice area by a guillotine door. The choice area is divided in half (perpendicular to the start area) by a Plexiglas barrier. The objects to be discriminated are placed in the choice area, one object on each side of the barrier. A small well in the floor of the choice area behind each object provides reinforcement, 0.1 cc of 10% (by volume) sucrose, for the correct choice.

Each rat is given one *test session* of 40 trials each day to a criterion of 39 consecutive correct responses in one day. Eight pairs of objects are used, with one object of each pair designated as the correct object. Each pair is presented for 5 trials in each test session. The order of presentation of the pairs and the right-left position of the correct object is determined in a pseudorandom counterbalanced manner with the following constraint: In each block of two test sessions, each object is on the right for half of the trials and on the left for half of the trials.

For each *trial*, one pair of objects is placed on the maze. The rat is placed in the start box. The guillotine door is lifted and the rat runs behind one of the objects to the well. A *correct response* is recorded if the front paws of the rat cross a line at the back of the rewarded object, and sucrose is injected into the reinforcement well. An *incorrect response* is recorded if the front paws of the rat cross the back of the unrewarded object, and no sucrose is injected into the cup. After making a choice, the rat returns to the start box. The appropriate objects are inserted for the next trial.

Control rats are learning the task to criterion after about 30 days of training. Testing is still in progress for the lesion groups, so the data are incomplete. This experiment shows that rats can perform another object discrimination that has been used to test monkeys, and will give valuable data for comparing the effects of lesions in rats and in monkeys.

Conclusions

Cognitive neuroscience seeks to relate cognitive and neural descriptions of behavior. In this endeavor, experiments with animals are critical. Although the continued development of biomedical technology provides us with increasing amounts of information about the neural mechanisms of cognition in people, these types of experiments are inherently limited by primary consideration for the therapeutic rather than the scientific benefits of the investigation. The accuracy of the integration of neural and cognitive descriptions is limited by the degree of accuracy in the least accurate description. Thus, a very sophisticated psychological analysis is of little use in neural-cognitive comparisons if the neural analysis is more limited. Consequently, animal models are critical to obtain the detailed neural information that is necessary to relate neural processes to cognitive processes. Furthermore, in the future, as our ability to obtain even more details of neural function in animals is increased, animal models will become even more important.

In the context of animal models, experiments with monkeys have contributed a great deal. However, these are inherently limited by the high cost and scarcity of primates. Some kind of mammalian model is necessary. Although experiments with invertebrates have provided us with a great deal of information about the neural mechanisms involved in plasticity, the nervous system of invertebrates is so different from that of vertebrates that invertebrates are not able to answer questions about localization of function within different neural systems. For the time being, rodents appear to be the animal of choice.

If rodents are to play such a pivotal role in cognitive neuroscience, some information about the validity of the generalization from rodents to other animals is important. If experiments are done correctly, of course, then the information obtained will certainly tell us something about the neural-cognitive relationships in rodents. However, some broader applicability is desirable, if it is valid. Assessing validity requires experiments that are as similar as possible for both rodents and the other species in question. The results we have obtained here, although limited, strongly suggest that at least some of the neural-cognitive links in rodents accurately generalize to monkeys, and may be characteristic of mammals in general. If further studies validate this conclusion, then rodent models provide all the power of sophisticated cognitive as well as neural investigations.

In terms of cognitive neuroscience, these results suggest that additional emphasis should be placed on experiments with rodents. The developments of comparative cognition have provided new behavioral tools and conceptual frameworks to investigate cognition in rats and other animals, and the field continues to provide new tools and new theories (Roitblat, 1987). Thus, rodent models provide a unique set of qualities: sophisticated cognitive analyses, sophisticated neural analyses, mammalian brain, ready availability, and low cost. If cognitive neuroscience is to develop to its maximum promise, it must make effective use of these advantages.

In terms of the specific issues relating brain mechanisms to memory, these experiments confirm the importance of the hippocampus and neuroanatomically related structures in memory, and provide the means to pursue more detailed questions about the neuroanatomical and behavioral dissociations that will permit particular neural systems to be related to particular mnemonic functions. Certainly, these data are consistent in demonstrating the role of the hippocampus and fornix in some types of memory. The experiments in progress should provide us additional information about the role of hippocampus in nonspatial, object discriminations, and the interactions between hippocampus and amygdala in memory.

Acknowledgments

The authors thank A. Durr for help in the preparation of the manuscript.

References

Aggleton, J. P., Hunt, P. R., & Rawlins, J. N. P. (1986). The effects of hippocampal lesions upon spatial and non-spatial tests of working memory. *Behavioral Brain Research, 19,* 133–146.

Bachevalier, J., Saunders, R. C., & Mishkin, M. (1985). Visual recognition in monkeys: Effects of transection of fornix. *Experimental Brain Research, 57,* 547–553.

Barnes, C. A. (1988). Spatial learning and memory processes: The search for their neurobiological mechanisms in the rat. *Trends in Neurosciences, 11,* 163–169.

Becker, J. T., & Olton, D. S. (1980). Object discrimination by rats: The role of frontal and hippocampal systems in retention and reversal. *Physiology and Behavior, 24,* 33–38.

Corkin, S. (1984). Lasting consequences of bilateral medial temporal lobectomy: Clinical course and experimental findings in H. M. *Seminars in Neurology, 4,* 249–259.

Eichenbaum, H., Fagan, A., & Cohen, N. J. (1986). Normal olfactory discrimination learning set and facilitation of reversal learning after medial-temporal damage in rats: Implications for an account of preserved learning abilities in amnesia. *Journal of Neuroscience, 6,* 1876–1884.

Gaffan, D. (1974). Recognition impaired and association intact in the memory of monkeys after transection of the fornix. *Journal of Comparative Physiological Psychology, 86,* 1100–1109.

Gaffan, D. (1977). Monkeys' recognition memory for complex pictures and the effect of fornix transection. *Quarterly Journal of Experimental Psychology, 29,* 505–514.

Gaffan, D., Gaffan, E. A., & Harrison, S. (1984). Effects of fornix transection on spontaneous and trained non-matching by monkeys. *Quarterly Journal of Experimental Psychology, 36B,* 285–303.

Gaffan, D., & Harrison, S. (1989). Place memory and scene memory: Effects of fornix transection in the monkey. *Experimental Brain Research, 74,* 202–212.

Gaffan, D., & Saunders, R. C. (1985). Running recognition of configural stimuli by fornix-transected monkeys. *Quarterly Journal of Experimental Psychology, 37B,* 61–71.

Mahut, H. (1972). A selective spatial deficit in monkeys after transection of the fornix. *Neuropsychologia, 10,* 65–74.

Mahut, H., & Moss, M. (1986). The monkey and the sea horse. In R. L. Isaacson & K. H. Pribram (Eds.), *The hippocampus* (Vol. 4, pp. 241–279). New York: Plenum Press.

Mahut, H., & Zola, S. M. (1973). A non-modality specific impairment in spatial learning after fornix lesions in monkeys. *Neuropsychologia, 11,* 255–269.

Mahut, H., Zola-Morgan, S., & Moss, M. (1982). Hippocampal resections impair associative learning and recognition memory in the monkey. *Journal of Neuroscience, 2,* 1214–1229.

Markowska, A. L., Olton, D. S., Murray, E. A., & Gaffan, D. (1989). A comparative analysis of the role of fornix and cingulate cortex in memory: Rats. *Experimental Brain Research, 74,* 187–201.

Milner, B. (1959). The memory deficit in bilateral hippocampal lesions. *Psychiatric Research Reports, 11,* 43–58.

Moss, M., Mahut, H., & Zola-Morgan, S. (1981). Concurrent discrimination learning of monkeys after hippocampal, entorhinal, or fornix lesions. *Journal of Neuroscience, 1,* 227–240.

Murray, E. A., Davidson, M., Gaffan, D., Olton, D. S., & Suomi, S. J. (1989). Effects of fornix transection and cingulate cortical ablation on spatial memory in rhesus monkeys. *Experimental Brain Research, 74,* 173–186.

O'Keefe, J., & Nadel, L. (1978). *The hippocampus as a cognitive map.* Oxford, England: Oxford University Press.

Olton, D. S., Becker, J. T., & Handelmann, G. E. (1979). Hippocampus, space and memory. *Behavioral and Brain Science, 2,* 313–365.

Olton, D. S., Walker, J. A., & Gage, F. H. (1978). Hippocampal connections and spatial discrimination. *Brain Research, 139,* 295–308.

Olton, D. S., Walker, J. A., & Wolf, W. A. (1982). A disconnection analysis of hippocampal function. *Brain Research, 233,* 241–253.

Raffaelle, K., & Olton, D. S. (1988). Differential involvement of the amygdala and hippocampus in learning. *Behavioral Neuroscience, 102,* 349–355.

Roitblat, H. L. (1987). *Introduction to comparative cognition.* New York: W. H. Freeman.

Rothblatt, L. L., & Hayes, L. L. (1987). Short-term recognition memory in the rat: Nonmatching with trial unique junk stimuli. *Behavioral Neuroscience, 101,* 587–590.

Scoville, W. B., & Milner, B. (1957). Loss of recent memory after bilateral hippocampal lesions. *Journal of Neurology, Neurosurgery, and Psychiatry, 20,* 11–21.

Squire, L. R. (1987). *Memory and brain.* New York: Oxford University Press.

Squire, L. R., & Cohen, N. J. (1984). Human memory and amnesia. In J. L. McGaugh, G. Lynch, & N. M. Weinberger (Eds.), *The neurobiology of learning and memory* (pp. 3–64). New York: Guilford Press.

Wible, C. G., Shiber, J. R., & Olton, D. S. (in press). Hippocampus, fimbria/fornix, amygdala, and memory: Object discrimination in rats. *Behavioral Neuroscience.*

Zola-Morgan, S., & Mahut, H. (1973). Paradoxical facilitation of object reversal learning after transection of the fornix in monkeys. *Neuropsychologia, 11,* 271–284.

Zola-Morgan, S., & Squire, L. R. (1986). Memory impairment in monkeys following lesions limited to the hippocampus. *Behavioral Neuroscience, 100,* 155–160.

12

Processing Subsystems of Memory

MARCIA K. JOHNSON / WILLIAM HIRST

Most students of cognition assume that human memory is not an undifferentiated system; at issue is how to conceptualize the parts of this system. We suggest that one fundamental division is between memories for the consequences of perceptual processing, such as seeing and hearing, and memories for the consequences of reflective processing, such as planning, comparing, and imagining. Cognition typically mixes both perceptual and reflective processing with such artful coordination that both phenomenal experience and memories for experiences possess an integrated, holistic quality. Consequently, the line between perceptual processing and reflective processing is often difficult to draw. Nevertheless, to understand the blend one must understand the ingredients—the separate perceptual and reflective mechanisms contributing to mental experience.

It should be emphasized at the outset that the perceptual processes we refer to include what are commonly called *top-down* as well as *bottom-up* processes (e.g., Palmer, 1975). People may make unconscious inferences in order to perceive depth, or their perception of a single word may be affected by their expectations and beliefs. In either case, the processing that yields depth perception or lexical access would not have occurred at that time if information had not impinged upon the sensorium in the first place. Such processing, which is *dependent* on the external world, constitutes perceptual processing. Other processing, of course, can occur independent of sensory stimulation. People can imagine, plan, develop beliefs, and solve problems without continuous guidance from external cues. Such independent internal processes are what we are referring to as reflective.

Memory here is assumed to record the processing underlying cognitive activities, not some further consequence of the processing (Kolers & Roediger, 1984). Memory is treated as an integral part of certain cognitive processes and not a separate mechanism. Recent work on connectionist networks illustrates the point. In a connectionist network, perception or categorization is a function of the weights of the connections, but the act of perceiving or categorizing also changes the weights (Rumelhart & McClelland, 1986). Thus, the processes that yield a perception or categorization also change the nature of subsequent processing by changing the weights on which this processing depends. A close tie between process and memory suggests that it may not make much sense to talk separately about process and memory. To claim that two processes involve different processing systems is equivalent to claiming that two different memory subsystems exist. Therefore, if we want to understand the nature of different memory subsystems, we must understand the processing these memory systems record.

A Working Framework

The framework we describe is an expansion of Johnson's multiple entry, modular memory system, or MEM (Johnson, 1983, 1990, 1991a). In developing this framework, we have been guided by introspection, a large body of evidence and theorizing from experimental cognitive psychology, and neuropsychological findings. We adopt the following terms to refer to different levels of analysis: The term *component* refers to the most primitive concepts in the present framework (these may be further decomposed in the future or for other purposes). *Subsystem* refers to the coordinated activities of two or more component processes; a *system* coordinates activities from two or more subsystems; and *memory* is a summary term for the coordinated activity of all memory systems. Of course, which level of analysis in this scheme is identified as a memory "system" is somewhat arbitrary. For example, even individual components may be considered systems in the "weak" sense (see Sherry & Schacter, 1987).

According to MEM, memory contains distinguishable perceptual and reflective memory systems. Positing these distinct memory systems has heuristic value, in that it has provided a means of organizing empirical facts obtained from cognitive/behavioral studies of normal memory (Johnson, 1983), a framework for interpreting observations about patients suffering from amnesia (Johnson, 1990), delusions (Johnson, 1988a), and confabulation (Johnson, 1991a), and has also generated new research (Hirst, Johnson, Kim, Phelps, Risse, & Volpe, 1986; Hirst, Johnson, Phelps, & Volpe, 1988; Johnson & Kim, 1985; Johnson, Kim, & Risse, 1985; Johnson, Peterson, Chua-Yap, & Rose, 1989; Weinstein, 1987; see also Johnson, 1990). The division between perceptual and reflective memories may capture functional organizations within the nervous system as well. Several behavioral dissociations support this claim. For example, reflective processing develops later than perceptual processes (e.g., Flavell, 1985; Moscovitch, 1985; Perlmutter, 1984; Schacter & Moscovitch, 1984). Moreover, reflective processes are more likely to be disrupted by stress, depression, aging, and the use of alcohol and other drugs than are perceptual processes (Craik, 1986; Eich, 1975; Hasher & Zacks, 1979, 1984; Hashtroudi & Parker, 1986). Furthermore, the breakdown in memory functioning found in patients with anterograde amnesia appears to fall disproportionately on reflective memory (Johnson, 1983; chapters in Cermak, 1982; Hirst, 1988). Conversely, reflective processes may be relatively intact while perceptually guided processes are disrupted, as in agnosias or certain cases of disrupted perceptual–motor skill learning. For example, there is evidence that Huntington's disease patients are impaired relative to Alzheimer's disease patients on a perceptual–motor skill task but Alzheimer's patients show the greater impairment on a recall task (Heindel, Butters, & Salmon, 1988).

Perceptual Memory

A closer analysis of perceptual memory suggests several divisions within the perceptual memory system. Perceptual processing can be divided along the lines of different perceptual modalities (seeing, hearing, etc.), but it can also be divided along amodal dimensions. For example, much of what people learn, regardless of modality, requires the processing of invariants in a complex stimulus array. People are not necessarily aware of this processing or of these invariants. In order to understand a speaker with an unusual accent, people must learn to distinguish cues in the speech signal that specify various phonemes, yet they may be unable to consciously isolate the relevant information or to tell someone what it is. Similarly, in learning to catch a ball, people have to coordinate their activity with changes in aspects of the stimulus array such as the rate of change in the size of the stimulus as a function of time. The memory system involved in recording these kinds of perceptual learning is called P-1.

On the other hand, other products of perceptual processing, namely, the phenomenal experiences of objects and events, are consciously accessible. Listeners may not know how they have adapted to a speaker's accent, but they know what words they have just heard. Ballplayers may not know what aspects of the changing stimulus array allowed them to catch a ball, but they know it is a ball they have caught. Such differences suggest that memory for perceptual phenomenal experience may involve a system other than P-1, which we call P-2.[1]

The postulation of P-1 and P-2 memory subsystems corresponds to the claim that there are interesting, functionally important differences in the memorial consequences of P-1 and P-2 perceptual processing. Recent neuropsychological studies provide dramatic examples of such a division in perceptual processes. Several studies have demonstrated that P-1 processing may take place although P-2 processing is disrupted. For instance, agnosic patients have a conscious awareness that an object is in front of them, but cannot identify it. Rather they claim that they see individual features without being able to assemble them into a whole (Luria, 1973; Marcel, 1983). In studies of other patients with brain damage, some patients can make perceptual discriminations without being aware of the object itself. Patients suffering from prosopagnosia may show a galvanic skin response to familiar faces even when they cannot consciously recognize the face (Damasio,

1985). Volpe, LeDoux, and Gazzaniga (1979) found that patients with extinction, who cannot identify an object if simultaneously presented with another object in the contralateral field, could nevertheless indicate whether
the two simultaneously presented objects were the same. Blindsight patients
can point to objects that they claim not to see (Weiskrantz, 1986).

Reflective Memory

Reflective processing can be sustained by internal events and can be independent of external events. The evolutionary development of reflective capabilities produced an explosion of mental possibilities: People can
consider analogues of perception (images) in the absence of the corresponding object or event; they can plan, solve problems, generate alternative
futures, concoct beliefs, and come to doubt these beliefs. Although we cannot yet list all the component subprocesses involved in reflection, much less
the relations among them, a modest and, it is hoped, tractable beginning is
to consider a minimum set of components that could, in principle, yield
reflection approaching the complexity we observe in normally functioning
adult humans.

A reasonably powerful reflective system would include at least four types
of component subprocesses, which might be called *noting, shifting, refreshing,*
and *reactivating* (Johnson, 1990). These processes activate or affect the
activation of already established memories or concepts and thereby establish new memories. *Noting* involves identifying relations among activated
objects or events; *shifting* involves a change in perspective that activates
alternative aspects of objects or events; *refreshing* prolongs ongoing activation through attentional processes; *reactivating* brings information that has
dropped out of consciousness back to an active (though not necessarily
conscious) state. To these four components of reflection, we need to add
supervisor functions that set agendas or goals and monitor outcomes (Miller,
Galanter, & Pribram, 1960; Nelson & Narens, 1990).

These four processes can vary in the extent to which they are strategically
driven or deliberate. To mark this deliberative dimension of reflection,
when noting, shifting, refreshing, and reactivating are carried out under
strategic control, they might be called *discovering, initiating, rehearsing,* and
retrieving, respectively. The control processes that set agendas and monitor
outcomes in the case of strategic reflection might be called *executive functions.*

In sum, a simple but powerful reflective system has to have several
component processes, including supervisor and executive functions, for
sustaining, organizing, and reviving events. These processes must be represented by at least two levels of reflection that differ in degree of deliberation. The two organized groupings of reflective component processes corresponding to these two levels can be thought of as two reflective subsystems,
R-1 and R-2 (see Figure 12.1). The component processes within and between

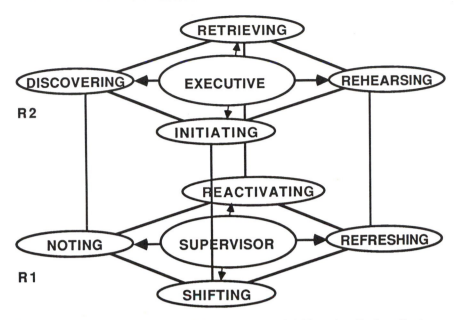

Figure 12.1 Reflective processes differ in degree of deliberation. Basic reflective processes (noting, shifting, refreshing, reactivating, and supervisor functions) are represented on the bottom of the cube and corresponding but more strategic reflective functions (discovering, initiating, rehearsal, retrieval, and executive functions) are represented on the top of the cube.

levels work together to allow mental activity to go beyond immediate perception.

Like the perceptual memory system, the reflective memory system is a record of prior processing. The distinction between R-1 and R-2 processes yields distinguishable reflective memory subsystems, also called R-1 and R-2. In theory, manipulations of reflective memory could differentially affect the various component processes in R-1 and R-2.

Activation of Memories

The processing at any point in time is recorded in the perceptual and reflective memory subsystem(s) that are active. At some future time, these records may be activated. By activation, we mean that the processing record is being replayed, so to speak (Kolers & Roediger, 1984). Exactly what is activated will depend on the kind of task probing memory. Complex experiences typically elicit a host of processes, and these complex mental events will yield memorial representations distributed across various memory subsystems. A task that probes memory may activate all or only some of this multiple and complex representation. Evidence of memory depends on an appropriate match between earlier processing and processing produced by the memory probe (e.g., Morris, Bransford, & Franks, 1977; Tulving &

Thompson, 1973). A perceptual identification task, for instance, may rely relatively more on representations formed by P-1, and a recognition task relatively more on representations formed by P-2 (e.g., Jacoby & Dallas, 1981). Recall, on the other hand, may depend critically on the reflective system (e.g., Hirst, Johnson, Kim, Phelps, & Volpe, 1986). A full understanding of different memory tasks would involve a clear articulation of the different subsystems the tasks call upon (also see Moscovitch, Winocur, & McLachlan, 1986).

The extent to which an individual is aware of the activation of a process will vary as a function of what kinds of processes and corresponding memory subsystems are involved and the kinds of interactions among activated processes. Thus consciousness is an emergent property of activation. Not all activated processes result in consciousness (e.g., see Kihlstrom, 1984). For instance, people are more likely to be conscious of R-2 processes than of P-1 processes even if both are activated. Moreover, one activated process may prevent another from becoming conscious, as in a divided attention experiment. In this regard, an activated process of which we ordinarily are not conscious may become conscious if activation of other processing is eliminated. Finally, consciousness may be determined by such factors as the degree of mutual activation among related elements, success in recruiting attention, and inhibition among patterns involving common elements or processing structures (e.g., Norman & Shallice, 1986).

Interaction Between Perceptual and Reflective Memory

In addition to specifying the component processes of perceptual and reflective memory, how perceptual and reflective subsystems interact must be specified. For example, representations from one subsystem may directly activate related representations from another, or interactions between perceptual and reflective memory may take place through supervisor and executive components (see Figure 12.2). A goal such as *look for a restaurant* might activate relevant perceptual schemas from perceptual memory as well as reflective plans adapted to the current situation (find a parking place, check the restaurant guide for this part of town, etc.). Typically, supervisor and executive functions have greater access to reflective memory than to perceptual memory, and greater access to P-2 than to P-1 subsystems. Also, through supervisor and executive functions, other component reflective processes (e.g., refreshing, reactivating) can be applied to representations in P-2, and perhaps in P-1.

It is important to emphasize that we do not define subsystems in terms of their complete independence or lack of interaction with each other. Rather, it seems more likely that subsystems must interact to support complex cognition and action. Characterizing this interaction remains a major challenge.

EXECUTIVE

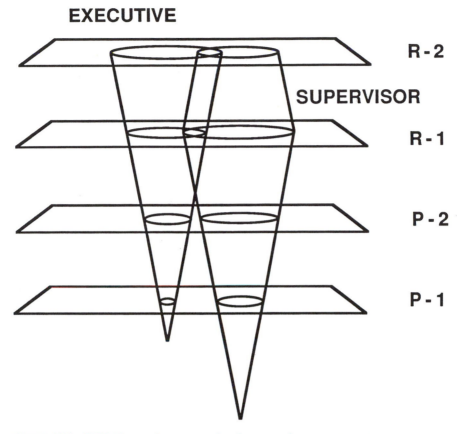

R-2

SUPERVISOR

R-1

P-2

P-1

Figure 12.2 Reflective and perceptual subsystems interact through supervisor and executive processes, which have relatively greater access to and control over reflective than perceptual subsystems.

Relation of MEM to Other Concepts

The conception of memory outlined in the last section draws on (and in some cases expands) a number of related ideas in the memory literature. For example, the concepts of *shifting* and *noting* are based on the demonstrated critical importance of organizational processes for recall (e.g., Mandler, 1967; Tulving, 1962). To further clarify the present framework, a brief discussion of its relation to several other ideas follows:

Supervisor and Executive Processes

One could represent monitoring and control functions as a single executive system that is separate from memory (e.g., Stuss & Benson, 1986), but in the present conceptualization, supervisor and executive functions are embedded within the reflection subsystems because learning and remembering depend critically on such functions. Supervisor and executive processes direct and monitor other processes (e.g., rehearsal) that have memorial consequences and new combinations of supervisor and executive processes themselves can be learned and remembered (e.g., a new learning strategy).

Automatic versus Effortful Processes

The distinction between automatic and effortful, or automatic and controlled processes (e.g., Hasher & Zacks, 1979, 1984; Shiffrin & Schneider, 1977) easily fits within the present framework. Automatic processes are those that seem to require no (or only minimal) executive functions; the most controlled processes require R-2 functions involving deliberate, conscious activity. Varying degrees of control would depend on how many or which reflective components are involved. The distinction between P-1 and P-2 and between R-1 and R-2 memory subsystems emphasizes that controlled processes are intimately involved in memory (Shiffrin & Schneider, 1977), yet also allow for the possibility that some types of memory depend less than others on deliberate processing (e.g., Eich, 1984; Hasher & Zacks, 1979; Jacoby, Woloshyn, & Kelley, 1989).

Short-Term versus Long-Term Memory

The distinction between short-term memory (STM) and long-term memory (LTM) can also be represented in MEM. In MEM, short-term memory (or working memory) as "capacity" could be identified either with the set of activated information with functional consequences or with the even more restricted subset of activated information of which we are aware. STM as "process" could be identified with the refreshing and rehearsing components of reflection. In any event, in MEM there is not a single STM buffer; rather, STM or working memory is distributed throughout the subsystems (Baddeley, 1983; Baddeley & Hitch, 1974). LTM would be the records of prior processing represented in all subsystems.

Episodic versus Semantic (or Generic) Memory

It has been proposed that general knowledge or semantic memory and specific incidents or episodic memory are different subsystems of memory (Tulving, 1983). In contrast, in MEM, knowing and remembering reflect attributions made on the basis of subjective qualities of mental experiences (Johnson, 1988a, 1988b; Klatsky, 1984). Remembering is not *either* autobiographical *or* nonautobiographical (i.e., semantic); rather, while remembering, we experience degrees of specificity, clarity, confidence in veridicality, and so on, and the greater the clarity and specificity, the more likely a memory will seem to refer to a distinct episode. Time and place information (Tulving, 1983) contribute greatly to this specificity, but probably should not be taken as defining features of autobiographical memory. For example, an especially important factor determining whether remembered information is felt to be autobiographical is whether it was the object of earlier reflective activity that would tie it in with other personal experiences. For example, anticipating an event, and reflecting back on it, create supporting memories that become evidence for the specificity and personal relevance of the event (Johnson, Foley, Suengas, & Raye, 1988, Study 2).

Similarly, there is no separate store for semantic memory; generic knowledge is represented in all subsystems. Furthermore, the types of generic knowledge represented in various subsystems might be quite different from each other (for example, that involved in learned eye movements for reading in P-1 and learned strategies for taking notes in R-2). Given that there are potential differences in appropriate cognitive models of the representation of different types of generic knowledge, and in the corresponding underlying neurobiological systems, the idea of a single generic memory system may be misleading.

Procedural versus Declarative Memory

It has also been proposed that memory consists of procedural and declarative memory systems (Cohen & Squire, 1980). In MEM, procedural knowledge (like generic knowledge) is distributed throughout the subsystems, but different types of skills or procedures (or components of complex skills or procedures) are very likely supported by different subsystems. Thus, we might not expect the procedure for threading a needle and that for counterbalancing lists in a learning experiment to be represented in the same way or supported by exactly the same structures. In addition, some procedures can be learned without strategic intervention or declarative representation; the learning and remembering of others may require strategic intervention or declarative representation (in fact, Anderson, 1982, suggests that procedural knowledge may start out as declarative knowledge). Furthermore, as learning occurs, control may pass from reflective to perceptual subsystems and vice versa for what might superficially appear to be the same task. Acquiring the skills necessary to complete a task efficiently is not always just a matter of learning to do the same skill "automatically" (Hirst, 1986). Skill acquisition can involve the use of different cues and subsystems as the task becomes restructured. Finally, not only may "declarative" information be implicated in learning and reactivating procedures, but the reverse is sometimes the case as well. For example, in order to tell someone a phone number, you may need to start to dial it as a cue to revive its declarative representation.

Various Combinations of Reflective Subprocesses

Useful insights and data have been generated by thinking about memory in terms of dichotomies such as STM/LTM, episodic/semantic, automatic/ effortful, procedural/declarative. Here we are suggesting that new insights and data might be generated by considering memory as the consequence of perceptually initiated and reflectively generated processes, and by attempting to specify the subsystems and component processes involved in establishing memory for perception and reflection. To illustrate the potential usefulness of this framework for thinking about similarities and differences among various cognitive activities, consider different combinations of reflective subprocesses. In Figure 12.3, the components of the reflective system are represented as either active, as indicated by solid lines, or as inactive, as indicated by dotted lines (to simplify the discussion, supervisor and executive functions have been omitted). This simple schema allows us to characterize various cognitive activities involving memory, as well as certain deficits.

For instance, as illustrated in Figure 12.3A, when discovering, initiating, rehearsing, and retrieving act in combination, the resulting activity is often characterized as intentional (i.e., strategy-driven) learning. If all R-1 processes are working as well, a wide range of memory tasks can be accomplished with no apparent cognitive deficiencies. However, as illustrated in Figure 12.3B, mental activity consisting only of noting, shifting, refreshing, and reactivating would yield unintentional learning alone—or, with guidance by R-1 supervisor processes, only relatively simple intentional learning. Consequently, the disruption of R-2 processing would severely limit the complexity of memory activities and thus would limit the complexity of what could be learned and remembered.

The pattern in Figure 12.3C, which highlights the combination of shifting, initiating, refreshing, and rehearsing, illustrates how the phenomenal experience of something like free association or stream of consciousness is realized. Goal-directed rote rehearsal and, if poorly controlled, perseveration or even compulsions, may arise from the combination of refreshing, rehearsing, reactivating, and retrieving (Figure 12.3D). The combination of noting, discovering, reactivating, and retrieving (Figure 12.3E) could produce a rigid type of rote rehearsal as well, resulting in a well-rationalized, well-remembered, but inflexible product. Shifting, initiating, noting, and discovering operating together would produce creative organizational possibilities useful in, for example, problem solving or brainstorming (Figure 12.3F).

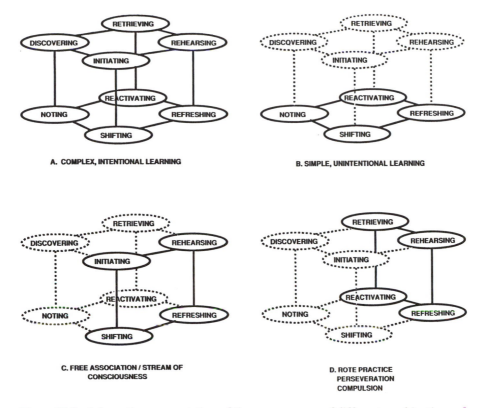

Figure 12.3 Schematic representation of the consequences of different combinations of reflective component processes.

A short-term memory deficit could arise when all components but re-freshing and rehearsing are intact (Figure 12.3G). When all components are intact but reactivating and retrieving, a long-term memory deficit very much like "core" anterograde amnesia might be observed (Figure 12.3H; see next section). Disruption of R-1 supervisor processes (Figure 12.3I) would eliminate the relatively "automatic" reflective control of activities, such as keeping active a set to note certain kinds of relations while engaged in some other memory task controlled by R-2. The consequence would be that reflective activities that usually seem "automatic" would require "effort." Alternatively, the complete deactivation of executive functions in R-2 (Figure 12.3J) may lead to an inability to voluntarily initiate schemas in P-1 or P-2. Here such disruption may prevent people from willfully undertaking well-learned, perceptually based actions (such as demonstrating how to use a hammer) in the absence of external cues (such as the hammer itself). Disruption of the relation between supervisor and executive processes

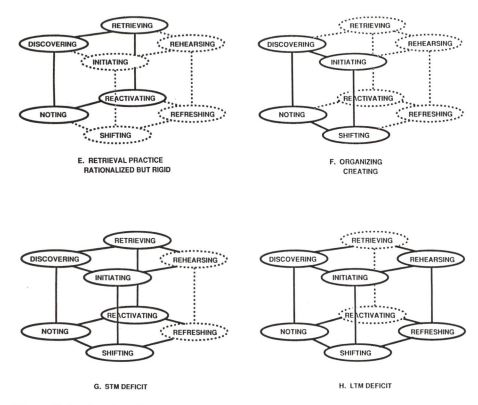

Figure 12.3   (Continued)

associated with R-1 and R-2 subsystems (Figure 12.3K) would, among other things, affect a person's ability to monitor the thoughts generated in R-1. When such disruptions occur, events imagined as a consequence of R-1 processes, for instance, might seem to have come from P-2 (see Johnson, 1991a).

Amnesia as a Reactivation Deficit

This section focuses on the way in which a particular memory deficit, "classic" anterograde amnesia, might be conceptualized in MEM. Available evidence suggests that perceptual memory processes (P-1 and P-2) are relatively intact in amnesics whereas some aspect(s) of reflection are disrupted (Johnson, 1983, 1990). A number of observations support this

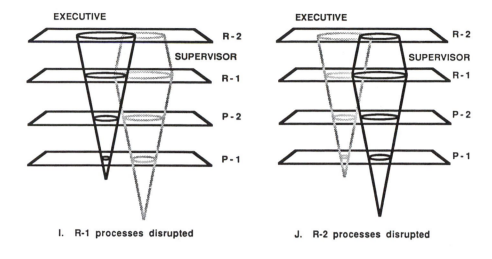

I. R-1 processes disrupted

J. R-2 processes disrupted

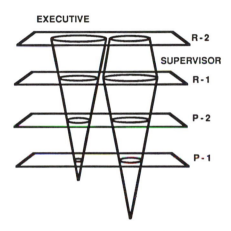

K. R-1 and R-2 do not exchange information
    about executive functions

Figure 12.3 (Continued)

characterization. In amnesics, some cognitive/perceptual and motor skills appear preserved (Cohen & Squire, 1980; Corkin, 1968), and these skills seem to be largely under perceptual control of P-1 and P-2 processes. The recall of amnesics is always profoundly disrupted, but recognition memory appears to be less disrupted (Hirst et al., 1986, 1988; Johnson & Kim, 1985; Weinstein, 1987). Such a pattern is predicted in the MEM framework because recognition often draws on entries resulting from perceptual processes whereas recall draws more on reflective entries. Furthermore, the degree of disruption that amnesics show in recognition memory (Weinstein, 1987), and in the acquisition of affective responses (Johnson, Kim, & Risse, 1985), depends on the degree to which reflection is required in the task.

As we have highlighted, reflection is a complex system involving a number of different components that interact in various ways, depending on task demands and on an individual's reflective capabilities. Many components of the reflective system appear to be largely intact in amnesics. The fact that amnesics' intellectual functions seem to be remarkably intact (Butters & Cermak, 1980; Milner, Corkin, & Teuber, 1968) suggests that executive functions may often be unaffected. Many amnesics can plan for the future, adopt and use complex mnemonics (Hirst & Volpe, 1988), and generally appear to be able to recruit what mnemonic resources they have to accomplish a task. There are exceptions to this characterization. For instance, Korsakoffs may have deficient metamemory (Hirst & Volpe, 1988; Shimamura & Squire, 1986). Nevertheless, in many amnesic patients, severe memory problems can be observed in the presence of what appears to be intact supervisor and executive functions.

Observations of normal short-term memory in at least some amnesics (Baddeley & Warrington, 1970) suggest that refreshing may be intact. Amnesics profit from contexts that clarify meaning (e.g., BAGPIPES—*The notes went sour when the seams split,* Johnson et al., described in Johnson, 1990; McAndrews, Glisky, & Schacter, 1987), suggesting that noting relations among simultaneously active concepts is intact. As for shifting, although some amnesics, particularly Korsakoffs, show perseveration (usually taken to be a sign of frontal lobe damage), other amnesics do not show signs of perseveration (Moscovitch, 1982). They easily shift from one perspective to another. Nevertheless, even with refreshing, noting, shifting, and supervisor component processes intact, severe amnesia could result from a disruption in the reactivating component of reflection.

Reactivating

Reactivating provides an opportunity for noting relations between perceptually noncontiguous elements. It is essential for establishing relational information that bridges individual items or events. These bridges can initiate further reactivations. And with each reactivation, the strength of the memory improves, thereby increasing the availability of a memory and extending the retention interval over which the memory can be detected (Ebbinghaus, 1885/1964; Linton, 1978). The resulting memory is a complex, cohesive representation of past experience and consists of both relational and item information (e.g., Hunt, Ausley, & Schultz, 1986; Mandler, 1980).

Memories become reactivated with the appropriate externally provided and internally generated cues (e.g., McGeoch, 1942; Tulving, 1983). Cues can reactivate relational information, item information, or both. Different underlying mechanisms may be responsible for item and relational reactivation. Furthermore, cues from different sources combine so that a memory may be reactivated when no single cue would be sufficient. One especially important kind of cue combination is when specific external and internal cues combine with "agendas" to increase the probability of reactivation of a memory. In MEM, *agendas* are goals or plans that govern through supervisor and executive functions the actions and mental activities of an individual. An agenda might be fairly global ("Eat dinner") or more specific ("Get the waitress's attention and order a meal"). In many instances, an agenda might specifically call on the use of memory, as in "Remember list A." These memory agendas can also be quite general ("Remember what you did when you were a kid") or quite specific ("Remember the animal words in the list that you studied five minutes ago"). Whether an agenda explicitly probes memory or only makes an implicit use of memory (Schacter, 1987), it will interact with memory by activating relevant information. An agenda about "ordering a meal" may activate a restaurant script, whereas an agenda about eating may activate a script about etiquette and, in the appropriate context, a restaurant script as well. This activated information need not be "conscious." However, it may facilitate the reactivation of a memory so that it does become conscious in the appropriate circumstances. This capacity of ongoing agendas to keep information in a state of increased susceptibility to appropriate cues may account not only for phenomena such as incubation effects in problem solving, but also for why explicit retrieval cues can often elicit memories when more general cues could not. A general probe such as "Remember the list of words studied five minutes ago" may activate a great deal of information relevant to the spatio-temporal frame "five minutes ago." This activation, however, may not be sufficient to raise the desired information to consciousness. With an additional, more explicit cue of "animal," the convergence of activations may become sufficient to produce a conscious recollection.

Reactivation can produce a special kind of "consolidation" of memories. *Consolidation* has been used to refer to a variety of types of processes (see Squire, 1987, and chapters in Weingartner & Parker, 1984). Often what investigators seem to mean by consolidation is an automatic, time-dependent, endogenously driven process that is nonselective in that all memories undergo consolidation unless a disruptive event occurs (see, e.g., discussions about this issue by Gold & McGaugh, 1984, Keppel, 1984, and Spear & Mueller, 1984). In contrast, in the present formulation, reactivation plays a consolidative role that is neither automatic nor nonselective. Reactivation is not a "stand alone" process, but rather an aspect of an integrated reflective system. Reactivation is partially driven by supervisor and executive functions. Moreover, the noting, shifting, and refreshing initiated by the reactivation of information contribute to the impact of reactivation on memory and the probability of further cycles of reactivation. Information that does not satisfy an ongoing agenda is unlikely to be reactivated unless prompted by a specific, strong external cue. Agenda-relevant information is more likely to be reactivated, and this newly activated information is likely to lead to the reactivation of additional agenda-relevant information. This idea of reactivation is similar to Spear and Mueller's (1984) notion of retrieval-based consolidation (see also Squire, 1987).

If the reactivating component of reflective memory is deficient in amnesics, then several things should follow. Amnesics may not be able to form and subsequently use relational information beyond that contained in concurrently available elements (i.e., through noting and refreshing). The resulting impoverished memory should make it difficult for agendas to activate and access information related to but not directly described in the agenda. In particular, an agenda such as "Recall the list that you just studied" provides little information about the material to be remembered. One must access this information through the relations connecting the present circumstances with the previous study session and the spatio-temporal memories of the study session with the particular items learned in the study session. If relational information is not formed and cannot be activated by the cue "the list you just studied," then recall of the list should be difficult, if not impossible, for an amnesic. On the other hand, if the agenda is "Determine whether you saw the word 'hat' in the list just studied," the cues provide quite specific item information, which is often enough to produce a familiarity response. Of course, such item-specific reactivation cannot alone make up for the absence of reflectively generated reactivation. Consequently, both recognition and recall are depressed with amnesia, although recall is disproportionately depressed when compared with recognition.

Summary

According to the present framework, memory is created by an intricate interplay of processes that are organized at the most global functional level into perceptual and reflective systems. Each of these systems consists of more specific functional subsystems that, in turn, are made up of yet more specific functional components. Perceptual and reflective subsystems are differentially involved in various learning and memory tasks: for example, the perceptual subsystems are more important in certain types of skill acquisition, and the reflective subsystems are more important in strategic memorization for the purposes of later recall. Our goal is to begin to specify the component processes in perceptual and reflective subsystems, to understand tasks in terms of the contributions of these component processes, and, eventually, to link such processes to neurobiological systems.

As a step in this direction, we have described some possible components and organizations of two reflective subsystems, R-1 and R-2. These subsystems are critical for what is sometimes called *declarative, episodic,* or *direct* memory, and for problem solving and other forms of productive thinking. Certain cognitive activities as well as deficits in cognitive functioning can be described in terms of the operation of components or combinations of components within the R-1 and R-2 subsystems. For example, we suggest that "classic" anterograde amnesia could result from the disruption of reactivation and retrieval components within the reflective subsystems. These components are critical for creating and strengthening connections between cognitive/behavioral agendas and events and establishing relations among perceptually noncontiguous events. Various other memory or cognitive deficits, and a variety of distinguishable normal cognitive activities (e.g., stream of consciousness, strategic learning) as well, can be specified in terms of patterns of reflective components that are active and disrupted or suppressed. Hopefully, efforts to more clearly specify processes at a cognitive level of description will make it easier to link cognitive functions to underlying brain structures and systems.

Note

1. P-1 and P-2 correspond to what was called "sensory" and "perceptual" subsystems in previous papers (e.g., Johnson, 1983, 1990). P-1 and P-2 can be further broken down into component subprocesses (see Johnson, 1991b).

Acknowledgments

Preparation of this chapter was supported in part by National Science Foundation grant BNS 8510633 to Marcia K. Johnson.

References

Anderson, J. R. (1982). Acquisition of cognitive skill. *Psychological Review, 89,* 369–406.
Baddeley, A. D. (1983). Working memory. *Philosophical Transactions of the Royal Society, B302,* 311–324.
Baddeley, A. D., & Hitch, G. J. (1974). Working memory. In G. H. Bower (Ed.), *The psychology of learning and motivation* (Vol. 8, pp. 47–89). New York: Academic Press.
Baddeley, A. D., & Warrington, E. K. (1970). Amnesia and the distinction between long-term and short-term memory. *Journal of Verbal Learning and Verbal Behavior, 9,* 176–189.

Butters, N., & Cermak, L. S. (1980). *Alcoholic Korsakoff's syndrome: An information processing approach to amnesia.* New York: Academic Press.

Cermak, L. S. (Ed.). (1982). *Human memory and amnesia.* Hillsdale, NJ: Lawrence Erlbaum.

Cohen, N. J., & Squire, L. R. (1980). Preserved learning and retention of pattern-analyzing skill in amnesia: Dissociation of knowing how and knowing that. *Science, 10,* 207–210.

Corkin, S. (1968). Acquisition of motor skill after bilateral medial temporal-lobe excision. *Neuropsychologia, 6,* 255.

Craik, F. I. M. (1986). A functional account of age differences in memory. In F. Klix & H. Hagendorf (Eds.), *Human memory and cognitive capabilities* (pp. 409–422). Amsterdam: North-Holland.

Damasio, A. R. (1985). Disorders of complex visual processing: Agnosias, achromatopsia, Balint's syndrome, and related difficulties of orientation and construction. In M. M. Mesulam (Ed.), *Principles of behavioral neurology* (pp. 259–288). Philadelphia: F. A. Davis.

Ebbinghaus, H. (1885/1964). *Memory: A contribution to experimental psychology.* New York: Dover.

Eich, J. E. (1975). State-dependent accessibility of retrieval cues in the retention of a categorized list. *Journal of Verbal Learning and Verbal Behavior, 14,* 408–417.

Eich, E. (1984). Memory for unattended events: Remembering with and without awareness. *Memory and Cognition, 12,* 105–111.

Flavell, J. H. (1985). *Cognitive development* (2nd ed.). Englewood Cliffs, NJ: Prentice Hall.

Gold, P. E., & McGaugh, J. L. (1984). Endogenous processes in memory consolidation. In H. Weingartner & E. S. Parker (Eds.), *Memory consolidation: Psychobiology of cognition* (pp. 65–83). Hillsdale, NJ: Lawrence Erlbaum.

Hasher, L., & Zacks, R. T. (1979). Automatic and effortful processes in memory. *Journal of Experimental Psychology, 108,* 356–388.

Hasher, L., & Zacks, R. T. (1984). Automatic processing of fundamental information: The case of frequency of occurrence. *American Psychologist, 39,* 1372–1388.

Hashtroudi, S., & Parker, E. S. (1986). Acute alcohol amnesia: What is remembered and what is forgotten. In H. D. Cappell, F. B. Glaser, Y. Israel, H. Kalant, W. Schmidt, E. M. Sellers, & R. Smart (Eds.), *Research advances in alcohol and drug problems* (Vol. 9, pp. 179–209). New York: Plenum.

Heindel, W. C., Butters, N., & Salmon, D. P. (1988). Impaired learning of a motor skill in patients with Huntington's Disease. *Behavioral Neuroscience, 102,* 141–147.

Hirst, W. (1986). Aspects of divided and selective attention. In J. E. LeDoux & W. Hirst (Eds.), *Mind and brain: Dialogues in cognitive neuroscience* (pp. 105–141). New York: Cambridge University Press.

Hirst, W. (1988). On consciousness, recall, recognition, and the architecture of memory. In K. Kirsner, F. Lewandowsky, & J. C. Dunn (Eds.), *Implicit memory.* Hillsdale, NJ: Lawrence Erlbaum.

Hirst, W., & Volpe, B. T. (1988). Memory strategies with brain damage. *Brain and Cognition, 8,* 379–408.

Hirst, W., Johnson, M. K., Kim, J. K., Phelps, E. A., & Volpe, B. T. (1986). Recognition and recall in amnesics. *Journal of Experimental Psychology: Learning, Memory, and Cognition, 12,* 445–451.

Hirst, W., Johnson, M. K., Phelps, E. A., & Volpe, B. T. (1988). More on recognition and recall in amnesics. *Journal of Experimental Psychology: Learning, Memory, and Cognition, 14,* 758–762.

Hunt, R. R., Ausley, J. A., & Schultz, E. E. (1986). Shared and item-specific information in memory for event descriptions. *Memory and Cognition, 14,* 49–54.

Jacoby, L. L., & Dallas, M. (1981). On the relationship between autobiographical memory and perceptual learning. *Journal of Experimental Psychology, 110,* 306–340.

Jacoby, L. L., Woloshyn, V., & Kelley, C. (1989). Becoming famous without being recognized: Unconscious influence of memory produced by dividing attention. *Journal of Experimental Psychology: General, 118*, 115–125.

Johnson, M. K. (1983). A multiple-entry, modular memory system. In G. H. Bower (Ed.), *The psychology of learning and motivation* (Vol. 17, pp. 81–123). New York: Academic Press.

Johnson, M. K. (1988a). Discriminating the origin of information. In T. F. Oltmanns & B. A. Maher (Eds.), *Delusional beliefs: Interdisciplinary perspectives* (pp. 34–65). New York: Wiley.

Johnson, M. K. (1988b). Reality monitoring: An experimental phenomenological approach. *Journal of Experimental Psychology: General, 117*, 390–394.

Johnson, M. K. (1990). Functional forms of human memory. In J. L. McGaugh, N. M. Weinberger, & G. Lynch (Eds.), *Brain organization and memory: Cells, systems and circuits* (pp. 106–134). New York: Oxford University Press.

Johnson, M. K. (1991a). Reality monitoring: Evidence from confabulation in organic brain disease patients. In G. Prigatano & D. L. Schacter (Eds.), *Awareness of deficit after brain injury* (pp. 175–197). New York: Oxford University Press.

Johnson, M. K. (1991b). Reflection, reality monitoring, and the self. In: R. G. Kunzendorf (Ed.), *Mental imagery* (pp. 3–16). New York: Plenum.

Johnson, M. K., Foley, M. A., Suengas, A. G., & Raye, C. L. (1988). Phenomenal characteristics of memories for perceived and imagined autobiographical events. *Journal of Experimental Psychology: General, 117*, 371–376.

Johnson, M. K., & Kim, J. K. (1985). Recognition of pictures by alcoholic Korsakoff patients. *Bulletin of the Psychonomic Society, 23*, 156–458.

Johnson, M. K., Kim, J. K., & Risse, G. (1985). Do alcoholic Korsakoff patients acquire affective reactions? *Journal of Experimental Psychology: Learning, Memory, and Cognition, 11*, 22–36.

Johnson, M. K., Peterson, M. A., Chua-Yap, E., & Rose, P. M. (1989). Frequency judgments and the problem of defining a perceptual event. *Journal of Experimental Psychology: Learning, Memory and Cognition, 15*, 126–136.

Keppel, G. (1984). Consolidation and forgetting theory. In H. Weingartner & E. S. Parker (Eds.), *Memory consolidation: Psychobiology of cognition* (pp. 149–161). Hillsdale, NJ: Lawrence Erlbaum.

Kihlstrom, J. F. (1984). Conscious, subconscious, unconscious: A cognitive perspective. In K. S. Bowers, & D. Meichenbaum (Eds.), *The unconscious reconsidered* (pp. 149–211). New York: Wiley.

Klatsky, R. L. (1984). Armchair theorists have more fun. *The Behavioral and Brain Sciences, 7*, 244.

Kolers, P. A., & Roediger, H. L., III (1984). Procedures of mind. *Journal of Verbal Learning and Verbal Behavior, 23*, 425–449.

Linton, M. (1978). Real world memories after six years: An in vivo study of very long-term memory. In M. M. Gruneberg, P. E. Morris, & R. N. Sykes (Eds.), *Practical aspects of memory* (pp. 3–24). New York: Academic Press.

Luria, A. R. (1973). *The working brain: An introduction to neuropsychology*. New York: Basic Books.

Mandler, G. (1967). Organization and memory. In K. W. Spence & J. T. Spence (Eds.), *The psychology of learning and motivation* (Vol. 1, pp. 327–372). New York: Academic Press.

Mandler, G. (1980). Recognizing: The judgment of previous occurrence. *Psychological Review, 87*, 252–271.

Marcel, A. J. (1983). Conscious and unconscious perception: An approach to the relations between phenomenal experience and perceptual processes. *Cognitive Psychology, 15*, 238–300.

McGeoch, J. A. (1942). *The psychology of human learning*. New York: Longmans.

McAndrews, M. P., Glisky, E. L., & Schacter, D. L. (1987). When priming persists: Long-lasting implicit memory for a single episode in amnesic patients. *Neuropsychologia, 25,* 497–506.

Miller, G. A., Galanter, E., & Pribram, K. A. (1960). *Plans and the structure of behavior.* New York: Holt, Rinehart and Winston.

Milner, B., Corkin, S., & Teuber, H. L. (1968). Further analysis of the hippocampal amnesic syndrome: Fourteen-year follow-up study of H. M. *Neuropsychologia, 6,* 215–234.

Morris, C. D., Bransford, J. D., & Franks, J. J. (1977). Levels of processing versus transfer appropriate processing. *Journal of Verbal Learning and Verbal Behavior, 16,* 519–533.

Moscovitch, M. (1982). Multiple dissociations of function in amnesia. In L. S. Cermak (Ed.), *Human memory and amnesia* (pp. 337–370). Hillsdale, NJ: Lawrence Erlbaum.

Moscovitch, M. (1985). Memory from infancy to old age: Implications for theories of normal and pathological memory. *Annals of the New York Academy of Sciences, 444,* 78–96.

Moscovitch, M., Winocur, G., & McLachlan, D. (1986). Memory as accessed by recognition and reading time in normal and memory-impaired people with Alzheimer's disease and other neurological disorders. *Journal of Experimental Psychology: General, 115,* 331–347.

Nelson, T. O., & Narens, L. (1990). Metamemory: A theoretical framework and some new findings. In G. H. Bower (Ed.), *The psychology of learning and motivation* (pp. 125–173). New York: Academic Press.

Norman, D. A., & Shallice, T. (1986). Attention to action: Willed and automatic control of behavior. In R. J. Davidson, G. E. Schwartz, & D. Shapiro (Eds.), *Consciousness and self-regulation* (pp. 1–18). New York: Plenum Press.

Palmer, S. E. (1975). Visual perception and world knowledge: Notes on a model of sensory-cognitive interaction. In D. A. Norman & D. E. Rumelhart (Eds.), *Explorations in cognition* (pp. 279–307). San Francisco: W. H. Freeman.

Perlmutter, M. (1984). Continuities and discontinuities in early human memory paradigms, processes, and performance. In R. Kail & N. E. Spear (Eds.), *Comparative perspectives on the development of memory* (pp. 253–284). Hillsdale, NJ: Lawrence Erlbaum.

Rumelhart, D. E., & McClelland, J. L. (1986). *Parallel distributed processing: Explorations in the microstructure of cognition: Vol 1. Foundations.* Cambridge, MA: MIT Press.

Schacter, D. L. (1987). Implicit memory: History and current status. *Journal of Experimental Psychology: Learning, Memory and Cognition, 13,* 501–518.

Schacter, D. L., & Moscovitch, M. (1984). Infants, amnesics, and dissociable memory systems. In M. Moscovitch (Ed.), *Infant memory* (pp. 173–216). New York: Plenum Press.

Sherry, D. F., & Schacter, D. L. (1987). The evolution of multiple memory systems. *Psychological Review, 94,* 439–454.

Shiffrin, R. M., & Schneider, W. (1977). Controlled and automatic human information processing: II. Perceptual learning, automatic attending, and a general theory. *Psychological Review, 84,* 127–190.

Shimamura, A. P., & Squire, L. R. (1986). Memory and metamemory: A study of the feeling of knowing phenomenon in amnesic patients. *Journal of Experimental Psychology: Learning, Memory, and Cognition, 12,* 452–460.

Spear, N. E., & Mueller, C. W. (1984). Consolidation as a function of retrieval. In H. Weingartner & E. S. Parker (Eds.), *Memory consolidation: Psychobiology of cognition* (pp. 111–147). Hillsdale, NJ: Lawrence Erlbaum.

Squire, L. R. (1987). *Memory and brain.* New York: Oxford University Press.

Stuss, D. T., & Benson, D. F. (1986). *The frontal lobes.* New York: Raven Press.

Suengas, A. G., & Johnson, M. K. (1988). Qualitative effects of rehearsal of memories for

perceived and imagined complex events. *Journal of Experimental Psychology: General,*
117, 377–389.

Tulving, E. (1962). Subjective organization in free recall of "unrelated" words. *Psychological Review, 69,* 344–354.

Tulving, E. (1983). *Elements of episodic memory.* New York: Clarendon Press/Oxford University Press.

Tulving, E., & Thompson, D. M. (1973). Encoding specificity and retrieval processes in episodic memory. *Psychological Review, 80,* 352–373.

Weingartner, H., & Parker, E. S. (1984). *Memory consolidation: Psychobiology of cognition.* Hillsdale, NJ: Lawrence Erlbaum.

Volpe, B. T., LeDoux, J. E., & Gazzaniga, M. S. (1979). Information processing of visual stimuli in an "extinguished" field. *Nature, 282,* 722–724.

Weinstein, A. (1987). Preserved recognition memory in amnesia. Unpublished doctoral dissertation, State University of New York at Stony Brook.

Weiskrantz, L. (1986). *Blindsight.* Oxford, England: Clarendon Press.

13

The Emergence of Multidimensional Approaches to the Structural Organization of Memory

RAYMOND P. KESNER

Memory is a concept that is invoked in order to account for a complex set of behavioral phenomena. Because of this complexity, theoreticians have proposed many schemas that describe and organize the memory domain. Similarly, because of the complexity of the central nervous system circuitry that subserves memory, many neural systems have been identified as critically involved in memory.

On a psychological level there are two schemas that organize the structure of memory. The first schema proposes that the basic unit of memory is composed of specific associative elements [stimulus-response (S-R), stimulus-stimulus (S-S), or stimulus-reinforcement] and interrelationships among these associative elements. The second schema proposes that the basic unit of memory is composed of higher order associative elements (attributes, working and reference memory, recognition memory, declarative and procedural memory) and the interrelationships among these higher order associative elements. On a neurobiological level the neural memory structure is represented by a set of elements (synapses, neural regions, and neural interconnections) that can subserve the specific associative or higher order psychological units.

Single Memory Analysis

Researchers who emphasize the importance of the first schema often assume that they are studying the neural contribution to a single memory based on the study of a single associative unit. Two separate research strategies have been used in studying the memory representation of single associative units. In the first research strategy, the emphasis is on neural regions and circuits that subserve memory as defined by a task (e.g., taste aversion learning, active or inhibitory avoidance learning, heart-rate conditioning). This approach is exemplified by researchers who, for example, have studied the neural circuits that mediate heart-rate conditioning in rabbits and pigeons as well as conditioned emotional response learning in rats (Cohen, 1980; Kapp, Pascoe, & Bixler, 1984; LeDoux, 1987). In the second research strategy, the emphasis is on cellular and molecular levels of analysis (e.g., single neurons, membrane channels, second messenger systems). This analytic approach is exemplified by researchers who have selected to analyze model systems using single stimulus-response units in the form of classical conditioning of a gill-withdrawal response in *Aplysia* or an eyelid closure response in the rabbit (Hawkins & Kandel, 1984; Thompson, 1980).

Dual Memory Analysis

In recent years, there has been a great emphasis on dual memory systems, or two classes of memory organization, with a specific emphasis on neural regions or circuits as neural units of analysis. The schemas for dividing memory differ among the various theoreticians and are often based on different sets of criteria. As an example, Mishkin (1982) and Mishkin, Malamut, and Bachevalier, (1984) have organized memory into a habit memory system based on S-R association links and an associative or representational memory (e.g., recognition memory) system based on a higher order level of organization of many associative elements. Furthermore, they assume that the two memory systems can operate independently. Based on extensive empirical work with monkeys, they propose that recognition memory, as one form of associative or representational memory, is stored and represented in higher order sensory areas of the cortex and involves active interaction with limbic-thalamic and cortical neural circuits. A different approach was taken by Olton (1983) and Olton, Becker, and Handlemann (1979). They organize the elemental associative structural units into two types of higher order memories, labeled *working* and *reference,* based on the differentiation of new informational aspects of a memory (working memory) from existing aspects of a memory (reference memory). The specific, personal, and temporal context of a situation is coded in working memory. This would translate into memory for events that occur on a specific trial in a task, biasing mnemonic coding towards the processing of incoming data. In contrast, general information concerning rules and procedures (general knowledge) of specific situations is coded in reference memory. This would translate into memory for events that happen on all trials in a task, biasing mnemonic coding toward the processing of expectancies based on the organization of the extant memory. One would expect that in any new task to be learned there would be a somewhat greater emphasis on working memory, but that after learning the emphasis would shift toward reference memory, unless the task requires the processing of new information on every trial. In this latter case both working and reference memory systems would be activated. This distinction between working and reference memory is closely akin to the distinction between episodic and semantic memory proposed by Tulving (1972).

Olton further suggests that both working and reference memory can operate independently of each other. Based on a large number of experiments, Olton (1983) has proposed that the hippocampus and its interconnections mediate working memory, while some other system, such as the neocortex, mediates reference memory. Using this approach, the psychological unit of analysis can be represented by the dual system of working and reference memory, while the neural unit of analysis might be represented by large interconnected neural systems.

A somewhat different approach has been taken by Squire and Cohen (1984). They have suggested that there are two types of memory, which they named *declarative* and *procedural,* with each being characterized by a specific set of operations within or between tasks (Cohen, 1984; Squire, 1983). Declarative memory is based on explicit information that is easily accessible and is concerned with specific facts or data. Procedural memory, on the other hand, is based on implicit information that is not easily accessible and is concerned with procedures and skills. In both types of memories there is a higher order level of organization of elemental associative units. Squire and Cohen also assume that the two types of memory systems are independent of each other. Furthermore, they propose that the medial temporal cortex, which includes the hippocampus, and diencephalic human brain areas mediate declarative, but not procedural, memory.

Each of these dual memory approaches emphasizes the importance of specific neural regions in coding memory, but each approach emphasizes a slightly different aspect of the organization of memory.

Multidimensional Memory Analysis

This approach recognizes the complexity of memory organization and proposes a multidimensional analysis of memory. Squire and Zola-Morgan (1988) have proposed a taxonomy of memory based on a distinction between declarative and nondeclarative memory. The declarative form of memory includes episodic and semantic representations of propositions and images. Nondeclarative forms of memory include unaware representations of motor, perceptual, and cognitive skills, priming, and simple classical conditioning. Furthermore, it is assumed that the hippocampus and related structures mediate declarative forms of memory. Kesner and DiMattia (1987) proposed that memory organization is based on functionally separate but interdependent attributes of memory, such as space, time, sensory perception, response, affect, and language. Furthermore, interactions between attributes can provide for a higher order structural organization, perhaps in the form of cognitive maps and environmental contexts. The psychological unit of analysis is represented by an attribute and interaction between attributes, while neural regions represent the neural level of analysis. I will elaborate in more detail on the theoretical assumptions that underlie this level of analysis in the next section.

Attribute Model of Memory: Psychological Aspects

Nature of Memory Representation

Based on earlier suggestions by Underwood (1969) and Spear (1976), Kesner has proposed that any specific memory is represented by a set of features or attributes that are specific and unique for each learning experience (Kesner, 1980; Kesner & DiMattia, 1987). Embedded within this multidimensional attribute framework are some of the explicit associations (e.g., S-S, S-R, S-Reward) mentioned earlier.

Kesner has suggested that in most animal experiments a set of at least five salient attributes characterizes mnemonic information. In humans one would also add a linguistic attribute. These are labeled *space, time, affect, sensory perception,* and *response.* A spatial attribute within this framework involves the coding and storage of specific stimuli representing places or relationships between places, which are usually independent of the subject's own body schema. It is exemplified by the ability to encode and remember maps and to localize stimuli in external space.

A temporal attribute involves the encoding and storage of specific stimuli or sets of spatially or temporally separated stimuli as part of an episode marking or tagging its occurrence in time—that is, separating one specific episode from previous or succeeding episodes.

An affect attribute involves the encoding and storage of reinforcement contingencies that result in positive or negative emotional experiences.

A sensory-perceptual attribute involves the encoding and storage of a set of sensory stimuli that are organized in the form of cues as part of a specific experience.

A response attribute involves the encoding and storage of information based on feedback from responses that occur in specific situations as well as the selection of appropriate responses.

The organization of these attributes can take many forms and probably is organized heterarchically. There are interactions between attributes that are very useful and can aid in identifying specific neural regions that might subserve a critical interaction. For example, the interaction between spatial and temporal attributes can provide for the external context of a situation, which is important in determining when and where critical events occurred. Another important interaction involves the temporal and affective attributes. In this case the interaction can provide important information concerning the internal context (internal state of the organism), which is important in evaluating emotional experiences. There is also a possible interaction between spatial and response attributes; this would result in the encoding and storage of responses that depend upon accurate assessment of one's body orientation in space (egocentric localization). This interaction is influenced by vestibular and kinesthetic input, which aids navigation in space relative to the animal. It is exemplified, for instance, by the ability to encode and remember right-left responses. Finally, there is an interaction between sensory-perception and response attributes. This interaction has been labeled *S-R associations*. In some tasks S-R associations are the most easily distinguishable attribute of a memory.

The attribute theoretical framework emphasizes the importance of (a) multiple measures of memory for any specific task, with the aim of assessing the contribution of each attribute, (b) finding tasks that accentuate the contribution of a single attribute or interaction between specific attributes, and (c) scaling of the difficulty of a task along a single dimension. For example, one can vary the temporal attribute by increasing list length, the spatial attribute by increasing the number of locations to be remembered, and the affect attribute by varying the magnitude of reinforcement. These manipulations are very important, because one often finds a reciprocal relationship between task difficulty and lesion size of specific neural regions.

Mechanics of Memory Representation

In this attribute model it is assumed that any specific memory is not only composed of a set of attributes, but is further organized into a data-based memory and an expectancy-based memory system (see Figure 13.1). The data-based memory system is biased toward the coding of incoming data concerning the present, with an emphasis on facts, data, and events that are usually personal and that occur within specific external and internal environmental contexts. To borrow a term from contemporary information processing theory, the emphasis is on bottom-up processing. During initial learning there is a great emphasis on the data-based memory system, which will continue to be of importance even after initial learning in situations where trial-unique or novel information needs to be remembered. The data-based memory system is akin to Olton's working memory or Tulving's episodic memory.

The expectancy-based memory system is biased toward previously stored information and can be thought of as one's general knowledge of the world. It can operate in the abstract in the absence of critical incoming data. From an information-processing view, the emphasis is on top-down processing. The expectancy-based memory system would tend to be of greater importance *after* a task has been learned if the situation is invariant and familiar. In most situations, however, one would expect a contribution of both systems with a varying proportion of involvement of one relative to the other.

Figure 13.1 Psychological and neural organization of data-based and expectancy- or knowledge-based memory.

SENSORY-PERCEPTION

SENSORY-PERCEPTION (Secondary sensory association cortex)

RESPONSE (autonomic) (Orbito-frontal cortex)

RESPONSE (somatic) (Premotor and supplementary motor cortex)

RESPONSE

SPATIAL (egocentric) (Dorsolateral or medial prefrontal cortex)

AFFECT (Orbito-frontal cortex)

AFFECT

SPATIAL (allocentric) (Posterior parietal cortex)

SPACE

TEMPORAL (Dorsolateral or medial prefrontal cortex)

TIME

S-R ASSOCIATIONS (Primary sensory cortex, motor cortex, brainstem, and cerebellum)

SENSORY-PERCEPTION

EGOCENTRIC LOCALIZATION (Caudate)

RESPONSE

INTERNAL CONTEXT (Amygdala)

AFFECT

SPACE

EXTERNAL CONTEXT (Hippocampus)

TIME

ATTRIBUTES

COGNITIVE MAP

EXPECTANCY-BASED MEMORY

HIGHER-ORDER ATTRIBUTES

ATTRIBUTES

DATA-BASED MEMORY

Memories within the expectancy-based memory system are assumed to be organized as a set of cognitive maps and their interactions that are unique for each memory. The exact nature and organization of knowledge structures within each cognitive map needs to be determined. The cognitive maps are labeled *spatial (allocentric)*, *spatial (egocentric)*, *temporal, affect, response (somatic)*, *response (autonomic)*, and *sensory-perceptual,* and are influenced by a set of attributes such as space, time, affect, response, and sensory perception as well as interactions between attributes such as space and response. Note that the same attributes are also associated with the data-based memory system. Of the many interactions between cognitive maps a few appear to be of critical importance. They are labeled *scripts,* representing the interaction between temporal and spatial cognitive maps; *schemas,* representing the interaction between spatial and sensory-perceptual cognitive maps; *moods,* representing the interaction between temporal and affect-laden cognitive maps; and *skills,* representing the interaction between sensory-perceptual and response cognitive maps. Within the expectancy-based memory system the combination of moods, scripts, and schemas can be thought of as similar to Olton's reference memory and Tulving's semantic memory, whereas skills can be thought of as akin to Mishkin's habit memory and Squire and Cohen's procedural memory. The structural organization of these attributes and their interactions with higher-order levels of organization (e.g., internal and external contexts egocentric localization and S-R associations as well as cognitive maps, scripts, schemas, moods, and skills) needs to be determined.

Dynamics of Memory Representation

How is critical attribute information processed within this multidimensional network? It is assumed that existing memory within the expectancy-based memory system is normally in an inactive state representing long-term memory. When critical neural regions are activated in the presence of a corresponding set of attributes, the resultant memory representation is labeled short-term memory. This short-term memory is a result of time-limited neural activation of critical neural systems. The duration, extent, and rate of short-term activation is in part influenced by selective attention and arousal processes. The level of attention can vary from automatic, requiring little effort, to controlled, requiring a great deal of effort. Automatic attention results in minimal neural activation, whereas attention requiring effort (e.g., rehearsal and mnemonic strategies) is likely to increase the duration of activation as well as the number of activated critical neural circuits. Empirical support for these latter assumptions has been presented elsewhere (Kesner & DiMattia, 1987).

The data-based memory system is assumed to operate in parallel with the expectancy-based memory system. The level of neuronal activation in these two systems varies depending upon the requirements of the situation. The data-based memory system differs from the expectancy-based memory system in that information can remain in an active state for a much longer time period. Some theoreticians have called this time period intermediate memory, a temporary holding mechanism, or long-term memory (Kesner, 1973; McGaugh, 1966; Rawlins, 1985). The data-based system is also influenced by attentional and arousal processes, which can affect the duration, extent, and rate of information processing. Prolonged activation of neuronal activity within activated neural regions subserving a corresponding set of attributes can trigger consolidation processes. Such consolidation processes could increase the probability of a structural change in the organization of the neural network mediating long-term memory within the expectancy-based memory system. Since this data-based memory system is likely to be activated when information is novel or in situations requiring processing of varied information, one is more likely to be aware of the dynamics of operation reflecting the explicit nature of this system.

Attribute Model of Memory: Neural Aspects

Which neural regions can be considered as candidates for the mediation of the attribute-based structure of memory? Within the model presented in Figure 13.1, it is proposed that for the data-based memory system the hippocampus mediates the external context, the amygdala the internal context, the caudate nucleus egocentric localization, and the primary-sensory cortex and motor cortex, brainstem, cerebellum circuit the S-R association system. Within the expectancy-based memory system it is proposed that the dorsolateral or medial prefrontal cortex mediates the temporal and spatial (egocentric) cognitive maps, the posterior parietal cortex the spatial (allocentric) cognitive maps, the orbito-frontal cortex the affect and response (autonomic) cognitive maps, the secondary sensory association cortex the sensory-perceptual cognitive maps, and the premotor and supplementary motor cortex the response (somatic) cognitive maps. The interactions between the dorsolateral or medial prefrontal cortex and posterior parietal cortex serve as the neural substrate for scripts; the posterior parietal cortex and secondary sensory association cortex subserve schemas; the dorsolateral or medial prefrontal and orbito-frontal cortex mediate moods; and the secondary sensory association cortex and premotor, supplementary motor cortex subserve skills.

It is assumed that interactive patterns of independently operating attributes provide the organizational framework for the existence of each unique memory. At the neurobiological level it is assumed that specific brain regions code or store the attributes and cognitive maps mentioned earlier as well as their specific interactions. In order to provide important information about the functional separability or independence of attributes and their underlying neural mechanisms, one can employ a double dissociation paradigm. Given that the independence assumption has empirical support, this attribute model would provide for both a psychological and a neural foundation for neural network models, because a large number of neural network models assume a parallel-distributed memory system.

Finally, relative amounts of neural activity within each critical brain region that codes a critical attribute, a cognitive map, or a set of attributes and cognitive maps provides for the total neuronal substrate associated with each unique memory representation.

Thus, different neural systems are assumed to mediate specific attributes. Furthermore, each attribute is processed by different neural systems (data or expectancy-based memory), depending upon the mechanics of information processing. When specific attributes trigger the activation of the underlying neuronal systems, the duration, extent, and rate of activation are affected by levels of attention and arousal as well as by the dynamics of maintaining information for longer periods of time. It is assumed that consolidation processes are most likely to be a property of the neuronal systems that mediate data-based memory systems. For example, the hippocampus is assumed to code the external context within the data-based memory system. Since the external context is very often an important component of a memory, the activation of consolidation processes in the hippocampus is likely to occur rather frequently. One possible neural mechanism that could mediate consolidation is long-term potentiation, which has been observed in the hippocampus (Teyler & Discenna, 1984). Since the cognitive maps within the expectancy-based memory system are assumed to be the most likely place for storage of information, consolidation processes can alter neurons within the neocortex—via transfer of information from the hippocampus, for example, or directly by altering neurons within the neocortex.

Finally, it is assumed that there is evolutionary continuity between animals and humans not only in terms of mnemonic function but also in terms of brain-memory function relationships. Given that this assumption has empirical support, one can devise neural networks that in animals have validity for human brain function and memory. The attribute model differs from previously proposed models in that it is more comprehensive and incorporates most of the critical features of the other models. There is also a suggestion of greater anatomical functional specificity as it relates to memory organization.

Empirical support and more detailed description of each component of this multidimensional model of memory has been presented elsewhere (Kesner & DiMattia, 1987).

Summary

Memory is a complex phenomenon due to the large number of potential interactions that are associated with the organization of memory at the psychological and neurological level. As a result, the study of the neu-robiological basis of the structure of memory lends itself readily to multiple theoretical views.

In this chapter an attempt is made to integrate some of the previously proposed views by presenting a comprehensive model of the structural organization of memory. This model is based on an attribute as the psycho-logical unit of memory and a neural system as the neurobiological unit of memory. At the psychological level it is proposed that each memory is composed of temporal, spatial, affect, sensory-perceptual, and response (somatic and autonomic) attributes and their interactions with higher order levels of organization (e.g., internal and external contexts, egocentric local-ization, and S-R associations as well as a variety of cognitive maps). The mechanics of processing these different attributes is carried out by a data-based memory system, which is biased toward the coding of incoming data concerning the present, and an expectancy-based memory system, which is biased toward the utilization of previously stored knowledge information in the form of cognitive maps.

Different neural systems mediate or subserve each attribute, interaction between attributes, or cognitive map. For the data-based memory system, the hippocampus mediates the interaction between temporal and spatial attributes (external context); the amygdala, the interaction between tem-poral and affect attributes (internal context); the caudate nucleus, the interaction between spatial and response attributes (egocentric localiza-tion); and the primary sensory cortex, motor cortex, brainstem, and cere-bellum circuit, the interaction between sensory-perceptual and response attributes (S-R association). Within the expectancy-based memory system it is proposed that the dorsolateral or medial prefrontal cortex mediates the temporal and egocentric spatial cognitive maps; the posterior parietal cor-tex, the allocentric spatial cognitive maps; the orbito-frontal cortex, the affect and response (autonomic) cognitive maps; the secondary sensory association cortex, the sensory-perceptual cognitive maps; and the premo-tor and supplementary motor cortex, the response (somatic) cognitive maps. Finally, dynamic processes, such as short-term memory, attention, arousal, and consolidation can modulate the operation of memory repre-sentations within the attribute organization of memory. It is hoped that this comprehensive multidimensional model will have heuristic value in a search for possible dissociations or critical interactions and the design of new experiments that will provide a better understanding of the neurologi-cal basis of memory.

References

Cohen, D. H. (1980). The functional neuroanatomy of a conditioned response. In R. F. Thompson, L. H. Hicks, & V. B. Shvyrkov (Eds.), *Neural mechanisms of goal-directed behavior and learning* (pp. 283–302). New York: Academic Press.

Cohen, N. (1984). Preserved learning capacity in amnesia: Evidence for multiple memory systems. In L. R. Squire & N. Butters (Eds.), *Neuropsychology of memory* (pp. 83–103). New York: Guilford Press.

Hawkins, R. D., & Kandel, E. R. (1984). Steps toward a cell-biological alphabet for elementary forms of learning. In G. Lynch, J. L. McGaugh, & N. M. Weinberger (Eds.), *Neurobiology of learning and memory* (pp. 385–404). New York: Guilford Press.

Kapp, B. S., Pascoe, J. P., & Bixler, M. A. (1984). The amygdala: A neuroanatomical systems approach to its contributions to aversive conditioning. In N. Butters & L. R. Squire (Eds.), *The neuropsychology of memory* (pp. 473–488). New York: Guilford Press.

Kesner, R. P. (1973). A neural system analysis of memory storage and retrieval. *Psychological Bulletin, 80,* 177–203.

Kesner, R. P. (1980). An attribute analysis of memory: The role of the hippocampus. *Physiological Psychology, 8,* 189–197.

Kesner, R. P., & DiMattia, B. V. (1987). Neurobiology of an attribute model of memory. In A. R. Morrison & A. N. Epstein (Eds.), *Progress in psychobiology and physiological psychology* (pp. 207–277). New York: Academic Press.

LeDoux, J. E. (1987). Emotion. In V. B. Mountcastle, F. Plum, & S. R. Geiger (Eds.), *Handbook of physiology: Section 1: The nervous system* (pp. 419–459). Bethesda, MD: American Physiological Society.

McGaugh, J. L. (1966). Time-dependent processes in memory storage. *Science, 153,* 1351–1358.

Mishkin, M. (1982). A memory system in the monkey. *Philosophical Transactions of the Royal Society of London. Series B: Biological Sciences (London), 298,* 85–95.

Mishkin, M., Malamut, B. L., & Bachevalier, J. (1984). Memories and habits: Two neural systems. In G. Lynch, J. L. McGaugh, & N. M. Weinberger (Eds.), *Neurobiology of learning and memory* (pp. 65–77). New York: Guilford Press.

Olton, D. S. (1983). Memory functions and the hippocampus. In W. Seifert (Ed.), *Neurobiology of the hippocampus* (pp. 335–373). New York: Academic Press.

Olton, D. S., Becker, J. T., & Handlemann, G. E. (1979). Hippocampus, space and memory. *Behavioral and Brain Sciences, 2,* 313–365.

Rawlins, J. N. P. (1985). Associations across time: The hippocampus as a temporary memory store. *The Behavioral and Brain Sciences, 8,* 479–496.

Squire, L. R. (1983). The hippocampus and the neuropsychology of memory. In W. Seifert (Ed.), *Neurobiology of the hippocampus.* New York: Academic Press.

Squire, L. R., & Cohen, N. J. (1984). Human memory and amnesia. In G. Lynch, J. McGaugh, & N. Weinberger (Eds.), *Neurobiology of learning and memory* (pp. 3–64). New York: Guilford Press.

Squire, L. R., & Zola-Morgan, S. (1988). Memory: Brain systems and behavior. *Trends in Neurosciences, 11,* 170–175.

Spear, N. F. (1976). Retrieval of memories: A psychobiological approach. In W. K. Estes (Ed.), *Handbook of learning and cognitive processes: Vol. 4. Attention and memory.* Hillsdale, NJ: Lawrence Erlbaum.

Teyler, T. J., & Discenna, P. (1984). Long-term potentiation as a candidate mnemonic device. *Brain Research Reviews, 7,* 15–28.

Thompson, R. F. (1980). The search for the engram, II. In D. McFadden (Ed.), *Neural mechanisms in behavior: A Texas symposium.* New York: Springer-Verlag.

Tulving, E. (1972). Episodic and semantic memory. In E. Tulving & W. D. Donaldson (Eds.), *Organization of memory.* New York: Academic Press.

Underwood, B. J. (1969). Attributes of memory. *Psychological Review, 76,* 559–573.

14

Conscious and Nonconscious Aspects of Memory: A Neuropsychological Framework of Modules and Central Systems

MORRIS MOSCOVITCH / CARLO UMILTA

It is commonplace to assert that we are conscious of only some of our mental processes. By this we typically mean both of the following: We can be conscious of only some mental processes, the others being forever out of the reach of our conscious experience, and, of those processes of which we can be conscious, only a small portion occupies our awareness at any given time. Although there is near-universal agreement about this general statement, there is much dispute about identifying the processes of which we can or cannot be conscious in principle, and about the mechanisms and processes involved in bringing that information to conscious awareness. There is even little agreement about what we mean by consciousness. Natsoulas (1978) discusses six common-sense meanings of consciousness in modern psychology, and we are sure that the enterprising theoretician can identify a few others. For our purposes, consciousness is defined as awareness of information so that a verbal or nonverbal description of it can be provided or a voluntary response can be made to it that is equivalent to the description.

To illustrate the difference between conscious and nonconscious forms of knowledge, consider the behavior of amnesic patients. Typically such patients perform extremely poorly on explicit tests of memory, such as recognition and recall, that require conscious recollection of the target items. Often the patients cannot even remember having studied any items. Despite this profound amnesia, performance on implicit tests of memory for the very same material may be relatively spared, and sometimes normal. In contrast to explicit tests, implicit tests make no reference to the past. Instead, memory for an item is inferred from changes in performance with respect to that item. Having studied a particular item, the amnesic patient processes it more accurately or efficiently when it is repeated, even though conscious recollection of the item or even the study episode, may be (virtually) absent (see Moscovitch, 1982a, 1984).

There is now an extensive literature documenting the dissociations between performance on implicit and that on explicit tests of memory

(Richardson-Klavehn & Bjork, 1988; Schacter, 1987). As a result, amnesia is now best described as an impairment only of conscious recollection of recently acquired information, not as a global failure to retain it (Moscovitch, Winocur, & McLachlan, 1986).

Although extensively documented, anmesia is not the only neuropsychological syndrome in which dissociations between performance on implicit and on explicit tests are found. In reviewing the literature on the topic, Schacter, McAndrews, and Moscovitch (1988) determined that such dissociations are found in a variety of syndromes—including neglect, aphasia, dyslexia, blindness, anesthesia, prosopagnosia, and possibly visual object agnosia. Indeed, it would not be stretching the truth to say that such dissociations are found for knowledge related to all input systems. That is, in each input system, a form of disorder can be found that is best described as a loss of access to consciousness rather than an absolute loss of knowledge related to that input system. Put in more neutral terms, performance on implicit tests is relatively normal despite apparent loss of function on explicit tests. Thus, *dyslexic patients* who cannot read when tested explicitly nonetheless guess correctly in choosing drawings that the words denote (e.g., Landis, Regard, & Serrant, 1980; Shallice & Saffran, 1986). *Prosopagnosic patients* who cannot match individual faces with their occupations or names when tested explicitly, nonetheless show higher GSRs (Galvanic Skin Responses) to faces that are paired with the appropriate occupation (e.g., Bauer, 1984) than with fabricated ones, and they read names more quickly when the names are accompanied by the matching face than when they are accompanied by a lure (De Haan, Young, & Newcombe, 1987). *Aphasic patients* who fail on explicit tests of comprehension show normal semantic priming and semantic context effects on lexical decision tasks (Milberg & Blumstein, 1981; Milberg, Blumstein, & Dworetsky, 1987). Patients with hemineglect who cannot attend to information on the neglected side and therefore fail to report it on explicit tests are nonetheless influenced by the neglected information when responding to input on the attended side (Behrmann, Moscovitch, Black, & Mozer, 1990; Marshall & Halligan, 1989). *Cortically blind* patients are influenced by visual events in the blind field: Their eye and hand movements indicate that they process visual information about location, direction of motion, and even orientation and shape, though they claim to have no concomitant, phenomenological experience of sight (Goodale, Milner, Jakobson, & Carey, 1991; Weiskrantz, 1986).

The observation that the explicit/implicit dissociation is characteristic of a number of such diverse syndromes suggests that there are basic and fundamental similarities in the organization of input systems. What are those similarities, and how do they relate to conscious and nonconscious processes? Why should comparable dissociations also be found for functions like memory and attention, even though neither is tied directly to input system? How is the organization of input systems related to the organization of memory and attentional systems? These questions will serve as a backdrop to the theoretical and empirical discussions in this paper and will lead to the presentation of a neuropsychological model of memory whose relation to input systems and attention is specified. The model is embedded in a conceptual framework of modules and central processes that grew out of a critical examination of Fodor's (1983, 1985) ideas on that topic (Moscovitch & Umilta, 1990). A brief summary of the main aspects of that framework as it pertains to modules follows.

Input Modules

Input modules are computational devices that have informational or propo-sitional content and are distinguished from other such devices by the following characteristics: *domain specificity, informational encapsulation* or *cognitive impenetrability*, and *shallow output*. Domain specificity entails that each module process information from only a restricted domain. For example, face-recognition modules, if they exist, would process information only about faces and no other visual stimulus. Informational encapsulation refers to the resistance of modules to higher order influences and to influ-ences from other modules. Thus, they are cognitively impenetrable, deliver-ing only a shallow output to higher order, central processes. An output is considered to be *shallow* if it is not semantically interpreted and if it provides no details about its derivation. The interpretation of the output is the responsibility of central systems that relate the output to a store of general knowledge (semantic memory).

These characteristics of modules ensure that the cognitive work they perform is not distorted by the beliefs, motivation, and expectancies of the organism. They are special-purpose devices suited to picking up informa-tion within their restricted domains, processing it efficiently and automat-ically, and delivering a precise, but narrow and presemantic, message to central systems for interpretation. In short, they are "stupid" but efficient systems, necessary for "representing the world veridically, and making it accessible to thought" (Fodor, 1983, p. 40).

Neuropsychological Evidence That Satisfies Criteria of Modularity

In the neuropsychological literature, the concept of modularity has, until recently, been considered much more loosely. A module is any area with a specialized function as determined by double (or sometimes even single) dissociation (Shallice, 1981, 1988). In this approach the term *module* is interchangeable with "component or component process." Indeed, the method of double dissociation is not adequate for distinguishing between modules, as we have defined them, and other nonmodular systems. In an attempt to provide greater precision and clarity in discussions of modu-larity at the neuropsychological level, we indicate what kind of neuro-psychological evidence would satisfy each of the criteria of modularity.

Domain specificity is satisfied by evidence of specialization of function: It is

the processing of only certain information that is impaired by a circumscribed lesion to one area and not by lesions to other areas of comparable
size. *Informational encapsulation* is satisfied by evidence that the function in
question remains intact in the face of gross intellectual decline caused by
degenerative or other neuropathological processes that do not affect the
area under investigation. Thus, if a function is unimpaired despite evidence
of dementia, this implies that the function is immune from the influence of
higher order processes that contribute to general intellectual functions.
The criterion of *shallow output* is satisfied by evidence of normal, domain-
specific performance without the ability to interpret semantically the information pertaining to that specific domain. For example, a patient of ours
was able to match different unusual views of the same object, without
having any notion of what the object was. This patient could be said to have
intact modules for perceptually representing objects in three dimensions.
The shallow output of the module is a "pictorial," nonsemantic, yet highly
specific, representation—a particular instance or token of the object, a
certain lamp or dog but not a generic lamp or dog.

   In a departure from Fodor (1983, 1985) that was motivated by an examination of the neuropsychological evidence, Moscovitch and Umilta (1990)
proposed the existence of three kinds of modules that differed from each
other in complexity and composition.

   1. Type I (basic) module. The basic modules are those that carry out a
single type of computation. They probably evolved to deal only with highly
relevant and predictable environmental stimuli. Among these would be
modules for the perception of basic sensory features in each modality and,
perhaps, some configurational stimuli such as faces, emotional expressions,
and phonemes that have invariant relational properties.

   2. Type II modules (innately assembled) consist of a collection of basic
modules whose organization is innately given and whose output is integrated or synthesized by a devoted, nonmodular processor. We use the term
devoted to indicate that this processor, though central, can only deal with
information coming from a particular group of modules and no other. In
some sense, this processor can be said to satisfy the criteria of domain
specificity and shallow output, but not that of informational encapsulation.
The organization is similar to one proposed by Turvey for vision (1973;
Michaels & Turvey, 1979) in which a devoted central processor integrates
input from modular feature analyzers. Like basic modules, Type II modules
are domain specific, though their domain is much broader. They correspond to the units or mechanisms in Luria's secondary zone (1966). Examples are modules that provide structural descriptions of objects (Riddoch &
Humphreys, 1987) and whose damage would lead to visual object agnosia.

   3. Type III modules (experientially assembled) are similar to Type II
modules except that nondevoted central systems are involved in assembling
basic modules and Type II modules. Once assembled, their functions be-

come modular with practice. In the cognitive literature, acquired automatic processes (Schneider, Dumais, & Shiffrin, 1977) would qualify as examples of the operation of Type III module. The organization of a word-form system is guided and formed by experience so that recognition of word forms becomes progressively more automatic, though it is a matter of debate whether it is ever completely effortless and mandatory (Behrmann, Moscovitch, & Mozer, in press; Egeth, 1989; Posner, 1989). A number of studies have suggested that attentional or central processes are necessary for assembling subroutines that are then run off automatically (Duncan, 1986; Logan, 1978, 1985; Norman & Shallice, 1986).

Input Modules and Consciousness

CONSCIOUS AWARENESS WITHOUT SEMANTIC KNOWLEDGE. The criterion of informational encapsulation and the neuropsychological evidence that satisfies it indicate that the information represented in modules and the operations performed on that information are not subject to conscious inspection. All that can be available to consciousness is the module's shallow output. Once it reaches consciousness, that shallow output can be interpreted and can guide voluntary action and thought. It participates in the control processes of behavior. Before being interpreted, however, shallow output can guide action and thought automatically (see Shallice's (1982) ideas regarding the inputs that work at the level of "contention scheduling"). In normal people, shallow output typically is interpreted immediately, and with little effort, but this does not necessarily mean that shallow output automatically receives a semantic interpretation once it reaches consciousness. The fate of that shallow output depends on the control processes that are brought to bear on it. As we have seen, some demented patients lack the means to interpret that output, yet can use that output, at its own level, to guide at least some of the behaviors in their repertoire. For example, they can read words they do not understand (Schwartz, Saffran, & Marin, 1980), match atypical views of objects they cannot classify (Moscovitch & Umilta, 1990; Warrington & Taylor, 1976), and so on. In short, they appear to display explicit knowledge only at a shallow level. Consciousness of shallow output does not entail semantic knowledge of even an elementary sort.

SHALLOW OUTPUT WITHOUT CONSCIOUS AWARENESS. We have noted that in a number of neuropsychological disorders, there is a dissociation between performance on implicit and that on explicit tests of knowledge (Schacter et al., 1988). In terms of our conceptual framework of modularity, these disorders can be viewed as arising from a disconnection of the module's shallow output from conscious awareness, while at the same time the output is made available to other systems that can affect behavior. Although this is a reasonable way to describe the disorders, there are still a number of problems that have to be resolved. One set of problems concerns the nature of the system mediating conscious awareness. Is there a single system or are there multiple systems, each associated with a specific module or sets of modules? Although we favor a single system (see Moscovitch & Umilta, 1990; Schacter, 1990; Schacter et al., 1988; Umilta, 1988), more empirical evidence is needed to decide conclusively between these alternatives.

The second set of problems concerns the shallow output itself. An alternative to the disconnection hypothesis is that brain damage has led to a degraded shallow output that is sufficient to drive performance on implicit, but not on explicit, tests of knowledge. Conscious awareness of the module's output requires a much stronger signal. This alternative interpretation falters on evidence that performance on at least some implicit tests of knowledge is equivalent in brain-damaged patients who show the dissociation and in normal people in whom the "signal" is presumably being delivered at full strength (see De Haan et al., 1987; Schacter et al., 1988).

A more serious question is whether the output is truly shallow. Critics of Fodor's theory of modularity frequently raise this question (see comments to Fodor, 1985). In defending his ideas, Fodor (1985) stated that the propositional content represented in modules can be quite rich, so that its computations and the resulting shallow output can appear to be deceptively "deep." What makes the output "shallow" is that it is not strategically derived or inferred, but is rather mandatorily driven by domain-specific input in a system that is informationally encapsulated from "deep knowledge." This argument is not an easy one to defend if one relies only on evidence from studies of normal people. In those studies, it is difficult to establish acceptable criteria for deciding whether the outputs are truly "shallow" or whether modules are not as informationally encapsulated as Fodor believes. If the latter turns out to be the case, higher order semantic knowledge can influence the module's operation, thus making its output truly deep. The problem is made more tractable by considering neuropsychological patients in whom there is evidence that they have no recourse to higher order knowledge that can affect the operation of the modules. A couple of examples will illustrate our point.

Patients with Wernicke's aphasia and Alzheimer's disease often perform at chance on explicit tests of comprehension for some common words or line drawings. Despite their gross deficits on explicit tests of comprehension, these patients show normal or even supranormal semantic priming effects on a lexical decision test for these very same words (Chertkow & Bub, 1990; Milberg & Blumstein, 1981). Moreover, Milberg, Blumstein, and Woretzky (1987) found disambiguation effects in the same type of lexical decision task. Taken together, these results suggest that input modules for words (word-form system [Warrington & Shallice, 1980]) or higher order lexical modules contain sufficient information for distinguishing a word from a nonword but also for producing context-dependent semantic priming effects—all without conscious awareness of the word's meaning.

A similar effect has been observed in face recognition. Bauer (1984) and Tranel and Damasio (1985) reported that prosopagnosic patients who cannot distinguish familiar from unfamiliar faces on explicit tests nonetheless display differential (GSR) or skin conductance responses (SCR) to familiar and unfamiliar faces. In Bauer's study, differential GSR responding was also noted when the face was paired with professions that truthfully or falsely described the occupations of the people. The latter finding suggests that the module (or modules) comprising the face-recognition system (see Bruce & Young, 1986) also contain some associative information about the face that can influence performance on tests of face recognition without conscious awareness.

De Haan et al. (1987), using a different procedure, had a prosopagnosic patient read names that were presented simultaneously with a face. When the names and faces were congruent, reading latencies were shorter than when they were not. The congruency effect, however, was not observed when higher order categories, such as professions, were paired with the face. The discrepancy between this finding and Bauer's merits further investigation. What is clear from both studies, however, is that an input module's shallow output contains not only structural, perceptual information about the stimulus but also some associative, "semantic" information. More studies are needed to corroborate these early findings, to determine whether such semantic, associative information is stored in other closely linked modules (Bruce & Young, 1986), and to gauge how extensive or deep is the network of association that is represented in a module.

Input Modules, Modular Representation, and Memory Without Awareness

The observation that input modules can project information about specific individual faces, words, and possibly objects implies that such modules can be modified by experience to store information. How else can they come to represent such specific, as opposed to generic, knowledge? Although logically necessary, there is little direct empirical evidence that modules can store new information. Such evidence would consist in showing that a patient whose behavior in a specific domain depends only on the shallow, uninterpreted output of the module can nonetheless acquire new knowledge in the specific domain of that module. For example, it would require demonstrating that prosopagnosic patients who show only implicit, nonconscious recognition of faces can acquire implicit knowledge of new faces. De Haan et al. (1987) claim that this is true of their patients. More recently, their finding was corroborated. Greve and Bauer (1990) found normal repetition priming effects (see below) for new faces in another prosopagnosic patient.

In fact, the growing literature on repetition (priming) effects may provide evidence, albeit less direct, in favor of the hypothesis that input modules are modifiable by experience and that it is this modification that supports performance on many implicit tests of memory (Moscovitch & Umilta, 1990; Schacter, 1990; Tulving & Schacter, 1990). Repetition effects refer to the phenomenon that stimuli are processed more accurately or more efficiently when they are repeated than when they are presented for the first time. Because conscious recollection of the past is not required, and is often absent, tests of repetition effects also serve as implicit or nonconscious tests of memory for the stimuli that are being repeated. Repetition effects in amnesic patients are often normal and can sometimes last for weeks or months, even when conscious recollection of the target event may have dissipated within minutes (for review see Moscovitch, 1984; Schacter, 1987a; Squire, 1987). There are now a substantial number of studies to show that repetition effects, some lasting for months, can be obtained for words, faces, and objects using a variety of techniques that include lexical decision (Moscovitch, 1982b; Scarborough, Cortese, & Scarborough, 1977), naming (Mitchell & Brown, 1988), speeded reading (Cohen & Squire, 1982; Kolers, 1976; Moscovitch et al., 1986), perceptual identification (Cermak, Talbot, Chandler, & Wolburst, 1985; Jacoby & Dallas, 1981), word completion (Graf & Schacter, 1985; Graf, Shimamura, & Squire, 1985), and face classification (Bentin & Moscovitch, 1988; Ellis, Young & Flude, 1990). For performance to be altered with repetition, there must be a lasting record or representation of the initial stimulus and/or the processes involved in picking it up. Repetition effects, therefore, can serve as a useful technique for determining the form of the stored information and, from our point of view, whether that information is represented in input modules. The current debate in the literature on repetition effects revolves around this very issue.

There is no consensus as to whether repetition effects involve reactivation of an abstract representation of the item (Clarke & Morton, 1983; Graff, & Mandler, 1984; Warren & Morton, 1982) or whether they depend on the storage of specific, structural features of the input (Hayman & Tulving, 1989). A third alternative is that repetition effects reactivate the processes used to pick up the initial information (Roediger, 1990). According to our conception of input modules, the information that is stored should be at the level of abstraction specific to the domain in which the module operates. As yet, we have only incomplete information in that regard, but it is sufficient to set some limits on the kind of theorizing that is possible. Neuro-psychological evidence on reading indicates that the visual word-form module represents information in a code that is sensitive to the visual (graphe-mic) properties of the word but not to the detailed specific features of the letters such as their font, size, script, and so on (Schwartz, Saffran, & Marin, 1980). This is consistent with findings of repetition effects on some tests such as lexical decision, reading, and perceptual identification where varia-tions in input modality makes a difference but variation in physical features does not (Carr, Brown, & Charalambous, 1989; Jacoby, 1983; Scarborough et al., 1977). Repetition effects on other tasks, such as word-stem completion and fragment completion, differ in that they show a degree of hyper-specificity to variation in physical features as well as some sensitivity to cross-modal effects (Hayman, & Tulving, 1989; Tulving & Schacter, 1990; Schwartz, 1990). Attempts to account for these variations across tasks are already being undertaken and the next few years should provide a resolution to the problem (see pp. 237–249).

There are far fewer studies of repetition effects in nonverbal domains, but the results obtained are similar. Repetition effects for line drawings of common objects (Jacoby et al., 1989; Warren & Morton, 1982) and for faces are modality specific as well as specific to the particular token rather than the class. With respect to that token, however, changes in viewpoint of the face (Young, & Flude, 1990; Ellis, Young, Flude, & Hay, 1987) and in size and orientation of the drawing (Jolicoeur, 1985; Jolicoeur & Milliken, 1987) have little influence on the repetition effect, indicating that what is represented is the individual face or particular object rather than its size or image from a particular viewpoint. Because repetition effects are not obtained, or are not as large, across different exemplars of the same class that share the same name (different cars, chairs, cats, or dogs, Bartram, 1974; Jacoby et al., 1989; Warren & Morton, 1982), the structural information that is represented in the input modules is item specific, a result that is consistent with the behavior of patients that retain perceptual, but not semantic, knowledge of objects (Moscovitch & Umilta, 1990; Warrington & Taylor, 1978).

In a series of studies Cooper, Schacter, Balleskros, and Moore (1990) make a similar point regarding repetition effects for line drawings of possible and impossible objects. Their observation that only line drawings of possible objects produce repetition effects led them (see also Schacter, 1990; Tulving & Schacter, 1990) to propose that repetition effects arise from a modification of perceptual representation systems (or input modules, in our terminology) whose internal structure enables them to represent only possible objects. Changes in size and orientation of the object seem to have no influence on repetition effects (Cooper et al., 1990).

If repetition effects are mediated by storage of domain-specific information in input modules, as we and others have suggested, it follows that similar repetition effects should also be observed for unfamiliar or new items after an initial exposure to them. Until recently, such repetition effects have been difficult to obtain, although there are reports of repetition effects for orthographically permissible nonwords (Carr, Brown, & Charalambous, 1989; Musen, Shimamura, & Squire, 1990) and for line drawings of unfamiliar, but possible, objects (see Cooper et al., 1990; Musen & Treisman, 1990). The failure to obtain consistent repetition effects for unfamiliar material may arise because the module that typically encodes such material is somewhat resistant to change. A single presentation may suffice for storing information about a familiar item, but either multiple presentation or single, but extended, presentation may be necessary for producing repetition effects for new material. The results of studies on face classification (Bentin & Moscovitch, 1988), word recognition (Dagenbach, Horst, & Carr, 1990) and object recognition (Schacter et al., in press) are consistent with this interpretation.

The most troublesome evidence against the hypothesis that repetition effects are mediated by input modules comes from studies showing that repetition effects can sometimes also be influenced by higher order, conceptual knowledge. To use Roediger and Blaxton's (1987) phrase, many repetition effects are not data driven, as would be expected if they were mediated only by input modules, but rather are conceptually driven. Thus, Blaxton (1989) found conceptually driven repetition effects that were as strong as the data-driven ones. Logan (1990), Masson and Freedman (1990), Oliphant (1983), and Ratcliff, Hockley, and McKoon, (1985), showed that repetition effects for visually presented words were drastically reduced or eliminated when the context in which the word appeared changed or when the classification scheme changed from a lexical decision test on the first presentation to a pronunciation decision on repetition. Last, in a series of experiments on perceptual identification, MacLeod & Masson (1990) showed that generating words mentally or even generating synonyms produced large and reliable repetition effects. If repetition effects depend merely on data-driven reactivation of a stored information in a visual word-form module, then changing the sentence context should still produce repetition effects, whereas internally generating a word should have little effect on the system.

It is not clear how these difficulties are to be resolved. One possible solution suggested by Moscovitch and Umilta (1990) and Tulving and Schacter (1990), and which we will explore in the section on central semantic systems, is that conceptual repetition effects are mediated by nonmodular systems. Another possibility is that conceptual repetition effects are mediated by input modules whose propositional context is deceptively deep. Yet a third possibility is that conceptual classification schemes preempt or mask the repetition effects that are mediated by modules. To resolve the issue, studies of conceptual repetition effects need to be conducted in certain agnosic or dyslexic patients, who show a disconnection of domain-specific knowledge from conscious awareness. Because the influence of higher order knowledge is severely reduced (if not altogether absent) in such patients, they form the definitive cases on which these various hypotheses regarding repetition effects can be tested. In the meantime, the hypothesis that input modules are modified by perceptual experience and mediate perceptual repetition effects is still viable and productive.

Central Systems

By contrast to modules, information received by central systems and used in their computations can be infinitely diverse, can come under voluntary control, and be open to both higher order and lower order influences.To quote Fodor, "the higher the cognitive process, the more it turns on the integration of information across superficially dissimilar domains" (1983,

p. 4). In principle, central systems are neither domain specific nor informationally encapsulated (though, under special circumstances, they may have those characteristics [Schwartz & Schwartz, 1984; Shannon, 1989]. As a result, it makes sense to characterize central systems by the function(s) they serve rather than by the information they compute.

To illustrate the variety of central systems, we identified four different types defined according to their function (for more details, see Moscovitch & Umilta, 1990).

1. Function 1: Forming Type II modules. In this instance we believe that devoted systems are associated with each modality and that each is located in different cortical regions, usually in close proximity to the sensory region that delivers modular input to them. Thus, though the domain may be large, its boundaries are typically confined to a single modality.

2. Function 2: Forming and maintaining Type III modules. Damage to the central systems having this function leaves the components of Type III modules intact, but impairs their organization into an operational unit (this applies also to function 1).

3. Function 3: Relating information to general knowledge. Two aspects of the central system's role can be distinguished: one is receiving informational content from modules; the other is the process that relates semantic knowledge to modular output. By combining both aspects, central processes assign meaning to modular output. Deficits in this function can arise either because the knowledge base, the semantic core, is "depleted" as in cases of Alzheimer's disease (Schwartz et al., 1980) or because the process that delivers (some of) the necessary semantic information is impaired as in patients with focal lesions who have lost catetory-specific information (see Damasio, 1990; McCarthy & Warrington, 1988; Warrington & Shallice, 1983, 1984).

4. Function 4: Planning. In planning, goals have to be set, strategies adopted, action sequences selected; the processes then have to be monitored and the outcomes verified against the internal representations of the goals that are to be achieved. These various components then need to be coordinated by a central processing device whose operations are assumed to be effortful, slow, and serial. This central processing device goes by a variety of names: *central executive* (Baddeley, 1986), *operating system* (Johnson-Laird, 1988), *attentional supervisory system* (Norman & Shallice, 1986), *central processor* (Umilta, 1988), and *central monitor* (Weiskrantz, 1988). Although strong evidence is lacking, our hunch is that the different aspects of planning, rather than being executed by a single processor, are instead relegated to different processing components that are interrelated through *the central processor* (discussed later), but that can be selectively impaired. More research is needed to distinguish between these possibilities.

Deficits in planning are often associated with frontal lesions, which produce both negative symptoms—such as a failure to plan and monitor responses in relation to previous events or future goals (Luria, 1973; Petrides, 1989) and positive symptoms such as perseveration or breaking of rules (Corkin, 1965; Milner, 1964; Mishkin, 1964). Progress has been made in isolating some deficits to different frontal regions of the prefrontal cortex (Goldman-Rakic, 1987; Milner, 1964; Mishkin, 1964; Petrides, 1989), but the description of these deficits is often task specific and fails to capture the general kind of impairment in planning that is seen in patients with frontal lesions. Some investigators locate the central processor or executive in the frontal lobe (Baddeley, 1986; Shallice, 1982, 1988), whereas others assume that only separate components of planning are localized in the frontal cortex—the central processor being associated, instead, with the midline reticular activating system (Baars, 1988; Crick, 1984). Our preference is to distinguish between the executive functions and operations of a *central processor* related to conscious awarness (see below).

   5. Function 5: Making information available to conscious awareness.

Conscious Awareness, Working Memory, and the Central Processor

Conscious awareness of mental representations and operations can be identified with the phenomenal experience of the contents and operation of a limited-capacity central processor (Umilta, 1988). The central processor is not equivalent to any of the central systems that we have identified by their functions in the previous sections; instead, it is the recipient of the output of modules and central systems and coordinator of their activity. Its base of operation is working memory, or, more properly, working memory grows out of the central processor in interaction with modular output and central systems.

   According to Baddeley (1986), working memory refers to "the temporary storage of information that is being processed in any range of cognitive tasks" (p. 34) *of which we are consciously aware* (our italics and addition). Initially, it was proposed that there is a single working memory system with a storage capacity limited to a fixed number of items (Baddeley, 1986). This idea has been challenged recently by a number of studies showing that the capacity of working memory (measured in items) varies with the functional domain (e.g., spatial versus verbal), and the type of task (reading span versus speaking span) used to assess capacity (for review see Daneman & Tardif, 1987). Faced with these data, one alternative was to posit multiple working memories, each linked to a particular domain (and task?) and each with its own unique, but fixed, capacity. This alternative, however, would still retain the idea that working memory is a type of store with a fixed capacity.

   There is a third alternative, which we favor. Working memory is a reflection of whatever operations and representations engage, or are engaged by, a central processor with limited resources. Thus, there is neither a single working memory in Baddeley's sense, nor are there multiple working memories. An estimate of the limited resources of the central processor, and therefore the capacity of working memory, is determined by the joint interactions of the information being held and the operations that are performed. The greater the demands of the operations on resources, the fewer the bits of information that can be held simultaneously in working memory. Assuming that different domains and tasks require different resources for their operation, the estimated capacity of working memory will vary. Working memory, then, is an operating system with a limited, but variable, capacity. Its components are also not fixed, but vary according to which ones currently interact with the central processor. In other words, working memory coincides with the cognitive complex (central systems and shallow outputs from modules) that at any given time is interacting with the central processor.

At the structural level, working memory is best associated with the operation of a central mechanism that is reciprocally linked to cortical and subcortical regions mediating the operation of central systems and the output of modules. The midline nuclei of the reticular activating system and thalamus meet these requirements (Crick, 1984; Heilman & Valenstein, 1979; Heilman, Watson, & Valenstein, 1985; Rizzolatti, & Camarda, 1987), but further research is needed to confirm whether the operation of these mechanisms is correlated with some of the characteristics of working memory. Whatever mechanism ultimately is found, working memory is the emergent construct that arises from the operation of central mechanisms and the limited array of cortical and subcortical structures that it can activate, or be activated by, at any given time.

Conscious Awareness and Central Systems

Engaging the central processor confers the quality of consciousness on the various central systems and outputs. For modules, only the shallow output can engage the central processor and thus gain entry to working memory and lead to conscious awareness. For central systems, on the other hand, even the intermediate steps or outcomes, as well as some aspects of the computations involved, may be included in working memory and thus become accessible to consciousness.

Can central systems, like modules, operate without engaging consciousness? For example, can planned behaviors (Function 4) be maintained without a degree of conscious awareness? Is semantic interpretation of modular output (Function 3) possible without conscious awareness? Anecdotal evidence of purposeful behavior during sleepwalking, as well as controlled observations of the phenomenon (for reviews see Bootin, Kihlstrom, & Schacter, 1990), suggests that central systems need not always be accompanied by full conscious awareness (though some minimal level of conscious awareness may be present even while the individual is sleepwalking). Even while awake, individuals conduct a number of habitual activities, such as fixing coffee or typing, with little conscious awareness. As a demonstration, Spelke, Hirst, & Neisser (1976) showed that individuals can type a manuscript while simultaneously shadowing a verbal message without loss of efficiency in *either* activity. Even monitoring for semantic classes was possible under these conditions of divided attention. It is open to debate whether in these instances it is the operation of a central system or that of Type III modules that is governing behavior. The complicated, and often novel, activity involved in typing a new manuscript, however, suggests that it is unlikely to be modular. Similarly, the assimilation of relatively complex information during deep anesthesia (see Kulli & Koch, 1990, for a review of effects of different anesthetics) also suggests that central systems, particularly those mediating Function 3, can operate when conscious awareness is minimal at best. Kihlstrom, Schacter, Cork, Hurt, & Behr (1990) showed that words heard by subjects during anesthesia were more likely to be emitted in response to a category cue than a control set of words. This finding lent credence to anecdotal reports of patients being affected by conversations they heard in the operating room while they were anesthetized.

Those observations are controversial and are not universally accepted, but the recent publication of well-controlled and well-documented studies suggests that such claims must be taken seriously.

General or Specific Deficits in Working Memory

According to the model we have proposed, deficits in working memory can arise from two sources: (a) damage or dysfunction of the central processor itself or all the connections to it; (b) damage to any of the central systems that interact with the central processor or any single projection to it. In the first instance, the deficit in resources will be general and will affect all domains and cognitive functions that are effortful. In the extreme case, it will result in a reduction in consciousness. But as long as resources remain above a certain threshold, the nature of the cognitive operations will not change, though efficiency will suffer. In contrast, the deficits that result from (b) will be highly specific and restricted only to the single system affected by the damage. Thus, rather than have an overall reduction, working memory capacity will be reduced only in a specific domain or for a particular cognitive function. In addition, the operation of the system mediating that function will be abnormal or, in the extreme case, lost entirely. If it is just the projection of the central system to the central processor that is lost, then the function mediated by the system will be normal but not conscious.

The differences between general versus specific working memory deficits are highlighted by contrasting the effects of working memory deficits in patients with Alzheimer's dementia, where a general loss of resources is believed to occur, and patients with conduction aphasia, where only a specific mechanism is believed to be damaged. Citing evidence from studies by Vallar and Baddeley (1984a, 1984b) and Baddeley, Lewis, and Vallar (1984) on conduction aphasia, Moscovitch and Umilta (1990) noted that in these patients the deficit is restricted only to the phonological store, a slave component or module of working memory. The deficit is not only quantitative but actually qualitative, in that patients lose their responsiveness to variables such as word length or articulatory suppression that depend on an intact phonological store in normal people. Patients with Alzheimer's disease, on the other hand, retain their responsiveness to such factors despite having reduced verbal spans and reading spans and lowered retention on the Brown–Peterson test (Morris, 1986). According to our model, the Alzheimer's patients' deficits on tests believed to be dependent on intact working memory arise because of the demands that such tests make on the resources of the central processor and the operation of the central systems that interact with it. The widespread neural degeneration in Alzheimer's disease is likely to disrupt the broad associative and integrative communication pathways among central systems and the central processor, as well as causing deterioration in the central systems themselves. As a result, Alzheimer's disease deficits are found even at the level of Type II modules, where sensory features are integrated into higher level representations by a devoted central system (Saffran & Coslett, 1990; Schlotterer, Moscovitch, & Crapper-McLachlan, 1983).

Long-Term Episodic Memory

In dealing with episodic memory, Fodor followed Gall's lead by making memory part of every module. Memory, for Fodor, is the stored information, or engrams, that are specific to the module to which they belong. At the same time, Fodor conceived of memory as a horizontal faculty whose properties are similar to those of other central systems. This duality, we believe, is an important characteristic of episodic memory. Fodor never explored the implications of his view in any depth, however, nor did he expend much effort in trying to reconcile these seemingly opposite aspects of memory. In the end, he lumped memory together with other functions of central systems that he believed would not yield to fruitful scientific investigation.

We, too, believe that memory has these dual aspects to which Fodor alluded, but we do not share his pessimistic conclusions. Rather, we attempt to reconcile these dual aspects in a model that shows that normal and pathological memory are best understood as a reflection of the joint operation, and breakdown, of modules and central systems.

Long-Term Episodic Memory: The Hippocampal Associative System

Episodic memory is memory for autobiographic episodes or events that retains a spatiotemporal context. Explicit tests of memory, such as recall and recognition, require conscious recollection of the target event. Recollection of the event may be accompanied by a memory of the context or only by a sense of familiarity indicating that the event had indeed been previously experienced. As we noted earlier, conscious recollection plays no part when memory is tested implicitly. We will have more to say about implicit tests of memory later. The present discussion concerns only explicit tests of episodic memory.

The model of explicit memory that we developed (Moscovitch, 1989; Moscovitch & Umilta, 1990) has much in common with other models (Mishkin, 1982; Schacter, 1990; Squire, 1987; Tulving & Schacter, 1990). Its unique feature, however, is the clear distinction between a hippocampal system that we propose is modular in its organization and operation and a frontal system that we believe interacts with the output of the hippocampal module, much as other central systems interact with shallow, modular output (see Figure 14.1).

The model that we outline is more a framework than a fully specified theory. It is useful primarily as a heuristic for organizing data and for suggesting new experiments. The critical assumption of the model is that events are picked up by input modules. As we noted earlier, the input modules, at least those that are Type II and III, are capable of being modified by the stimuli that activate them. As a result, they store information about the activating event. The central systems involved in interpreting the event may also be modified by stimulation, although this view has yet to

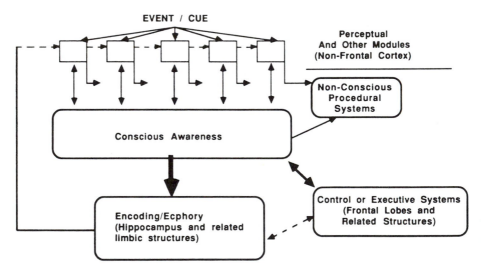

Figure 14.1 A neuropsychological model of memory. The dashed lines indicate that the interaction is optional. The cortical modules that interact with the hippocampal system will vary depending on the information about the event that is available to consciousness when the event is initially experienced and when it is being remembered. (See text for details from Moscovitch, 1989.)

be tested widely (but see Blaxton, 1990). Following Kirsner and Dunn (1985) we use the term *record* to refer to a modification of the neural circuitry of input modules (and perhaps central systems) in response to stimulation. The modification contains information about the stimulating event and has processing consequences so that identical and perhaps related stimuli can be processed more quickly by the module. We reserve the term *engram* to refer to the informational content of the record that can be accessed or reactivated to contribute to recollection. We use *memory trace* (Hayman & Tulving, 1989) to refer to a collection of bound engrams that contribute to recollection of an experienced event and its context. Whether the record is strictly perceptual or whether it also has semantic content is currently disputed.

The output of the modules, which are presumed to be located in the posterior and midlateral neocortex, is delivered both to working memory whose representations are consciously apprehended and to other procedural systems that can affect behavior but whose representations and operations are not available to conscious inspection. Information that is consciously apprehended, that is, information in working memory, and only that information, is picked up automatically by the hippocampus and its related structures. In other words, *the specific domain of the hippocampal module is information that is consciously apprehended.* The hippocampus then binds or integrates the records or engrams that gave rise to the conscious experience to form a memory trace. This, we believe, is achieved by the reciprocal pathways connecting the hippocampus, via the entorhinal cortex (hippocampal gyrus) to the cortical modules (Hyman, Damasio, Van Hoesen, Barnes, 1984; Hyman, Von Hoesen, & Damasio, 1990). *Memory consolidation* refers to this process. The hippocampus concurrently encodes the memory trace as a file entry or index within its structure. Conscious recollection of an event occurs when a cue, either externally presented or internally generated, gains access to working memory and activates the hippocampal index. That automatically causes the information stored in the hippocampus to interact with the collection of bound engrams or memory trace. The output of that interaction is then delivered to working memory. Once initiated, the process is rapid, obligatory, informationally encapsulated, and cognitively impenetrable. We do not have access to the process that delivered the memory to us. It is consistent with our introspections that a memory merely "popped" into our mind, much as perceptual events pop out of the background preattentively (Treisman, 1988). This automatic, mandatory process by which retrieval information (a cue) is brought into interaction with stored information is called *ecphory*, a term along with engram, that was first coined by Semon (1921, cited in Schacter, Eich, & Tulving, 1978). The information thus retrieved as a result of cue–engram interaction may or may not be veridical. It is the "shallow" output of the hippocampal module.

Because it is modular, and because its representations consist of indexes to bound, or associated engrams, we refer to the hippocampal system as *associative*.

IS THE HIPPOCAMPAL SYSTEM MODULAR? NEUROPSYCHOLOGICAL CRITERIA. Like the input systems, the hippocampal system seems to satisfy the criteria of modularity. In examining this claim, it should be borne in mind that identical interpretations of the criteria cannot be applied to the hippocampus and to input modules. In contrast to input modules, the hippocampal system typically receives semantically interpreted information. Its output, therefore, is also expected to be of a similar order of complexity or depth. When we judge whether the hippocampal system is informationally encapsulated and whether its output is shallow, we should take the unique properties of the hippocampal system into account.

1. *Domain specificity*:   The domain of the hippocampal system is information represented in consciousness that is then registered or retrieved from memory. Double dissociation experiments have shown that damage to the hioppocampus and its related structures produces global amnesia in isolation from other symptoms, whereas damage to other structures, such as the frontal or parietal lobes, can produce a range of other cognitive deficits while leaving memory relatively unaffected (Milner, 1966, 1974). In short, there is ample evidence of double dissociation between hippocampal memory function and cognitive functions of other cortical regions.

2. *Informational encapsulation*:   According to this criterion, once the hippocampal system receives its input it is resistant to higher order influences until it emits its output, that is, until consolidation processes at study and ecphoric processes at retrieval, are complete. These operations are mandatory. Conscious awareness of them is impossible. Because the input to the hippocampal system typically consists of highly processed and interpreted information, and because the output is necessarily similar, it makes it difficult to determine whether this criterion is satisfied. However, one can apply the same criterion test to the hippocampal memory as was applied to other input modules and see if it passes: Is memory spared even in the face of obvious dementia? The answer seems to be "Yes." Memory loss is not a prominent feature of dementia if the neuropathological process causing the dementia does not attack the hippocampal system directly (Neary, Snowden, Northern, & Goulding, 1988). Morever, there are reports of patients with semantic memory loss whose episodic memory is nonetheless spared (De Renzi, Liotti, & Nichelli, 1987). Finally, Geschwind, Quadfasel, and Segarra (1966) report a patient with isolation of the speech zone from other cortical regions, but not from the hippocampal limbic system. That patient could repeat but not understand speech. Nonetheless, she was able to learn phrases and sentences that she heard on the radio. These kinds of evidence suggest that the hippocampal system is informationally encapsulated, but clearly other experiments are necessary to make the case more convincing.[1]

3. *Shallow output*:   The criterion of shallow output is satisfied as long as the output is not semantically richer than the input. We propose that this

criterion is satisfied if a memory of an event is not placed in its proper temporospatial context; that is, it is recognized as a memory of the past but it cannot be interpreted properly within the context of other past or current events. There is no temporal order in the hippocampal system beyond that which occurs as a result of association between events. As with input modules, interpretation is dependent on the involvement of central systems, in this case the frontal lobes and perhaps other structures, that make "sense of" and organize the output of the hippocampal system and the input to it.

Studies of patients with large bilateral lesions of the frontal lobes and related structures provide a dramatic demonstration of how memory might operate if it relied only on the shallow output of the hippocampal system. Patients with such lesions, caused by aneurysms of the anterior communicating artery, confabulate a great deal (for review see Berlyne, 1972; Moscovitch, 1989). Their confabulations are not pure fabrications but seem to consist of "memories" that ecphoric processes deliver automatically to consciousness. These "memories" are then combined in a seemingly haphazard fashion without regard to their temporal order or spatial context, or to internal consistency or even plausibility. The "content" of the memory, the uninterpreted output of the hippocampal module, may have a basis in the individual's past experience, but the context in which it is placed may be so outlandish that only someone with knowledge of the patient's past can discern the elements of truth in the narrative. We will say more about the contribution of central systems, especially the frontal lobes, to memory in the next section.

Strategic Processes in Memory: The Frontal Lobes and Related Structures as Central Systems

Memory that relies only on the automatic reactivation of stored engrams is reduced to a reflex. That holds as much for memory with conscious awareness that depends on reactivation of memory traces through the hippocampal system as it does for memory without awareness that depends on reactivation of engrams directly.

Memory, however, is also a far more complex act or skill, elements of which are often under voluntary control. Consider the processes involved in figuring out where you had met a person who looks familiar but whom you can't place, or in trying to recall with whom you had seen a movie a year or two ago, or in remembering what you did during the summer of your fifteenth year. In each of these cases it is unlikely that a memory merely pops into mind in response to the cue. Remembering in such cases is an intelligent, rather than a reflexive, act that relies on strategic processes as much as problem solving does. What confers "intelligence" on memory are the strategic processes associated with prefrontal and other central systems. "The processes coordinate, interpret, and elaborate the information in working memory to provide the hippocampal-associative memory system with the appropriate information that it takes as its input at encoding and at retrieval" (Moscovitch & Umilta, 1990, p. 39). Comparable processes are involved in evaluating the hippocampal system's shallow output and placing those retrieved memories in a proper spatio-temporal context.

As a result, damage to the frontal lobes and related areas (the frontal system) typically does not affect memory for the target item or the content of an event, but rather memory for the spatio-temporal context (Milner, Petrides, & Smith, 1985; Moscovitch, 1982a; Schacter, 1987b). Frontal patients performed poorly on tests of recency (Milner, 1974), temporal order (Vriezen & Moscovitch, 1990; Shimamura, Janowsky, & Squire, 1990), list differentiation (Moscovitch, 1982a), conditional associative learning and self-ordered pointing (Petrides & Milner, 1982), frequency estimation (Smith & Milner, 1984), metamemory (Shimamura & Squire, 1986), and source memory (Janowsky, Shimamura, & Squire, 1989; Schacter, 1987b; Shimamura & Squire, 1989). Even performance on tests of recognition and recall may sometimes be impaired following frontal lesions if organization at encoding or directed search or monitoring at retrieval is a prominent factor (Delberg-Desrouesne, Beauvois, & Shallice, 1990; Mayes, 1988; Moscovitch, 1989; della Rochetta, 1986).

Similar deficits are also observed in patients with damage to structures that are closely linked to the frontal lobes and may be said to form part of the frontal system or loop (Alexander, De Long, & Strick, 1986; Brown & Marsden, 1990). Thus patients with neostriatal damage or dysfunction, such as those with Parkinson's or Huntington's disease, are also impaired on tests of temporal order (Sagar, Sullivan, Gabrieli, Corbin, & Growdon, 1988; Vriezen & Moscovitch, 1990), conditional associative learning (Canavan et al., 1989; Vriezen & Moscovitch, 1990), and free recall of categorized lists (St. Cyr, 1989). Some patients with Parkinson's disease and many patients with Huntington's disease will also be impaired on free, but not cued recall (Butters, Heindel, & Salmon, 1990; Butters, Salmon, Heindel & Granholm, 1989; Huberman, Freedman, & Moscovitch, 1988).

Even elderly people, without frank damage to the frontal lobe but with evidence of frontal dysfunction, also have noticeable deficits on "frontal-sensitive" tests such as release from P.I. (Moscovitch, 1982b; Moscovitch & Winocur, 1983; Winocur & Moscovitch, 1990), list differentiation (Dywan, Moscovitch & Jacoby, 1990), and source amnesia (Craik, Morris, Morris, & Loewen, 1990).

Two other differences should be noted between the effects of frontal and hippocampal damage that arise from our conception of them as central and modular systems, respectively. Hippocampal deficits are domain specific— they should affect memory and no other cognitive processes. Memory deficits associated with frontal system damage, on the other hand, do not occur in isolation but are invariably accompanied by related cognitive deficits in other domains. Indeed, the type of memory disorder seen in patients with frontal system damage seems to be analogous to their cognitive deficits on tests of sequencing (Kolb & Milner, 1981), planning (Shallice, 1982, 1988), and monitoring (Luria, 1966).

Even if we restrict our analysis only to memory, there is yet another critical difference between frontal and hippocampal deficits. Memory loss following hippocampal lesions is temporally graded with the most recently acquired memories being the most severely affected (Albert, Butters, & Levin, 1979; Cermak, 1976; Korsakoff, 1889a; 1889b; Milner, 1966; Squire & Cohen, 1982; but see Warrington & Sanders, 1971, for a dissenting view). The deficits are also typically restricted to episodic memory. In keeping with the view of the frontal lobes as central system structures, the memory deficits associated with frontal system damage seem to affect all memories, remote and recent, semantic and episodic, in the same way. Memory for temporal order is as impaired for recent as for remote events (Sagar, Cohen, Sullivan, Corkin, & Growdon, 1988; Shimamura et al., 1990), as is directed memory search when measured by the Crovitz word-prompt test (Moscovitch, 1989). Moscovitch (1989) also observed that deficits in temporal ordering and directed search occurred even with regard to historical events that took place before the subject was born and that are more appropriately considered to be part of semantic, rather than episodic, memory.

Although the deficits observed after frontal system damage are consistent with our view that it constitutes a central system, it is important not to lose sight of the fact that the frontal lobe, let alone the frontal system, is not a homogeneous structure (Jones & Powell, 1970; Pandya & Barnes 1987). There is ample evidence of functional and structural differentiation within the frontal lobes, even with respect to memory (Goldman-Rakic, 1987; Milner, Petrides, & Smith, 1985; Petrides, 1989). It remains to be seen, however, whether subsystems within the frontal lobes will continue to behave like central system structures, but with more circumscribed functions all of which bear a family resemblance to each other (Teuber, 1972), or whether, as smaller and smaller subsystems are identified, they will begin to appear more and more like modules (for discussion, see Moscovitch & Umilta, 1990).

Consciousness and Memory

Because frontal lobes are central system structures, the processes they mediate are accessible to consciousness. We are aware of some of the strategies we use in remembering how we solved a memory problem, but rarely, if ever, do we know how a particular memory popped into mind. We may be aware of the cue that gave rise to it, but not of the ecphoric process, mediated by the modular hippocampal system, that delivered the memory to consciousness. In short, we are not aware of the workings of the hippocampal-associative system but only of the operations and the representations of central systems that occupy working memory and of the shallow output of the hippocampal system. We are conscious of the memory strategies that are adopted, the questions that are delivered as cues to the hippocampal system, the answers that are automatically elicited from it, and the evaluation of those answers, all of which are associated with the frontal system, but we are not conscious of the ecphoric operations of the hippocampal system itself.[2]

Memory Without Conscious Awareness

THE CONTRIBUTION OF INPUT MODULES: THE PERCEPTUAL RECORD AND NONCONSCIOUS MEMORY. The hippocampus receives as its input information of which we are consciously aware and delivers its output automatically to consciousness. The function of the hippocampus is to make available recently acquired information to consciousness. As we noted earlier, the engrams are not themselves stored in the hippocampus but in those input structures that are initially involved in processing the information as well as in the structures that may be necessary for interpreting them. The hippocampus binds these engrams and stores an index that, when triggered by the appropriate cue, reactivates those engrams and interacts with them.

Damage to the hippocampus and its related structures impairs conscious recollection and, as a result, performance on explicit tests of memory. The "engrams," however, are not lost but exist as "perceptual" records in the input modules and as "conceptual" or semantic records in central systems (see below). Memory without awareness may be possible by direct sensory or conceptual reactivation of the engrams. This is not to say that the subject retrieves an entire memory episode but fails to recognize it as a memory. Rather, by observing changes in the subject's performance with experience we can infer that an engram exists—that is, that memory for the experience is retained. Implicit tests of memory capitalize on this fact. Repetition priming effects, a measure of performance on implicit tests of memory, refer to the facilitation in processing that occurs when an item is repeated, whether or not the subject consciously recollects the initial presentation of the item.

As we noted earlier, in line with similar observations made by other investigators (see Richardson-Klavehn & Bjork, 1988; Roediger, 1990; Tulving & Schacter, 1990), the characteristics of perceptually based repetition effects are those that would be expected if they are dependent on changes in input modules. Perceptually based repetition effects are data-driven, modality specific and not influenced by higher order, semantic processes. The repetition effect is such that it preserves the form of representation characteristic of the input module that is being altered. Thus, repetition effects for faces, words, and objects are abstract, yet visual, representations of the "form" of the stimulus, not its particular instantiation so long as there is a structural correspondence among the various tokens. For example, repetition effects for faces occur regardless of the view (full face, three-quarter, or profile) between the initial and repeated presentation; repetition effects for words also are insensitive to font and letter size; and repetition effects for objects are insensitive to size, viewpoint, or even alterations in the tokens (two different dresses), so long as a common structural description referring to basic features (Biederman, 1987; Warrington & James, 1986) is maintained across repetitions. This accords well with evidence from other sources about the characteristics of the word-form systems (word modules) Schwartz et al., 1980; Warrington & Shallice, 1980), the face-recognition system (face modules) (Bruce & Young, 1986), and the object-recognition system (visual object module) (Farah, 1990; Humphreys & Riddoch, 1987; Warrington & Taylor, 1978). We think that the extreme hyperspecificity of repetition priming effects advocated by Tulving and Schacter (1987) is peculiar to only some implicit tests of memory such as stem and fragment completion. The "fragmented" nature of the stimuli provide incomplete information to the input module and therefore more complete (specific) information is required to reactivate the engram. As we shall see below, it is also doubtful that either test is dependent only on the operation of input modules; more than likely, central systems are necessary for efficient performance.

THE SEMANTIC RECORD AND NONCONSCIOUS MEMORY. Not all implicit tests of memory are data driven (i.e., not all tests involve the reactivation of perceptual records in input modules). Some tests are more conceptual. For example, providing exemplars when given the taxonomic category as a cue, answering conceptual questions, and even completing words when given letters as cues, may all require some degree of conceptual processing. All these tests show some cross-modal influences, and are affected by semantic factors at processing or at retrieval (Graf & Ryan, 1990; Macleod & Masson, 1990; Roediger, 1990). As with data-driven implicit tests of memory, performance on conceptually driven tests is independent of performance on explicit tests. Amnesic patients with damage to the hippocampus or related structures, but who are otherwise cognitively unimpaired, continue to perform as well on conceptual implicit tests of memory as they do on perceptual tests (Gardner, Boller, Moreines, & Butters, 1973; Graf et al., 1985; Jacoby & Witherspoon, 1982). This suggests that structures involved in interpreting the shallow output of input modules can themselves be modified by experience. The stored engrams of semantic structures (the semantic record) can be reactivated by appropriate semantic input to mediate conceptual repetition effects. As we noted earlier, under some circumstances, central systems can operate without accompanying conscious awareness. Studies of conceptual repetition effects indicates that they can also contribute to memory without conscious awareness.

If this hypothesis is correct, conceptual, but not perceptual, repetition effects should be absent in patients with damage to structures believed to mediate central system functions. A number of recent studies support this hypothesis.

The most dramatic effect was obtained by Blaxton (1990) in a study of patients with unilateral, anterior, but lateral, temporal-lobe excisions. Damage to that region of the cortex impairs performance on verbal and visual semantic judgment tasks (Hebb's anomalies, picture completion, Mooney figures, word and object classification [for review see Kolb & Whishaw, 1990]) and leads to a reduction in category fluency (Newcombe, 1969). According to Blaxton, patients with left temporal lobectomies failed to show conceptual repetition effects on a word-association test though they had normal perceptual repetition effects on stem completion.

If conceptual repetition effects depend on central systems, then it would be expected that patients with Alzheimer's dementia (which affects the temporal lobes) would also show poor conceptual repetition effects but normal perceptual effects. Indeed, Butters, Heindel, & Salmon (1990) found that repetition effects on a semantic word-association test was absent in patients with Alzheimer's disease, although such patients perform normally on speeded reading of transformed script (Moscovitch et al., 1986) and perceptual identification of words (Keane, Gabrielli, Fennema, Growdon, & Corkin, 1991) both of which are more perceptually based. Performance on stem completion as an implicit test of memory was intermediate, consistent with the observation in normal people that this test has aspects of both conceptual and perceptual repetition effects (Gabrielli, 1989). Some studies of Alzheimer patients found little or no repetition effects on a test of stem completion (Shimamura, Salmon, Squire, & Butters, 1987), whereas others found a reduced, but significant, effect (Huberman et al., 1988).

FUNCTIONAL, STOCHASTIC, AND STRUCTURAL INDEPENDENCE AMONG IMPLICIT TESTS OF MEMORY. The foregoing analysis suggests that insofar as different neural structures mediate performance on implicit tests of memory (what we call structural independence), then performance on those tests should also be functionally and stochastically independent of each other, much as performance on implicit tests and explicit tests are independent of each other. Functional independence between performance on two tests of memory refers to the differential effects that some independent variables, such as modality and depth of processing, have on performance on different tests of memory. Stochastic independence between two tests states that correctly "remembering" an item on one test is unrelated to the probability that the same item will be correctly retained on another test. Witherspoon and Moscovitch (1989) were able to confirm this prediction by showing that stochastic independence can be found between two implicit tests: perceptual identification and fragment completion. They interpreted this finding as supporting a multiple-component view of memory, which states that any memory test, implicit or explicit, consists of a variety of component processes. Tests of stochastic independence reflect the extent to which the critical components or structures in each test are common to the tests. The greater the commonality, the greater the dependence. Interpreted in the framework of modules and central systems, it refers not only to the modules and central systems that may mediate each test, but also to the operations involved in gaining access to the engrams.

Levels of Processing, Incidental Learning, and Conscious Recollection: The Hippocampal System

One of the major insights of the levels of processing theory was that memory, as conscious recollection, is a natural byproduct of cognitive processing (Craik & Lockhart, 1972). The intention to commit information to memory is less important for later recollection than is the depth to which the target is initially processed. Although in general this principle seems to hold, no mechanism was suggested to account for this effect. In addition, two other findings were cited as evidence against the levels of processing theory. The first was that the mere presentation of a target, such as a word, seems to activate automatically its deep semantic associates. Countless experiments on semantic priming and spreading activation can be cited in support of this claim (for review see Neely, 1990). More recently, a number of investigators have shown that semantic activation of a target can occur even without subjective perceptual awareness of the target (Cheesman & Merikle, 1984, 1985; Marcel, 1983). Yet it is known, in seeming contradiction of the levels of processing theory, that recollection of these items is poor (Forster, Booker, Schacter, & Davis, 1990; Forster & Davis, 1984).

In our model, the hippocampus is driven mandatorily and without effort by information that enters working memory. No intention to memorize or remember is necessary. The information is picked up by the hippocampus and "encoded" as soon as it is in working memory. The more deeply processed the information, the more unique or distinctive is the hippocampal index. This, in turn, makes access to the information easier because it is uniquely specified by its retrieval cue (Klein & Salz, 1976; Moscovitch & Craik, 1976). If the cue is not unique, recollection for the target will suffer despite its being processed deeply. Mere rote rehearsal of the target does not lead to better recollection (Craik & Watkins, 1973) because the item is maintained in a phonological buffer (Baddeley, 1986) that is a modular slave system to working memory and not an actual part of it. Similarly, the semantic information that is automatically activated by the target item does not gain access to working memory. That it remains largely, if not completely, unconscious, and therefore not in working memory, is shown by the fact that the magnitude of semantic priming can be as large for items that fall outside subjective perceptual awareness as for items that fall within perceptual awareness (Cheesman & Merikle, 1984, 1985). To be in working memory implies conscious awareness. If a property or feature of an object is not in conscious awareness, it cannot activate the hippocampal system, and consequently, it cannot lead to proper conscious recollection. The level to which an item is processed has to be conscious to have an impact on conscious recollection.

Conversely, conscious processing to a deep level will lead to good recollection only if the hippocampus is activated. Cermak and Reale (1978)
confirmed anecdotal reports (viz. Milner, 1966) that memory in amnesic
patients remains poor despite consciously processing information to a deep
level. This finding was cited as evidence against a strong version of level-of-
processing theory (Baddeley, 1978). The inadequacy of the level-of-
processing theory for dealing with evidence from amnesic patients is understandable in the framework of our model. In most amnesic patients the
hippocampus and/or its related structures are damaged. The information
available in working memory cannot be consolidated into long-term memory and consequently leaves a poor memory trace regardless of how deeply
processed it is.

Attention, Encoding, and Memory:
The Hippocampal System and Input Modules

If the specific domain of the hippocampal system is consciously apprehended information, it follows that nonattended information could not
be consolidated, and conscious recollection would suffer. This is common
knowledge that is even supported by well-controlled experiments (Craik &
Byrd, 1982). What is not obvious is whether performance on implicit tests
for the nonattended input would be relatively spared. If it were, it would
imply that input modules can pick up and store information even when
attention is not directed at the stimulus.

Although only a few relevant studies have been conducted on this topic,
the results consistently indicate that attentional manipulations have relatively little effect, if any, on performance on perceptually driven implicit
tests of memory. Parkin, Reid, & Russo (1990) showed that word-stem completion was unaffected by performance of an attention-demanding concurrent task at input, although conscious recollection on cued recall tests was
severely diminished in the concurrent task condition. Using a different
technique, Merikle and Reingold (1991) obtained comparable results. They
showed subjects two words, one above the other, and subjects were instructed to attend to only one. Again, recognition memory for the ignored
word dropped to chance, but repetition effects for it were above baseline
and normal. Eich (1984) had subjects shadow a message in one ear while
targets were presented in the other. Although conscious recollection for the
target, as measured by recognition, was at chance, repetition effects, as
measured by free association to a cue, were well above baseline. Similar, but
even more dramatic, effects were reported by Kihlstrohm, Schacter, Cork,
Hurt, & Behr (1990) in subjects who heard the target while they were
anesthetized. Not only do these results suggest that nonattended information is retained, but the conceptual nature of the implicit test suggests that
the targets were processed to a deep semantic level. That is, they left a
semantic record behind.

More studies like these are needed both to substantiate the phenomena and to determine whether there are any limits to the retention of unattended information. In addition, implicit tests of memory, if properly combined with techniques derived from the attention literature, could be used to settle questions about the fate of unattended information that have occupied the attention and perception literature for at least a century (Allport, 1989; James, 1890).

Attention, Retrieval, and Memory:
The Frontal and Hippocampal Systems

To impair conscious recollection by dividing attention at encoding is easy; to do so by dividing attention at retrieval appears to be much more difficult (see Baddeley, Lewis, Eldridge, & Thomson 1984). Why should that be the case? Our introspections are not very helpful. Consider the effort many of us have expended in trying to remember the name of a movie, a restaurant, a childhood friend, or an answer on a final exam. Consider also how annoying we found distractions at those times. "I had it on the tip of my tongue until you interrupted" or "If I could just concentrate a little longer" are common complaints.

How could such effortful activity not be disrupted by diverting attention away from it? Why should it take heroic measures to demonstrate the effect in the laboratory? Perhaps a clue to the answer lies in the strategies that some people recommend should be used in this situation. One is to forget about it and the answer will sooner or later "pop" into mind. The other is to keep providing cues to yourself until the right one causes the answer to "pop" into mind. These strategies suggest that the effortful process is the one involved in discovering the appropriate cue; once the cue is available, recovering the memory is effortless, almost automatic.

In the framework of our model, strategic search or retrieval is mediated by the frontal lobes (see Moscovitch, 1989, for review), which are central system structures. The recovery of the memory, ecphory, is an automatic process, mediated by the hippocampal system, which is modular. As we noted earlier (see the section on working memory), processes associated with central systems make greater demands on resources of the central processor than do modular processes, which typically require little or no resources for their operation.

A possible explanation for the difficulty investigators have had in showing memory impairment when attention is divided at retrieval is that they used memory tests that primarily tap the hippocampal, rather than the frontal, system. Typically, tests of recognition and recall are used, and performance on those is most often affected by hippocampal damage, but much more rarely by frontal damage. If a concurrent task is to have its most deleterious effect at retrieval, it must interfere with resource-demanding frontal functions. Examination of the literature lends some support to this

conjecture. Baddeley et al. (1984) found that the greatest effect of a concurrent task at retrieval was on word fluency, a test that is sensitive to frontal
damage. More recently, Jacoby and his colleagues (Dywan & Jacoby, 1990;
Jacoby, Woloshyn, & Kelly, 1989) have shown strong interference effects on
a recognition test that effectively requires subjects to distinguish between
lists of famous and nonfamous people, a test that likely requires the strategic
retrieval processes associated with the frontal lobes. This hypothesis is
supported by evidence that performance on this test in elderly people is
significantly correlated with their performance on the Wisconsin Card
Sorting Test, which is sensitive to frontal dysfunction. Similarly, Park, Smith,
Dudley, and Lafonza (1989), found that interference at output affected
memory for categorized lists which also is affected by frontal lesions (della
Rochetta, 1986).

Encouraged by the evidence in the literature, Moscovitch designed a
number of experiments to test the hypothesis that concurrent tasks at
retrieval in normal people will interefere more with performance on memory tests sensitive to frontal than to hippocampal damage. The concurrent
interference test that was chosen was a sequential finger-tapping test. The
subject had to tap the fingers of the right hand in the sequence, index, ring,
middle, small, and continue to do so throughout the encoding or retrieval
phase of the memory test, or during both phases. This sequential motor test
was chosen because we believed that it might also be mediated by the
frontal lobes. In addition, it was not too demanding, so its effects, if any,
would be selective.

The first test we chose was release from proactive interference, because
previous work had shown that damage to the left frontal cortex prevented
release from proactive interference but did not influence overall recall
during the buildup of PI. Hippocampal damage, on the other hand, lowered
overall recall but had no effect on release from PI.

The results showed that concurrent interference at study, or at test, had
little effect on overall performance. Interference at both study and test,
however, significantly reduced release from PI.

As a second test, Moscovitch looked at the effects of the same concurrent
task on performance on the California Verbal Learning Test, which is a test
of free recall of a 16-item list that consists of four items from each of four
categories. Recall performance on such categorized lists is impaired in
patients with damage to the frontal lobes (della Rochetta, 1986) or to
related structures in the neostriatum (Brown & Marsden, 1990; St. Cyr,
1989). The impairment arises because these patients fail to take full advantage of the organization inherent in the list. Clustering by category at recall
is reduced in these patients.

As in the previous test, concurrent interference at both study and test, but
at neither alone, reduced overall recall and clustering. When a 16-item list
of random words was substituted for the categorized list, the interference
task had no effect. Recall of the list of random words presumably depends

more on hippocampal, rather than frontal, mediation because it offers little opportunity for organized (strategic) retrieval.

As a final test, Moscovitch had subjects tap sequentially during a test of letter, and of category, fluency. The letter fluency test requires subjects to name as many words as they can in a minute that begin with a given letter, whereas the category fluency test requires that subjects name as many exemplars in a given category as they can in the same amount of time. Performance on the letter fluency test is impaired following damage to the orbital region of the frontal cortex (Milner, 1964), whereas performance on the category fluency test is affected more by damage to the anterior temporal lobes (Newcombe, 1969). As predicted, the sequential tapping test reduced output by about 25% on the letter fluency test but by less than 1% on the category fluency test.

The results of these studies are consistent with the hypotheses that strategic retrieval processes that are mediated by the frontal lobes are resource demanding and are susceptible to interference produced by a concurrent task. Associative retrieval processes that are mediated by the hippocampus make fewer demands on cognitive resources and so are not as susceptible to interference. These findings are consistent with the view that the hippocampal system is modular and that the frontal lobes are central system structures. Further studies are needed, however, to support and extend these conclusions. In particular, it is important to determine whether similar results could be obtained with interfering tasks that engage mechanisms in the posterior neocortex rather than in the frontal lobes.

Notes

1. Reports of hysterical or functional amnesia would seem to constitute evidence that the hippocampal system is not informationally encapsulated. The individual's expectancies and motivations seem to influence the output of the system. First, documented evidence of hysterical amnesia is difficult to obtain. Second, even if we accept the claim, it is not clear that the hippocampal system is not functioning. The information may be properly stored and ecphoric output may be normal but prevented from being interpreted properly. Alternatively, the input to the system through consciousness may be distorted so that proper retrieval cues are not provided. Once appropriate retrieval cues are received, memories return automatically and seemingly even against the will of the individual. Last, hysterical amnesia as in fugues, may affect remote, rather than recent, memories and those may not be handled by the hippocampus at all.

2. For both central systems and modular ones, however, the internal working of the algorithms that act on the representation never become conscious. In short, the workings of the devices themselves are cognitively impenetrable and computationally autonomous.

Acknowledgments

Preparation of this manuscript was supported by National Science and Engineering Research Councils of Canada Grant A8347 to Morris Moscovitch. The ideas reported in this paper were inspired by a year spent at the Institute for Advanced Studies of the Hebrew University in Jerusalem in 1985–1986 (see Moscovitch & Umilta, 1990). We also thank Martene Behrman and Gordon Winocur for their comments.

References

Albert, M. S., Butters, N., & Levin, J. (1979). Temporal gradients in the retrograde amnesia of patients with alcoholic Korsakoff's disease. *Archives of Neurology, 36,* 211–226.

Alexander, G. E., De Long, M. R., & Strick, P. L. (1986). Parallel organization of functionally segregated circuits linking basal ganglia and cortex. *Annual Review of Neuroscience, 9,* 357–381.

Allport, A. (1989). Visual attention. In M. I. Posner (Ed.), *Foundation of cognitive science.* Cambridge, MA: MIT/Bradford.

Baars, B. J. (1988). *A cognitive theory of consciousness.* Cambridge, England: Cambridge University Press.

Baddeley, A. D. (1978). The trouble with levels: A re-examination of Craik and Lockhart's framework for memory research. *Psychological Review, 85,* 708–729.

Baddeley, A. D. (1986). *Working memory.* Oxford, England: Oxford University Press.

Baddeley, A., Lewis, V., Eldridge, M., & Thomson, N. (1984). Attention and retrieval from long-term memory. *Journal of Experimental Psychology: General, 113,* 518–540.

Baddeley, A. D., Lewis, V. J., & Vallar, G. (1984). Exploring the articulatory loop. *Quarterly Journal of Experimental Psychology, 36,* 233–252.

Bartram, D. (1974). The role of visual and semantic codes in object naming. *Cognitive Psychology, 10,* 325–356.

Bauer, R. M. (1984). Autonomic recognition of names and faces in prosopagnosia: A neuropsychological application of the guilty knowledge test. *Neuropsychologia, 22,* 457–469.

Behrmann, M., Moscovitch, M., Black, S. E., & Mozer, M. (1990). Perceptual and conceptual mechanisms in neglect dyslexia: Two contrasting case studies, *Brain, 113,* 1163–1183.

Behrmann, M., Moscovitch, M., & Mozer, M. (in press). Directing attention to words and nonwords in normal subjects and in a computational model: Implications for neglect dyslexia. *Cognitive Neuropsychology.*

Bentin, S., & Moscovitch, M. (1988). The time course of repetition effects for words and unfamiliar faces. *Journal of Experimental Psychology: General, 117,* 148–160.

Bentin, S., & Moscovitch, M. (1990). Psychophysiological indices of implicit memory performance. *Bulletin of the Psychonomic Society, 28,* 346–352.

Berlyne, N. (1972). Confabulation: *British Journal of Psychiatry, 120,* 31–39.

Biederman, I. (1987). Recognition by components: A theory of human image understanding. *Psychological Review, 94,* 115–147.

Blaxton, T. A. (1989). Investigating dissociations among memory measures: Support for a transfer appropriate processing framework. *Journal of Experimental Psychology: Learning, Memory, and Cognition, 15,* 657–668.

Blaxton, T. A. (1990). *Dissociations among memory measures in both normal and memory impaired subjects.* Manuscript submitted for publication.

Bootzin, R. R., Kihlstrom, J. F., & Schacter, D. L. (1990). *Sleep and cognition.* Washington, DC: American Psychological Association.

Brown, R. G., & Marsden, C. D. (1990). Cognitive function in Parkinson's disease: From description to theory. *Trends in Neurosciences, 13,* 21–29.

Bruce, F., & Young, A. (1986). Understanding face recognition. *British Journal of Psychology, 77,* 305–327.

Butters, N., Heindel, W. C., & Salmon, D. P. (1990). Dissociation of implicit memory in dementia: Neurological implications. *Bulletin of the Psychonomic Society, 28*, 359–366.

Butters, N., Salmon, D. P., Heindel, W., & Granholm, E. (1989). Episodic, semantic, and procedural memory: Some comparisons of Alzheimer's and Huntington's disease patients. In R. Terry (Ed.), *Aging and the brain*. New York: Raven Press.

Canavan, A. G. M., Passingham, R. E., Marsden, C. D., Quinn, N., Wyke, M., & Polkeg, C. E. (1989). The performance on learning tasks of patients in the early stages of Parkinson's disease. *Neuropsychologia, 17*, 141–156.

Carr, T. H., Brown, J. S., & Charalambous, A. (1989). Repetition and reading: Perceptual encoding mechanisms are very abstract but not very interactive. *Journal of Experimental Psychology: Learning, Memory, and Cognition, 15*, 763–779.

Cermak, L. S. (1976). The encoding capacity of a patient with amnesia due to encephalitis. *Neuropsychologia, 14*, 311–326.

Cermak, L. S., & Reale, L. (1978). Depth of processing and retention of words by alcoholic Korsakoff patients. *Journal of Experimental Psychology: Human Learning and Memory, 4*, 165–174.

Cermak, L. S., Talbot, N., Chandler, K., & Wolburst, L. R. (1985). The perceptual priming phenomenon in amnesia. *Neuropsychologia, 23*, 615–622.

Cheesmen, J., & Merikle, P. M. (1984). Priming with and without awareness. *Perception and Psychophysics, 36*, 387–395.

Cheesman, J., & Merikle, P. M. (1985). Word recognition and consciousness. In D. Besner, T. G. Waller, & G. E. MacKinnon (Eds.), *Reading research: Advances in theory and practice* (Vol. 5). New York: Academic Press.

Chertkow, H., & Bub, D. (1990). Semantic memory loss in dementia of the Alzheimer type. In M. F. Schwartz (Ed.), *Modular deficits in dementia*. Cambridge, MA: MIT/Bradford.

Clarke, R., & Morton, J. (1983). Cross modality facilitation in tachistoscopic word recognition. *Quarterly Journal of Experimental Psychology, 35A*, 79–96.

Cohen, N. J., & Squire, L. R. (1982). Preserved learning and retention of pattern analysing skill in amnesia: Dissociation of "knowing how" and "knowing that." *Science, 210*, 207–209.

Cooper, L. A., Schacter, D. L., Ballesteros, S., & Moore, C. (1990). Priming of structural representations of three-dimensional objects. Paper presented at the meeting of the Psychonomic Society, New Orleans, Lousiana.

Corkin, S. (1965). Tactually-guided maze learning in man: Effects of unilateral cortical excisions and bilateral hippocampal lesions. *Neuropsychologia, 3*, 339–351.

Craik, F. I. M., & Byrd, M. (1982). Aging and cognitive deficits: The role of attentional resources. In F. I. M. Craik, & S. Trehub (Eds.), *Aging and cognitive processes*. New York: Plenum Press.

Craik, F. I. M., & Lockhart, R. S. (1972). Levels of processing: A framework for memory research. *Journal of Verbal Learning and Verbal Behavior, 11*, 671–684.

Craik, F. I. M., Morris, L. W., Morris, R. G., & Loewen, E. R. (1990). Aging, source amnesia, and frontal lobe functioning. *Psychology and Aging, 5*, 148–151.

Craik, F. I. M., & Watkins, M. J. (1973). The role of rehearsal in short-term memory. *Journal of Verbal Learning and Verbal Behavior, 12*, 599–607.

Crick, F. (1984). Functions of the thalamic reticular complex: The searchlight hypothesis. *Proceedings of the National Academy of Sciences, U.S.A., 81*, 4586–4593.

Dagenbach, D., Horst, S., & Carr, T. (1990). Adding new information to semantic memory: How much learning is enough to produce automatic priming? *Journal of Experimental Psychology: Memory, Learning, and Cognition, 16*, 581–591.

Damasio, A. R. (1990). Category-related recognition defects as a clue to the neural substrates of knowledge. *Trends in Neurosciences, 13*, 95–98.

Daneman, M., & Tardif, T. (1987). Working memory and reading skill re-examined. In M. Coltheart (Ed.), *Attention and performance XII: The psychology of reading*. Hillsdale, NJ: Lawrence Erlbaum.

De Haan, E. H. F., Young, A., & Newcombe, F. (1987). Face recognition without awarness. *Cognitive Neuropsychology, 4,* 385–415.

Delberg-Derouesne, J., Beauvois, M. F., & Shallice, T. (1990). Preserved recall versus impaired recognition: A case study. *Brain, 113,* 1045–1074.

De Renzi, E., Liotti, M., & Nichelli, P. (1987). Semantic amnesia with perseveration of autobiographic memory: A case report. *Cortex, 23,* 575–597.

Duncan, J. (1986). Disorganization of behaviour after frontal-lobe damage. *Cognitive Neuropsychology, 3,* 271–290.

Dywan, J., & Jacoby, L. L. (1990). Effects of aging on source monitoring: Differences in susceptibility to false fame. *Psychology and Aging, 5,* 379–387.

Eich, E. (1984). Memory for unattended events: Remembering with and without awareness. *Memory and Cognition, 12,* 105–111.

Ellis, A. W., Young, A. W., & Flude, B. M. (1990). Repetition priming and face processing: Priming occurs within a system that responds to the identity of a face. *Quarterly Journal of Experimental Psychology, 42A,* 495–512.

Ellis, A. W., Young, A. W., Flude, B. M., & Hay, D. C. (1987). Repetition priming of face recognition. *The Quarterly Journal of Experimental Psychology, 39A,* 193–210.

Farah, M. J. (1990). *Visual agnosia.* Cambridge, MA: MIT/Bradford.

Fodor, J. (1983). *The modularity of mind.* Cambridge, MA: MIT Press.

Fodor, J. A. (1985). Multiple review of *The modularity of mind. Behavioral and Brain Sciences, 8,* 1–42.

Forster, K., Booker, J. Schacter, D. L., & Davis, C. (1990). Masked repetition priming: Lexical activation or novel memory trace. *Bulletin of the Psychonomic Society, 28,* 341–345.

Forster, K. I., & Davis, C. (1984). Repetition priming and frequency attenuation in lexical access. *Journal of Experimental Psychology, 10,* 680–698.

Gabrieli, J. D. (1989). Paper presented to the Memory Disorders Research Society. Boston, MA.

Gardner, H., Boller, F., Moreines, J., & Butters, N. (1973). Retrieving information from Korsakoff patients: Effects of categorical cues and references. *Cortex, 9,* 165–175.

Geschwind, N., Quadfasel, F. A., & Segarra, J. M. (1966). Isolation of the speech area. *Neuropsychologia, 6,* 327–340.

Goldman-Rakic, P. S. (1987). Circuitry of primate prefrontal cortex and regulation of behavior by representational memory. In F. Plum (Ed.), *Handbook of physiology—The nervous system* (Vol. 5). Bethesda, MD: American Physiological Society.

Goodale, M. A., Milner, A. D., Jakobson, L. S., & Carey, D. P. (1991). Ancerological dissociation between perceiving objects and grasping them. *Nature, 349,* 154–156.

Graf, P., & Mandler, G. (1984). Activation makes words more accessible but not necessarily more retrievable. *Journal of Verbal Learning and Verbal Behavior, 23,* 405–424.

Graf, P., Mandler, G., & Haden, M. (1982). Simulating amnesic symptoms in normal subjects. *Science, 218,* 1243–1244.

Graf, P., & Ryan, L. (1990). Transfer-appropriate processing for implicit and explicit memory. *Journal of Experimental Psychology: Learning, Memory, and Cognition, 16,* 978–992.

Graf, P., & Schacter, D. L. (1985). Implicit and explicit memory for new associations in normal and amnesic subjects. *Journal of Experimental Psychology: Learning, Memory, and Cognition, 11,* 501–518.

Graf, P., Shimamura, A. P., & Squire, L. R. (1985). Priming across modalities and priming across category levels: Extending the domain of preserved function in amnesia. *Journal of Experimental Psychology: Learning, Memory, and Cognition, 11,* 385–395.

Greve, K. W., & Bauer, R. M. (1990). Implicit learning of new faces in prosopagnosia: An application of the mere-exposure paradigm. *Neuropsychologia, 28,* 1035–1042.

Hayman, C. A. G., & Tulving, E. (1989). Is priming in fragment completion based on a "traceless" memory system? *Journal of Experimental Psychology: Learning, Memory, and Cognition, 15,* 941–956.

Heilman, K. M., & Valenstein, E. (1979). Mechanisms underlying hemispatial neglect. *Annals of Neurology, 5,* 166–170.

Heilman, K. M., Watson, R. T., & Valenstein, E. (1985). Negelct and related disorders. In K. M. Heilman & E. Valenstein (Eds.), *Clinical neuropsychology* (2nd ed.). Oxford, England: Oxford University Press.

Huberman, M., Freedman, M., & Moscovitch, M. (1988). Performance on implicit and explicit tests of memory in patients with Parkinson's and Alzheimer's disease. Paper presented at the International Society for the Study of Parkinson's Disease, Jerusalem, Israel.

Hyman, B. T. , Damasio, A. R., Van Hoesen, G. W., & Barnes, C. L. (1984). Alzheimer's disease: Cell specific pathology isolates the hippocampal formation. *Science, 225,* 1168–1170.

Hyman, B. T., Van Hoesen, G. W., & Damasio, A. R. (1990). Memory-related neural systems in Alzheimer's disease: An anatomic study. *Neurology, 40,* 1721–1730.

Jacoby, L. L. (1983). Remembering the data: Analyzing interactive processes in reading. *Journal of Verbal Learning and Verbal Behavior, 22,* 485–508.

Jacoby, L. L., & Dallas, M. (1981). On the relationship between autobiographical memory and perceptual learning. *Journal of Experimental Psychology: General, 110,* 306–340.

Jacoby, L. L., & Witherspoon, D. (1982). Remembering without awareness. *Canadian Journal of Psychology, 32,* 300–324.

Jacoby, L. L., Woloshyn, V., & Kelley, C. M. (1989). Becoming famous without being recognized: Unconscious influences of memory produced by dividing attention. *Journal of Experimental Psychology: General, 118,* 115–125.

Humphreys, G. W., & Riddoch, M. J. (1987). *To see but not to see: A case study of visual agnosia.* London: Lawrence Erlbaum.

James, W. (1898/1950). *The principles of psychology.* New York: Holt and Company (1890); Dover Press (1958).

Janowsky, J. S., Shimamura, A. P., & Squire, L. R. (1989). Source memory impairment in patients with frontal lobe lesions. *Neuropsychologia, 27,* 1043–1056.

Johnson-Laird, P. N. (1988). A computational analysis of consciousness. In A. J. Marcel & E. Bisiach (Eds.), *Consciousness in contemporary science* (pp. 357–368). Oxford, England: Oxford University Press.

Jolicoeur, P. (1985). The time to name disoriented natural objects. *Memory and Cognition, 13,* 289–303.

Jolicoeur, P., & Milliken, B. (1989). Identification of disoriented objects: Effects of context of prior presentation. *Journal of Experimental Psychology: Learning, Memory and Cognition, 15,* 200–210.

Jones, E. G., & Powell, T. P. S. (1970). An anatomical study of converging sensory pathways within the cerebral cortex of the monkey. *Brain, 93,* 793–820.

Keane, M. M., Gabrieli, J. D., Fennema, A. C., Growdon, J. H., & Corkin, S. (1991). Evidence for a dissociation between perceptual and conceptual priming in Alzheimer's disease. *Behavioral Neuroscience, 105,* 326–342.

Kihlstrom, J. F., Schacter, D. L., Cork, R. C., Hurt, C. A., & Behr, S. E. (1990). Implicit memory following surgical anaesthesia. *Psychological Science, 1,* 303–306.

Kirsner, K., & Dunn, D. (1985). The perceptual record: A common factor in repetition priming and attribute retention. In M. I. Posner & O. S. M. Marin (Eds.), *Attention and performance XI* (pp. 547–566). Hillsdale, NJ: Lawrence Erlbaum.

Klein, K., & Saltz, E. (1976). Specifying the mechanisms in a levels-of-processing approach to memory. *Journal of Experimental Psychology: Human Learning and Memory, 2,* 671–070.

Kolb, B., & Milner, B. (1981). Peformance of complex arm and facial movements after focal brain lesions. *Neuropsychologia, 19,* 491–504.

Kolb, B., & Whishaw, I. Q. (1990). *Fundamentals of human neuropsychology* (3rd ed.). New York: Freeman.

Kolers, P. A. (1975). Memorial consequences of automatized encoding. *Journal of Experimental Psychology: Human Learning and Memory, 1*, 689–701.

Kolers, P. (1976). Reading a year later. *Journal of Experimental Psychology: Human Learning and Memory, 2*, 554–556.

Korsakoff, S. S. (1889a). Etude medico-psychologique sur une forme des maladies de la mémoire. *Revue Philosophique, 5*, 501–530.

Korsakoff, S. S. (1889b). Uber eine besondere Form psychischer Storung combiniert mit multipler Neuritis. *Archiv fur Psychiatrie und Nervenkrankheiten, 21*, 669–704. (Translation by M. Victor & P. I. Yakovlev (1955). *Neurology, 5*, 394–406.)

Kulli, J., & Koch, C. (1991). Does anaesthesia cause loss of consciousness? *Trends in Neurosciences, 14*, 6–10.

Landis, T., Regard, M., & Serrant, A. (1980). Iconic reading in a case of alexia without agraphia caused by a brain tumor: a tachistoscopic study. *Brain and Language, 11*, 45–53.

Logan, G. D. (1978). Attention in character classification: Evidence for the automaticity of component states. *Journal of Experimental Psychology: General, 107*, 32–63.

Logan, G. D. (1985). Executive control of thought and action. *Acta Psychologica, 60*, 193–210.

Logan, G. D. (1990). Repetition priming and automaticity: Common underlying mechanisms? *Cognitive Psychology, 22*, 1–35.

Luria, A. R. (1966). *Higher cortical functions in man.* New York: Basic Books.

Luria, A. R. (1972). *The man with a shattered world: The history of a brain wound.* New York: Basic Books.

MacLeod, C. M., & Masson, M. E. J. (1990). Priming in perceptual identification relies on a context-sensitive interpretation. Paper presented at the meeting of the Psychonomic Society, New Orleans, Louisiana.

Marcel, A. J. (1983). Conscious and unconscious perception: Experiments on visual masking and word recognition. *Cognitive Psychology, 15*, 197–237.

Marshall, J. C., & Halligan, P. W. (1989). Blindsight and insight into visuo-spatial neglect. *Nature, 336*, 766–767.

Masson, M. E. J., & Freedman, L. (1990). Fluent identification of repeated words. *Journal of Experimental Psychology: Learning, Memory, and Cognition, 16*, 355–373.

Mayes, A. R. (1988). *Human organic memory disorders.* Cambridge, England: Cambridge University Press.

McCarthy, R. A., & Warrington, E. K. (1988). Evidence for modality specific meaning systems in the brain. *Nature, 334*, 428–430.

Merikle, P. M., & Reingold, E. M. (1991). Comparing direct (explicit) and indirect (implicit) measures to study unconscious memory. *Journal of Experimental Psychology: Learning, Memory and Cognition, 17*, 224–233.

Michaels, C. F., & Turvey, M. T. (1979). Central sources of visual masking: Indexing structures supporting seeing at a single, brief glance. *Psychological Research, 41*, 1–61.

Milberg, W., & Blumstein, S. E. (1981). Lexical decision and aphasia: Evidence for semantic processing. *Brain and Language, 14*, 371–385.

Milberg, W., Blumstein, S. E., & Dworetzky, B. (1987). Processing of lexical ambiguities in aphasia. *Brain and Language, 31*, 138–150.

Milner, B. (1964). Some effects of frontal lobectomy in man. In J. M. Warren & K. Akert (Eds.), *The frontal granular cortex and behavior* (pp. 313–331). New York: McGraw-Hill.

Milner, B. (1966). Amnesia following operation on the temporal lobe. In C. W. M. Whitty & O. L. Zangwill (Eds.), *Amnesia.* London: Butterworth.

Milner, B. (1974). Hemispheric specialization: Scope and limits. In F. O. Schmitt & F. G. Worden (Eds.), *The neurosciences: Third research program.* Cambridge, MA: MIT Press.

Milner, B., Petrides, M., & Smith, M. L. (1985). Frontal lobes and the temporal organization of memory. *Human Neurobiology, 4*, 137–142.

Mishkin, M. (1964). Preservation of central sets after frontal lesions in monkeys. In J. M. Warren & K. Akert (Eds.), *The frontal granular cortex and behavior*. New York: McGraw-Hill.

Mishkin, M. (1982). A memory system in the monkey. *Philosophical Transactions of the Royal Society of London, B298*, 85–96.

Mitchell, D. B., & Brown, A. S. (1988). Persistent repetition priming in picture naming and its dissociation from recognition memory. *Journal of Experimental Psychology: Learning, Memory, and Cognition, 14*, 213–222.

Morris, R. G. (1986). Short-term forgetting in senile dementia of the Alzheimer's type. *Cognitive Neurospsychology, 3*, 77–97.

Moscovitch, M. (1982a). Multiple dissociations of function in amnesia. In L. S. Cermak (Ed.), *Human memory and amnesia*. Hillsdale, NJ: Lawrence Erlbaum.

Moscovitch, M. (1982b). A neuropsychological approach to perception and memory in normal and pathological aging. In F. I. M. Craik & S. Trehub (Eds.), *Memory and cognitive processes in aging*. New York: Plenum Press.

Moscovitch, M. (1984). The sufficient conditions for demonstrating preserved memory in amnesia: A task analysis. In N. Butters & L. R. Squire (Eds.), *The neuropsychology of memory*. New York: Guilford Press.

Moscovitch, M. (1985). Memory from infancy to old age: Implications for theories of normal and pathological memory. *Annals of the New York Academy of Sciences, 444*, 78–96.

Moscovitch, M. (1989). Confabulation and the frontal system: Strategic vs. associative retrieval in neuropsychological theories of memory. In H. L. Roediger III & F. I. M. Craik (Eds.), *Varieties of memory and consciousness: Essays in honor of Endel Tulving*. Hillsdale, NJ: Lawrence Erlbaum.

Moscovitch, M., & Craik, F. I. M. (1976). Depth of processing, retrieval cues, and uniqueness of encoding as factors in recall. *Journal of Verbal Learning and Verbal Behavior, 15*, 447–458.

Moscovitch, M. & Umilta, C. (1990). Modularity and neuropsychology: Implications for the organization of attention and memory in normal and brain-damaged people. In M. F. Schwartz (Ed.), *Modular processes in dementia*. Cambridge, MA: MIT/Bradford.

Moscovitch, M., Winocur, G., & McLachlan, D. (1986). Memory as assessed by recognition and reading time in normal and memory impaired people with Alzheimer's disease and other neurological disorders. *Journal of Experimental Psychology: General, 115*, 331–347.

Moscovitch, M., & Winocur, G. (1983). Contextual cues and release from proactive inhibition in young and old people. *Canadian Journal of Psychology, 37*, 331–344.

Musen, G., & Treisman, A. (1990). Implicit and explicit memory for visual patterns. *Journal of Experimental Psychology: Learning, Memory, and Cognition, 16*, 127–137.

Natsoulas, T. (1978). Consciousness. *American Psychologist, 33*, 906–914.

Neary, D., Snowden, J. S., Northern, B., & Goulding, P. (1988). Dementia of frontal lobe type. *Journal of Neurology, Neurosurgery and Psychiatry, 51*, 353–361.

Neely, J. H. (1990). Semantic priming effects in visual word recognition: A selective review of current findings and theories. In D. Besner & G. Humphreys (Eds.), *Basic processes in reading: Visual word recognition*. Hillsdale, NJ: Lawrence Erlbaum.

Newcombe, F. (1969). *Missile wounds of the brain: A study of psychological deficits*. London: Oxford University Press.

Norman, D. A., & Shallice, T. (1986). Attention to action: Willed and automatic control of behavior. In R. J. Davidson, G. E. Schwartz, & D. Shapiro (Eds.), *Consciousness and self-regulation: Advances in research* (Vol. 4, pp. 1–18). New York: Plenum Press.

Oliphant, G. W. (1983). Repetition and recency effects in word recognition. *Australian Journal of Psychology, 35*, 393–403.

Pandya, D., & Barnes, C. L. (1987). Architecture and connections of the frontal lobe. In E. Perecman (Ed.), *The frontal lobes revisited*. New York: The IRBN Press.

Park, D. C., Smith, D. A., Dudley, W. N., & Lafronza, V. N. (1989). Effects of age and a divided attention task presented during encoding and retrieval on memory. *Journal of Experimental Psychology: Learning, Memory, and Cognition, 15*, 1185–1191.

Parkin, A. J., Reid, T., & Russo, R. (1990). On the differential nature of implicit and explicit memory. *Memory and Cognition, 18*, 307–314.

Petrides, M. (1989). Frontal lobes and memory. In F. Boller & J. Grafman (Eds.), *Handbook of neuropsychology* (Vol. 3). North Holland: Elsevier Science Publishers B.V. (Biomedical Division).

Posner, M. I. (1989). Structures and function of selective attention. In T. Boll & B. Bryant (Eds., The Master Series), *Clinical neuropsychology and brain function: Research, measurement and practice*. Washington, DC: American Psychological Association.

Ratcliff, R., Hockley, W., & McKoon, G. (1985). Components of activation: Repetition and priming effects in lexical decision and recognition. *Journal of Experimental Psychology: General, 114*, 435–450.

Richardson-Klavehn, A., & Bjork, R. A. (1988). Measures of memory. *Annual Review of Psychology, 39*, 475–543.

Riddoch, M. J., & Humphreys, G. W. (1987). Visual object processing in optic aphasia: A case of semantic access agnosia. *Cognitive Neuropsychology, 4*, 131–186.

Rizzolatti, G., & Camarda, R. (1987). Neural circuits for spatial attention and unilateral neglect. In M. Jennerod (Ed.), *Neurophysiological and neuropsychological aspects of spatial neglect*. North Holland: Elsevier Science Publishers B.V. (Biomedical Division).

della Rochetta, A. I. (1986). Classification and recall of pictures after unilateral frontal or temporal lobectomy. *Cortex, 22*, 189–211.

Roediger, H. L. (1990). Implicit memory: Retention without remembering. *American Psychologist, 45*, 1043–1056.

Roediger, H. L., & Blaxton, T. A. (1987). Effects of varying modality, surface features, and retention interval on priming in word fragment completion. *Memory and Cognition, 15*, 379–388.

Saffran, E., Fitzpatrick-De Salme, E. J., & Coslett, H. B. (1990). Visual disturbances in dementia. In M. F. Schwartz (Ed.), *Modular deficits in Alzheimer's disease*. Cambridge, MA: MIT/Bradford.

Sagar, J. J., Cohen, N. J., Sullivan, E. V., Corkin, S., & Growdon, J. H. (1988). Remote memory in Alzheimer's disease and Parkinson's disease. *Brain, 111*, 185–206.

Sagar, J. J., Sullivan, E. V., Gabrieli, J. D. E., Corkin, S., & Growdon, J. H. (1988). Temporal ordering and short-term memory deficits in Parkinson's disease. *Brain, 111*, 525–540.

Scarborough, D. L, Cortese, C., & Scarborough, H. S. (1977). Frequency and repetition effects in lexical memory. *Journal of Experimental Psychology: Human perception and Performance, 3*, 1–17.

Schacter, D. L. (1987a). Memory, amnesia, and frontal lobe dysfunction. *Psychobiology, 15*, 21–36.

Schacter, D. L. (1987b). Implicit memory: History and current status. *Journal of Experimental Psychology: Learning, Memory, and Cognition, 13*, 501–518.

Schacter, D. L. (1990). Perceptual representational systems and implicit memory: Toward a resolution of the multiple memory systems debate. In A. Diamond (Ed.), *Development and neural bases of higher cognition: Annals of the New York Academy of Sciences*.

Schacter, D. L, Cooper, L. A., Delaney, S. M., Peterson, M. A., & Tharan, M. Implicit memory for possible and impossible objects: Constraints on the construction of structural descriptions. *Journal of Experimental Psychology: Learning, Memory, and Cognition, 17,* 3–19.

Schacter, D. L., Eich, J. E., & Tulving, E. (1978). Richard Semon's theory of memory. *Journal of Verbal Learning and Verbal Behavior, 17,* 721–743.

Schacter, D. L., McAndrews, M. P., & Moscovitch, M. (1988). Access to consciousness: Dissociations between implicit and explicit knowledge in neuropsychological syndromes. In L. Weiskrantz (Ed.), *Thought without language* (pp. 242–278). Oxford, England: Oxford University Press.

Schlotterer, G., Moscovitch, M., & Crapper-McLachlan, D. (1983). Visual processing deficits as assessed by spatial frequency contrast sensitivity and backward masking in normal aging and Alzheimer disease. *Brain, 107,* 309–325.

Schneider, W., Dumais, S. T., & Shiffrin, R. M. (1984). Automatic and control processing and attention. In R. Parasuraman & D. R. Davies (Eds.), *Varieties of attention.* New York: Academic Press.

Schwartz, B. L. (1989). Effects of generation indirect measures of memory. *Journal of Experimental Psychology: Learning, Memory, and Cognition, 15,* 1119–1128.

Schwartz, M. F., Saffran, E. M., & Marin, O. S. M. (1980). Fractionating the reading process in dementia: Evidence for word-specific print-to-sound associations. In M. Coltheart, K. E. Patterson, & J. C. Marshall (Eds.), *Deep dyslexia.* London: Routledge and Kegan Paul.

Schwartz, M. F., & Schwartz, B. (1984). In defence of organology. *Cognitive Neuropsychology, 1,* 25–42.

Shallice, T. (1981). Neurological impairment of cognitive processes. *British Medical Bulletin, 37,* 187–192.

Shallice, T. (1982). Specific impairments of planning. *Philosophical Transactions of the Royal Society of London, B298,* 199–209.

Shallice, T. (1988). *From neuropsychology to mental structure.* Cambridge, England: Cambridge University Press.

Shallice, T., & Saffran, E. M. (1986). Lexical processing in the absence of explicit word identification: Evidence from a letter-by-letter reader. *Cognitive Neuropsychology, 3,* 429–458.

Shannon, B. (1989). Consciousness. Working paper No. 25 of the Rotman Centre for Cognitive Science in Education, the Hebrew University of Jerusalem.

Shimamura, A. P., Janowsky, J. S., & Squire, L. R. (1990). Memory for temporal order in patients with frontal lobe lesions and patients with amnesia. *Neuropsychologia, 28,* 803–813.

Shimamura, A. P., Salmon, D. P., Squire, L. R., & Butters, N. (1987). Memory dysfunction and word priming in dementia and amnesia. *Behavioral Neuroscience, 101,* 347–351.

Shimamura, A. P., & Squire, L. R. (1989). A neuropsychological study of fact memory and source amnesia. *Journal of Experimental Psychology: Learning, Memory, and Cognition, 13,* 464–473.

Spelke, E., Hirst, W., & Neisser, U. (1976). Skills of divided attention. *Cognition, 4,* 215–230.

Squire, L. R. (1987). *Memory and brain.* New York: Oxford University Press.

Squire, L. R., & Cohen, N. (1982). Remote memory, retrograde amnesia, and the neuropsychology of memory. In L. S. Cermak (Ed.), *Human memory and amnesia.* Hillside, NJ: Lawrence Erlbaum.

Stuss, D. T., & Benson, D. F. (1986). The frontal lobes. New York: Raven Press.

St. Cyr, J. A. (1989). Paper presented at the Society for Research on Memory Disorders, Boston, Massachusetts.

Teuber, H.-L. (1972). Unity and diversity of frontal lobe functions. *Acta Neurobiologica Experimentalis, 32,* 615–656.

Tranel, E., & Damasio, A. R. (1985). Knowledge without awareness: An autonomic index of facial recognition by prosopagnosics. *Science, 228,* 1453–1454.

266   DISSOCIATIONS AND MODELS

Treisman, A. (1988). Features and objects: The fourteenth Bartlett memorial lecture. *Quarterly Journal of Experimental Psychology, 40A*, 201–237.

Tulving, E., & Schacter, D. L. (1990). Priming and human memory systems. *Science, 247*, 301–306.

Turvey, M. T. (1973). On peripheral and central processes in vision: Inferences from an information-processing analysis of masking with patterned stimuli. *Psychological Review, 80*, 1–52.

Umilta, C. (1988). The control operations of consciousness. In A. J. Marcel & E. Bisiach (Eds.), *Consciousness in contemporary science* (pp. 334–356). Oxford, England: Oxford University Press.

Vallar, G., & Baddeley, A. D. (1984a). Fractionation of working memory: Neuropsychological evidence for a phonological short-term store. *Journal of Verbal Learning and Verbal Behavior, 23*, 151–161.

Vallar, G., & Baddeley, A. D. (1984b). Phonological short-term store, phonological processing and sentence comprehension: A neuropsychological case study. *Cognitive Neuropsychology, 1*, 121–141.

Vriezen, E., & Moscovitch, M. (1990). Temporal ordering and conditional associative learning in Parkinson's disease. *Neuropsychologia, 28*, 1283–1294.

Warren, C., & Morton, J. (1982). The effects of priming on picture recognition. *British Journal of Psychology, 73*, 117–129.

Warrington, E. K., & James, M. (1986). Visual object recognition in patients with right hemisphere lesions: Axes or features? *Perception, 15*, 355–366.

Warrington, E. K., & Sanders, H. I. (1971). The fate of old memories. *Quarterly Journal of Experimental Psychology, 23*, 432–442.

Warrington, E. K., & Shallice, T. (1980). Word-form dyslexia. *Brain, 103*, 99–112.

Warrington, E. K., & Shallice, T. (1984). Category specific semantic impairments. *Brain. 107*, 829–853.

Warrington, E. K., & Taylor, A. M. (1978). Two categorical stages of object recognition. *Perception, 7*, 695–705.

Weiskrantz, L. (1986). *Blindsight*. Oxford, England: Clarendon Press.

Winocur, G., & Moscovitch, M. (1990). A comparison of cognitive function in community dwelling and institutionalized old people of normal intelligence. *Canadian Journal of Psychology, 44*, 435–444.

Witherspoon, D., & Moscovitch, M. (1989). Independence of the repetition effects between word fragment completion and perceptual identification. *Journal of Experimental Psychology: Memory, Learning, and Cognition, 15*, 22–30.

Commentary on Part II

Weiskrantz opens his chapter by pointing out the dual justification for the existence of neuropsychology. On the one hand, it has a clinical utility (discussed in detail by Lezak in Part IV). On the other hand, he argues, it is a rich source of information for drawing inferences about normal cognitive functioning. He emphasizes that it is dissociations rather than associations of functions that are most crucial for understanding cognition. However, demonstrating that two processes are dissociable is an extraordinarily difficult task. He argues that even one of the most persuasive phenomena, that of double dissociation, does not provide incontrovertible evidence (see also Dunn & Kirsner, 1988). A double dissociation involves the following set of findings. A manipulation M (either behavioral or biological) alters performance in task A but has no effect in task B. A second manipulation N alters performance in task B but not in task A. Weiskrantz points out that examples of M and N have been found to strengthen proposed distinctions between long- and short-term memory. Other examples come from the recent literature distinguishing implicit from explicit memory (Schacter, 1987; Tulving & Schacter 1990). Weiskrantz examines the specificity of an impairment exquisitely at the behavioral level. This detailed behavioral analysis is crucial for understanding the effects of any lesion or treatment.

The task of determining whether two processes are distinct can be made a little easier by applying the methods of neuroscience. Weiskrantz argues that the convergence of data from clinical neuropsychology, neuroanatomy, neurochemistry, theoretical experimental psychology, and neural networking are helpful in uncovering dissociations in cognition. The techniques that are currently available are, perhaps, better than those of 20 years ago. For example, a detailed analysis of cognitive behavior used in conjunction with imaging techniques can provide valuable data concerning behavioral dissociations (Posner, Petersen, Fox, & Raichle, 1988).

Schacter and Nadel provide an excellent example of how one can approach the problem of dissociations in memory from an interdisciplinary perspective. They draw on evidence from cognitive psychology, neuropsychology, and psychobiology to argue that there are different forms of spatial memory. Spatial memory has been well studied in many different species and has clear evolutionary significance. There is almost a century of cumulative data on how spatial memory can be impaired in patients with lesions. For example, we know that the dominant and nondominant hemispheres play different roles in spatial information processing. Schacter and Nadel conclude that spatial memory can be represented in two different forms. These are labeled primary and secondary spatial memory. Primary spatial memory refers to spatial information that is acquired through direct experience, such as navigating through an area. In contrast, secondary spatial memory is acquired indirectly—for example, by processing symbols as in reading a map. Studies in patients with memory impairments provide evidence for a double dissociation of these two types of spatial memory. That is, primary and secondary spatial memory can be selectively disrupted and spared. Schacter and Nadel also discuss how various lesions in rodents can produce specific spatial memory deficits that parallel those described in humans.

Olton et al. also present a detailed discussion of spatial processing and compare the performance and effects of lesions in different species. Schacter and Nadel argued that such studies provide evidence for their distinction between primary and secondary spatial memory. However, it can be difficult to extrapolate results obtained from studies in rodents to humans. Olton et al. note that there may be a number of reasons why lesion studies in different species may lead to different behavioral effects. Often the tasks used are different. The function of the neural pathways mediating the behavior may be different. Further, tasks that are relevant for one species may not be relevant for another. An extensive discussion of this issue can be found in MacPhail (1982). The work of Olton et al. illustrates ways to resolve these issues and the types of experiments that need to be performed.

The experiments performed in both rats and monkeys were consistent in showing that performance in the T-maze delayed conditional discrimination test can be dissociated from performance in a conditional place-object discrimination task. In both species, lesions of the fornix impaired performance in the former but not in the latter. These results, therefore, further support Schacter and Nadel's notion that spatial memory is not unitary.

Even though some of the neuroanatomical structures that are important in learning and memory have been extensively investigated, and the importance of areas such as the hippocampus and amygdala have been noted, their exact role in mediating and modulating memory functions still remains to be determined after more than 100 years of research. The chapters by Olton et al. and Schacter and Nadel in this part, and by McNaughton and Smolensky in Part I, go a considerable way toward clarifying this role.

Johnson and Hirst are also trying to identify and describe the different features of distinguishable memory systems. Their approach goes beyond the tradition in which experiments in normal subjects are used to develop models of memory, and biological issues are ignored. Johnson and Hirst point out the value of studying the differentiated nature of memory both developmentally and through the study of clinical phenomena. This leads to an interest in developing an understanding of central nervous system organization from the differentiated nature of memory. This direction is increasingly evident in cognitive studies by traditional experimental psychologists.

Having outlined their model based upon the distinction between perceptual and reflective processes, they then show how it relates to distinctions that have formed the basis of other theories of memory dissociations. These include short- versus long-term memory, episodic versus semantic memory, and procedural versus declarative memory. Johnson and Hirst's distinction between perceptual and reflective processes has some similarities to Schacter and Nadel's distinction between primary and secondary spatial processing. Primary spatial memory would appear to be based more on perceptual than on reflective processes. The problem of relating the many concepts of various psychological models to one another is a major difficulty confronting neuroscientists and is discussed in more detail below. An important feature of Johnson and Hirst's approach is that they attempt to clarify the component processes that make up the subsystems they propose and to describe a variety of both normal and disrupted cognitive activities in terms of these component processes.

Kesner begins his chapter by pointing out that some researchers examine one form of memory in great detail. The discussion of associative learning by the various contributors to Part I is an example. He then notes that others make binary distinctions between component cognitive processes (e.g., Bailey's analysis of short- and long-term memory and Schacter and Nadel's analysis of primary and secondary spatial memory). Kesner provides a synthesis of the many types of cognitive function. He presents a model derived from many binary-type studies that provides a synthesis of all this research. He suggests, primarily on the basis of research in laboratory animals, that five attributes characterize mnemonic information: spatial, temporal, affect (related to reinforcement contingencies), sensory-perceptual, and response. He further divides the memory system into a data-based system and an expectancy-based system. The data-based system is involved in the coding of essentially autobiographical information and is similar to Tulving's episodic memory. The expectancy-based system revolves around previously stored knowledge and resembles Tulving's semantic memory. Kesner goes on to speculate about the neural systems that may mediate interactions between the various attributes of memory. For example, he sees the hippocampus as mediating the interaction between temporal and spatial attributes and the amygdala as mediating the interaction between temporal and affect attributes.

Clearly, the full model cannot be tested in a single experiment. However, different experiments can test various parts of the model, such as the role of the caudate nucleus in egocentric localization. The advantage of a model that is as extensive as the one provided by Kesner is that it allows many problems to be cast within its framework.

Moscovitch and Umilta, like Kesner, take a broad view of cognition and develop a scheme for discussing all types of cognitive behavior. Their chapter is organized around the theme of consciousness, a theme that has intrigued philosophers for many years but has only recently reemerged as one worthy of study by cognitive scientists. They begin with a discussion of dissociations between processes that require conscious recollection (explicit processes) and those that do not (implicit processes). For example, amnesic patients show dissociations between performance in tests of free recall and recognition, and performance in tests of repetition priming. That is, these patients process information more efficiently as a result of prior exposure, although they have no awareness of that exposure. In discussing such dissociations, Moscovitch and Umilta distinguish between modular and central processes. Modular processes are domain specific and occur in three different forms. Type I modules are those that are hardwired, autonomous, and highly specific and inflexible in their function. Examples of such modules are those involved in the perception of color or acoustic frequency. Type II modules consist of clusters of Type I modules and are also hardwired and domain specific. Such modules might be involved in object perception. Type III modules are those that have been "experientially assembled" by a central processor. With repeated use their function becomes automatic. In contrast, central processes have a more general purpose. Whereas modules are defined by the information they compute, central processes are better defined by the function they serve. Moscovitch and Umilta distinguish five central functions: Forming Type II modules, forming Type III modules, relating information to general knowledge, planning, and making information available to conscious awareness. Central processes, therefore, have general utility in a wide range of information processing situations. Intelligent behavior is presumably mediated by central processes.

Having defined the difference between central and modular processes, Moscovitch and Umilta then use this scheme to discuss attention, episodic memory, working memory, and implicit and explicit memory. They argue that the hippocampus functions as a modular system, providing evidence of domain specificity, information encapsulation, and shallow output. The frontal lobes, in contrast, function as a central system. Damage to these structures produces deficits in a number of cognitive domains beyond those involved in explicit episodic memory. For example, frontal lesions cause impairments in sequencing, planning, and monitoring. The reader will note the extensive use of clinical material by Moscovitch and Umilta to support their model. Likewise, Milberg and Albert provide a discussion of central versus modular cognitive functions from a clinical perspective in Part IV.

All the contributors to Part II are interested in exploring distinctions between cognitive behaviors. This approach is not unique to cognitive neuroscience. Much of science involves a reductionist approach to problems. For this reason, most disciplines are divided, and subdivided, into different areas. The aim is to understand the whole based on the functioning of the component parts. Classification and categorization are basic to all kinds of scholarship. In attempting to understand the artistic process, for example, art works have been classified by artist, country of origin, period, style, or medium. Moreover, these classificatory schemes are used to evaluate the significance of a given piece of art. For example, without such schemes it can be difficult to appreciate a Jackson Pollock or the impact of postimpressionists on contemporary art. Similarly, we need appropriate classificatory schemes to evaluate a piece of cognitive behavior. As the contributors to Part II demonstrate, cognitive behavior, like art, can be divided in many different ways. For example, behavioral engineers often use an operational basis for their taxonomies, whereas cognitive scientists may use a more theoretical "process" base.

At the very least, classification provides a framework within which to discuss phenomena and generate hypotheses. However, it does more than just that. Every explicit classificatory scheme makes an implicit statement about mechanism. It implies that there is something different about the mechanisms underlying phenomena that are allocated to different classes.

Two broadly defined classificatory schemes are used to organize a set of phenomena. One can begin by listing all the phenomena to be classified and then try to find commonalities. This method of *upward classification by empirical grouping* (Mayr, 1982) contrasts with that of *downward classification by logical division*, in which differences are sought between apparently identical phenomena. The current emphasis in cognitive neuroscience seems to be on the second approach (but see Fleishman & Quaintance, 1984; Tulving, 1985), and this is reflected in the chapters in Part II. That is, most of the contributors begin by suggesting that functions A and B may be distinct aspects of some form of cognitive behavior. They test this by examining whether various treatments differentially alter A and B. If so, this would be considered evidence for their binary distinction. In this way cognition gets broken up into component processes.

Is attention to biology essential for providing conclusive evidence for particular dissociations, or have biological manipulations such as lesions (either induced experimentally or caused by accidents of nature) just been more successful than behavioral ones in providing robust and dramatic effects? In some areas it is difficult to find examples where an experimental psychologist could not argue that the study of normals using the traditional methods of experimental psychology was capable of providing just as strong evidence for a dissociation between two cognitive functions as any other approach (Crowder, 1982). For example Schacter and Nadel's review of the experimental psychology literature alone might, perhaps, be deemed sufficient to justify their proposed distinction between primary and secondary spatial memory. However, their suggestion is considerably strengthened by the convergent evidence provided by neurobiological studies. Similarly, Moscovitch and Umilta's use of clinical material considerably strengthens their discussion of the dissociations between various implicit and explicit processes. In some cases we feel that neuropsychologists have been better able than experimental psychologists to provide evidence for distinct cognitive processes. For example the evidence furnished by Rapp and Caramazza (in Part IV) for distinct aspects of language functions derived from neuropsychological studies is much more compelling than that derived from any studies of which we are aware in normals.

An assumption made by all the contributors in Part II is that dysfunction can tell us about normal brain function. In particular there is a belief that an analysis of the behavioral changes resulting from a lesion in the brain can provide valuable information about normal cognitive behavior. Although caution must certainly be exercised in making interpretations using this engineering-like approach to behavior (see, e.g., Gregory 1961) the contributors provide convincing evidence to support their belief.

As Weiskrantz points out, the study of cognitive dysfunction is of value not only because it yields information about normal cognitive function. It is also important for clinicians who diagnose and treat cognitive dysfunction. This is considered in detail in Part IV. The reader should note that several of the contributions in that part could equally well be included here (those of Rapp and Caramazza, and of Milberg and Albert).

Another strategy for examining dissociations involves the administration of drugs to normal volunteers. Such treatments can produce relatively specific cognitive impairments. These impairments disappear when the drug is eliminated from the body or following the administration of a pharmacological antagonist. The reversibility of these impairments offers advantages over the irreversible lesions observed in many patient populations. For example benzodiazepines (e.g., compounds such as diazepam [Valium]) can cause a marked anterograde amnesia. However, the impairment appears to be limited to tests of explicit memory (Lister, 1985). In this way benzodiazepine-treated normal volunteers resemble (undrugged) amnesic patients. They differ, however, from memory-impaired dementia patients. Normal subjects treated with drugs that interfere with cholinergic functioning show impairments that more closely resemble those of patients with dementia (Weingartner, 1985; Lister & Weingartner, 1987).

Both Schacter and Nadel and Johnson and Hirst note that developmental studies can also provide evidence for distinctions between cognitive processes. By charting the ontogeny of various cognitive behaviors, evidence has been provided that one form of cognitive behavior is different from another (Nadel & Zola-Morgan, 1984; Schacter & Moscovitch, 1984).

Yet another strategy uses an evolutionary perspective (see Sherry & Schacter, 1987). Such an approach may be particularly useful for providing evidence concerning the distinctiveness of modules like those described by Moscovitch and Umilta. An evolutionary analysis suggests that the memory systems responsible for song learning in birds or for imprinting are probably distinct from those involved in learning the location of food caches.

One reason providing an integrated view of cognitive neuroscience is difficult is that there are so many unrelated dichotomies. It is usually left to the reader to consider how one dichotomy is related to another. Weiskrantz mentions evidence for dissociations between long- and short-term memory. Schacter and Nadel dissociate primary and secondary spatial memory. Johnson and Hirst discuss perceptual and reflective processes. Kesner mentions data-based and expectancy-based memory systems. Moscovitch and Umilta discuss implicit and explicit memory. To these dissociations can be added distinctions between episodic and semantic memory (Tulving, 1983), procedural and declarative memory (e.g., Cohen & Squire, 1980), habits versus memories (Mishkin, Malamut, & Bachevalier, 1984), and automatic versus effortful processes (Hasher & Zacks, 1979). Many more are surely waiting in the wings to be discovered in the coming years.

Newell (1973) has questioned the value of experiments designed around putative binary oppositions arguing, "You can't play 'Twenty Questions' with nature and win." His concern was that a dichotomy based upon two experimenter-devised tests that does not relate to the rest of the literature is not going to be helpful. Dichotomies must be generalizable. Newell's uneasiness about the Twenty Questions approach to cognition has led to his recent strategy of developing a unified theory of cognition (Waldrop, 1988a, 1988b).

It is certainly logical to try to develop cognitive models following the development of classificatory schemes. However, can all the distinctions be put together to form a coherent theory of cognition? Johnson and Hirst, Kesner, and Moscovitch and Umilta all attempt to do this. Their schemes are considerably enhanced by the discussions of how they relate to those of others.

How should we judge the value of a model? Should it be on the basis of its ability to make predictions or to provide us with a framework within which to think about cognition? Many theories and models may provide aids to discovery without assigning permanent significance to scientific theory (Hesse, 1966). For example, a number of questions could be raised using an astrological perspective. Sheldon's theories of body type in relation to personality (Sheldon, 1940) suggest questions concerning endocrine function and behavior and genetic mechanisms underlying individual differences in cognition. The answers to these questions are of interest even to those who reject Sheldon's theories. How do we choose between the models of Johnson and Hirst, Kesner, and Moscovitch and Umilta and the many others in the cognitive literature?

Popper (1968) argues that we should choose the most simple and falsifiable. He also argues that there are three requirements for the growth of knowledge in a field. First, a new theory should proceed from some *simple*, *new*, and *powerful*, *unifying idea* about some connection or relation between hitherto unconnected things or facts or new "theoretical entities." Second, the theory should be *independently testable*—that is, not only should it explain all currently known facts, it should have new and testable consequences. Finally, the theory should pass some new and severe tests (Popper, 1963, pp. 241–242). That is, the independently testable hypotheses should be confirmed in the laboratory.

All the models or theories in this section satisfy the first of Popper's criteria—they provide useful, new, and unifying frameworks for organizing facts about cognition. It is less clear whether they all satisfy the second requirement. Kesner's model clearly makes predictions about underlying neurobiology, but it is less clear whether the model is capable of making predictions at the purely behavioral level. While the first two of Popper's criteria can be assessed on purely formal grounds, the third criterion can only be fulfilled by empirical testing and therefore requires further experimentation.

Part II was put together with the idea of asking how we can identify the component processes that make up higher mental functions. We believe that the contributors have clearly shown that an interdisciplinary approach has made significant progress toward achieving this goal. Partially successful attempts have been made to incorporate our knowledge of the component processes into models that account for complex cognitive behaviors. However, the continued proliferation of models attests to the fact that no one of these schemes is either detailed or precise enough.

References

Cohen, N. J., & Squire, L. R. (1980). Preserved learning and retention of a pattern analyzing skill in amnesia: Dissociation of knowing how and knowing that. *Science, 210,* 207–210.

Crowder, R. G. (1982). General forgetting theory and the locus of amnesia. In L. S. Cermak (Ed.), *Human memory and amnesia.* Hillsdale, NJ: Lawrence Erlbaum.

Dunn, J. C., & Kirsner, K. (1988). Discovering functionally independent mental processes: The principle of reversed association. Psychological Review, *95,* 91–101.

Fleishman, E. A., & Quaintance, M. K. (1984). *Taxonomies of human performance.* San Diego: Academic Press.

Gregory, R. L. (1961). The brain as an engineering problem. In W. H. Thorpe & O. L. Zangwill (Eds.), *Current problems in animal behavior* (pp. 307–330). Cambridge, England: Cambridge University Press.

Hesse, M. B. (1966). Models and analogies in science. South Bend, IN: University of Notre Dame Press.

Hasher, L., & Zacks, R. T. (1979). Automatic and effortful processes in memory. *Journal of Experimental Psychology: General, 108,* 356–388.

Lister, R. G. (1985). The amnesic action of benzodiazepines in man. *Neuroscience and Biobehavioral Reviews, 9,* 87–94.

Lister, R. G., & Weingartner, H. J. (1987). Neuropharmacological strategies for understanding psychobiological determinants of cognition. *Human Neurobiology, 6,* 119–127.

MacPhail, E. M. (1982). Brain and intelligence in vertebrates. Oxford, England: Clarendon Press.

Mayr, E. (1982). The growth of biological thought. Cambridge, MA: Belknap Press.

Mishkin, M., Malamut, B., & Bachevalier, J. (1984). Memories and habits: Two neural systems. In G. Lynch, J. L. McGaugh, & N. M. Weinberger (Eds.), *Neurobiology of learning and memory* (pp. 65–77). New York: Guilford Press.

Nadel, L., & Zola-Morgan, S. (1984). Infantile amnesia: A neurobiological perspective. In M. Moscovitch (Ed.), *Infant memory* (pp. 145–171). Hillsdale, NJ: Lawrence Erlbaum.

Newell, A. (1973). You can't play 20 questions with nature and win. In W. G. Chase (Ed.), *Visual information processing* (pp. 283–308). New York: Academic Press.

Popper, K. R. (1963). *Conjectures and refutations*. London: Routledge and Kegan Paul.

Popper, K. R. (1968). *The logic of scientific discovery*. London: Hutchinson.

Posner, M. I., Petersen, S. E., Fox, P. T., & Raichle, M. E. (1988). Localization of cognitive operations in the human brain. *Science, 240*, 1627–1631.

Schacter, D. L. (1987). Implicit memory: History and current status. *Journal of Experimental Psychology: Learning, Memory and Cognition, 13*, 501–522.

Schacter, D. L., & Moscovitch, M. (1984). Infants, amnesics and dissociable memory systems. In M. Moscovitch (Ed.), *Infant memory* (pp. 145–172). Hillsdale, NJ: Lawrence Erlbaum.

Sheldon, W. H. (1940). *The varieties of human physique*. New York: Harper and Brothers.

Sherry, D. F., & Schacter, D. L. (1987). The evolution of multiple memory systems. *Psychological Review, 94*, 439–454.

Tulving, E. (1983). *Elements of episodic memory*. New York: Oxford University Press.

Tulving, E. (1985). On the classification problem in learning and memory. In L. G. Nilsson & T. Archer (Eds.), *Perspectives in learning and memory* (pp. 67–91). Hillsdale, NJ: Lawrence Erlbaum.

Tulving, E., & Schacter, D. L. (1990). Priming and human memory systems. *Science, 247*, 301–306.

Waldrop, M. M. (1988a). Toward a unified theory of cognition. *Science, 241*, 27–29.

Waldrop, M. M. (1988b). Soar: A unified theory of cognition. *Science, 241*, 296–298.

Weingartner, H. J. (1985). Models of memory dysfunctions. *Annals of the New York Academy of Sciences, 444*, 359–369.

III

Modulation of Cognition

If cognitive scientists were asked to alter an individual's cognitive functions, they would probably suggest changing rehearsal strategies or altering how information is presented or retrieved. However, a number of factors not normally associated with cognitive functioning have a major impact on how we think, learn, and remember. These include variables such as arousal, stress, emotion, and motivation. The contributors in Part III consider some of these *modulators* of cognition.

In Part I we obtained multifaceted pictures of neural systems that may be directly involved in establishing a record of experience, maintaining that record, and retrieving experience from memory. Many of the contributors to Part I provided an empirical/theoretical model of the "intrinsic" neurobiologically "hardwired" components of memory, together with the software that would make those components operational. These memory systems were described either in terms of analyses of well-defined neurobiological "circuits" or alternatively by isolating and identifying simple types of learning and memory, which then spurred an examination of the neurobiology that is necessary for that type of learning to take place.

In contrast, the contributors to Part III consider how various factors modulate learning and memory. Such factors as emotional state, arousal, and reinforcement contingencies can alter how well something is learned and the allocation of resources involved in selectively attending to and processing events. There is a parallel between this modulation in the cognitive domain and the concept of neuromodulation as discussed by neuroscientists. A neuromodulator is a chemical that does not directly alter neuronal function, but instead modifies the action of neurotransmitters that do have direct effects on neuronal activity. Similarly, if information is not being processed, a modulating factor such as an increase in motivation will not, on its own, change cognitive functioning. In contrast, if some experience is being processed, then motivational changes can influence cognitive functioning (such as how well something is learned). This distinction between modulating and mediating factors allows us to consider how nominally noncognitive functions such as emotion play a role in cognitive processing.

McGaugh opens the section by making a persuasive case for the importance of modulation in cognitive functions such as memory. He also lays out the groundwork for understanding how this modulatory system interacts with systems that may mediate memory. Koob follows with a discussion of the neurobiology of stress and arousal. His analysis fits with McGaugh's approach to memory modulation and also serves as a logical bridge to the contribution of Eysenck, who presents a discussion of how anxiety alters cognition from a psychological perspective. Eysenck discusses the methodology that can be productively used to research this complex issue. His methods and metaphors are derived from cognitive science rather than neurobiology. This is also the case for Leventhal, who provides a general overview of emotion and cognition, with a strong developmental emphasis. He develops a broad theory of emotion and cognition derived from biology, social psychology, and cognitive science.

15

Neuromodulation and the Storage of Information: Involvement of the Amygdaloid Complex

JAMES L. MCGAUGH

We do not know whether all of our experiences leave a neural record. But it seems clear from experimental findings as well as experiential fact that all experiences are not remembered equally well (or poorly). Some of our memories are vivid and lasting, whereas others are but faint and fleeting. What accounts for the enormous variation in the strength of our memories? The central hypothesis guiding the research findings summarized in this chapter is that neuromodulatory systems activated by experiences serve to modulate the storage of newly acquired information (Gold & McGaugh, 1975). According to this view, the strength of memory for an event depends, at least in part, on the degree to which the experience activates modulatory systems.

The view that neuromodulatory systems play a role in regulating memory storage is based on the extensive and well-known evidence indicating that the retention of recently acquired information can be influenced by a variety of posttraining treatments (McGaugh & Herz, 1972). Susceptibility to posttraining modulating influences appears to be a common feature of human as well as animal memory (McGaugh, 1989b). Of particular relevance are the findings, first reported over 25 years ago (e.g., Breen & McGaugh, 1961; McGaugh, Westbrook, & Thomson, 1962; Westbrook & McGaugh, 1964), indicating that, in rats tested a day or longer following training, the retention of recently learned tasks is enhanced by low doses of CNS stimulants administered shortly after training. The drugs were ineffective when administered several hours following training. Comparable effects were obtained in a variety of training tasks, including a latent learning task (McGaugh, 1966; Westbrook & McGaugh, 1964). The findings of these early experiments provided compelling evidence that the drugs affected retention by altering memory processing initiated by training (McGaugh, 1973, 1989c; McGaugh & Petrinovich, 1965).

279

Hormonal Modulation of Memory Storage

The evidence indicating that retention can be enhanced (McGaugh, 1973) as well as impaired by posttraining administration of many drugs known to affect brain functioning suggests the possibility that memory storage may be regulated by endogenous physiological systems that are activated by learning experiences (Gold & McGaugh, 1975). That is, posttraining suscep-tibility to modulation of memory storage provides an opportunity for the physiological consequences of experiences to modulate the strength of memory. It is well known that experiences such as those resulting from the training tasks typically used in studies of learning and memory in animals can activate the release of a variety of hormones, and there is now consider-able evidence indicating that, in rats and mice, retention of information is affected by posttraining administration of hormones that are normally released by training experiences (McGaugh, 1983; McGaugh & Gold, 1989). Experiments examining the influences of hormones on learning and mem-ory have shown that retention can be altered by posttraining injections of a number of hormones, including epinephrine, ACTH, vasopressin, opioid peptides, CCK, substance P, and other peptide hormones (for reviews, see McGaugh, 1983, 1989a; McGaugh & Gold, 1989).

Studies of the effects of the adrenal medullary hormone epinephrine on retention in rats and mice provide strong support for the view that this hormone has a modulatory influence on memory storage. In the first study examining the memory-enhancing effect of epinephrine, Gold and van Buskirk (1975) reported that, when injected peripherally following training in an inhibitory (passive) avoidance task, epinephrine produced dose-dependent and time-dependent effects on subsequent retention. Retention was enhanced by low and moderate doses and impaired by high doses. Injections of epinephrine administered several hours following training were ineffective. Such findings are consistent with the view that the epi-nephrine effects seen on the retention test are due to influences of the hormone on the storage of the information provided by the training experi-ence.

The finding that the effects of posttraining epinephrine on memory are long lasting provides additional support for this view. Figure 15.1 shows the findings of an experiment examining the effects, on retention, of high and low doses of epinephrine administered to mice immediately following train-ing on a Y-maze position discrimination (Introini-Collison & McGaugh, 1986). Escape from footshock was used as motivation. On the retention test given to independent groups 1 day, 1 week, or 1 month later, the location of the correct (i.e., safe) alley was reversed and the mice received 6 retraining trials. Errors made on the discrimination reversal trials were used as the index of retention of the original training. As can be seen, the effects of the posttraining injections of the two doses of epinephrine produced compar-able effects at all retention intervals: As was found with the inhibitory

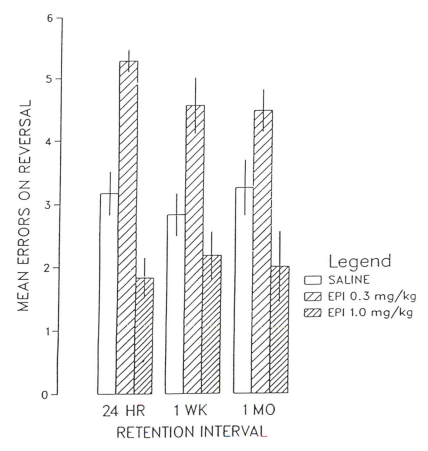

Figure 15.1 Effects of posttraining epinephrine (i.p.) on retention of a Y-maze position discrimination as tested by discrimination reversal training. Independent groups of mice were tested at the intervals indicated following training. See text for details. (From Introini-Collison & McGaugh, 1986.)

avoidance task, retention was enhanced by the lower dose and impaired by the higher dose. The finding that epinephrine modulated the retention of a response requiring choice provides additional support for the view that the effects were due to influences on the storage of information rather than to nonassociative influences on mobility.

While most studies of the effects of epinephrine have used aversively motivated tasks, including inhibitory and active avoidance as well as discrimination learning (McGaugh, 1989a; McGaugh & Gold, 1989), epinephrine also enhances retention of an appetitively motivated task (Sternberg, Isaacs, Gold, & McGaugh, 1985). These and many other studies of the effects of epinephrine on memory in rats and mice have provided strong evidence that retention in a variety of tasks is enhanced by low doses of epinephrine. The finding that the hormone is effective only when administered shortly after training is consistent with the view that endogenous epinephrine normally released by training serves to modulate the storage of the training experience (Gold & McGaugh, 1975; McGaugh, 1983; 1989a; McGaugh & Gold, 1989).

Involvement of the Amygdaloid Complex

Much recent research has been motivated by an interest in determining how ephinephrine affects brain systems involved in memory. As epinephrine does not readily enter the brain, it seems unlikely that this hormone directly affects brain processes involved in memory storage. Evidence from a number of recent studies in my laboratory suggests that epinephrine probably acts peripherally at receptors on visceral afferents projecting to the brain. More specifically, the evidence strongly suggests that the memory-modulating effects of posttraining epinephrine involve, at one step in the process, the activation of a noradrenergic system within the amygdaloid complex (Liang, Juler, & McGaugh, 1986; McGaugh, 1989b).

The first evidence suggesting the involvement of the amygdala in the memory modulating effects of epinephrine came from experiments examining retrograde amnesia produced by posttraining electrical stimulation of the amygdala. The amnesia typically produced by this treatment was not obtained in adrenal demedullated rats (Bennett, Liang, & McGaugh, 1985). Retrograde amnesia did result, however, if epinephrine was administered after the training and immediately prior to the brain stimulation (Liang, Bennett, & McGaugh, 1985). These findings strongly suggested that amygdala functioning is influenced by peripheral epinephrine. The findings of other experiments examining the role of the amygdala indicated that lesions of the stria terminalis (ST), a major amygdala pathway, blocked the amnesic effect of amygdala stimulation (Liang & McGaugh, 1983a) as well as the memory-enhancing effect of posttraining epinephrine (Liang & McGaugh, 1983b). More recent evidence (discussed further below) argues that ST lesions interfere primarily with outputs from the amygdala. On the basis of these findings it seems unlikely that the amygdala serves as a permanent storage site for the type of learning assessed in these experiments. Rather, the findings suggest that the amygdala affects memory through influences on other brain regions. Our finding that retention of one task used in these studies, inhibitory avoidance, is unaffected by amygdala lesions made several days after training provides further support for this view. (Liang et al., 1982).

Involvement of Norepinephrine Within the Amygdala

Evidence from several recent studies suggests that the memory-modulation function of the amygdala involves activation of norepinephrine (NE) receptors within the amygdaloid complex. In the first study examining the role of amygdala NE in memory Gallagher and her colleagues found that retention is impaired by posttraining intraamygdala injections of the $\beta_{1,2}$-adrenergic receptor antagonist propranolol (Gallagher, Kapp, Pascoe, & Rapp, 1981). Further, findings from my laboratory suggest that epinephrine effects on memory involve activation of NE receptors within the amygdala. In rats trained on an inhibitory avoidance task, intraamygdala injections of propranolol block the memory-enhancing effect of posttraining intraperitoneal injections of epinephrine. In addition retention is enhanced by posttraining intraamygdala injections of NE (Liang, Juler, & McGaugh, 1986).

Studies of the effects of opioid peptide agonists and antagonists provide additional evidence arguing that amygdala NE is involved in the modulation of memory storage, as opioid peptides are known to inhibit the release of NE (Arbilla & Langer, 1978; Montel, Starke, & Weber, 1974; Nakamura, Tepper, Young, Ling, & Groves, 1982; Werling, Brown, & Cox, 1987). There is extensive evidence indicating that retention is enhanced by posttraining systemic (Introini-Collison & McGaugh, 1987; Izquierdo, 1979; Messing et al., 1979) as well as intraamygdala (Gallagher & Kapp, 1978; Introini-Collison, Nagahara, & McGaugh, 1989) injections of opiate antagonists such as naloxone and naltrexone. The possibility that such effects involve NE is supported by the findings (Izquierdo & Graudenz, 1980) that the memory-enhancing effects of systemically injected naloxone are blocked by propranolol. The effects of naloxone on memory, in mice, are also blocked in animals treated with DSP4, a neurotoxin that produces a relatively specific reduction in brain NE (Introini-Collison & Baratti, 1986). Evidence from several studies suggests that the effects of naloxone on memory involve amygdala NE. In an extensive series of studies, Gallagher and her colleagues demonstrated that, in rats given posttraining intraamygdala injections, retention is enhanced by naloxone and impaired by opiate agonists (Gallagher, 1982, 1985; Gallagher & Kapp, 1978; Gallagher et al., 1981). Further, the memory-enhancing effects of both peripheral as well as intraamygdala injections of naloxone are blocked in rats with neurotoxic-induced (6-OHDA) lesions of the dorsal noradrenergic pathway (Fanelli, Rosenberg, & Gallagher, 1985; Gallagher, Rapp, & Fanelli, 1985).

In further support of the view that naloxone effects on memory involve amygdala NE we have found that posttraining intraamygdala injections of propranolol block the memory-enhancing effects of systemically as well as intraamygdally administered naloxone (Introini-Collison et al., 1989b; McGaugh, Introini-Collison, & Nagahara, 1988). The findings of the experiment examining the effects of posttraining intraamygdala injections of propranolol are shown in Figure 15.2 The rats in this experiment were trained in an inhibitory avoidance task as well as the Y-maze position discrimination task described earlier. Propranolol or a buffer solution was injected intraamygdally via implanted cannulae immediately after training. Then, the animals received either saline or naloxone i.p. Retention was tested 1 week later. Comparable effects were obtained with the two tasks.

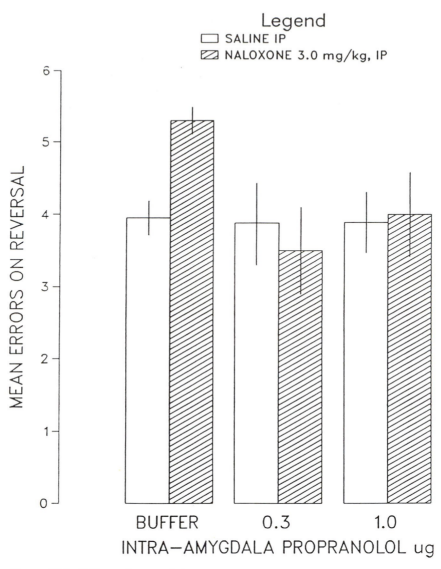

Figure 15.2 Effects of posttraining naloxone (i.p.) on retention of rats given bilateral intraamygdala injections of propranolol immediately prior to the intraperitoneal injections. The animals were trained on a Y-maze position discrimination task and tested for retention one week later with errors made on discrimination reversal training as the index of retention of the original training. See text for details. (From McGaugh, Introini-Collison, & Nagahara, 1988.)

Naloxone significantly enhanced retention in the rats given the buffer solution intraamygdally. However, the naloxone effect on retention was blocked in animals given intraamygdala injections of propranolol prior to the i.p. naloxone injections.

The findings of other experiments indicate that the memory-enhancing effects of systematically administered naloxone are also blocked by intra-amygdala injections of the β_1-antagonist atenolol as well as the β_2-antagonist zinterol. In contrast, α_1- (prazosin) and α_2- (yohimbine) antagonists were ineffective. The β-antagonists were effective only when administered to the amygdala: Injections into either the caudate putamen or cortex dorsal to the amygdala did not block the memory-enhancing effects of naloxone. Thus, the findings of the studies using β-adrenergic blockers suggest that both epinephrine and naloxone effects on memory involve the activation of NE receptors within the amygdala.

Involvement of the ST and VAF Pathways

As was summarized above, lesions of the ST block the memory-impairing effects of electrical stimulation of the amygdala as well as the memory-enhancing effects of systematically administered epinephrine (Liang & McGaugh, 1983a, 1983b). The results of other recent studies indicate that ST lesions also block the memory-enhancing effects of posttraining nalox-one i.p. as well as the memory-impairing effects of posttraining β-endor-phin i.p. (McGaugh, Introini-Collison, Juler, & Izquierdo, 1986). In addition, ST lesions block cholinergic influences on memory: Both the memory-enhancing effects of oxotremorine and the memory-impairing effects of atropine are blocked in animals with ST lesions (Introini-Collison, Arai, & McGaugh, 1989a). These findings provide strong support for the general hypothesis that all of these treatments modulate memory storage through influences involving the amygdala.

As the ST carries both afferent and efferent fibers, these studies do not reveal whether the effects of ST lesions are due to loss of amygdala afferents or to an impairment of amygdala influences on other brain regions. A recent study comparing the effects of lesions of the ST with those produced by lesions of another amygdala pathway, the VAF (ventral amygdalofugal pathway), provides some clarification of this issue (Liang, McGaugh, & Yao, 1990). As is shown in Figure 15.3 (top), ST lesions blocked the effects, on memory (examined in an inhibitory avoidance task), of posttraining intra-amygdala injections of NE: The lesions blocked both the memory-enhancing effects of a low dose of NE as well as the memory-impairing effects of a high dose. However, as is shown in Figure 15.3 (bottom), VAF lesions did not block the memory-modulating effects of intraamygdala NE.

These results clearly argue that the effects, on memory, resulting from activation of the amygdala are due to outputs to other brain regions medi-

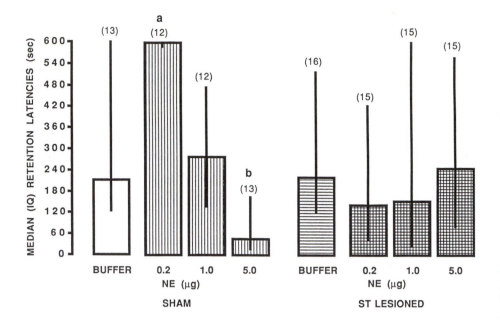

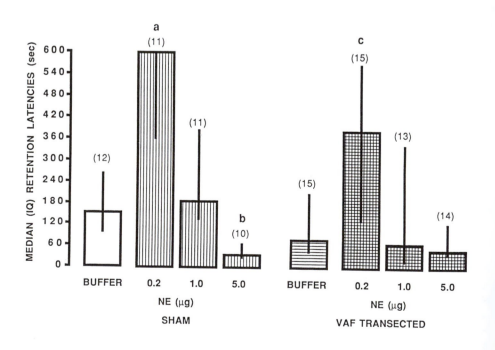

ated by the ST. Additional evidence indicated that VAF lesions attenuated but did not block the memory-enhancing effects of systemically administered epinephrine. That is, in VAF-lesioned animals, a higher dose of epinephrine was required for enhancement of memory. These findings, considered together with the previous finding that ST lesions block the memory-modulating effects of systemically administered epinephrine, suggest that the VAF may mediate the central influences of peripheral epinephrine on amygdala functioning, whereas the ST mediates amygdala influences on memory storage in other brain regions. It is clear from the findings of these experiments that lesions of these amygdala pathways do not prevent learning. While these pathways are not critical for learning or memory, an intact ST appears to be required for the regulation of memory storage by posttraining neuromodulatory systems.

Memory-Modulating Influences of GABAergic Agonists and Antagonists

The findings summarized above suggest the interesting possibility that the amygdala may serve to integrate the interaction of neuromodulatory influences on memory storage. The evidence indicating that ST lesions, as well as intraamygdala injections of β-adrenergic antagonists, block the memory-modulating effects of posttraining treatments affecting adrenergic, noradrenergic, and opioid peptidergic systems is consistent with this suggestion (Introini-Collison et al., 1989b; Liang & McGaugh, 1983b; McGaugh et al., 1986, 1988). The finding that ST lesions also block cholinergic influences on memory provides additional support for this possibility (Introini-Collison, et al., 1989a).

As a further examination of this general hypothesis, we have recently investigated the effects, on memory storage, of treatments affecting GABAergic systems. As is summarized below, the findings of these experiments indicate that posttraining treatments affecting GABAergic systems may also influence memory storage through effects involving the amygdaloid complex.

An early experiment from my laboratory (Breen & McGaugh, 1961) indicated that posttraining systemic administration of the GABAergic antagonist picrotoxin enhanced appetitively motivated maze learning in rats. Comparable effects have been obtained in experiments using a wide variety

---

Figure 15.3 (Top) Stria terminalis (ST) lesions block the memory-enhancing effect of posttraining intraamygdala injections of norepinephrine (NE). (Bottom) VAF lesions do not block the effect of intraamygdala injections of NE. NE was administered via implanted cannulas to sham-lesioned, ST-lesioned, and VAF-lesioned rats immediately following training in an inhibitory avoidance task. Retention was examined 1 day later. (From Liang, McGaugh, & Yao, 1990.)

of training tasks (Bovet, McGaugh, & Oliverio, 1966; Brioni & McGaugh, 1988; McGaugh, Castellano, & Brioni, 1990). In recent studies using aversively motivated training tasks we have provided extensive evidence indicating that posttraining systemic administration of GABAergic antagonists, including picrotoxin and bicuculline, enhance retention while GABAergic agonists, including GABA A agonist muscimol and the GABA B agonist baclofen, impair retention (Brioni & McGaugh, 1988; Castellano & McGaugh, 1989; Castellano, Introini-Collison, Pavone, & McGaugh, 1989b). The findings of one experiment examining the effects of posttraining i.p. injections of picrotoxin on 1-day retention of an inhibitory avoidance task are shown in Figure 15.4 (Castellano & McGaugh, 1989). When administered immediately posttraining, the drug produced dose-dependent enhancement of retention. The drug did not affect retention latencies that received no footshock on the training trial. Further, the effects were time dependent: Injections of picrotoxin administered 2 hours following training were ineffective. We have obtained highly comparable results in studies of the effects of bicuculline on memory in this task as well as the effects of both picrotoxin and bicuculline on retention of the Y-maze discrimination task (Brioni & McGaugh, 1988).

Although aversively motivated tasks have been used extensively in examining the memory-modulating influences of GABAergic antagonists on memory, the effects are not due simply to drug potentiation of fear. The

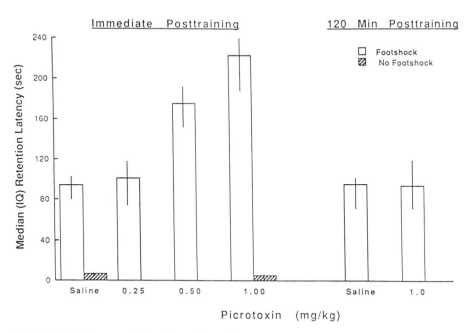

Figure 15.4 Dose and time-dependent effects of posttraining picrotoxin (i.p.) on retention of an inhibitory avoidance response. See text for details. (From Castellano & McGaugh, 1989.)

evidence that picrotoxin enhances the retention of appetitively motivated learning tasks (e.g., Breen & McGaugh, 1961) clearly argues against that suggestion. Further, as is shown in Figure 15.5, posttraining systemic injections of picrotoxin enhance the learning of learned safety (McGaugh et al., 1990). In this experiment mice first received a series of tones, and for half of the animals the tone was paired with footshock. Two days later, animals in four groups were tested for exploratory locomotor activity in a Y maze in the presence of the tone (left four bars in the figure). As can be seen, the number of alley entries (the measure of activity) was significantly reduced in animals that had received footshock. Mice in the four other groups received extinction training (a series of tones alone, without shock) on the day following training and were then tested, as were the other groups, for locomotor activity a day later. Immediately following the extinction training the mice received i.p. injections of either saline or picrotoxin. As can be seen, the extinction session was effective: The activity scores of extinguished controls were higher than those of nonextinguished controls. Further, the postextinction picrotoxin enhanced extinction: In the groups given extinction training the mean activity scores of picrotoxin-injected mice were higher than those of saline controls and comparable to those of animals that received only tones on the original training session. The other two groups received picrotoxin injections but no extinction training on the second day. As can be seen, the drug alone (i.e., without the extinction training) did not affect subsequent performance.

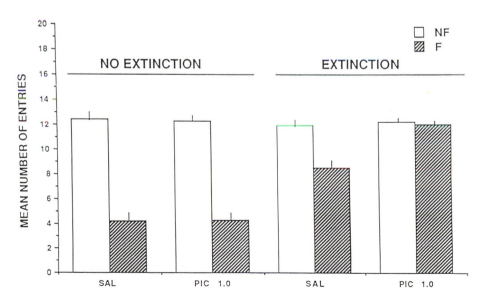

Figure 15.5 Effects of postextinction administration of picrotoxin (1.0) on extinction of a conditioned fear response. Bars indicate mean (± SEM) alley entries on the test session. See text for details. (From McGaugh, Castellano, & Brioni, 1990.)

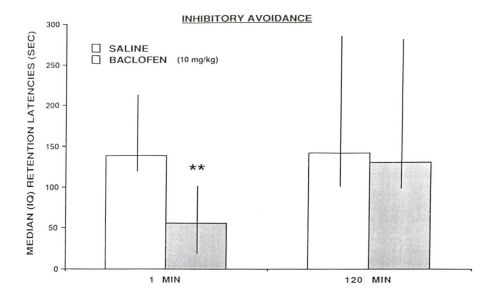

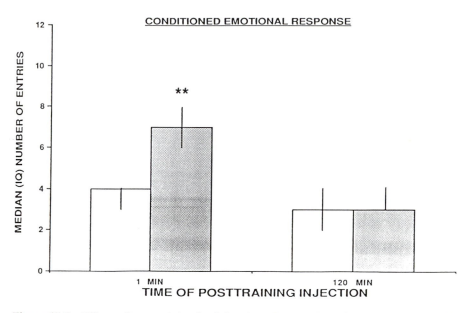

Figure 15.6 Effects of posttraining baclofen (i.p.) on retention of (upper figure) inhibitory avoidance and (lower figure) conditioned fear tasks. See text for details. Bars indicate mean and interquartile ranges. $**p < 0.01$ versus saline-injected controls. See text for details. (From Castellano, Brioni, Nagahara, & McGaugh, 1989a.)

Thus, the findings of these studies provide strong evidence that retention can be enhanced by posttraining injections of GABAergic antagonists. Other recent findings indicate that retention is impaired by posttraining injections of GABAergic agonists, including muscimol (Brioni, Nagahara, & McGaugh, 1989; Castellano & Pavone, 1988; Castellano et al., 1989b) and baclofen (Castellano, Brioni, Nagahara, & McGaugh, 1989a; Swartzwelder, Tilson, McLamb, & Wilson, 1987). Figure 15.6 shows the time-dependent memory-impairing effects of posttraining administration of baclofen i.p., in mice trained in an inhibitory avoidance response (upper figure) and a conditioned fear (conditioned emotional response) task (lower figure) (Castellano et al., 1989b). The conditioned fear task was the same as that used for the picrotoxin study summarized earlier (except that there was no extinction session). As can be seen, baclofen produced comparable dose-dependent memory-impairing effects in the two tasks. As was indicated above alley entries on the Y-maze retention test were used as the index of retention of conditioned fear: Mice given posttraining baclofen displayed less fear than did controls and mice given injections 2 hours posttraining.

Examination of State Dependency in GABAergic Influences on Memory

In studies of the effects of posttraining treatments on memory it is, of course, important to include controls for a variety of possible factors, other than influences on information storage, that might affect performance on the retention test. Groups given delayed posttraining treatments as well as groups given posttraining treatments but no footshocks (in inhibitory avoidance training) are typically included in such studies. Izquierdo (1984) has reported that the retention impairment produced by some posttraining treatments may be based on state dependency. That is, newly acquired information may be stored in a state induced by the posttraining treatment and, thus, not be accessible when retention is assessed while the animal is in a normal state. These findings raise the question of whether retention enhancement produced by posttraining treatments such as those found with GABAergic antagonists might be due to the induction of state dependency. However, as the animals in such studies are usually not injected prior to the retention testing session, a state-dependent interpretation of drug-induced enhancement of retention would require the ad hoc assumption that the animal's state normally occurring at the time of the retention test is congruent with the state induced following the training. If that is the case, administration of the same treatment prior to the retention test should be expected to decrease the congruency and, as a consequence, attenuate the retention enhancement.

The findings shown in Figure 15.7 are based on an experiment examining this issue (Castellano & McGaugh, 1989). Mice were injected with saline or picrotoxin (0.5 or 1.0 mg/kg) immediately posttraining. Animals in different groups then received injections of saline or picrotoxin at one of several intervals (30, 10, or 3 min) prior to the retention test. As can be seen, the posttraining injections produced dose-dependent effects on retention. However, the drug injections prior to the retention test did not affect retention performance. We have obtained highly comparable effects in a study examining the memory-enhancing effects of bicuculline (Castellano & McGaugh, 1990). Further, we have also found that the memory-impairing effects of posttraining muscimol are not altered by muscimol administered prior to the retention test (Castellano & McGaugh, 1990). These findings clearly argue that the memory-modulating effects of GABAergic drugs are not based on state dependency. In other recent studies we have found no evidence of state dependency in the memory-impairing effects of the opiate agonists morphine and dynorphin (Introini-Collison, Cahill, Baratti, & McGaugh, 1987). Although there is compelling evidence that some posttraining treatments induce effects on retention that are state dependent

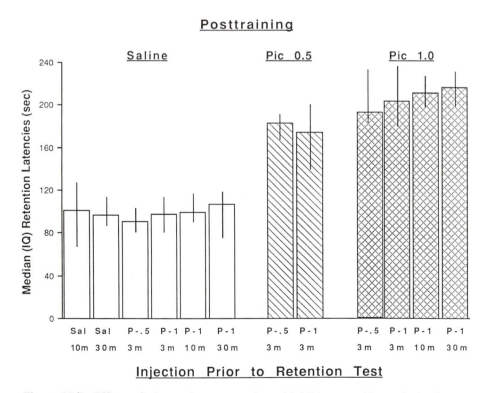

Figure 15.7 Effects of picrotoxin on retention of inhibitory avoidance: lack of state-dependency. Picrotoxin and/or saline was administered (i.p.) posttraining and prior to testing. See text for details. (From Castellano & McGaugh, 1989.)

(Izquierdo, 1984), we have obtained no evidence that enhancement and impairment of retention produced by posttraining drugs and hormones are due to state dependency. Rather, our findings are consistent with the view that the drugs and hormones we have studied affect retention by modulating posttraining processes involved in information storage.

Involvement of the Amygdala in GABAergic Influences on Memory Storage

As GABAergic drugs readily pass the blood-brain barrier, it seems clear that their effects on memory are due to direct influences on brain GABAergic systems. To examine this issue more explicitly, we have studied the effect of bicuculline methiodide, a GABAergic antagonist that does not readily pass the blood-brain barrier. When injected i.p. immediately posttraining, this drug does not affect retention (Brioni & McGaugh, 1988). However, as is shown in Figure 15.8, when injected intraamygdally, bicuculline methiodide produces dose-dependent enhancement of retention of inhibitory avoidance. The effects appear to be initiated by GABAergic influences within

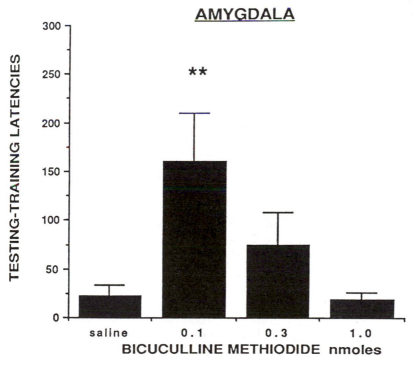

Figure 15.8 Effect of posttraining bilateral intraamygdala injections of bicuculline on retention of an inhibitory avoidance response. Bars indicate means ($\pm$SEM). $**p < 0.01$ versus saline-injected controls. (From Brioni, Nagahara, & McGaugh, 1989.)

the amygdala, as injections dorsal to the amygdala are ineffective. Further, retention is impaired by intraamygdala injections of the GABAergic agonists baclofen and muscimol (Brioni et al., 1989, Castellano et al., 1989a).

It is of particular interest that, as we previously found with the adrenergic and opioid peptidergic systems, GABAergic influences on memory appear to involve activation of NE receptors. In other recent experiments we have found that propranolol blocks the memory-enhancing effects of bicuculline. This effect was obtained with intraperitoneal as well as intraamygdala injections of bicuculline and propranolol (McGaugh, Introini-Collison, Brioni, & Nagahara, unpublished findings).

Neuromodulatory Systems and Memory

The findings summarized in this chapter leave little doubt that the storage of new information is subject to the modulating influences of hormonal and neurotransmitter systems activated by learning experiences. The finding that is perhaps most critical for this conclusion is that hormones (and treatments affecting hormonal systems) influence retention when administered within a short period following training. This is precisely what would be expected on the assumption that the hormones alter retention by modulating the storage of new information (Gold & McGaugh, 1975). This view assumes, of course, that memories are stored, or consolidated over time following training (McGaugh, 1966; McGaugh & Herz, 1972).

Several other key findings provide additional support for this hypothesis. First, the memory-modulating effects of posttraining hormone treatments are long lasting. Enhancing as well as impairing effects of posttraining injections of epinephrine on retention are seen even 1 month following training. Second, the effects cannot be readily accounted for by nonspecific or nonmemorial influences. For example, when rats are trained in an inhibitory avoidance task, posttraining injections of epinephrine do not affect retention latencies of controls that receive no footshock on the training trial (Liang & McGaugh, 1983b). Third, within the range of tasks examined to date the effects of posttraining administration of hormones can be obtained with a variety of tasks and procedures, including inhibitory avoidance, active avoidance, appetitive learning, and discrimination learning. It remains to be determined whether there are forms of memory that are not subject to the modulating influences of posttraining treatments affecting hormonal systems. Finally, although this chapter focuses on treatments affecting adrenergic and opioid peptidergic hormones, comparable effects have been obtained with other hormones, including ACTH and vasopressin, that are known to be released following training in learning tasks comparable to those used in the studies summarized in this chapter (McGaugh, 1983, 1989a; McGaugh & Gold, 1989). All of this evidence strongly points to a role for hormones in influencing the strength of memory for the experiences leading to their release.

There are, of course, many ways in which hormones might act to regulate the storage of information. There is evidence that hormones may act on both peripheral and central systems that influence memory. For example, Gold (1988) has obtained evidence suggesting that influences of epinephrine on retention may be mediated, at least in part, by the release of glucose. Vasopressin may affect memory through peripheral influences on blood pressure as well as through direct influences on the brain (Burbach, Kovacs, de Wied, van Nispen, & Greven, 1983; De Wied, 1984; Koob, this volume). The findings from my laboratory summarized in this chapter provide extensive evidence indicating that the amygdaloid complex is involved in mediating the influences of several neuromodulatory systems on memory. Among the critical findings supporting this conclusion is the evidence that lesions of the stria terminalis and intraamygdala administration of β-adrenergic receptor antagonists block the memory-influencing effects of a variety of treatments, including adrenergic, noradrenergic, cholinergic, opioid peptidergic and GABAergic treatments affecting hormonal and neurotransmitter systems. Further, memory-modulating effects can be produced by intraamygdala injections of NE, as well as opiate and GABAergic receptor agonists and antagonists. Such effects, like those found with peripheral injections, are dose dependent as well as time dependent, and are not obtained with injections administered to the caudate putamen and the cortex dorsal to the amygdaloid cortex.

It is well known that the amygdala is activated by many kinds of sensory stimulation, and extensive evidence from lesion studies suggests that the nuclei of the amygdaloid complex are involved in a variety of forms of learning, particularly those that involve retention of aversively motivated tasks and integration of sensory information (Gaffen & Harrison, 1987; Gentile, Jarrell, Teich, McCabe & Schneiderman, 1986; Hitchcock & Davis, 1987; Iwata, LeDoux, Meeley, Arneric, & Reis, 1986; Kapp, Frysinger, Gallagher, & Haselton, 1979; LeDoux, 1986; Murray & Mishkin, 1985; Saunders, Murray, & Mishkin, 1984). The findings of studies examining the effects of posttraining treatments on memory suggest the interesting possibility that the amygdaloid complex may be a site at which interactions of hormonal and transmitter systems orchestrate endogenous influences on the storage of information. It may be that information is stored in temporary neural alterations located within the amygdala. However, as lesions of the amygdala do not appear to impair the retention of old, well-established memories (Liang et al., 1982; Milner, 1966), it would seem likely that the long-term effects resulting from activation of the amygdala are due to influences on storage in other brain regions.

The most general conclusion suggested by the findings considered in this chapter is that a complete account of the neurobiological systems mediating learning and memory must include an understanding of the role of neuromodulatory systems activated by experiences in regulating the cellular neurobiological processes underlying the storage of information.

References

Arbilla, S., & Langer, S. Z. (1978). Morphine and β-endorphin inhibit release of noradrenaline from cerebral cortex but not of dopamine from rat striatum. *Nature, 271*, 559–561.

Bennett, C., Liang, K. C., & McGaugh, J. L. (1985). Depletion of adrenal catecholamines alters the amnestic effect of amygdala stimulation. *Behavioural Brain Research, 15*, 83–91.

Bovet, D., McGaugh, J. L., & Oliverio, A. (1966). Effects of posttrial administration of drugs on avoidance learning of mice. *Life Sciences, 5*, 1309–1315.

Breen, R. A., & McGaugh, J. L. (1961). Facilitation of maze learning with posttrial injections of picrotoxin. *Journal of Comparative and Physiological Psychology, 54*, 498–501.

Brioni, J. D., & McGaugh, J. L. (1988). Posttraining administration of GABAergic antagonists enhance retention of aversively motivated tasks. *Psychopharmacology, 96*, 505–510.

Brioni, J. D., Nagahara, A. H. & McGaugh, J. L. (1989). Involvement of the amygdala GABAergic system in the modulation of memory storage. *Brain Research, 487*, 105–112.

Burbach, J. P. H., Kovacs, G. L., de Wied, D., van Nispen, J. W., & Greven, H. M. (1983). A major metabolite of arginine vasopressin in the brain is a highly potent neuropeptide. *Science, 221*, 1310–1312.

Castellano, C., Brioni, J. D., Nagahara, A. H., & McGaugh, J. L. (1989a). Posttraining systemic and intra-amygdala administration of the GABA-b agonist baclofen impairs retention. *Behavioral and Neural Biology, 52*, 170–179.

Castellano, C., Introini-Collison, I. B., Pavone, F., & McGaugh, J. L. (1989b). Effects of naloxone and naltrexone on memory consolidation in CD1 mice: Involvement of GABAergic mechanisms. *Pharmacology, Biochemistry and Behavior, 32*, 563–567.

Castellano, C., & McGaugh, J. L. (1989). Retention enhancement with posttraining picrotoxin: Lack of state dependency. *Behavioral and Neural Biology, 51*, 165–170.

Castellano, C., & McGaugh, J. L. (1990). Effects of post-training bicuculline and muscimol on retention: Lack of state dependency. *Behavioral and Neural Biology, 54*, 156–174.

Castellano, C., & Pavone, F. (1988). Effects of ethanol on passive avoidance behavior in the mouse: Involvement of GABAergic mechanisms. *Pharmacology, Biochemistry and Behavior, 29*, 321–324.

De Wied, D. (1984). Neurohypophyseal hormone influences on learning and memory processes. In G. Lynch, J. L. McGaugh, & N. M. Weinberger (Eds.), *Neurobiology of learning and memory* (pp. 289–312). New York: Guilford Press.

Fanelli, R. J., Rosenberg, R. A., & Gallagher, M. (1985). Role of noradrenergic function in the opiate antagonist facilitation of spatial memory. *Behavioral and Neural Biology, 50*, 112–119.

Gaffan, D., & Harrison, S. (1987). Amygdalectomy and disconnection in visual learning for auditory secondary reinforcement by monkeys. *Journal of Neuroscience, 7*, 2285–2292.

Gallagher, M. (1982). Naloxone enhancement of memory processes: Effects of other opiate antagonists. *Behavioral and Neural Biology, 35*, 375.

Gallagher, M. (1985). Re-viewing modulation of learning and memory. In N. M. Weinberger, J. L. McGaugh, & G. Lynch (Eds.), *Memory systems of the brain: Animal and human cognitive processes* (pp. 311–334). New York: Guilford Press.

Gallagher, M., & Kapp, B. S. (1978). Manipulation of opiate activity in the amygdala alters memory processes. *Life Sciences, 23*, 1973–1978.

Gallagher, M., Kapp, B. S., Pascoe, J. P., & Rapp, P. R. (1981). A neuropharmacology of amygdaloid systems which contribute to learning and memory. In Y. Ben-Ari, *The amygdaloid complex* (pp. 343–354). Amsterdam: Elsevier/North-Holland.

Gallagher, M., Rapp, P. R., & Fanelli, R. J. (1985). Opiate antagonist facilitation of time-dependent memory processes: Dependence upon intact norepinephrine function. *Brain Research, 347*, 284–290.

Gentile, C. G., Jarrell, T. W., Teich, A., McCabe, P. M., & Schneiderman, N. (1986). The role of amygdaloid central nucleus in the retention of differential Pavlovian conditioning of bradycardia in rabbits. *Behavioral Brain Research, 20*, 263–273.

Gold, P. E. (1988). Plasma glucose regulation of memory storage processes. In C. D. Woody, D. L. Alkon, & J. L. McGaugh (Eds.), *Cellular mechanisms of conditioning and behavioral plasticity* (pp. 329–342). New York: Plenum Press.

Gold, P. E., & McGaugh, J. L. (1975). A single-trace, two-process view of memory storage processes. In D. Deutsch & J. A. Deutsch (Eds.), *Short-term memory* (pp. 145–158). New York: Academic Press.

Gold, P. E., & van Buskirk, R. B. (1975). Facilitation of time-dependent memory processes with posttrial epinephrine injections. *Behavioral Biology, 13*, 145–153.

Hitchcock, J. M., & Davis, M. (1987). Fear-potentiated startle using an auditory conditioned stimulus: Effect of lesions of the amygdala. *Physiology and Behavior, 39*, 403–408.

Introini-Collison, I. B., Arai, Y., & McGaugh, J. L. (1989a). Stria terminalis lesions attenutate the effects of posttraining atropine and oxotremorine on retention. *Psychobiology, 17*, 397–401.

Introini-Collison, I. B., & Baratti, C. M. (1986). Opioid peptidergic systems modulate the activity of b-adrenergic mechanisms during memory consolidation processes. *Behavioral and Neural Biology, 46*, 227–241.

Introini-Collison, I. B., Cahill, L., Baratti, C. M., & McGaugh, J. L. (1987). Dynorphin induces task specific impairment of memory. *Psychopharmacology, 15*, 171–174.

Introini-Collison, I. B., & McGaugh, J. L. (1986). Epinephrine modulates long-term retention of an aversively-motivated discrimination task. *Behavioral and Neural Biology, 45*, 358–365.

Introini-Collison, I. B., & McGaugh, J. L. (1987). Naloxone and β-endorphin alter the effects of posttraining epinephrine on memory. *Psychopharmacology, 92*, 229–235.

Introini-Collison, I. B., Nagahara, A. H., & McGaugh, J. L. (1989b). Memory-enhancement with intra-amygdala posttraining naloxone is blocked by concurrent administration of propranolol. *Brain Research, 476*, 94–101.

Iwata, J., LeDoux, J. E., Meeley, M. P., Arneric, S., & Reis, D. J. (1986). Intrinsic neurons in amygdaloid field projected to by the medial geniculate body mediate emotional responses conditioned to acoustic stimuli. *Brain Research, 383*, 195–214.

Izquierdo, I., & Graudenz, M. (1980). Memory facilitation by naloxone is due to release of dopaminergic and β-adrenergic systems from tonic inhibition. *Psychopharmacology, 67*, 265–268.

Kapp, B. S., Frysinger, R. C., Gallagher, M., & Haselton, J. R. (1979). Amygdala central nucleus lesions: Effect on heart rate conditioning in the rabbit. *Physiology and Behavior, 23*, 1109–1117.

LeDoux, J. E. (1986). Sensory systems and emotion: A model of affective processing. *Integrated Psychiatry, 4*, 237–248.

Liang, K. C., Bennett, C., & McGaugh, J. L. (1985). Peripheral epinephrine modulates the effects of posttraining amygdala stimulation on memory. *Behavioural Brain Research, 15*, 93–100.

Liang, K. C., Juler, R., & McGaugh, J. L. (1986). Modulating effects of posttraining epinephrine on memory: Involvement of the amygdala noradrenergic system. *Brain Research, 368*, 125–133.

Liang, K. C., & McGaugh, J. L. (1983a). Lesions of the stria terminalis attenuate the amnestic effect of amygdaloid stimulation on avoidance responses. *Brain Research, 274*, 309–318.

Liang, K. C., & McGaugh, J. L. (1983b). Lesions of the stria terminalis attenuate the

enhancing effect of post-training epinephrine on retention of an inhibitory avoidance response. *Behavioural Brain Research, 9,* 49–58.

Liang, K. C., McGaugh, J. L., Martinez, J. L., Jr., Jensen, R. A., Vasquez, B. J., & Messing, R.
B. (1982). Posttraining amygdaloid lesions impair retention of an inhibitory avoidance response. *Behavioural Brain Research, 4,* 237–249.

Liang, K. C., McGaugh, J. L., & Yao, H.-Y. (1990). Involvement of amygdala pathways in the
influence of posttraining amygdala norepinephrine and peripheral epinephrine
on memory storage. *Brain Research, 104,* 262–265.

McGaugh, J. L. (1966). Time-dependent processes in memory storage. *Science, 153,* 1351–
1358.

McGaugh, J. L. (1973). Drug facilitation of learning and memory. *Annual Review of
Pharmacology, 13,* 229–241.

McGaugh, J. L. (1983). Hormonal influence on memory. *Annual Review of Psychology, 34,*
297–323.

McGaugh, J. L. (1989a). Involvement of hormonal and neuromodulatory systems in the
regulation of memory storage. *Annual Review of Neuroscience, 12,* 255–287.

McGaugh, J. L. (1989b). Modulation of memory storage processes. In P. R. Solomon, G. R.
Goethals, C. M. Kelley, & B. R. Stephens (Eds.), *Memory: Interdisciplinary approaches*
(pp. 33–64). New York: Springer Verlag.

McGaugh, J. L. (1989c). Dissociating learning and performance: Drug and hormone
enhancement of memory storage. *Brain Research Bulletin, 23,* 339–345.

McGaugh, J. L., Castellano, C., & Brioni, J. (1990). Picrotoxin enhances latent extinction
of a conditioned emotional response. *Behavioral Neuroscience, 104,* 262–265.

McGaugh, J. L., & Gold, P. E. (1989). Hormonal modulation of memory. In R. B. Brush & S.
Levine (Eds.), *Psychoendocrinology* (pp. 305–339). New York: Academic Press.

McGaugh, J. L., & Herz, M. J. (1972). *Memory consolidation.* San Francisco: Albion Publishing Company.

McGaugh, J. L., Introini-Collison, I. B., Juler, R. G., & Izquierdo, I. (1986). Stria terminalis
lesions attenuate the effects of posttraining naloxone and β-endorphin on retention. *Behavioral Neuroscience, 100,* 839–844.

McGaugh, J. L., Introini-Collison, I. B., & Nagahara, A. H. (1988). Memory-enhancing
effects of posttraining naloxone: Involvement of β-noradrenergic influences in the
amygdaloid complex. *Brain Research, 446,* 37–49.

McGaugh, J. L., & Petrinovich, L. F. (1965). Effects of drugs on learning and memory.
International Review of Neurobiology, 8, 139–196.

McGaugh, J. L., Westbrook, W. H., & Thomson, C. W. (1962). Facilitation of maze learning
with posttrial injections of 5-7-diphenyl-1-3-diazadamantan-6-ol (1757 I.S.). *Journal
of Comparative and Physiological Psychology, 55,* 710–713.

Messing, R. B., Jensen, R. A., Martinez, J. L., Jr., Spiehler, V. R., Vasquez, B. J., Soumireu-
Mourat, B., Liang, K. C., & McGaugh, J. L (1979). Naloxone enhancement of
memory. *Behavioral and Neural Biology, 27,* 266–275.

Milner, B. (1966). Amnesia following operation on the temporal lobes. In C. W. M. Whitty
& O. L. Zangwill (Eds.), *Amnesia* (pp. 109–133). London: Butterworths.

Montel, H., Starke, K., & Weber, F. (1974). Influence of morphine and naloxone on the
release of noradrenaline from rat brain cortex slices. *Naunyn-Schmiedeberg's Archives
of Pharmacology, 283,* 357–369.

Murray, E. A., & Mishkin, M. (1985). Amygdala impairs crossmodal association in
monkeys. *Science, 228,* 604–606.

Nakamura, S., Tepper, J. M., Young, S. J., Ling, N., & Groves, P. M. (1982). Noradrenergic
terminal excitability: Effects of opioids. *Neuroscience Letters, 30,* 57–62.

Saunders, R. C., Murray, E. A., & Mishkin, M. (1984). Further evidence that the amygdala
and hippocampus contribute equally to recognition memory. *Neuropsychologia, 22,*
786–796.

Sternberg, D. B., Isaacs, K., Gold, P. E., & McGaugh, J. L. (1985). Epinephrine facilitation of appetitive learning: Attenuation with adrenergic receptor antagonists. *Behavioral and Neural Biology, 44*, 447–453.

Swartzwelder, H. S., Tilson, H. A., McLamb, R. L., & Wilson, W. A. (1987). Baclofen disrupts passive avoidance retention in rats. *Psychopharmacology, 92*, 398–401.

Werling, L. L., Brown, S. R., & Cox, B. M. (1987). Opioid receptor regulation of the release of norepinephrine in brain. *Neuropharmacology, 26*, 987–996.

Westbrook, W. H., & McGaugh, J. L. (1964). Drug facilitation of latent learning. *Psychopharmacology, 5*, 440–446.

16

Arousal, Stress, and Inverted U-shaped Curves: Implications for Cognitive Function

GEORGE F. KOOB

Arousal

Arousal as a term is widely used but difficult to define. According to Berlyne (1960), arousal was defined as "alertness or intensity of attention" and was largely reflected in activity of the reticular arousal system. Arousal has also been described as a nonspecific facilitation of cortical transmission that results in alertness, responsiveness, wakefulness, and vigilance (Hebb, 1972). Finally, arousal has been characterized as a "nonspecific response to a great many different independent variables and can be measured by a number of dependent variables" (Hennessy & Levine, 1979).

Berlyne (1960) outlined three major determinants of arousal. Intensive variables reflect information about the intensity of stimulation. For example, some stimuli are more intense and elicit more arousal. Affective variables reflect emotional state, such as fear, or even rewarding conditions. The third determinant was collative variables such as novelty. For example, the amount of arousal produced by a stimulus increases with its novelty. Other collective variables included "suprisingness," suddeness of change, or incongruous stimulus patterns.

The dependent variables associated with arousal included physiological measures—such as autonomic, neuroendocrine, and electrophysiological signs—and numerous behavioral measures. Behavioral measures of arousal included locomotor activity, the sleep/wake cycle, and motivational measures.

The relationship between arousal and performance has classically been described as an inverted U-shaped function where low to moderate levels of arousal can facilitate performance, but higher levels of arousal can disrupt performance. While some have hypothesized more than one mechanism of arousal (see Broadbent, 1971; Robbins, 1984), it is the thesis of this chapter that the right side of the inverted U-shaped arousal function represents the hypothetical construct of "stress." A further thesis of this chapter is that the different behavioral manifestations of stress may reflect an overactivation of endogenous brain arousal systems, each of which may have specific and independent actions, but also act on cognitive function via a common substrate.

Stress

Stress was defined by Selye (1980) as "the nonspecific (common result) of any demand upon the body (usually, but not always noxious)." More recent definitions emphasize the critical importance of psychological variables as determinants of stress, as elegantly outlined by Mason (1971). In such a conceptual framework, stress is defined as "anything which causes an altera-tion of psychological homeostatic processes" (Burchfield, 1979). The most sensitive and most frequently used dependent variable for a state of "stress" has been an increase in the production of adrenocorticotropin (ACTH) hormone. However, such definitions and the use of this neuroendocrine dependent variable suggest an important overlap between the concepts of stress and arousal as intervening variables or hypothetical constructs. As elegantly elaborated by Hennessy and Levine (1979) in terms of psychoen-docrine measures, the pituitary adrenal system is also a sensitive and reli-able measure of arousal as well as stress. These authors argued that although the concept of arousal involves a wide range of independent and dependent variables, stress primarily refers to the excess consequences of arousal that produce medical pathology such as tissue damage and adrenal hypertrophy. However, this hypothesis can be extended to behavioral responses and psychological states in which excess arousal is reflected in deterioration of adaptive responses and a deterioration in performance, ultimately leading to psychopathology.

Neurobiological Substrates of Arousal

While activation of the pituitary-adrenal axis is clearly a reflection of arousal and stress, the importance of psychological variables in the stress response requires a neurological substrate to process the interaction of sensory stimuli with the ultimate neuroendocrine response. The presumed substrate for the afferent limb of the stress response is the brain system involved in the processing of emotion (i.e., the limbic system). Under this conceptualization, activation of the pituitary-adrenal axis should reflect activation of the limbic system. What, however, is the neurobiologic sub-strate for the "state" of arousal or stress? According to Berlyne (1960), the portion of the nervous system hypothesized to do with "alertness or inten-sity of attention" was the reticular arousal system (RAS). The RAS as con-ceptualized here was a system containing both ascending pathways to the cortex and descending pathways to the motor systems. Two projection systems were delineated. The specific projection system carried information about the exact location and quality of the stimulus all the way from a receptor to a cortical projection area. The diffuse projection system appeared to process the urgency of stimuli and ignored their finely discriminable properties. These conceptualizations about the RAS take on new meaning as knowledge of the function of some of the specific neurochemical pathways becomes available.

Catecholamines

Two major brain catecholamine systems have been implicated in the hypothetical constructs of arousal and stress, the midbrain dopamine neurons and the pontine noradrenergic system from the locus coeruleus. The forebrain dopamine systems originate in the midbrain and form two basic pathways, the mesocorticolimbic dopamine system and the nigrostriatal dopamine system. The mesocorticolimbic dopamine system projects from the ventral tegmental area to the nucleus accumbens, olfactory tubercle, septum, and frontal cortex, whereas the nigrostriatal dopamine system originates in the substantia nigra and projects to the corpus striatum. Both the mesocorticolimbic and nigrostriatal dopamine systems have been implicated in drug- and non-drug-induced motor behavior. 6-hydroxydopamine lesions of the nigrostriatal dopamine system blocked the focused stereotyped behavior induced by high doses of psychomotor stimulant drugs (Kelly, Seviour, & Iverson, 1975) and produced Parkinsonian impairment of performance in a sensitive reaction time task (Amalric & Koob, 1987). In contrast, 6-hydroxydopamine lesions of the mesocorticolimbic dopamine system blocked the locomotor activation induced by psychomotor stimulants (Kelly et al., 1975), produced decreased locomotor behavior in an open field (Joyce & Iversen, 1978), and abolished the development of adjunctive drinking (Robbins & Koob, 1980).

The mesocorticolimbic dopamine system has also been implicated in reward motivation and learning (Bloom, Schulman, & Koob, 1989). In particular the frontal cortex projection has been implicated in learning (Simon, Scatton, & Le Moal, 1980) and this system appears to be greatly activated during stress (Thierry, Tassin, Blanc, & Glowinski, 1976). Thus, it appears that the midbrain dopamine systems have an important role in locomotor activation and the facilitation of response initiation. Global deficits in this system result in global pathology of the ability to move, such as Parkinson's disease, and presumably in the overt manifestations of arousal. However, a high level of dementia associated with Parkinson's disease suggest that this system may also have some role in cognitive response initiation (see also specific cognitive deficits associated with dopamine receptor antagonists; Weingartner, Burns, Diebel, & LeWitt, 1984).

The major source of forebrain noradrenergic neurons can be found in the nucleus locus coeruleus located in the brainstem reticular formation at the level of the isthmus. The locus coeruleus contains only norepinephrine neurons in rats and primates and projects widely in the brain but notably to the cortex, hippocampus, and cerebellum. The locus coeruleus and its ascending forebrain projections have been implicated in a wide variety of theoretical constructs—such as reward, learning, memory, anxiety, attention, and stress. Recently, its putative role in stress and attention has led to new hypotheses regarding a role in arousal function important for cognitive function.

In early work, exposure to a variety of stressors produced decreases in brain and adrenal noradrenaline in rats (Barchas & Greedman, 1963). Exposure to stressors also increased turnover of norepinephrine (Thierry, Javoy, Glowinski, & Kety, 1968) and increased the brain levels of norepinephrine metabolism (Cesar, Hague, Sharman, & Werdinius, 1974). These results were all consistent with the hypothesis that stressors can activate central as well as peripheral noradrenergic neurons. More recent work has implicated the forebrain norepinephrine neurons in attentional mechanisms. In electrophysiological studies, increases in noradrenergic transmission produced increases in the signal to noise ratio of transmission through primary sensory afferents (Foote, Freedman, & Oliver, 1975; Segal & Bloom, 1976). Locus coeruleus neuronal activity varied during the sleep/wake cycle with almost no activity during REM sleep (Aston-Jones, 1985; Jacobs, 1987). Stimuli from various modalities stimulated locus coeruleus activity with stressful stimuli being particularly effective (Jacobs, 1987) and showing rapid habituation (Aston-Jones & Bloom, 1981; Jacobs, 1987).

In lesion studies destruction of the dorsal noradrenergic bundle has been shown to have a variety of behavioral effects. While the prolongation of extinction associated with dorsal noradrenergic bundle lesions (Mason & Iversen, 1979) has not been readily replicated (Pisa & Fibiger, 1983; Tombaugh, Pappas, Roberts, Vickers, & Szostak, 1983), the original learning deficits associated with destruction of the locus coeruleus (Anlezark, Cros, Greenway, 1973) have been replicated and exploited under special circumstances (Everitt, Robbins, & Sellen, 1989; Robbins & Everitt, 1982; Robbins, Everitt, Cole, Archer, & Mohamed, 1985; Sessions, Kant, & Koob, 1976). Dorsal bundle lesions impaired acquisition of a variety of tasks, particularly complex or stressful ones. For example, dorsal bundle lesions disrupted acquisition of conditioned suppression and a conditional discrimination but not simple, simultaneous visual discriminations (Everitt et al., 1989). More important perhaps, rats with dorsal bundle lesions showed attentional deficits in a variety of tasks. Lesion rats showed deficits in performance in a continuous performance test of vigilance when bursts of noise were presented immediately prior to expected visual stimuli or when the stimuli were made temporally unpredictable (Carli, Robbins, Evender, & Everitt, 1983).

These observations have led Robbins (1984) to propose that the dorsal noradrenergic bundle emanating from the locus coeruleus may be involved in cortical arousal or the upper mechanism of arousal as conceptualized by Broadbent (1971). According to this hypothesis, the locus coeruleus dorsal noradrenergic bundle may participate in cortical arousal by enhancing inhibitory effects and may keep the animal from becoming overly distractible. This type of mechanism would perhaps prevent the disruptive effects of arousal on discrimination (Robbins, 1984). This conceptualization is consistent with the hypothesized role of cortical norepinephrine in vigilance as generated from electrophysiological studies (Aston-Jones, 1985), and is consistent with the activation of cortical norepinephrine in stress (discussed earlier). Robbins (1984) also makes an important distinction between the role of dopamine and that of norepinephrine in arousal. Dopamine is hypothesized to be involved in activating responding, a lower mechanism of arousal. The mesocorticolimbic dopamine neurons may make responding faster in execution and increase the frequency of initiation, but they do not alter the accuracy of responding. Perhaps a similar conceptualization can be made for the role of two neuropeptides hypothesized to be involved in arousal, stress, and learning.

Corticotropin Releasing Factor

The discovery of a central nervous system activating role for corticotropin releasing factor (CRF) at the cellular, electroencephalographic, and behavioral levels has raised not only the possibility of a role for this peptide in mobilizing physiological function in the classical hypothalamic pituitary axis (Vale, Spiess, Rivier, & Rivier, 1981), but also the possibility of a neurotropic role for CRF as a primary mediator of the behavioral state of stress and the behavioral responses to stress. CRF has been localized to the central nervous system both in hypothalamus and extrahypothalamic structures. CRF cells and fibers have been found in the amygdala, bed nucleus of the stria terminalis, substantia inominata, parabrachial nucleus, and some neocortical areas (Sawchenko & Swanson, 1985).

When directly administered intracerebroventricularly (ICV) CRF produces a profound/dose-dependent activation of the electroencephalogram (EEG) (Ehlers et al., 1983). Doses of 0.015 to 0.15 nmol produce a long-lasting activation of EEG. At the cellular level, CRF produces increases in the firing rate of cells within the locus coeruleus (Valentino, Foote, & Aston-Jones, 1983). CRF also has been shown to produce a significant depolarization and excitation of hippocampal pyramidal cells (Aldenhoff, Gruol, Rivier, Vale, & Siggins, 1983).

The electrophysiological activation produced by central administration of CRF is paralleled by a dose-dependent activation of behavior. Both ovine and rat CRF are equally active in stimulating locomotor activity of rats in a familiar environment (Koob et al., 1984; Sherman & Kalin, 1987; Sutton,

Koob, Le Moal, Rivier, & Vale, 1982). These effects appear to be independent of direct mediation by the pituitary-adrenal system because they were observed in hypophysectomized and dexamethasone-treated rats (Britton, Lee, Dana, Risch, & Koob, 1986; Britton, Norela, Garcia, & Rivier, 1986; Eaves, Britton, Rivier, Vale, & Koob, 1985). The locomotor activation caused by CRF is not antagonized by the opiate antagonist, naloxone, or by low doses of a dopamine receptor antagonist (Koob et al., 1984). Nor is this activation reversed by 6-hydroxydopamine lesions of the region of the nucleus accumbens, lesions that reverse the locomotor-stimulant effects of indirect sympathomimetics such as amphetamine and cocaine (Swerdlow, Geyer, Vale, & Koob, 1986). The activating effects of CRF do appear to depend on forebrain CRF receptors in that a cold-cream plug of the cerebral aqueduct will block the locomotor activation produced by cisterna magna but not lateral ventricle injections of CRF (Tazi, Swerdlow, Le Moal, Rivier, Vale, & Koob, 1987).

CRF also appears to potentiate behavioral responses associated with exposure to stressors. For example, CRF decreases eating and the approach to food in deprived rats exposed to an open field with food available (Britton, Koob, Rivier, & Vale, 1982) and decreases exploratory behavior in nondeprived rats exposed to a variety of open field situations (Berridge & Dunn, 1986; Sutton et al., 1982; Takahashi, Kalin, Vandenburg, & Sherman, 1989). CRF also enhances shock-elicited "freezing" behavior (Sherman & Kalin, 1988), the acoustic startle response (Swerdlow et al., 1986) and shock-induced fighting (Tazi, Dantzer, Le Moal, Rivier, Vale, & Koob, 1987). In operant situations CRF produces effects in a conflict test opposite to those of benzodiazepines (Britton, Morgan, Rivier, Vale, & Koob, 1985) and enhances the suppression produced by the conditioned stimulus in a conditioned suppression test.

More recent work using a CRF antagonist, α-helical CRF 9-41 which has 10 times the affinity of CRF for the CRF-binding site, has shown effects opposite to these of CRF. The CRF antagonist can also block the effects of stressors on behavior. α-Helical CRF attenuates the suppression of feeding produced by restraint stress (Krahn, Gasnell, Grace, & Levine, 1986), attenuates shock-induced fighting (Tazi, Dantzer, Le Moal, Rivier, Vale, & Koob, 1987), reverses the suppression of exploratory behavior produced by restraint stress (Berridge & Dunn, 1987; Takahashi et al., 1989), and attenuates the acquisition of conditioned suppression (Cole et al., 1989).

The neural substrate for these stress-enhancing effects of CRF may involve, at least in part, activation of the locus coeruleus–cortex noradrenergic system. Systemic administration of propranolol reverses the effects of CRF on CS responding in a conditioned suppression test (Cole & Koob, 1988). Similar effects were found with central (ICV) administration of a nonlipophilic α₂-agonist, suggesting an effect on central noradrenergic systems. Such an agonist would presumably block the firing of the locus coeruleus. Other recent work suggests that the locus coeruleus (the site of origin of forebrain norepinephrine) may be particularly sensitive to the suppression of exploratory behavior observed with central administration of CRF (Butler et al., 1988). The reversal of response inhibition by blockade of noradrenergic receptors has some behavioral specificity in that propranolol does not reverse but actually potentiates CRF-induced locomotion (Cole & Koob, 1988).

The results describing neuronal activation, behavioral activation, and stress-enhancing actions of exogenous CRF all suggest a possible role for CRF as a fundamental activating system. However, the results with the CRF antagonist suggest that endogenous central CRF systems may become active in situations involving response inhibition, particularly when the situation contingencies are novel and where some degree of stress is involved. The possibility that endogenous CRF systems are involved in the activating properties of stress needs further exploration.

These results also indicate that the sites of action for CRF in the response activating versus the response inhibiting effects of stressors may be different. The response activating effects may involve an action in the forebrain (cold-cream plug experiment) whereas the response inhibiting properties of CRF may involve an action in the hindbrain (pons) possibly associated with activation of central cortical noradrenergic systems (propranolol and locus coeruleus experiments).

The functional significance of an endogenous central nervous system CRF neuronal network may have developed as a means for an organism to mobilize not only the pituitary adrenal system but also various levels of the central nervous system. Thus, CRF may act on both levels of arousal—not only mobilizing response activation but also activating systems necessary for response selection. Such a system would be well placed to provide a major contribution to the state of arousal and, one may speculate, would be a likely mechanism for the shift from a state of high-intensity arousal to a state of stress.

Vasopressin

Arginine vasopressin, or antidiuretic hormone, has been known for its two major peripheral hormonal actions: to retain water from the kidney, and to increase blood pressure. However, some of the earliest behavioral studies on peptides showed that arginine vasopressin (AVP) administered systemically could prolong extinction of active avoidance (De Wied, 1971), and could improve retention of inhibitory (passive) avoidance when injected post-training (Bohus, Kovacs, & De Wied, 1978). These "memory"-enhancing effects of systemic AVP, have now been replicated in numerous studies and by numerous groups (Koob et al., 1985), and these effects have been observed in appetitive as well as in aversively motivated tasks (Dantzer, Bluthe, Koob, & Le Moal, 1987). These effects were originally interpreted as suggesting that AVP has a physiological role in "memory," particularly memory consolidation, and that this action is mediated directly in the central nervous system (De Wied & Bohus, 1979).

More recent work with vasopressin antagonist peptides has questioned this original hypothesis. Most if not all of the behavioral actions of systemically administered AVP to date have been reversed by pressor AVP antagonist analogs of AVP (Koob, 1987). Based on these observations, the behavioral effects of systemically administered AVP were hypothesized to be mediated by the arousal changes (Sahgal, 1988) secondary to peripheral visceral changes such as increases in blood pressure (Le Moal et al., 1981). Support for this hypothesis has come from several studies showing that only ICV doses of the AVP antagonist large enough to reverse systemic blood pressure effects of AVP were effective in reversing the behavioral effects of systemic AVP (Ettenberg, 1984; Lebrun, Le Moal, Koob, & Bloom, 1985). While this position is still controversial (De Wied, Gaffori, Van Ree, & de Jong, 1984; Koob et al., 1989), there is some consensus that the systemic doses of AVP itself do produce effects on blood pressure that can be of behavioral significance.

There is also widespread consensus that AVP also has "memory"-enhancing effects when injected directly into the brain in nanogram quantities (De Wied, 1976; Kovacs, Bohus, & Versteeg, 1979), doses that do not increase systemic blood pressure (Koob et al., 1986). A physiological role for memory processes via direct action on the CNS was suggested by the observation that a specific vasopressin antiserum injected ICV posttraining in inhibitory avoidance produced deficits in subsequent retention trials (Van Wimersma Greidanus, Docterom, De Wied, 1975). Also, local intracerebral injection of AVP into areas such as the dorsal septal nucleus, dorsal hippocampus and parafascicular nucleus facilitates retention of inhibitory (passive) avoidance (Kovacs, Bohus, & Versteeg, 1979), and destruction of these structures, as well as of the noradrenergic projections to them, blocks the facilitatory effects of AVP on avoidance behavior (Kovacs, Bohus, Versteeg, De Kloct, & De Wied, 1979; Van Wimersma Greidanus & De Wied, 1976; Van Wimersma Greidanus, Croiset, Bakkar, & Bouman, 1979). Similar facilitatory effects on memory have been observed in an appetitively motivated social memory task with microinjections of AVP into the lateral septum (Dantzer, Koob, Bluthe, & Le Moal, 1988). In this study opposite effects ("memory" impairment) were observed with microinjections of a hydrophilic AVP antagonist into the lateral system.

These findings suggest that vasopressinergic innervation of certain limbic structures (i.e., septum, hippocampus, amygdala) may be involved in the modulation of "memory" consolidation. This effect is consistent with the actions of systemic AVP, but may result from completely different mechanisms. Where systemic AVP appears to involve actions of visceral afferent signals, central AVP may reflect a direct neurotropic role, possibly involving an action on central noradrenergic system (Kovacs, Bohus, Versteeg, De Kloot, & De Wied, 1979).

Implications for Cognitive Function

Four major brain neurotransmitter systems have been discussed here in the context of the arousal-stress continuum and performance. Although they have been studied under different conceptual frameworks and with different hypotheses under test, these systems are drawn together by some common threads, both behaviorally and neurobiologically. First, there is the ubiquitous phenomenon of U-shaped dose-effect functions in behavioral pharmacology, particularly as related to performance in more cognitive tasks. Characteristically, drugs or peptides alter performance in a dose-dependent monotonic function up to some optimum dose, and then the effectiveness of the treatment declines in a similar monotonic function with ever-increasing doses. With treatments that have clear physiological effects, such as systematically administered vasopressin, where high doses greatly increase blood pressure, one could easily consider the descending part of the dose-effect curve as a nonspecific debilitory effect superimposed upon the normal facilitatory effect of AVP acting on a memory substrate.

However, this explanation will not explain the many cases of U-shaped functions that are observed even after central injections of very small amounts of what appear to be endocrinologically inert analogs of neurotransmitters. An alternative explanation centers on the relationship between performance and arousal described above, which is classically also a U-shaped function. Here, poor performance is observed both at low levels of arousal and at very high states of arousal. The best performance is generated at some optimal arousal state in between. However, this explanation begs the question of the actual mechanism by which these neurotransmitters alter cognitive processes. This explanation also begs the question of what is the role of arousal in cognitive function.

The thesis of this paper is that the central nervous system may contain multiple arousal systems that operate on both of the previously hypothesized mechanisms (upper and lower). Clearly dopamine forms a prototype for a lower arousal mechanism involved in response initiation. Norepinephrine systems in the cortex and hippocampus may be a prototypic attentional system involved in upper arousal mechanisms. Perhaps vasopressin can modulate this noradrenergic mechanism, particularly in the septum and hippocampus. Interestingly, both the noradrenergic and vasopressinergic systems appear most active in tasks involving aversive contingencies or stressors.

CRF, however, appears to impact on both lower and upper mechanisms of arousal. CRF can enhance motor activity as stimulant drugs do, but it can also apparently enhance behavioral responses to stress that involve possibly noradrenergic mechanisms. Under this conceptualization, CRF systems in the brain form a basic arousal system in which overactivity may contribute to what is often described as a "state of stress." What is unknown at this time is whether endogenous CRF systems have more of a role in the performance-enhancing or in the performance-decreasing actions of arousal (i.e., stress) or even whether these actions can be separated at a neurobiologic level.

How these three systems are interrelated can only be addressed at this time by preliminary speculation. Stressors and CRF activate noradrenergic systems, and it may be that the inhibition of behavior associated with stressors may involve CRF activation of forebrain norepinephrine systems. This stressor-induced behavioral inhibition may also be responsible for the right side (descending) of the U-shaped function relating arousal versus performance.

Vasopressin, however, may be more involved in the performance-enhancing aspects of noradrenergic function, or the left side (ascending) of the U-shaped function relating arousal versus performance. Unfortunately, to date little work has been done comparing both systems in the same tasks. It would be of some interest to know if AVP could improve performance in tasks sensitive to dorsal noradrenergic bundle lesions.

These three systems, norepinephrine, CRF, and vasopressin, have another common possible site of biologic interaction. Each of these systems has a peripheral hormonal role that may be related to its central nervous system function. Norepinephrine is a sympathetic hormone involved in autonomic activation and autonomic responses to stress. CRF is the hypothalamic releasing factor thought to be primarily involved in the pituitary adrenal response to arousal and stress. Vasopressin acts as antidiuretic hormone and has been hypothesized to have a physiological role in blood pressure regulation in situations of stress. It is interesting to speculate that these systems may have evolved from basic hormonal regulators of homeostasis to a similar role in the central nervous system, a role that at a functional level may be homologous to their peripheral role in activation and arousal.

In summary, arousal is conceptualized as an intensity of attention and in its extreme transforms antecedents and consequences to the hypothetical construct of stress. There may be multiple levels of arousal, and at least two are reflected in what Robbins (1984) has described as a noradrenergic-cortical upper level mechanism and a dopaminergic subcortical lower level mechanism. Two neuropeptides, AVP and CRF, may interface with the upper level noradrenergic mechanism and act at different levels of the arousal-stress continuum. AVP is hypothesized to modulate the performance-enhancing effects of norepinephrine activation, whereas CRF is hypothesized to drive the performance-inhibiting effects of norepinephrine activation. CRF may also act at lower level arousal mechanisms, and as a basic arousal system may play a key role in behavioral responses to stress.

Acknowledgments

Preparation of this manuscript was supported in part by Grant DK 26741 from NIADDKD and the Catherine T. and John MacArthur Foundation.

References

Aldenhoff, J. B., Gruol, D. L., Rivier, J., Vale, W., & Siggins, G. R. (1983). Corticotropin releasing factor decreases post-burst hyperpolarizations and excites hippocampal pyramidal neurons in vitro. *Science, 221*, 875–877.

Amalric, M., & Koob, G. F. (1987). Depletion of dopamine in the caudate nucleus but not nucleus accumbens impairs reaction time performance in rats. *Journal of Neuroscience, 7*, 2129–2134.

Anlezark. G. M., Cros, T. J., & Greenway, A. P. (1973). Impaired learning and decreased cortical norepinephrine after bilateral locus coeruleus lesions. *Science, 181*, 682–684.

Aston-Jones, G. (1985). Behavioral functions of locus coeruleus derived from cellular attributes. *Physiological Psychology, 13*, 118–126.

Aston-Jones, G., & Bloom, F. E. (1981). Norepinephrine-containing locus coeruleus neurones in behaving rats exhibit pronounced responses to non-noxious environmental stimuli. *Journal of Neuroscience, 1*, 887–900.

Barchas, J. D., & Greedman, D. X. (1963). Brain amines: Response to physiological stress. *Biochemical Pharmacology, 12*, 1232–1235.

Berlyne, D. E. (1960). *Conflict arousal and curiosity.* New York: McGraw-Hill.

Berridge, C. W., & Dunn, A. J. (1986). Corticotropin-releasing factor elicits naloxone sensitive stress-like alterations in exploratory behavior in mice. *Regulatory Peptides, 16*, 83–93.

Berridge, C. W., & Dunn, A. J. (1987). A corticotropin-releasing factor antagonist reverses the stress-induced changes of exploratory behavior in mice. *Hormones and Behavior, 21*, 393–401.

Bloom, F. E., Schulman, J. A., & Koob, G. F. (1989). Catecholamines and behavior. In *Catecholamines II. Handbook of experimental pharmacology* (Vol. 90/II, p. 27), Berlin: Springer Verlag.

Bohus, B., Kovács, G. L., & De Wied, D. (1978). Oxytocin, vasopressin and memory: Opposite effects on consolidation and retrieval processes. *Brain Research, 157*, 414–417.

Britton, D. R., Koob, G. F., Rivier, J., & Vale, W. (1982). Intraventricular corticotropin-releasing factor enhances behavioral effects of novelty. *Life Sciences, 31*, 363–367.

Britton, K. T., Lee, G., Dana, R., Risch, S. C., & Koob, G. F. (1986). Activating and "anxiogenic" effects of CRF are not inhibited by blockade of the pituitary-adrenal system with dexamethasone. *Life Sciences, 39*, 1281–1286.

Britton, K., Morgan, J., Rivier, J., Vale, W., & Koob, G. F. (1985). Chlordiazepoxide attenuates CRF-induced response suppression in the conflict test. *Psychopharmacology, 86*, 170–174.

Britton, D. R., Varela, M., Garcia, A., & Rivier, J. (1986). Dexamethasone suppresses pituitary-adrenal but not behavioral effects of centrally administered CRF. *Life Sciences, 38*, 211–216.

Broadbent, D. E. (1971). *Decision and stress,* New York: Academic Press.

Burchfield, S. (1979). The stress responses: A new perspective. *Psychosomatic Medicine, 41*, 661–672.

Butler, P. D., Weiss, J. M., Stout, J. C., & Nemeroff, C. B. (1990). Corticotropin-releasing factor produces anxiogentic and behavioral activating effects following microinfusion into the locus coeruleus. *Journal of Neuroscience, 10*, 176–183.

Carli, M., Robbins, T. W., Evenden, J. L., & Everitt, B. J. (1983). Effects of lesions to ascending noradrenergic neurons on performance of a 5-choice serial reaction time task in rats: Implications for theories of dorsal noradrenergic bundle function based on selective attention and arousal. *Behavioral Brain Research, 9*, 361–380.

Cesar, P. M., Hague, P., Sharman, D. F., & Wardinius, B. (1974). Studies on the metabolism of catecholamines in the central nervous system of the mouse. *British Journal of Pharmacology, 51*, 187–195.

Cole, B. J., Britton, K. T., Rivier, C., Rivier, J., Vale, W., & Koob, G. F. (1989). *Corticotropin releasing factor and conditioned fear.* Unpublished manuscript.

Cole, B. J., & Koob, G. F. (1988). Propranolol antagonizes the enhanced conditioned fear produced by corticotropin releasing factor. *Journal of Pharmacology and Experimental Therapy, 247*, 902–910.

Cole, B. J., & Koob, G. F. (1989). Low doses of corticotropin releasing factor potentiate amphetamine induced stereotyped behavior. *Psychopharmacology, 99*, 27–33.

Dantzer, R., Bluthe, R. M., Koob, G. F., & Le Moal, M. (1987). Modulation of social memory in male rats by neurohypophyseal peptides. *Psychopharmacology, 91*, 363–368.

Dantzer, R., Koob, G. F., Bluthe, R. M., & Le Moal, M. (1988). Septal vasopressin modulates social memory in male rats. *Brain Research, 457*, 143–147.

De Wied, D. (1971). Long term effect of vasopressin on the maintenance of a conditioned avoidance response in rats. *Nature, 232*, 58–60.

De Wied, D. (1976). Behavioral effects of intraventricularly administered vasopressin and vasopressin fragments. *Life Sciences, 19*, 685–690.

De Wied, D., & Bohus, B. (1979). Modulation of memory processes by neuropeptides of hypothalamic-neurohypophyseal origin. In M. A. B. Brazier (Ed.), *Brain mechanisms in memory and learning: From the single neuron to man* (pp. 139–149). New York: Raven Press.

De Wied, D., Gaffori, O., Van Ree, J. M., & de Jong, W. (1984). Central target for the behavioral effects of vasopressin neuropeptides. *Nature, 308*, 276–278.

Eaves, M., Britton, K. T., Rivier, J., Vale, W., & Koob, G. F. (1985). Effects of corticotropin releasing factor on locomotor activity in hypophysectomized rats. *Peptides, 6*, 923–926.

Ehlers, C. L., Henriksen, S. J., Wang, M., Rivier, J., Vale, W., & Bloom, F. E. (1983). Corticotropin releasing factor produces increases in brain excitability and convulsive seizures in the rat. *Brain Research, 278*, 332–336.

Ettenberg, A. (1984). Intracerebroventricular application of a vasopressin antagonist of vasopressin prevents both the "memory" and "aversive" actions of vasopressin. *Behavioral Brain Research, 14*, 201–211.

Everitt, B. J., Robbins. T. W., & Selden, N. R. W. (1990). Functions of the locus ceruleus noradrenergic system: A neurobiological and behavioural synthesis. In D. J. Heal & C. A. Marsden (Eds.), *The pharmacology of noradrenaline in the CNS.* New York: Oxford University Press.

Foote, S. L., Freedman, R., & Oliver, A. P. (1975). Effects of putative neurotransmitters on neuronal activity in monkey auditory cortex. *Brain Research, 86*, 229–242.

Hebb, D. O. (1972). *Textbook of psychology.* Philadelphia: Saunders.

Hennessy, J. W., & Levine, S. (1979). Stress, arousal, and the pituitary-adrenal system: A psychoendocrine hypothesis. In *Progress in psychobiology and physiological psychology* (8th ed., pp. 133–178). New York: Academic Press.

Jacobs, B. L. (1987). Brain monoaminergic unit activity in behaving animals. *Progress in psychobiology and physiological psychology* (Vol. 12, pp. 171–206).

Joyce, E. M., & Iversen, S. D. (1978). The effect of 6-hydroxydopamine lesions to mesolimbic dopamine terminals on spontaneous behaviour in the rat. *Neuroscience Letters (Supplement), 2*, 289.

Kelly, P. H., Seviour, P. W., & Iversen, S. D. (1975). Amphetamine and apomorphine responses in the rat following 6-OHDA lesions of the nucleus accumbens septi and corpus striatum. *Brain Research, 94*, 507–522.

Koob, G. F. (1987). Neuropeptides and memory. In L. L. Iversen, S. D. Iversen, & S. H. Snyder (Eds.), *Handbook of psychopharmacology. Vol. 19, Behavioral pharmacology, an update* (pp. 531–573). New York: Plenum Press.

Koob, G. F., Dantzer, R., Bluthe, R. M., Lebrun, C., Bloom, F. E., & Le Moal, M. (1986). Central injections of arginine vasopressin prolong extinction of active avoidance. *Peptides, 7*, 213–218.

Koob, G. F., Lebrun, C., Dantzer, R., & Le Moal, M. (1989). Role of neuropeptides in learning versus performance: Focus on vasopressin. *Brain Research Bulletin, 23*, 359–364.

Koob, G. F., Lebrun, C., Martinez, J. L., Jr., Dantzer, R., Le Moal, M., & Bloom, F. E.. (1985). Arginine vasopressin, stress and memory. In R. W. Schrier (Ed.), (pp. 195–201). New York: Raven Press.

Koob, G. F., Swerdlow, N., Seelingson, M., Eaves, M., Sutton, R., Rivier, J., & Vale, W. (1984). CRF-induced locomotor activation is antagonized by alpha flupenthixol but not naloxone. *Neuroondocrinology, 39*, 459–464.

Kovacs, G. L., Bohus, B., & Versteeg, D. H. G. (1979a). Facilitation of memory consolidation by vasopressin: Mediation by terminals of the dorsal noradrenergic bundle? *Brain Research, 172*, 73–85.

Kovacs, C. L., Bohus, B., Versteeg, D. H. G., De Kloet, E. R., & De Wied, D. (1979b). Effect of oxytocin and vasopressin on memory consolidation sites of action and catecholaminergic correlates after micro injection into limbic-midbrain structures. *Brain Research, 175*, 303–314.

Krahn, D. D., Gosnell, B. A., Grace, M., & Levine, A. S. (1986). CRF antagonist partially reverses CRF- and stress-induced effects on feeding. *Brain Research Bulletin, 17*, 285–289.

Lebrun, C., Le Moal, M., Koob, G. F., & Bloom, F. E. (1985). Vasopressin pressor antagonist injected centrally reverses peripheral behavioral effects of vasopressin but only at doses that reverse increases in blood pressure. *Regulatory Peptides, 11*, 173–181.

Le Moal, M., Koob, G. F., Koda, L. Y., Bloom, F. E., Manning, M., Sawyer, W. H., & Rivier, J. (1981). Vasopressin antagonist peptide: Blockade of pressor receptor prevents behavioral action of vasopressin. *Nature, 291*, 491–493.

Mason, J. W. (1971). A re-evaluation of the concept of "non-specificity" in stress specificity in stress theory. *Journal of Psychiatric Research, 8*, 323–333.

Mason, S. T., & Iversen, S. D. (1979). Theories of the dorsal bundle extinction effect. *Brain Research, 1*, 107–137.

Pisa, M., & Fibiger, H. C. (1983). Evidence against a role of the rats dorsal noradrenergic bundle in selective attention and place memory. *Brain Research, 272*, 319–329.

Robbins, T. W. (1984). Cortical noradrenaline, attention and arousal. *Psychological Medicine, 14*, 13–21.

Robbins, T. W., & Everitt, B. J. (1982). Functional studies of the central catecholamines. In *International Review of Neurobiology*, (Vol. 23, pp. 303–365). New York: Academic Press.

Robbins, T. W., Everitt, B. J., Cole, B. J., Archer, T., & Mohammed A. (1985). Functional hypotheses of the coeruleocortical noradrenergic projection: A review of recent experimentation and theory. *Physiological Psychology, 13*, 127–150.

Robbins, T. W., & Koob, G. F. (1980). Selective disruption of displacement behaviour by lesions of the mesolimbic dopamine system. *Nature, 285*, 409–412.

Sahgal, A. (1988). A critique of the vasopressin memory hypothesis. *Psychopharmacology, 83*, 215–228.

Sawchenko, P. R., & Swanson, L. W. (1985). Localization, co-localization and plasticity of corticotropin-releasing factor immunoreactivity in rat brain. *Federation Proceedings, 44*, 221–227.

Segal, M., & Bloom, F. E. (1976). The action of norepinephrine in the rat hippocampus. IV. The effects of locus coeruleus stimulation on evoked hippocampal unit activity. *Brain Research, 107*, 513–525.

Selye, H. (1980). In *Selye's guide to stress research* (p. v). New York: Van Nostrand Reinhold.

Sessions, G. R., Kant, G. J., & Koob, G. F. (1976). Locus coeruleus lesions and learning in the rat. *Physiology and Behavior*, *17*, 853–859.

Sherman, J. E., & Kalin, N. H. (1987). The effects of ICV-CRH on novelty-induced behavior. *Biochemistry and Biophysics Research Communication*, *26*, 699–703.

Sherman, J. E., & Kalin, N. H. (1988). ICV-CRM alters stress-induced freezing behavior without affecting pain sensitivity. *Pharmacology, Biochemistry and Behavior*, *30*, 801–807.

Simon, H., Scatton, B., & Le Moal, M. (1980). Dopaminergic A10 neurones are involved in cognitive functions. *Nature*, *286*, 150–151.

Sutton, R. E., Koob, G. F., Le Moal, M., Rivier, J., & Vale, W. (1982). Corticotropin-releasing factor (CRF) produces behavioral activation in rats. *Nature*, *297*, 331–333.

Swerdlow, N. R., Geyer, M. A., Vale, W. A., & Koob, G. F. (1986). Corticotropin releasing factor potentiates acoustic startle in rats: Blockade by chlordiazepoxide. *Psychopharmacology*, *88*, 147–152.

Takahashi, L. K., Kalin, N. H., Vandenburgt, J. A., & Sherman, J. E. (1989). Corticotropin-releasing factor modulates defensive withdrawal and exploratory behavior in rats. *Behavioral Neuroscience*, *103*, 648–654.

Tazi, A., Dantzer, R., Le Moal, M., Rivier, J., Vale, W., & Koob, G. F. (1987). Corticotropin-releasing factor antagonist blocks stress-induced fighting in rats. *Regulatory Peptides*, *18*, 37–42.

Tazi, A., Swerdlow, N. R., Le Moal, M., Rivier, J., Vale, W., & Koob, G. F. (1987). Behavioral activation of CRF: Evidence for the involvement of the ventral forebrain. *Life Science*, *41*, 41–49.

Thierry, A. M., Javoy, F., Glowinski, J., & Kety, S. S. (1968). Effects of stress on the metabolism of norepinephrine, dopamine and serotonin in the central nervous system of the rat. I. Modifications of norepinephrine turnover. *Journal of Pharmacology and Experimental Therapy*, *163*, 163–171.

Thierry, A. M., Tassin, J. P., Blanc, G., & Glowsinksi, J. (1976). Selective activation of the mesocortical dopaminergic system by stress. *Nature*, *263*, 242–244.

Tombaugh, T. N., Pappas, B. A., Roberts, D. C. S., Vickers, G. J., & Szostak, C. (1983). Failure to replicate the dorsal bundle extinction effect: Telencephalic norepinephrine depletion does not reliably increase resistance to extinction but does augment gustatory neophobia. *Brain Research*, *261*, 231–242.

Vale, W., Spiess, J., Rivier, C., & Rivier, J. (1981). Characterization of a 41 residue ovine hypothalamic peptide that stimulates the secretion of corticotropin and beta-endorphin. *Science*, *213*, 1394–1397.

Valentino, R. J., Foote, S. L., & Aston-Jones, G. (1983). Corticotropin-releasing factor activates noradrenergic neurons of the locus coeruleus. *Brain Research*, *270*, 363-367.

van Wimerama Greidanus, T. B., Crosiet, G., Bakker, I., & Bouman, H. (1979). Amygdaloid lesions block the effect of neuropeptides (vasopressin, ACTH[4-10]) on avoidance behavior. *Physiology and Behavior*, *22*, 291–295.

van Wimerama Greidanus, T. B., & De Wied, D. (1976). Dorsal hippocampus: A site of action of neuropeptides on avoidance behavior? *Pharmacology, Biochemistry and Behavior*, *5* (*Supplement 1*), 29–33.

van Wimerama Griedanus, T. B., Docterom, J., & De Wied, D. (1975). Intraventricular administration of antivasopressin serum inhibits memory consolidation in rats. *Life Sciences*, *16*, 637–644.

Weingartner, H., Burns, S., Diebel, R., & LeWitt, P. (1984). Cognitive impairments in Parkinson's disease: Distinguishing between effort-demanding and automatic cognitive processes. *Psychiatry Research*, *11*, 223–285.

17

Anxiety and Cognitive Functioning:
A Multifaceted Approach

MICHAEL W. EYSENCK

Historical Background

The topic of anxiety is one that has been considered from a variety of perspectives over the years. However, there would probably be general agreement with Lang (1971) that it is useful in terms of measurement to distinguish among behavioral, physiological, and verbal response systems. These different kinds of measure are clearly not equivalent, because there are numerous stressful situations in which these response systems fail to respond concordantly. Examples of discordance are provided by Craske and Craig (1984) and by Weinberger, Schwartz, and Davidson (1979).

While Lang's (1971) emphasis was on response systems, it is obviously necessary to consider in detail the underlying functional systems that produce the observable response measures. Of particular importance here is the interdependent functioning of the cognitive and physiological systems. However, an understanding of the major functions of the cognitive and physiological systems in anxiety considered separately will probably be required before their interdependent functioning can be understood. Accordingly, the major goal of the theoretical perspective presented in this chapter is to obtain a clearer understanding of the ways in which the cognitive system is involved in anxiety.

Anxiety and cognitive functioning have frequently been considered in terms of individual differences. The most common approach has been to compare the cognitive performance of extreme groups on the personality dimension of trait anxiety (cf. Spielberger, Gorsuch, & Lushene, 1970). These groups are sometimes compared under neutral conditions and sometimes under stressful conditions. An alternative approach (but one that has been used rarely) is to compare the cognitive performance of anxious patients with that of normal controls. Despite the paucity of relevant empirical evidence, Beck and Emery (1985) proposed an elaborate theory based on the assumption that there are numerous differences in cognitive functioning between patients with generalized anxiety disorder and normal controls.

314

This is not the place to document in detail the contributions that have been made using these approaches (see Eysenck, 1982, for a review). However, it is important to discuss some of the most consequential limitations of the previous approaches in order to understand why a new approach is required. Most researchers on trait anxiety in normals have addressed a rather narrow range of issues relating to the effects of anxiety on the efficiency of cognitive functioning. Although such research can provide information that is of relevance to academic achievement, it is rather unlikely to be informative about those aspects of cognitive functioning that play a role in the development of anxiety.

Research comparing the cognitive performance of currently anxious patients with that of normal controls is intrinsically limited, because it is not possible to provide an unequivocal account of any differences that are found between the two groups. In particular, it is not clear whether the direction of causality is from anxiety to cognition, or from cognition to anxiety.

A strange characteristic of research to date is that there have been essentially no attempts to integrate the findings from normals high and low in trait anxiety with those from clinically anxious patients. This is especially difficult to understand in view of the common assumption that normals who are high scorers on trait anxiety are vulnerable to anxiety neurosis.

Research Strategy

As we have already seen, a major limitation of research on the role of the cognitive system in anxiety disorders is that it provides only correlational evidence. It is thus not possible to decide whether nonnormal cognitive performance in anxious patients reflects their current mood state or whether it reflects stable characteristics associated with vulnerability to clinical anxiety. The optimal research strategy would involve the use of a prospective study in which normals who subsequently developed anxiety disorder were compared with current anxious patients and with normal controls. It could be concluded that cognitive functioning plays a part in the etiology of anxiety disorder if the functioning of the cognitive system in those normals who subsequently develop anxiety disorder resembled that of patients who had anxiety disorder at the time of testing more than did that of normals who did not later develp anxiety disorder. On the other hand, nonnormal cognitive functioning would not be regarded as either a cause or a precipitant of anxiety disorder if the premorbid group did not differ from the normal controls in their cognitive functioning.

Despite the value of prospective studies, they suffer from the disadvantages of being both expensive and time-consuming. As a consequence, the strategy adopted in the research program has involved the administration of various cognitive tasks to three groups of subjects: currently anxious patients (with a diagnosis of generalized anxiety disorder); recovered anxious patients (who previously had been diagnosed as having generalized anxiety disorder, but who have been recovered for at least six months); and normal controls. It is assumed that those cognitive measures that reflect stable characteristics associated with vulnerability to anxiety should distinguish the normal controls from both of the other two groups, whereas those cognitive measures that reflect current mood state should distinguish the currently anxious group from the other two groups.

This research strategy suffers from some potential problems of design and interpretation. "Recovery" is notoriously difficult to define unequivocally, and the findings that are obtained will probably be affected by the exact definition employed. At the interpretative level, recovered patients might exhibit the same nonnormal cognitive functioning as currently anxious patients because suffering from anxiety disorder permanently affects the cognitive system, rather than because nonnormal cognitive functioning reflects vulnerability. An additional possibility is that nonnormal cognitive functioning simply takes longer than other components of generalized anxiety disorder to revert to normal. However, the requirement that all of our recovered anxious patients must have been recovered for at least six months prior to testing reduces the plausibility of this explanation for nonnormal functioning in recovered patients.

The problems with the research strategy just discussed mean that the data obtained need to be interpreted with caution. However, there is no doubt that the inclusion of recovered anxious patients in the experimental design is of great value in attempting to resolve the complex causality issue. This is especially true if the basic strategy is extended in various ways, all of which we have either already used or intend to use in the future. First, it must be recognized that it is very difficult (or even impossible) to establish the null hypothesis that the cognitive functioning of recovered anxious patients is identical to that of normal controls. In order to achieve this, it would be essential to demonstrate that the cognitive performance of the recovered patients was the same as that of normal controls across a wide range of tasks under both stressful and neutral conditions.

Second, the basic strategy measures only the end points in the recovery process—that is, before any recovery has occurred and six months after recovery is apparently complete. Much more information can be obtained by conducting a longitudinal study in which cognitive functioning, mood state, and degree of recovery are all assessed repeatedly during the treatment period and beyond. It is, of course, exceptionally difficult to determine causality in this area, but it is clearly relevant to discover whether cognitive functioning reverts to normal before or after other measures of mood state and recovery indicate that recovery has occurred. If some aspect of cognitive functioning reverts to normal before other measures do, then it is at least possible that altered cognitive functioning has played some part in the overall recovery process.

Third, it is assumed that high levels of trait anxiety predispose to clinical anxiety. There is little directly relevant evidence. However, McKeon, Roa, and Mann (1984) considered groups of obsessive-compulsive patients having either highly anxious or nonanxious premorbid personalities. The highly anxious group had experienced only half as many life events as had the latter group during the 12 months preceding the onset of the disorder, which suggests that those high in trait anxiety are more vulnerable to stress than other people are, and therefore that relatively few life events may be necessary to precipitate anxiety disorder. If normals high in trait anxiety are, indeed, vulnerable to clinical anxiety, then it is possible to relate findings from normal subjects high in trait anxiety to those from recovered anxious patients. In essence, if there is some aspect of cognitive functioning that is nonnormal in recovered anxious patients, it might be anticipated that that aspect would also be nonnormal in normals having high trait anxiety. If those results were obtained, they would certainly strengthen the argument that the aspect of cognitive functioning in question reflected a vulnerability factor.

Throughout the research program, there has been an emphasis on investigating individual differences in the processing of threat-related stimuli. The reason for this is that it is assumed that clinically important differences between anxious patients and normal controls in cognitive functioning are more likely to be obtained when threat-related stimuli rather than neutral stimuli are presented. Since anxious patients typically report physical health and/or social concerns, most of the experiments in the research program have involved the presentation of threatening words relating to both physical health (e.g., *mutilated*) and social concerns (e.g., *inferior*).

The final ingredient in the research strategy concerns the selection of cognitive tasks to be used. Two major considerations have guided this selection. First, it has been assumed that there are major differences between anxious patients and normal controls in preattentive and attentional processes. As Beck and Emery (1985) pointed out, "The [anxious] patient is hyper-vigilant, constantly scanning the environment for signs of impending disaster or personal harm. . . . The anxious patient selectively attends to stimuli that indicate possible danger and becomes oblivious to stimuli that indicate that there is no danger." Second, it is a reasonable working hypothesis that those aspects of cognitive functioning that reflect a vulnerability factor are likely to involve overlearned or automatic processes, whereas those aspects of cognitive functioning that reflect current mood state will tend to involve controlled processes of which there is conscious awareness. In order to test this hypothesis, and to ensure a reasonably comprehensive assessment of cognitive functioning in clinical anxiety, we have made use of some tasks that assess automatic processes and other tasks that assess controlled processes.

The selection of tasks can be contrasted with those typically used in research on cognitive factors in clinical depression. Such research, whether using premorbid subjects or recovered depressed patients, has relied very heavily on self-report questionnaires. Since questionnaires do not assess automatic processing, it is not altogether surprising that this research has consistently failed to identify a cognitive vulnerability factor in depression (see, for example, Eaves & Rush, 1984; Lewinsohn, Steinmetz, Larson, & Franklin, 1981; Wilkinson & Blackburn, 1981). A typical conclusion was reached by Wilkinson and Blackburn (1981): "Depressed patients did not show any cognitive distortions after recovery, their scores, on all measures, being equivalent to those of normal and recovered other subjects. Hence, cognitive distortions would appear to be specific to the illness phase of depression and not to depression-prone individuals" (pp. 289–290).

Empirical Evidence

One of the major paradigms investigated in the research program is one that was first used by a student of mine, C. Halkiopoulos, and the results from which are reported in Eysenck, MacLeod, and Mathews (1987). In essence, pairs of words were presented concurrently, one to each ear. All of the words presented to one of the ears had to be shadowed, and the words presented to that ear were a mixture of threatening and neutral words. In contrast, only neutral words were presented to the other ear. Occasionally, the subject was required to respond as rapidly as possible to a tone presented to the shadowed or to the unattended ear shortly after a pair of words had been presented. What was of critical interest was the speed of responding to the tone on those trials on which a threatening and a neutral word were presented concurrently. It was assumed that this response speed would provide an approximate measure of the allocation of processing resources.

Normals scoring high on the Facilitation–Inhibition Scale (Ullmann, 1962), a questionnaire essentially measuring trait anxiety, responded very rapidly when the tone followed a threat-related word in the same ear, but rather slowly when it followed a threat-related word in the other ear. The implication is that they allocated extra processing resources to the ear to which a threat-related word had been presented. Low scorers on the Facilitation–Inhibition Scale had the opposite pattern of response latencies, indicating avoidance of the channel on which a threat-related word had just been presented.

A visual analog of the preceding paradigm has been used a number of times with patients suffering from generalized anxiety disorder (e.g., MacLeod, Mathews, & Tata, 1986). Subjects were required to read the upper word in each pair of visually presented words and to respond as rapidly as possible to a small dot that could appear in the space just vacated by either one of the words. One threat-related and one neutral word were presented on the critical trials, and the threat-related word could appear in either location. Regardless of the location, it has been found that anxious patients respond faster to the dot when it replaces a threat-related word than they do when it replaces a neutral word. This suggests that anxious patients selectively allocate processing resources to threatening sources of information. On the other hand, normal controls show the opposite tendency.

It is not clear from studies on the selective allocation of processing resources whether an automatic process is involved, or whether there might be individual differences in some strategic process. Relevant evidence was obtained by Mathews and MacLeod (1986). They utilized a dichotic listening paradigm in which neutral stories were shadowed, while threat-related and neutral words were presented concurrently in the unattended channel. In addition, subjects performed a simple visual reaction-time task, which was synchronized with the presentation of words in the unattended channel. Anxious patients had slower reaction times when the visual signal coincided with threat-related rather than neutral words, whereas control subjects had comparable response latencies with both types of unattended material. This pattern of results suggests that anxious patients have a selective processing bias favoring threatening stimuli. Since no subject was able to recognize the threat-related words presented to the unattended channel at above chance level, it is probable that this selective processing bias is automatic, at least in the sense of being independent of awareness.

A second major line of research has been concerned with the interpretation of ambiguous stimuli having both a threatening and a neutral interpretation. The first study in the series was reported by Eysenck et al. (1987). Homophones (e.g., *gilt, guilt*) having both a threatening and a neutral meaning were presented auditorily, and subjects were instructed simply to write down the spelling of each word. Normal subjects were used, and there was a significant corrrelation of + .60 between trait anxiety and the number of threatening interpretations of homophones written down. In subsequent unpublished research it has been found that trait anxiety also correlates highly with the number of threatening interpretations of homophones among currently anxious patients. It has also been discovered that currently anxious patients write down significantly more threatening interpretations than do normal controls. Recovered anxious patients have also been tested on this homophone task, and their number of threatening interpretations was intermediate between those of currently anxious groups and those of normal control groups. This makes it unclear whether the tendency of currently anxious patients to provide threatening interpretations of homophones is part of a vulnerability factor, a reflection of current mood state, or a combination of the two.

A somewhat similar paradigm has been used in recent research. Ambiguous sentences (e.g., "The two men watched as the chest was opened") having both a threatening and a neutral interpretation are presented. On a subsequent recognition-memory test, sentences representing the gist of either the threatening or the neutral interpretation of each ambiguous sentence are presented, and the subjects' task is to decide whether each sentence means the same as one of the sentences presented at acquisition. Currently anxious patients differ from normal controls in being more likely to select threatening interpretations on the recognition-memory test, presumably because of their greater tendency to interpret the ambiguous sentences in a threatening fashion. Recovered anxious patients closely resemble normal controls on this task, suggesting that a bias toward interpreting ambiguous sentences in a threatening fashion reflects anxious mood state rather than being a vulnerability factor.

Another major part of the research program has been concerned with distractibility. There is evidence in normals (e.g., Dornic & Fernaeus, 1981, Eysenck & Graydon, 1989) that those high in trait anxiety are more distractible than these low in trait anxiety. Mathews, May, Mogg, and Eysenck (submitted) investigated distractibility in generalized anxiety disorder patients. They made use of both neutral and threatening distracting stimuli, and discovered that anxious patients were generally more distractible then were normal controls. They also found that this greater distractibility was more evident with threatening than with neutral distractors. Recovered anxious patients resembled normal controls when neutral distractors were presented, but they manifested significantly greater distractibility than controls when threatening distractors were presented. The implication is that the tendency for attention to be captured by threatening stimuli forms part of a cognitive vulnerability factor for generalized anxiety disorder.

Several studies have investigated memory bias in clinical depression. The usual finding (e.g., Bradley & Mathews, 1983) is that depressed patients recall relatively more negative adjectives than normal controls do, at least when the adjectives have been encoded with respect to oneself. This negative bias effect has proved remarkably difficult to obtain in patients suffering from generalized anxiety disorder. Mogg, Mathews, and Weinman (1987) carried out several experiments, and consistently failed to demonstrate any tendency among anxious patients to exhibit a recall bias favoring negative or threatening words, and the same was true of recognition memory.

However, the findings from normal groups high and low in neuroticism (a personality dimension that correlates highly with trait anxiety) indicate that there is a negative recall bias among normals high in neuroticism. Young and Martin (1981) discovered that those high in neuroticism recalled negative rather than positive information to a greater extent than those low in neuroticism. This finding was replicated by Martin, Ward, and Clark (1983), who suggested that those high in neuroticism selectively attended to negative information about themselves at acquisition. In a recent study (Mathews, Mogg, May, & Eysenck, 1989), groups of currently anxious, recovered anxious, and normal controls were compared on their cued recall of threatening, neutral, and positive words. The groups did not differ in terms of negative recall bias. However, there was a statistically significent correlation of +.30 between trait anxiety and recall bias for the whole sample, although the correlation was not significant within the normal control group.

Why is a negative recall bias found in normals high in neuroticism or trait anxiety but not in patients suffering from generalized anxiety disorder? Presumably normals high in trait anxiety have a rich network of associations to personally threatening stimuli, and this leads to elaboration of processing and to high levels of long-term memory. The absence of a negative recall bias in currently anxious patients is more puzzling. However, their attempts to prevent anxiety becoming intolerably intense may lead them to develop cognitive avoidance strategies (e.g., selective ignoring or failure to rehearse) when confronted by threat-related stimuli.

Graf and Mandler (1984) have drawn a distinction between one memory process (implicit memory) that involves activation and is relatively automatic, and a second memory process (explicit memory) that involves elaboration and affects free and cued recall. In those terms, Mathews et al. (1989) discovered that anxiety patients do not have a negative bias with explicit memory. However, on a word-completion measure of implicit memory, anxious patients produced a greater number of threatening word completions than did normal controls, with recovered anxious patients intermediate. These findings suggests that internal representations of threat words show greater activation in anxious patients but not greater elaboration. The fact that the pattern of results was quite different for explicit and implicit memory suggest that this conceptual distinction may prove important in the attempt to elucidate differences between anxious and nonanxious groups in memory functioning.

Theoretical Implications

It is not possible at present to provide a coherent theoretical account of cognitive functioning and anxiety. However, some trends are noticeable in the data. Anxious patients with generalized anxiety disorder typically differ from normal controls in their cognitive functioning in ways that correspond to the differences between normals high and low in trait anxiety. Thus, normals high in trait anxiety selectively allocate processing resources to threat-related rather than to neutral stimuli, they tend to interpret ambiguous stimuli in a threatening fashion, they are high in distractibility, and they have a negative recall bias. Generalized anxiety patients exhibit the same pattern of cognitive functioning, except that they do not have a negative recall bias. These findings are generally consistent with the notion that normals high in trait anxiety are more vulnerable to generalized anxiety disorder than are normals low in trait anxiety.

It is important to note that the overall similarity of cognitive functioning in normals high in trait anxiety and in patients with generalized anxiety disorder has not been established previously. As was discussed earlier, previous research has dealt with cognitive functioning either in normals varying in trait anxiety level or in anxious patients, and there has hitherto been remarkably little interest in comparing anxious normals with anxious patients. The high degree of similarity between the two groups at a cognitive level may prove to be of considerable theoretical interest in the search for vulnerability factors in anxiety neurosis.

However, consideration of the findings from recovered anxious patients suggests that much more research is needed in order to establish the nature of the cognitive vulnerability factor in anxiety neurosis. Recovered anxious patients do not appear to be biased in favor of threatening interpretations of ambiguous stimuli, and they do not show a negative recall bias. Accordingly, in terms of the logic of the experimental approach being adopted, neither of these aspects of cognitive functioning forms part of a vulnerability factor. However, the position is rather different so far as distractibility to threatening stimuli is concerned. The evidence so far indicates that high distractibility to threatening stimuli characterizes currently anxious patients, recovered anxious patients, and normals high in trait anxiety. It is thus possible that a tendency toward attentional capture by threatening stimuli forms part of a vulnerability factor in generalized anxiety disorder.

Why should it be that distractibility to threatening stimuli makes an individual vulnerable to generalized anxiety disorder? Someone whose attention is captured by mildly threatening environmental stimuli will obviously find the environment to be more threatening than will someone who does not have a bias in favor of processing such stimuli. It seems reasonable that finding the environment to be subjectively threatening could play some part in the development of generalized anxiety disorder.

At a more speculative level, we can consider why it is that so many of the cognitive features associated with anxiety involve the attentional system. This issue is of particular interest because much of the evidence on clinical depression suggests that depression is associated more with the functioning of the memory than with that of the attentional system (Eysenck, in preparation). These differences may be related to important findings reported by Finlay-Jones and Brown (1981). They discovered that depressed patients indicated that past losses played an important role in the onset of depression, whereas anxious patients more frequently indicated present or future dangers as being responsible.

There is a methodological problem with the study by Finlay-Jones and Brown (1981). The nature of the events causing anxiety and depression differed, with the consequence that there is a confounding between the nature of the events and their time orientation. However, Eysenck (unpublished) obtained very similar findings with normal individuals who were asked to identify the factors that had made them either anxious or depressed in the past. He also found that some events (e.g., an important examination) could be associated with either anxiety or depression, depending on whether they lay in the past or in the future at the time that the anxiety or depression was experienced.

If anxious individuals are especially concerned about present or future dangers, then a likely strategy is to monitor the environment in order to detect the first signs of danger as soon as possible. From this perspective, it makes sense that they would have a very active engagement with the external environment, and that the attentional system would be involved. It also makes sense that anxious individuals would engage in constant scanning of the environment combined with selective allocation of processing resources to threatening and threat-related stimuli.

In contrast, if depression stems from past events that cannot be altered, it is understandable that depressed individuals would adopt the strategy of passive disengagement from the external environment. Perhaps depressed patients have a greater bias toward focusing on internal rather than external sources of information (e.g., past losses) than do anxious individuals. As a consequence, depression is associated with the functioning of the memory system rather than with that of the attentional system.

In sum, we have seen that the evidence indicates very clearly that anxious patients differ substantially in their cognitive functioning from normal controls. There are at least two major implications. First, an adequate understanding of generalized anxiety disorder must necessarily consider the cognitive system, as Beck and Emery (1985) have already argued forcibly. Second, we now have relatively detailed information concerning some of the cognitive processes and biases that are found in anxious patients. This information should be of use in the treatment of anxiety neurosis by means of cognitive therapy. This form of therapy has typically not been based on experimental research and theory, and this may well have limited its efficacy. Some of our evidence indicates that anxious patients differ from normals in cognitive processes operating below the level of conscious awareness. This suggests a major limitation of cognitive therapy, which is concerned exclusively with processes that are consciously accessible. The ultimate goal of the research program is to increase our understanding of cognitive functioning in generalized anxiety disorder to a level at which significant improvements in the effectiveness of cognitive forms of therapy can be achieved.

Acknowledgments

Grateful thanks are offered to the Wellcome Trust, whose generous financial assistance to Professor Andrew Mathews and myself enabled the research discussed in this chapter to be carried out.

References

Beck, A. J., & Emery, G. (1985). *Anxiety disorders and phobias: A cognitive perspective*. New York: Basic Books.
Bradley, B., & Mathews, A. (1983). Negative self-schemata in clinical depression. *British Journal of Clinical Psychology, 22*, 173–181.

Craske, M. G., & Craig, K. D. (1984). Musical performance anxiety: The three-systems model and self-efficacy theory. *Behavior Research and Therapy, 22*, 267–280.

Dornic, S., & Fernaeus, S.-E. (1981). Individual differences in high-load tasks: The effect of verbal distraction. *Reports of the Department of Psychology, University of Stockholm*, No. 569.

Eaves, G., & Rush, A. J. (1984). Cognitive patterns in symptomatic and remitted unipolar depressives. *Journal of Abnormal Psychology, 93*, 31–40.

Eysenck, M. W. (1982). *Attention and arousal: Cognition and performance.* Berlin: Springer Verlag.

Eysenck, M. W. (in preparation). *Anxiety: The cognitive perspective.* London: Lawrence Erlbaum.

Eysenck, M. W., & Graydon, J. (1989). Susceptibility to distraction as a function of personality. *Personality and Individual Differences, 10*, 681–687.

Eysenck, M. W., MacLeod, C., & Mathews, A. (1987). Cognitive functioning and anxiety. *Psychological Research, 49*, 189–195.

Finlay-Jones, R. A., & Brown, G. W. (1981). Types of stressful life events and the onset of anxiety and depressive disorders. *Psychological Medicine, 11*, 803–815.

Graf, P., & Mandler, G. (1984). Activation makes words more accessible, but not necessarily more retrievable. *Journal of Verbal Learning and Verbal Behavior, 13*, 553–568.

Lang, P. (1971). The application of psychophysiological methods to the study of psychotherapy and behavior modification. In A. Bergin & S. Garfield (Eds.), *Handbook of psychotherapy and behavior change* (pp. 476–503). New York: Wiley.

Lewinsohn, P. M., Steinmetz, J. L., Larson, D. W., & Franklin, Y. (1981). Depression related cognitions: Antecedents or consequences. *Journal of Abnormal Psychology, 90*, 213–219.

MacLeod, C., Mathews, A., & Tata, P. (1986). Attentional bias in emotional disorders. *Journal of Abnormal Psychology, 95*, 15–20.

Martin, M., Ward, J. C., & Clark, D. M. (1983). Neuroticism and the recall of positive and negative personality information. *Behaviour Research and Therapy, 21*, 495–503.

Mathews, A., & MacLeod, C. (1986). Discrimination of the threat cues without awareness in anxiety states. *Journal of Abnormal Psychology, 95*, 1–8.

Mathews, A., May, J., Mogg, K., & Eysenck, M. W. (1990). Attentional bias in anxiety: Selective search or defective filtering? *Journal of Abnormal Psychology, 99*, 166–173.

Mathews, A., Mogg, K., May, J., & Eysenck, M. W. (1989). Implicit and explicit memory biases in anxiety. *Journal of Abnormal Psychology, 98*, 236–240.

McKeon, J., Roa, B., & Mann, A. (1984). Life events and personality traits in obsessive-compulsive neurosis. *British Journal of Psychiatry, 144*, 185–189.

Mogg, K., Mathews, A., & Weinman, J. (1987). Memory bias in clinical anxiety. *Journal of Abnormal Psychology, 96*, 94–98.

Spielberger, C. D., Gorsuch, R., & Lushene, R. (1970). *The state trait anxiety inventory (STAI) test manual.* Palo Alto, CA: Consulting Psychologists Press.

Ullmann, L. P. (1962). An empirically derived MMPI scale which measures facilitation–inhibition of recognition of threatening stimuli. *Journal of Clinical Psychology, 18*, 127–132.

Weinberger, D. A., Schwartz, G. E., & Davidson, R. J. (1979). Low-anxious, high-anxious, and repressive coping styles: Psychometric patterns and behavioral and physiological responses to stress. *Journal of Abnormal Psychology, 88*, 369–380.

Wilkinson, I. M., & Blackburn, J. M. (1981). Cognitive style in depressed and recovered depressed patients. *British Clinical Psychology, 20*, 283–292.

Young, G. C. D., & Martin, M. (1981). Processing of information about self by neurotics. *British Journal of Clinical Psycholaogy, 20*, 205–212.

18

Emotion: Prospects for Conceptual and Empirical Development

HOWARD LEVENTHAL

My primary goal in this chapter is to elaborate a framework for the study of emotion. This framework has three basic themes. The first is that the understanding of emotional processes requires a biopsychosocial approach (Engel, 1977). The central aspect of this theme is the recognition that a comprehensive model of emotion requires analysis and theory at social, psychological, and biological levels and the specification of linkages across levels. From this perspective, no level is more "fundamental" or important than another, and explanation at any one level will always be incomplete in the absence of understanding of the others. For example, neurobiological or neurochemical "explanations" of emotional behavior cannot proceed without a clear psychological analysis of emotional phenomenon at the psychological or behavioral level. The subjective states—the expressive and instrumental reactions associated with emotion, the interrelationships between them, and the psychological processes underlying them—specify the acts, structures and functions that the biological mechanisms must perform.

Similarly, psychological analyses or models that aim to describe the psychological or computational processes that convert a stimulus into a subjective experience and/or an expressive reaction must be consistent with biological knowledge. Biological mechanisms will both define limits and suggest modes of operation for psychological mechanisms. If a psychological algorithm for the transformation of a stimulus into an emotional state is incompatible with biological operations, the psychological model must be wrong. For example, Lashley (1950) rejected psychological models of skilled performance based upon response chaining because neural conduction is too slow to control such performance by feedback from a series of smaller, chained response units; he argued that acts such as playing musical scales were controlled by a centrally organized motor template. Biological models can also suggest which responses or organized response units are hardwired and which are acquired, establishing further guidelines and limits for psychological analysis. The recognition of such cross levels constraints places one in opposition to advocates of "virtual" computing—that is, of positions stating that the critical feature of a psychological model is its ability to predict and account for behavior, the specific way in which the computing is conducted and the physical or physiological substrate in which it is done being of no significance.

The second theme is that mental processes, even those which seem unitary, reflect the joint operation of a set of componential or modular units (Gazzaniga, 1989). From this perspective, emotions (as seen in others or felt in oneself) are the product of the information-processing activity of an organized set of componential units; emotions are not disorganized nor are they states that fill in the gaps or interrupts of so-called planful behavior (Fisher, Carnochan, & Shaver, in press). Thus, anger reflects the *integrated* activity of a different set of units than those active in the generation of joy or pleasure. A componential approach requires that theory and data at all levels be consistent with the notion that relatively independent functional units are organized and integrated to generate our basic emotional competency.

The third basic theme is developmental. It suggests that both experiential and maturational factors are involved in the addition of components and their changing organization over time, leading to increased elaboration of the emotional system. These themes will be elaborated on in a later section, where I spell out the features of a "perceptual motor framework" for the study of emotion.

Before describing this framework, I will briefly review a few of the basic functions of emotions, because a credible model, componential or otherwise, must be compatible with and able to "perform" these functions. I will then review three componential approaches that have been popular in contemporary psychology, for they show why a more elaborate componential model is necessary. After that I will describe the componential revision of the "sensory-motor model," which I have previously described (Leventhal, 1979; 1980; 1984), and then will discuss future directions for research and a few implications for treatment of emotional disorders.

Basic Functions of Emotions

It is generally recognized that emotional reactions have two broadly defined sets of functions, the first external (i.e., the communication and organization of social interactions), the second internal (i.e., the organizing of response systems and communicating to consciousness an awareness of the overall organization of the behavioral system). Internal functions set the stage for and sustain behavior that can meet homeostatic and motivational needs.

At least two different though related approaches are visible in contemporary views of the external functions of emotion, one having a Darwinian emphasis, the other a learning or socialization emphasis. The Darwinian theme links emotional expression and communication with group organization and survival. For example, expressions of positive affect are said to be associated with the formation of groups and with reproduction, the expression of aggression and submission with the establishment of territory and within-group dominance hierarchies, and the expression of fear with intraspecific signaling of predatory threat (Plutchik, 1984; Rowell, 1972). As Darwin (1904/1872) pointed out, the shape of these expressive behaviors reflects both the external, communicative need, that is the need to be discriminable, and the internal, homeostatic demand of preserving organismic function so as to allow for species propagation.

The socialization theme focuses upon emotions as a signal system in which positive and negative expressions play a critical role in social communication and are elaborated upon to form more complex subjective and expressive reactions through social learning. Although emotional expressions such as smiles and expressions of distress (and fear) may be prewired, their purpose is to ensure a learning process that leads to a mutual bonding (infant to mother and mother to infant) and successful parenting. Thus, the infant's ability to learn ensures the elaboration of these emotions when they are expressed and experienced within social relationships.

The contemporary focus on the internal functions of emotion (i.e., as conscious experiences involved in communication to self) has its origin with William James's (1950/1890) hypothesis that the subjective experience of emotion is a product of the experience of somatic (that is, autonomic) reactions. This theme is alive and well despite early (Cannon, 1927, 1931) and later efforts to destroy it (Schachter & Singer, 1962). James's hypothesis seemed reasonable when little was known about the less accessible areas of the brain, which we now judge important for the control of emotional reactions, and the central nervous system was viewed as a sensory and motor system with no functions other than decoding stimuli and encoding overt reactions (James, 1950/1890). The survival of his hypothesis appears to be due to its grounding in individual experience and "intuition."

A key problem remaining from James's legacy is whether the organism is designed to monitor the activity of specific autonomic (e.g., heart rate, blood pressure, electrodermal activation) and expressive channels (facial change, body tension), or whether the organism is designed to monitor collections of these reactions as experiences of general "states"—e.g., emotions, effort, fatigue, and so on. (Leventhal & Mosbach, 1983). The very weak association between direct measures of autonomic events and reports of these events such as for blood pressure (Baumann & Leventhal, 1985; Watson & Pennebaker, 1989), the presence of individual and gender differences in ability to track autonomic function (Katkin, 1985) and the difficulty in training individuals to engage in monitoring (Lacroix, 1986) suggest either that feedback from autonomic reaction is unnecessary for the experience of emotion, or that organismic design favors the accumulation of these internal signals into the relatively undifferentiated packages represented by emotional states. This latter solution allows the organism to dedicate its on-line, limited-capacity, information-processing system to various aspects of the external world.

A substantial body of research exists on the functions of emotion as an activator and integrator of phasic reactions of the neural, hormonal, and immune systems during adaptive activities such as fight or flight. Less work exists on the positive emotions accompanying interpersonal activities such as mutual sharing. Studies have also examined the relationship of individual differences in "tonic" affective orientation or chronic mood to the development of organic dysfunctions. Examples include studies of the linkage of repressed hostility to coronary disease (Matthews, 1988), and depression to immune function (Levy et al., 1989). Indeed, the vast array of material incorporated in the study of emotion suggests that the term *emotion* is best treated as a chapter heading and index for a set of phenomena rather than a unitary scientific construct.

Finally, a number of contemporary views of emotion focus on how the homeostatic processes communicate to consciousness to prepare the organism for adapting to environmental challenges. There seems little reason to doubt that emotional expressions and experience are differentiated into basic categories such as joy, anger, fear, sadness, and disgust (Izard, 1971; 1977; Tomkins, 1962, 1963). Although individuals do not always attend to them, these differences can serve as valuable cues for behavior—e.g., depression and the fatigue accompanying it cue that one has lost valuable resources and is in need of rest and recuperation; anxiety can cue avoidance of competitive and threatening situations; and anger can cue (and signal others) that one is ready to attack.

In summary, emotions have critical functions with regard to communication and the organization of action. The communicative functions involve both the interpersonal expressive systems (primarily intraspecific) and the intrapersonal subjective system, which communicates readiness for action to the self. The external and internal aspects of communication insure the organization of interpersonal processes ranging from bonding and reproduction to territoriality and group structure. The somatic, organizing functions involve the activation and integration of phasic (and tonic) reactions of the somatic system—for example, neural, hormonal, and immune functions concerned with adaptive activities such as fight or flight, rejection of pathogens, mutual sharing, and so on. What type of emotion theory can account for this complexity?

Emotion as Component Process

Because emotions serve multiple functions, we can expect theories of emotion to be complex. We can also expect particular theories to emphasize one or another set of functions and ignore the others. As we shall see, this is indeed the case.

Traditional Component Views

My view of three currently popular approaches to the analysis of emotional processes will help to clarify what componential approaches have and have not accounted for with respect to the functions of emotion. It will also serve as a platform for the more detailed componential anaysis which follows.

THE THREE-LEVELS ANALYSIS. The first theoretical approach handles the complexity of emotion by conceptualizing emotion as a tripartite response system, consisting of subjective (typically identified by verbal report), expressive, and autonomic components (Lang, 1979; Lang, Kozak, Miller, Levin, & McLean, 1980). The distinction is important for recognizing the independence of the three systems and the considerable degree of response specificity detectable in human behavior (Lacey, 1967)—that is, people reporting anger or fear, may or may not have autonomic activation, and when they have autonomic activation they may or may not report subjective reactions or show expressive behaviors of fear. The major deficit of this approach is the absence of any clear mechanism or hypotheses to suggest the conditions that would organize the systems in different ways, or of any hypotheses regarding the functional utility of different organizations. The model was also mute about the organizations that would characterize different emotional states, though data was generated suggesting that the persistence of phobic reactions such as fear of snakes was due in part to the asynchrony between autonomic and cognitive systems—that is, the absence of autonomic response in the presence of a subjective fear response (Lang et al., 1980).

In summary, the three-levels model did little to expand our understanding of the external functions of emotion (i.e., its role in interpersonal communication) and provided only limited insight into the internal, organizing functions of emotion. This model served primarily as a platform for the study of each of the three systems.

THE COGNITION-AROUSAL MODEL. The failure to detect differences in the pattern of autonomic events for different emotions and the evidence for the differentiation of emotion at a subjective level created a major problem for emotion theory. Efforts to resolve the inconsistency—that we feel distinct emotions in the absence of objective evidence for autonomic differentiation—led to theories that argue that emotions involve the integration of separate response components—such as autonomic activity and cognition of external events or autonomic activity plus patterned feedback—from more articulate motor systems such as the face. The cognition-arousal hypothesis proposed by Schachter and Singer (1962), currently the most popular componential view of emotion, states that emotional experience is generated by the integration of a generalized state of autonomic arousal with cognitions specific to the experienced quality of each emotion. The undifferentiated autonomic arousal is *necessary* to create the subjective experience of emotion, and the specific, situational cognitions (e.g., the presence of an attacker versus the presence of a loved one), *are necessary to generate the qualitative aspect of emotion* (i.e., that one feels fear, anger, sadness, etc.). In this hypothesis, cognition is not just a stimulus to emotional reactions, it is the mental event that *is*, and that is equivalent to the quality (color or hue) of emotion. From the perspective of the cognition-arousal hypothesis, subjective feelings are critical markers of emotion, though expressive behaviors are also used as an indicator.

The major contribution of the cognition-arousal model has been the exploration of the social-environmental or cognitive factors responsible for emotional experience and behavior. Studies generated by this framework have been especially valuable in demonstrating the importance of the social context for the definition or resolution of ambiguous emotional states, an important contribution given the ambiguity of many social encounters. The model runs into serious difficulty, however, in at least four areas. First, both human and animal data have cast doubt on the hypothesis that autonomic participation is necessary for emotional experience and behavior (see Leventhal, 1980; Reisenzein, 1983). Second, the model makes the very clear suggestion that a socialization history is essential to create the cognitions essential for the quality of emotional feelings. The data on the imitation of distinct, adult, emotional expressions in the newborn (Field, Woodson, Greenberg, & Cohen, 1982) and the data on the similarity of emotional expressions and judgments across widely separated cultures are clearly inconsistent with the socialization hypothesis (i.e., with its historical aspect), though they do not directly contradict it. There also seems little reason to identify the quality of emotion with cognition: The two seem quite different and the organization of the nervous system suggests the two sets of events are the product of separable processing components (Leventhal, 1970; Zajonc, 1980).

The model also fails to account for the many functions emotion is presumed to serve. While it clarifies the role of the social environment in shaping individual emotional experience, it does little to clarify the role of expressions and feelings in social behavior and the creation of social organization. The model also fails to address the internal, organizing functions of emotion. Thus, although it has stimulated research, there is little reason to believe it is a valid representation of the emotion mechanism.

MODELS BASED ON FACIAL EXPRESSIONS. A third set of theorists have suggested that expressive reactions hold the key to both the behavioral and subjective differentiation of emotional states. Studies of expressive behavior have focused primarily on two functions: the role of expression in interpersonal behavior and the role of expression in the generation of subjective feelings. With regard to the latter, the hypothesis that efference from specific patterns of facial expression generates the qualitative experience of emotion has generated considerable research (Izard, 1971; Laird, 1984; Strack, Martin, & Stepper, 1988; Tomkins, 1962, 1963). While some theorists in this camp suggest that feedback from expressive behavior is sufficient for emotional experience, none would argue that it is typical for emotions to be experienced on the basis of facial feedback alone.

There are at least three reasons for rejecting the facial feedback hypothesis. First, if feedback from facial muscles, the skin, or the vasculature (see Zajonc, 1985) were *the primary* source of emotional states (both their subjective quality and motivational push), one might expect that it would be relatively simple to demonstrate this primacy and that variations in expression would lead to very substantial changes in feeling; neither is the case (Leventhal, 1980). Second, these hypotheses do not distinguish between the consequences of feedback from the same expressive pattern, such as a smile, when the response is spontaneous or volitional. Neither logic nor data support the idea that these two origins of expressive change will lead to the same feeling state because they have the same efferent pattern. Third, while expressive feedback may make a small contribution to feeling, given what we now know about the nervous and somatic systems, it seems more reasonable to expect a central nervous system source for primary emotions such as anger, fear, disgust, and joy.

While there is reason to question the primacy of expressive feedback in the generation of feelings, there is no reason to question the importance of data on the role of expressive behaviors in social judgment. Paul Ekman and his colleagues have generated extremely important empirical contributions respecting facial behavior and emotion. They have developed a system to identify specific expressive patterns for different emotions (Ekman, Levenson, & Friesen, 1983); have obtained evidence for the universality of these patterns, as expressed and judged, across wide, cultural gaps (Ekman & Friesen, 1972); and have generated data suggesting that different facial patterns may be associated with unique patterns of autonomic (Ekman et al., 1983) and cerebral activity (Davidson, Ekman, Savon, & Friesen, 1987).

Studies of expressive behavior have made a substantial contribution to our understanding of the role emotions play in interpersonal communication and organizing group processes. They have done less, however, in accounting for internal, subjective experience, and have only recently begun to deal with the possible effect of emotion on organizing the internal, biological substratum.

In summary, all three of the approaches reviewed agree in their division of emotional reactions into subjective, autonomic, and expressive response systems. They also agree in assigning a key role to cognition in the elicitation of emotion, but only the cognition-arousal model postulates that feeling qualities, the color of emotional experience, are equivalent to cognition. In addition, facial response patterns have been identified for specific emotions, and very similar expressive patterns are assigned common labels across widely varying cultures. The significance of facial motor feedback for subjective feelings is less clear, and in my judgment, very likely a secondary issue.

While each of these three componential approaches has enhanced our understanding of emotional processes in particular domains, none of the three provide a satisfactory framework for the development of a comprehensive theory of emotion. Indeed, none of the three can account for the substantial array of data that has been generated by the others, as each works from a conceptual base that omits concepts needed to account for the empirical data of the others. In addition, insufficient attention is paid by all three to the role of central nervous system processes in the creation of emotional states, and none give explicit recognition to the hierarchical organization of mental and neurological processes. Finally, with the exception of investigators involved in the analysis of facial expressions, the concepts adopted in these models (autonomic arousal, cognition, etc.), are far too molar and ignore the modular structure of processes at each level of organization.

The Perceptual-Motor Model

Over the past decade I have proposed and updated a framework or model of emotion designed to help us understand how emotions are constructed and how the constructive process changes over the life span (Leventhal, 1979, 1980, 1984; Leventhal & Mosbach, 1983). Two sets of constraints were foremost in constructing the model: the empirical data generated in a host of laboratories over the prior 30 years, and the functions specified in the prior section. Both sets of constraints pointed to the presence of a core set of emotional reactions that are present at birth, and both clearly suggested that these core reactions were elaborated upon during development.

The very brief summary of this model that follows differs from prior presentations in two ways. First, increased emphasis is given to the componential or modular nature of the cognitive and expressive motor processes generating emotion. A componential analysis of emotional reactions forces us to recognize that both terms (i.e., emotion and cognition) *stand for complex sets of partially independent processes that are integrated in the generation of emotional or cognitive reactions* (Leventhal, 1970; Leventhal & Scherer, 1987). Second, emotion, qua emotion, is treated as a process that organizes componential operators in affecting behavior. A major challenge to emotion theory will be the specification of the factors responsible for this organization.

Componential Units for Emotional Processing

Before describing the model, it is necessary to say something about what is meant by an emotion-processing component. In this model, emotional reactions are seen as the product of a complex set of processors that generate specific emotional reactions to specific stimulus conditions. The varying structure of these processors includes what I will call *cognitive-expressive modules, action systems,* and *somatic receptor systems.* Cognitive-expressive modules are units sensitive to interpersonal, expressive cues (facial and vocal) that generate parallel reactions in response to such cues (these basic components of sensory-motor processing are discussed below). While direct evidence is lacking, I suspect that some expressive modules are closed units similar to the phonetic modules identified by Liberman and Mattingly (1989) for the perception and production of phonetic units in speech. While some of these modules are "unique" to emotional reactions, others may well be regarded as components of the cognitive system, though some operate so early in the processing hierarchy—that is, so prior to consciousness—that one might decline to call them cognitive.

Action systems organize approach and avoidance behavior, are supraordinate to expressive modules, and can subsume more than one expressive module. Action systems are open to a wide array of stimuli and produce a broader range of behaviors within the parameters of their function. Moreover, a broad array of memory structures will be accessed by each action system in the guidance of approach and/or avoidance responding.

Somatic receptor systems integrate both expressive modules and action systems with the organism's physiological system most broadly defined. These systems are general in nature and vary with both phasic (emotional) energy demands and tonic (mood) state. The systems are integrated by hormones that are manufactured in various places in the body and that communicate and integrate specific functions by acting upon receptor sites on target cells such as macrophages, neurons such as those regulating core temperature, and cells lining organ systems such as the gastrointestinal tract. Thus, these systems integrate the brain, that is, cognitive-expressive modules and action systems, with widely separated physiological processes involved in defense against pathogens and the expenditure of energy.

Whatever the eventual definition and positioning of these componential units, it is clear that interactions between cognition and emotion and the development and change in emotion over the life span takes place at these levels—that is, in the way these modules and systems are integrated with one another and with the informational base that is available to them. From this perspective, prior debates respecting the interaction or temporal priority of emotion and/or cognition become irrelevant, as these issues require the analysis of the interactions and temporal priorities among the componential units comprising the emotional and cognitive domains and not between the domains as unitary wholes (Leventhal & Scherer, 1987). Similarly, emotional development may involve the introduction of a componential unit into the emotion process or a change in the organization or relationship between the components; development does not refer to the addition of emotions comprised of entire new units.

Hierarchical Levels of the Perceptual-Motor Model

I have hypothesized that the components involved in the construction of emotion are organized at three levels: (1) sensory-motor, (2) schematic, and

(3) conceptual. I will briefly describe each level and the components presumed critical for the definition of each, and I will present hypotheses regarding the mode of action for each level and how they are involved in the generation of emotional episodes.

SENSORY-MOTOR PROCESSING. Recent data suggest that newborns are equipped with basic skills that allow them to *mirror* adult, expressive behaviors. Examples are seen in the report of Field, et al. (1982), who found that one- and two-day old infants mirrored adult smiles and adult expressions of surprise and distress. While the precise number (5 or 7) of basic emotions may never—and need never—be fully resolved (Ekman, Friesen, & Ellsworth, 1972; Izard, 1971), there is reason to believe they reflect the action of prewired stimulus-response units. Following Piaget, I have used the term *sensory-motor* to describe these prewired units (see also Fischer, et al., in press). These reaction patterns could be thought to reflect the action of a core set of relatively closed, cognitive-expressive modules. Viewed thus, they would satisfy Fodor's (1983) restrictive definition of a module as each of these reactions appears to mirror a highly specific set of facial patterns from adult caregivers. The infant's ability to mirror other facial expressions such as tongue protrusion and mouth opening (Meltzoff & Moore, 1977), raises questions as to whether these reactions reflect the operation of a set of relatively closed modules or the operation of a system for mirroring facial patterns. In any case, the data showing that these expressive patterns are preserved and elaborated (common patterns are recognized and produced across vastly different cultures; [Ekman, & Ellsworth, 1972]) are consistent with a modular interpretation.

We do not know if these core reactions are accompanied by their expected and appropriate subjective states. Although it is premature, therefore, to equate them with adult emotions, it appears reasonable to hypothesize that they form the core units or seeds around which adult emotions are elaborated. For example, the infant's automatic, sensory-motor smile is likely to elicit smiles from caregivers. The cognitive contents of the social-contextual factors surrounding these smiles, along with the autonomic reactions elicited by them, are likely to be joined together, forming the schemata for "joy" or "pleasure." I used the term *mirror* and avoided the terms *imitate* or *reflex* when describing sensory-motor processing for two reasons. First, *imitate* suggests a conscious, volitional copying of the adult expression. Although we do not know if the infant makes a conscious decision when copying the adult's expression, we do know that infants lack a concept of self until a later age, approximately 15 to 18 months (See Lewis, Sullivan, Stranger, & Weiss, 1989). This and questions about the development of volitional motor pathways suggest that early imitative behavior lacks a sense of self as a volitional, conscious copier. On the other hand, *reflex* implies that these infant expressions are simple, prewired reactions elicited by their necessary and sufficient stimulus with no broad implications for mental function.

In summary, the role assigned sensory-motor reactions in the domain of emotion is the same as that assigned phonetic modules for the perception and production of speech: They are central structures designed to perceive and create a basic vocabulary of emotional units, units that can be integrated and combined in larger emotional structures and that add to the structure of the self system. While these units are integrated with other components, they persist as modules throughout a lifetime.

SCHEMATIC EMOTIONAL PROCESSING. Emotional schemata represent memories *of* prior emotional experiences. Thus, schemata are integrations of core expressive modules into larger complexes essential for the experience, expression, and somatic conditions of emotion and the motivational push generated by emotion. The motivational *push* derives from the incorporation of the core expressive modules under larger, relatively open action systems designed to regulate behavioral *approach, avoidance,* and/or *response inhibition* in response to specific, environmental stimuli. These supraordinate action systems are critical components of the schematic memory structure, and they will focus attention on specific external stimuli and retrieve those semantic and motor memories needed to regulate approach and/or avoidant behavior in the current environment.

There is some risk in using the term *schema* to refer to this level of processing as memory schemata are often conceived as passive networks activated by an "arousal" system. Viewing the schema as an arrangement of action systems with their associated cognitive content and expressive modules stresses the dynamic nature of this memory structure. Thus, the activation of a schema brings into play a variety of emotion (and cognitive) processors that then draw upon external and internal information in elaborating a complete emotion. The term *schema* emphasizes the organization of the components such that activity in any one component tends to activate the others.

For example, the approach system, which appears to be coordinate with the three parts of the dopaminergic system (Deakin & Crow, 1986; Iversen, 1977), integrates motor and cognitive activities to allow for the focus of attention, the retrieval of spatial and object memories, emotional or evaluative reactions (possibly), and the generation of action procedures relevant to moving toward, observing, consuming and/or attacking objects. The avoidance system, possibly nonadrenergic, encourages rapid, widespread scanning biased toward dangerous "locations," retrieves information relevant to threat interpretations, and activates motor programs for escape (Tucker & Williamson, 1984). Response inhibition, on the other hand, directs attention away from environmental cues, retrieves information to facilitate distraction, and leads to nonresponse to the threatening agent and the possibility of death (Holt, 1986). As these supraordinate modules are action directed, they also initiate organized somatic receptor systems that have broad effects on broadly based physiological reactions—e.g., the autonomic and endocrine patterns associated with preparation for action (approach, fight, or flight) or inaction (pituitary adrenocortotropic activity). Finally, the direction of attention, the information retrieved, and the action readiness will also reflect the nature of the subsumed, expressive-core module. Thus, the direction of attention and the content of information retrieved from memory during approach activity will be moderated by the specific subsumed core—e.g., interpersonal joy, sexual anticipation, or anger.

As the supraordinate action systems are open, their activation and the activation of core modules subsumed under them will generate schemata or acquired organizations reflecting the repeated activation of the expressive and action components in relation to specific social contexts involving self and other. Thus, the typical trigger for schematic processing will be the social and contextual cues that have been associated with prior activation of these modules. Emotion schemata also can be activated by other entry points, such as the activation of an expressive module by physiognomic adjustments (Strack et al., 1988) or the activation of an action system through exercise, and so on. Because autonomic activity is linked to the action modules and is represented and integrated within the emotion schema, it too can be a cue to the activation of the schema and its component systems and their associated cognitive contents and expressive modules, leading to the generation of complex emotional states (Leventhal, 1986). For example, the depression of sympathetic activity and the suppression of motor activity during illness commonly leads to depressed mood and the activation of depressive expression and depressive thought.

CONCEPTUAL PROCESSING. Conceptual emotional processing involves the creation of symbolic representations *about* emotional events. Emotional episodes have a beginning, middle, and end, and this unfolding over time can be described as a series of emotion labels embedded in propositions about the conditions eliciting emotion, the duration of emotional experience, the likely behavioral and interpersonal consequences of emotional experience, and ways of reacting to emotional reactions in oneself and others. The temporal unfolding of an episode can be divided into subunits or frames that appear to identify key organizations or periods between transitions, and the assembly of frames can be described as the script for the episode. Thus, conceptual processing involves higher level conceptual organizations, which permit volitional or controlled activity with respect to the various components comprising emotional behavior.

Controlled activity at this level also implies the existence of a module for self-awareness that is the source of sense of self as the originator of action. Moreover, conceptual processing not only construes the self as the source of thoughts, feelings, and actions, it also represents the self as represented by others (Leventhal, 1984; Lewis, et al., 1989; Oatley & Johnson-Laird, 1987). Thus, conceptual processes represent mutuality—that is, I have feelings, I know that I have feelings, I know that my feelings are partially visible, I know that another person is aware of and reacting to my feelings, and I know that he knows he is aware of and reacting to my feelings. There will always be a margin of uncertainty in the overlap between these two domains of mutual awareness.

Conceptual elaborations are likely to serve as the primary framework for the integration and elaboration of core and schematic emotion into a sequence of events that comprise an emotional episode. This integrative and elaborative process in emotion management will play a critical role for the development of notions about the self and the development of skills for the selection of various programs for coping or defending against affective states (Lacroix, 1986). Thus, conceptual elaboration plays a critical role in very-long-term facets of emotional development.

While conceptual structures may be the key to longer term management of affect and the development of concepts of self, these structures also play a critical role in short-term adaptation. It is likely that conceptual rules guide coping activities, which are important for reviewing emotional experiences and reducing their intensity. For example, we can think, plan, and talk about emotion without necessarily having the subjective *feeling* or experience of emotion, or the somatic conditions and motivational pushes created by the presence of the emotional state. Conceptual processes may also direct defensive operations—for example, controlling the fluctuations between avoiding and taking in information.

In summary, the components or propositions and procedures that participate in conceptual processing play a critical role in storing information about emotional states (i.e., their elicitors, performance, and consequences), and interact with and exert a degree of control of emotional processes at "lower" levels. It is important, however, to avoid the all-too-frequent error that emotional states (i.e., being in the throes of constructing the expressive, subjective, and cognitive aspects of fear, anger, joy, etc.) are the same as thinking or talking about feelings, or deliberately enacting the expressive and instrumental acts typically associated with feeling.

Dynamics of the Perceptual-Motor System

The three-level division outlined here is clearly provisional; depending upon the questions raised, other slices may prove useful; as many as seven operational levels have been proposed by others (Carver & Scheier, 1981; Powers, 1973). The number of levels needed will vary depending upon what one learns about the dynamics of emotion systems (i.e., how the system operates to generate emotion and how it might change over time). Further analysis and experimentation should clarify the precise nature of the cross-levels interactions (e.g., establishing set points for lower levels, providing declarative or procedural information for upper levels, etc.) and indicate the need for additional levels. Four aspects of dynamics merit comment at this point: (1) parallel processing and dissociation of levels;

(2) levels interaction; (3) temporal organization; and (4) activators and emotions as goals.

PARALLEL PROCESSING. The first rule of system operation is that conceptual, schematic, and sensory-motor mechanisms operate *in parallel* in generating emotional reactions. While the system operates as a whole, it is clear that one or another of these levels may be more or less dominant in shaping the contours of an "emotional" reaction at a given point in time. Which level will dominate depends upon the individual's experiential history and the situational context in which the emotional reaction takes place. For example, sensory-motor processing would seem to be dominant for the creation of emotional responses in infancy and less important for adult emotion. But sensory-motor processing does not vanish with maturation and socialization, and it may play a powerful role in shaping reactions in adulthood depending upon the situational context. Because sensory-motor processing (i.e., activity of these expressive motor modules) is highly dependent upon the presence of interpersonal stimuli, the quality of emotional activity, both expressive and subjective, should be more distinctive in social settings. When interpersonal stimuli are absent, processing may be dominated by conceptual activity and may have a less distinctive, affective quality, though the intensity and distinctiveness of emotional experience could be enhanced by conceptual reverie which activates emotion schemata.

Because expressive modules and action systems are sensitive to different stimuli, it is not surprising that they may become dissociated from one another and form theoretically interesting emotional response syndromes that are not anticipated by commonsense views of emotion as discrete units such as fear, joy, anger, and so on. For example, the social environment fails to respond to and therefore discourages emotional expressions of distress in young males. As a consequence, it is not surprising to find expressive blandness associated with arousal of motor approach or avoidance modules. Dissociation of this sort has been documented among males in a variety of emotional "pathologies" such as phobias (Lang et al., 1980).

CROSS-LEVELS CONTROL. Second, there is continual *direct and indirect communication and control exerted between levels* of the system. Investigators generally think of top-down activity when discussing cross-levels control (e.g., cognitive control of autonomic events or voluntary control of facial expression). The tactics for executing control will vary, however, depending upon the components involved. For example, control of one motor system by another can be seen in the use of voluntary smiles to hide expressive displays of annoyance or disgust reflective of one's automatic, emotional response to another person in a social exchange; control in this case is direct, with the voluntary response inhibiting or overriding the automatic reaction. By contrast, the use of cognitive procedures to control autonomic motor system activity is likely to be indirect. For example, I can slow my heart rate by an abrupt change in position (i.e., standing up suddenly to induce postural hypotension), and I can condition this slowing to external events or to a thought or mental image by systematically pairing these cues to the postural change (Riley & Furedy, 1984). With extended training, indirect control may be replaced by direct control.

Cognitive models of emotion, such as Bower's (Bower & Cohen, 1982), Mandler's (1975), or Oatley and Johnson-Laird's (1987), give relatively little attention to bottom-up processing, though influence from "lower" to upper level modules may be the most pervasive and important aspect of emotional processing. For example, the dopaminergic and noradrenergic components of the central and autonomic nervous system appear to be the anatomic and functional units for the schematic processing of behavioral approach and avoidance. Each appears to be supraordinate to a separate set of core, expressive-motor modules, the dopamine system for positive affects (joy) and addictive urges (Wise, 1988), and the noradrenergic for negative emotions (fear and pain/distress). Indeed, data at the cellular level show that noradrenergic neurons of the central nervous system in mammals such as the cat are activated in response to a wide range of stressors ranging from the presence of a dog through white noise to fever induced by pyrogen injections. Moreover, the activity of these central, noradrenergic neurons tracks changes in heart rate and habituates to stressors (e.g., white noise), as does heart rate and the orienting response. Further analysis may suggest that this motor system may meet many of the criteria set by Fodor (1983) for a processing module, as it is sensitive to a particular range of inputs and has well-defined effects on response output—ranging from attentional changes through habituation and learning to instrumental, escape action (Tucker & Williamson, 1984). This system impacts on higher level, conceptual modules, influencing both momentary and long-term images of the self. Self-esteem, the feeling that one is capable and in control, is undercut by fear; at the moment that one is frightened, the self is perceived as out of control, and this can lead to the self perception that one is chronically fearful. In addition, motor modules are linked to widely dispersed, somatic receptor systems that impact upon immune activity and potential for illness. Thus, an action module can link higher cognitive processes to the soma, providing a bidirectional path for cognitive influence on health and the influence of health upon mood and cognition.

TEMPORAL ORGANIZATION: EMOTIONAL ACTS, FRAMES, AND SCRIPTS. The third principle in dynamic function concerns control of the unfolding or elaboration of emotional episodes in real time. This elaboration involves the integration of core, expressive-motor acts, with schematic processors (i.e., action systems and their associated declarative and procedural content) to form emotional frames. Sensory-motor processes provide the organized core of sensory cues and expressive motor activity. Schematic processes elaborate and connect the core to a more elaborate perceptual content and to subjective feelings and action tendencies and their associated autonomic reactions. The consequence is an emotional *frame*, (i.e., an expressive and autonomic motor reaction with associated subjective feeling) that is integrated with an elaborate set of contextual cues that are largely social in content (i.e., a cognized self and a cognized other). Finally, it is likely that conceptual operations sequence and organize the frames into scripts (Abelson, 1976), or meaningful, temporally extended, emotional themes.

ACTIVATORS AND EMOTIONAL GOALS. The models of emotion proposed by cognitivists tend to focus on the stimulus events and the cognitive content elaborating and shaping the emotional response. Thus, cognition creates the goals and groundwork for behavior whereas emotion is an event caused by the disruption of goal-directed cognitive activity (Mandler, 1975). Even the expanded view of the functions of emotion presented by Oatley and Johnson-Laird (1987) places emotion in a subordinate or assisting role to cognition; that is, emotion's functions are to sustain a particular direction of cognitive activity, to direct attention to plans and goals, to sustain figure-ground relationships of goals to background, and so on. While these views are consistent with data on the conditions for the arousal of a number of emotions (Ellsworth & Smith, 1985), they are limited as they encompass only some of the conditions that lead to emotion. They ignore the role of emotion as a goal—that is, as focusing attention, selecting responses, and sustaining ongoing instrumental activity whose purpose is to regulate (enhance or reduce) affect itself.

The reciprocal and intimate relationship of emotion to cognition emphasized in the perceptual-motor model makes clear that plans and behavior are often in the service of emotion. Emotion is not only important, therefore, for highlighting and motivating "cognitive" goals and procedures, as when anger focuses the mind on completing a manuscript to deal with an hostile critic, emotion can be a goal as well as the force sustaining action. The addictions to chemical substances, to gambling, to loved ones, and so on, are examples of behavioral systems where cognition and behavior are designed to enhance emotional states. The hypothesis that the generation and maintenance of emotional states is a critical objective of mammalian behavior calls for a thorough reexamination of the cognitivists' view. A consequence of this analysis is the recognition that the intensity and quality of emotional states cannot be fully accounted for to any reasonable degree by an analysis in which emotion is totally dependent upon stimulus conditions and their interpretation—for example, fear as a product of the perceived threat accompanying task disruption, depression a product of the extent of an interpersonal loss, or pain the product of the intensity and duration of exposure to a noxious stimulus. The emotions elicited by such disruptions are a product of an emotional mechanism whose action is not fully predictable from cognition. The loss of a spouse may provoke anger, anxiety, and/or depression, which is partially predictable from the cognitive content of the disruption, but which may also include the operation of an opponent, or homeostatic, emotional process, where the absence of positive affect due to loss of contact uncovers a latent, negative affect that originally served to moderate the intensity of the positive emotion generated by the close attachments (Solomon & Corbit, 1974). Habituation and inhibition of the motor and autonomic components of emotional reactions by repeated exposure to a noxious stimulus or with repeated use of an addictive substance such as nicotine, may have a profound impact on pain and the experienced value of the substance, and these changes are not readily predicted or understood from a purely cognitive framework. Cognitive processes per se cannot account for the full range of emotion dynamics. In summary, models of emotional processes that ignore the activation of emotion modules as both a goal and a motivator of planned activity are incomplete, and perhaps, fundamentally incorrect.

Implications for Research

A framework of the type I have outlined poses a number of directions for research, including new slants on existent problems. I will briefly discuss two: the relationship between emotion and cognition, which has been a central focus for argument and investigation in contemporary psychology, and the change of emotion with development.

The Relationship Between Cognition and Emotion

Leventhal and Scherer (1987) have suggested that a componential analysis would obviate the need for debates as to the relative primacy of emotion, as argued by Zajonc (1980, 1984), and of cognition, as argued by Lazarus (1982). Lazarus is clearly correct in arguing that situational perceptions and/or cognitions both provoke and shape emotional reactions, and Zajonc is correct in arguing that emotional reactions may indeed precede and shape cognition. They are both correct for at least two reasons. First, as should be clear from the framework I have outlined, there is a continuing interaction between cognitive and emotional processing, each serving as goals for and driving the other, cognitions creating emotions and emotions generating cognitive reorganizations.

Second, there is continuing interaction between the components (both psychological and biological) that are involved in the construction of cognitive and emotional reactions. Thus, to answer questions about the interaction of emotion and cognition, we must ask which particular level of cognitive and emotional activities is involved in the interaction and what components of each at that level are involved in the interaction. For example, if we are interested in the interaction of cognition with expressive behavior, we need at the very least to distinguish between automatic and volitional motor expressions on the one hand and perceptual cognition and conceptual or propositional cognition on the other. If our focus is the impact of emotional states on cognitive activity, we must specify whether we are interested in preattentive cognitive processes, attentional processes or consciously generated cognitive activity and the module and level of emotional processing involved in the interaction.

Consider the question of the interaction of emotion and memory. The approach to this problem follows that used in studies of state-dependent memory (Eich, 1980), where investigators examined the effects upon memory retrieval as a function of the match between emotional state during learning and the emotional state during retrieval. This search for state-dependent effects has been only partially successful (see Leventhal & Tomarken, 1987). Although some studies show effects and others do not, by far the largest body of data suggests that memory can alter attention, thus affecting what is learned and altering retrieval, though there is less evidence for the state-dependent hypothesis, which assumes that maximum effects will appear when the emotional state during retrieval matches that during learning.

A componential analysis suggests that investigators may have erred by considering emotion and memory as unitary constructs and by seeking memory differentials as a function of the match, for example, between happiness at learning and happiness or a negative mood state such as anger or fear at recall, and by using verbal material for studying the problem. If the match process is redefined as a match between levels and components, one must ask whether the same or a different component or module was present at learning and retrieval and not whether the emotional states were the same or different and whether the components involved had similar or differential access to the memory store for verbal material. If different emotions (e.g., anger and happiness) share the same left-frontal-lobe activation (Davidson, 1984), based upon the same or similar dopaminergic action system, and if that action system plays a prominent role in memory retrieval, these two apparently divergent emotions would have overlapping domains of declarative and procedural knowledge, and a subject might retrieve both aggressive and pleasurable content in these apparently contrasting states. Indeed, such overlap may have been the basis for psychoanalytic observations respecting object relations disorders and the basis for the more mundane example of a grandparent's loving, but aggressive and sometimes frightening assertion to his three-year-old grandchild: "I'm going to eat you up!"

The content of the memory material used in these studies should also affect emotion memory outcomes. For example, if noradrenergic pathways involved in avoidance or fear behaviors have more intimate connections with the right parietal than with the temporal lobe, the arousal of avoidance activity might have greater impact on attention to and memory for movement and location than memory for things. It might prove difficult indeed to find state-dependent (i.e., matching) effects for fear or depression on verbal memories, as the components constructing these emotions may not have special links to the verbal system (Davidson, 1984).

The above componential speculations are not inconsistent with hypotheses that emotions such as fear, joy, and depression could have a major impact on initial learning. A depressed or frightened person might be more attentive to, more completely rehearse, and better retain depressing or frightening content (Bower, Monteiro, & Gilligan 1978). But, although the mood relevant material would be better learned, the analysis suggests that it would not necessarily be more accessible if the individual's emotional state at the time of retrieval matched that at learning. Specifically, the interaction of emotion and memory is not fully described by a network model in which moods and emotions are nodes linked to nodes for cognitive content (Bower & Cohen, 1982). The network formulation seems to overlook the active, search process associated with schemata containing specific action systems. Action systems to approach or avoid, and, perhaps the expressive modules subsumed by them, are active processors: They direct attention and search for information. This active view resonates with Clarke and Isen's (1982) suggestion that emotions alter search heuristics and is less congenial to models postulating the activation of nodes by the flow of energy or activation along the paths of a passive network.

EMOTIONAL DEVELOPMENT. A componential framework suggests that developmental changes in emotion will involve both the addition of new components to existent ones, and changes in the organization and interaction among components. Thus, the expressions we call fear, anger, shame, happiness, and/or frustration in the infant will differ in substance from the emotions given the very same labels in adulthood. And as cognitivists have argued, it is clear that some emotions cannot be experienced until additional cognitive components have been added to the emotional processing system. As mentioned earlier, Michael Lewis has found that the sense of self, which may emerge from the infant's and toddler's intentional actions and the social interchange with caregivers, is a critical, cognitive component for the appearance of emotions such as shame. Cognitive additions of this type are clearly important for the development of new, complex emotions such as pride, guilt, and so on. The model I have presented, however, also suggests that many components may preserve their integrity, even when they are incorporated into new organizations.

A componential approach leaves considerable room for variation in the organization of emotion as a function of socialization practices in different cultures. Given the substantial degree of independence between the modules and levels, variation in socialization practices can generate different linkages and organizations of the components comprising the emotion system. For example, parents who encourage vigorous activity by tossing their two-year-old in the air when he or she is joyful can link intense physiological activation with joy, whereas parents who interact quietly with their infant when joyful may generate a quite different linkage between the expressive component of joy and motor and somatic systems. The flexibility created by the partial independence of emotion components may even allow a culture to delete or override a primary emotion such as anger (Levy, 1984). Limitations exist, however. Thus, the assumption that expressive modules are innate and nonerasable should permit reorganization of the system. It is theoretically possible for a therapist to enhance the salience of a hidden affect by creating conditions for the function of an inhibited, expressive module and linking it with new cognitions and action potentials. However, doing so may require simulating conditions appropriate to an earlier time in the individual's life history.

Concluding Remarks

While there is reason to believe that the componential theme will prove highly robust, it is clear that observational and experimental data will require alteration of many of the highly speculative ideas I offered here. The following are the features of the componential view that I expect to prove durable. First, emotions will prove to be multicomponent constructions. Second, multiple components will be found and demarcated at multiple levels. Third, many of the components will be found to enjoy a considerable degree of functional independence. Fourth, different socialization practices may link components in typical and/or unexpected ways. Many emotional organizations that we consider universal and innate may actually be products of learning, the expected organization being a virtual necessity given the environments in which these components emerge. Fifth, cognitive components, in particular those associated with the ever-evolving sense of self, will play a major role in the elaboration of emotion and be responsible for the appearance and disappearance of specific emotional states. Sixth, emotions will be given the status of goals as well as energizers of behavior in models designed to simulate behavioral processes. Seventh, the components will be represented as active processing units in emotional schemata. Finally, empirical research will identify components or modules for emotional processing that do fit a priori or commonsense categories, and elements we think unitary may often prove to be multicomponent. For example, it is not obvious that perceptual memories divide into separate component functions such as memory for faces, things and place, yet these specific component systems are likely to have quite different involvement with emotion. Similarly, the differentiation between volitional and spontaneous motor systems has been given insufficient attention with regard to the experience of emotion, though the distinction has been recognized as important in recent studies and in studies of faking emotional states.

In summary, the componential approach I have presented offers a framework for interdisciplinary research on emotion. Its two basic themes, that emotional reactions reflect the operation of organizations of componential or modular units, and that emotion and cognition are reciprocally intertwined, provide a basis for further analysis. Neither emotion nor cognition can be fully understood and no model of behavior can make sense of emotional activity with a view that is restricted to emotion as a product of behavioral interrupts or emotion as the energizer and organizer of an organism whose cognitive and procedural plans are the basis of its existence. Endocrines and endocrine communication, the somatic receptor system that some claim to be the heart of mood and emotion (Pert, 1986), define goals and execute procedures for single-celled organisms. In complex mammalian forms they are channeled in neurons and secreted by glands. Indeed, one might suggest that phylogenesis and individual development has elaborated perceptual, cognitive, and motor systems to better serve the execution of genetic plans by these chemical messengers—giving added emphasis to the metaphor, "The mind is in the body." The framework also makes clear how little we know and how much we have to learn.

References

Abelson, R. P. (1976). Script processing in attitude formation and decision-making. In U. S. Carroll & J. W. Payne (Eds.), *Cognitive and social behavior*. Hillsdale, NJ: Lawrence Erlbaum.

Baumann, L. J., & Leventhal, H. (1985). "I can tell when my blood pressure is up, can't I?" *Health Psychology, 4*, 203–218.

Bower, G. H., & Cohen, P. R. (1982). Emotional influences on learning and cognition. In M. S. Clarke & S. T. Fisker (Eds.), *Affect and cognition: The 17th Annual Carnegie Symposium on Cognition* (pp. 263–289). Hillsdale, NJ: Lawrence Erlbaum.

Bower, G. H., Monteiro, K. P., & Gilligan, S. G. (1978). Emotional mood as a contest for learning and recall. *Journal of Verbal Learning and Verbal Behavior, 17*, 573–585.

Cannon, W. B. (1927). The James-Lange theory of emotions: A critical examination and an alternative theory. *American Journal of Psychology, 34*, 106–124.

Cannon, W. B. (1931). Again the James-Lange and the thalamic theories of emotion. *Psychological Review, 38*, 281–295.

Cannon, W. B., & Scheier, M. F. (1982). Control theory: A useful conceptual framework for personality-social, clinical and health psychology. *Psychological Bulletin, 92*, 111–135.

Carver, C. S., & Scheier, M. F. (1981). *Attention and self-regulation: A control-theory approach to human behavior*. New York: Springer Verlag.

Clarke, M. S., & Isen, A. M. (1982). Toward understanding the relationship between feeling states and social behavior. In A. H. Hastorf & A. M. Isen (Eds.), *Cognitive social psychology* (pp. 73–108). North Holland: Elsevier.

Darwin, C. (1904/Original work published in 1872). *The expression of the emotions in man and animals*. London: Murray.

Davidson, R. J. (1984). Hemispheric asymmetry and emotion. In K. R. Scherer & P. Ekman (Eds.), *Approaches to emotion* (pp. 39–57). Hillsdale, NJ: Lawrence Erlbaum.

Davidson, R. J., Ekman, R., Saron, C., Senulis, J. A., & Friesen, W. V. (1990). Approach-withdrawal and cerebral asymmetry: Emotional expression and brain physiology I. *Journal of Personality and Social Psychology, 58,* 330–341.

Deakin, J. F. W., & Crow, T. J. (1986). Monoamines, rewards and punishments—The anatomy and physiology of the affective disorders. In J. F. W. Deakin (Ed.), *The biology of depression* (pp. 1–25). Oxford, England: Alden Press.

Eich, J. E. (1980). The cue-dependent nature of state-dependent retrieval. *Memory and Cognition, 2,* 157–173.

Ellsworth, P. C., & Smith, C. A. (1985). Patterns of cognitive appraisal in emotion. *Journal of Personality and Social Psychology, 48,* 813–838.

Ekman, P., & Friesen, W. V. (1972). Constants across culture in the face and emotion. *Journal of Personality and Social Psychology, 17,* 124–129.

Ekman, P., Friesen, W. V., & Ellsworth, P. (1972). *Emotion in the human face.* Elmsford, NY: Pergamon Press.

Ekman, P., Leveson, R. W., & Friesen, W. V. (1983). Autonomic nervous system activity distinguishes among emotions. *Science, 22,* 1208–1210.

Engel, G. L. (1977). The need for a new medical model: A challenge for biomedicine. *Science, 196,* 129–136.

Field, T. M., Woodson, R., Greenberg, R., & Cohen, D. (1982). Discrimination and imitation of facial expression by neonates. *Science, 218,* 179–181.

Fischer, K. W., Carnochan, P., & Shaver, P. R. (in press). How emotions develop and how they organize development. *Cognition and Emotion.*

Fodor, J. A. (1983). *The modularity of mind.* Cambridge, MA: MIT Press.

Gazzaniga, M. S. (1989). Organization of the human brain. *Science, 245,* 947–952.

Holst, D. V. (1986). Vegetative and somatic components of tree shrews' behavior. *Journal of the Autonomic Nervous System, Supplement,* 657–670.

Iversen, S. D. (1977). Brain dopamine systems and behavior. In L. L. Iversen, S. D. Iversen, & S. H. Snyder (Eds.), *Handbook of psychopharmacology: Drugs, neurotransmitters and behavior* (Vol. 8, pp. 333–384). New York: Plenum Press.

Izard, C. E. (1971). *The face of emotion.* New York: Appleton-Century-Crofts.

Izard, C. E. (1977). *Human emotions.* New York: Plenum Press.

James, W. (1950/Original work published in 1890). *The principles of psychology* (Vol. 1). New York: Dover (Original: New York, Henry Holt).

Katkin, E. S. (1985). Blood, sweat, and tears: Individual differences in autonomic self-perception. *Psychophysiology, 22,* 125–135.

Lacey, J. I. (1967). Somatic response patterning of stress: Some revisions of activation theory. In M. Appley & R. Trumbell (Eds.), *Psychological Stress* (pp. 14–39). New York: Appleton-Century-Crofts.

Lacroix, J. M. (1986). Mechanisms of biofeedback control: On the importance of verbal (conscious) processing. In R. J. Davidson, G. E. Schwartz, & D. Shapiro (Eds.), *Consciousness and self-regulation* (Vol. 4, pp. 137–162). New York: Plenum Press.

Laird, J. D. (1984). The real role of facial responses in the experience of emotion: A reply to Torrangeau and Ellsworth and others. *Journal of Personality and Social Psychology, 47,* 909–917.

Lang, P. J. (1979). Language, image and emotion. In P. Pliner, K. R. Blankstein, & J. M. Spigel (Eds.), *Perception of emotion in self and others* (Vol. 5). New York: Plenum Press.

Lang, P. J., Kozak, M. J., Miller, G. A., Levin, D. N., & McLean, A. (1980). Emotional imagery: Conceptual structure and pattern of somato-visceral response. *Psychophysiology, 17,* 179–192.

Lashley, K. S. (1950). In search of the engram. *Symposia of the Society for Experimental Biology, 4,* 454–482.

Lazarus, R. (1982). Thoughts on the relationship between emotion and cognition. *American Psychologist, 37,* 1019–1024.

Leventhal, H. (1979). A perceptual-motor processing model of emotion. In P. Pliner, K. Blankstein, & I. M. Spigel (Eds.), *Advances in the study of communication and affect: Perception of emotion in self and others* (Vol. 5, pp. 1–46). New York: Plenum Press.

Leventhal, H. (1980). Toward a comprehensive theory of emotion. *Advances in Experimental Social Psychology, 13*, 139–207.

Leventhal, H. (1986). Symptom reporting: A focus on process. In S. McHugh & T. M. Vallis (Eds.), *Illness behavior: A multidisciplinary model* (pp. 219–237). New York: Plenum Press.

Leventhal, H., & Mosbach, P. A. (1983). The perceptual-motor theory of emotion. In J. Cacioppo & R. Petty (Eds.), *Social psychophysiology* (pp. 353–388). New York: Guilford Press.

Leventhal H., & Scherer, K. R. (1987). The relationship of emotion to cognition: A functional approach to semantic controversy. *Cognition and Emotion, 1*, 3–28.

Leventhal, H., & Tomarken, A. J. (1986). Emotion: Today's problems. In M. R. Rosenzweig & L. W. Porter (Eds.), *Annual Review of Psychology, 37*, 565–610. Palo Alto, CA: Annual Reviews.

Levy, R. I. (1984). The emotions in comparative perspective. In K. R. Scherer & P. Ekman (Eds.), *Approaches to emotion* (pp. 397–412). Hillsdale, NJ: Lawrence Erlbaum.

Levy, S. M., Heberman, R. B., Simons, A., Whiteside, T., Lee, J., McDonald, R., & Beadle, M. (1989). Persistently low natural killer cell activity in normal adults: Immunological, hormonal, and mood correlates. *Natural Immunity and Cell Growth Regulation, 8*, 173–186.

Lewis, M., Sullivan, M. W., Stranger, C., & Weiss, M. (1989). Self-development and self-conscious emotions. *Child Development, 60*, 146–156.

Liberman, A. M., & Mattingly, I. G. (1989). A specialization for speech perception. *Science, 243*, 489–494.

Mandler, G. (1975). *Mind and emotion.* New York: Wiley.

Matthews, K. A. (1988). CHD and Type A behavior: Update on an alternative to the Booth-Kewley and Friedman quantitative review. *Psychological Bulletin, 104*, 373–380.

Meltzoff, A. N., & Moore, N. K. (1977). Imitation of facial and manual gestures by human neonates. *Science, 225*, 75–78.

Oatley, K., & Johnson-Laird, P. N. (1987). Towards a cognitive theory of emotion. *Cognition and Emotion, 1*, 29–50.

Pert, C. B. (1986). The wisdom of the receptors: Neuropeptides, the emotions, and bodymind. *Advances, 3*, 8–16.

Plutchik, R. (1984). Emotions: A general psycho-evolutionary theory. In K. R. Scherer & P. Ekman (Eds.), *Approaches to emotion* (pp. 197–219). Hillsdale, NJ: Lawrence Erlbaum.

Powers, W. T. (1973). *Behavior: The control of perception.* Chicago: Aldine.

Reisenzein, R. (1983). The Schachter theory of emotion: Two decades later. *Psychological Bulletin, 94*, 239–264.

Riley, D. M., & Furedy, J. J. (1984). Psychological and physiological systems: Mode operation and interaction. In S. R. Burchfield (Ed.), *Psychological and physiological interactions in the response to stress* (pp. 3–33). New York: Hemisphere.

Rowell, T. (1972). *The social behaviour of monkeys.* Middlesex, England: Penguin.

Schachter, S., & Singer, J. E. (1962). Cognitive, social, and physiological determinants of emotional state. *Psychological Review, 69*, 379–399.

Solomon, R. L., & Corbit, J. D. (1974). An opponent-process theory of motivation: Temporal dynamics of affect. *Psychological Review, 81*, 119–145.

Strack, F., Martin, L. L., & Stepper, S. (1988). Inhibiting and facilitating conditions of the human smile: A nonobtrusive test of the facial feedback hypothesis. *Journal of Personality and Social Psychology, 54*, 768–777.

Tomkins, S. S. (1962). *Affect-imagery-consciousness* (Vol. 1). New York: Springer Verlag.

Tomkins, S. S. (1963). *Affect-imagery-consciousness* (Vol. 2). New York: Springer Verlag.

Tucker, D. M., & Wiliamson, P. A. (1984). Asymmetric neural control systems in human self-regulation. *Psychological Review, 91*, 185–215.

Watson, D., & Pennebaker, J. W. (1989). Health complaints, stress, and distress: Exploring the central role of negative affectivity. *Psychological Review, 96*, 234–254.

Wise, R. A. (1988). The neurobiology of craving: Implications for the understanding and treatment of addiction. *Journal of Abnormal Psychology, 97*, 118–132.

Zajonc, R. B. (1980). Feeling and thinking: Preferences need no inferences. *American Psychologist, 35*, 151–175.

Zajonc, R. B. (1984). On the primacy of affect. *American Psychologist, 39*, 117–123.

Zajonc, R. B. (1985). Emotion and facial efference: A theory reclaimed. *Science, 228*, 15–21.

Commentary on Part III

McGaugh has documented in many studies that biological treatments that alter hormonal function or neurotransmission can affect the integrity of information in memory. His research has investigated the timing of these effects and the brain systems mediating them. He also shows that central effects of various treatments can be mediated through the peripheral nervous system. The modulatory effects of hormonal treatment are long lasting, sizable, can be specific, and play a role in different types of learning and memory. McGaugh provides evidence for the importance of noradrenergic, opiate, and GABAergic systems in the modulation of memory. He then integrates the roles of these various neurotransmitters by considering the functioning of the amygdala. The manipulations that McGaugh describes do not determine whether or not a memory is formed but rather alter the strength of memory traces. In this sense the treatments are modulatory.

McGaugh makes a strong case for the value of pharmacological tools for investigating memory processes. In the commentary on Part II we pointed out how drugs can be used to examine the differentiated nature of memory. McGaugh, however, uses them to focus on mechanisms of memory modulation. While he does examine their effects on different tasks, his primary agenda is not to discover dissociations in memory. McGaugh's work also highlights how neurochemical systems do not function independently. No behavioral function (e.g., type of memory) is likely to be dependent on the function of only one neurotransmitter. This does not mean that a pharmacological manipulation using a highly selective agent will not exert quite specific behavioral effects.

Koob provides an overview of research investigating the relationship between arousal and performance in terms of both behavioral and biological variables. *Arousal* is a term that is frequently used but not easily defined. Koob notes that it is not unidimensional. Performance changes with arousal according to an inverted U-shaped function. That is, at both low and high levels of arousal subjects' cognitive performance is relatively poor. Koob argues that the right side of that function represents the effects of stress. In

349

this way he describes stress not just in terms of arousal but in terms of performance.

Koob presents data showing how several different biological systems contribute to the stress response. One of the systems that has been shown to be particularly important is the hypothalamic pituitary adrenal axis. This system has been well characterized in terms of anatomy, physiology and chemistry. There is also a large body of data implicating the limbic system in the processing of emotional information. Koob notes that the reticular activating system, the catecholamines, the midbrain dopamine neurons, the locus coeruleus, corticotropin releasing factor (CRF), and vasopressin are also all involved in stress and arousal.

More specifically, and as elaborated by Robbins (1984), Koob supports the hypothesis that dopamine is involved in a low-level arousal system particularly concerned with response initiation. In contrast, the norepinephrine system in cortex and hippocampus is suggested to be involved in attention—an upper arousal mechanism. CRF is viewed as playing a role in both the lower and upper mechanisms of arousal. There is still much uncertainty as to how these systems interact. Moreover, they have peripheral as well as central determinants. At present, there is no complete integration of the working mechanisms of these different systems.

While Koob examines the neurobiology of stress in some detail, he pays less attention to behavior, particularly cognition. This does not imply a lack of interest in cognition. Instead, he appreciates that one of the principal difficulties facing research in this area is the issue of how to measure stress and arousal behaviorally. There remains some uncertainty as to what constitutes stress, and Koob notes this in his chapter.

The study of cognition would be incomplete without some appreciation of how affect modulates aspects of cognitive function. Eysenck addresses this issue by discussing how anxiety alters information processing. Much is known of the neurobiology of anxiety from studies both in humans and in lower animals (Last & Hersen, 1988; Marks, 1987). Eysenck argues that a psychobiological understanding of anxiety requires that we begin by examining the cognitive and physiological processes separately before we begin to integrate these two systems. This approach is certainly consistent with the history of research in the area. The extensive work on the neurophysiology, neuroanatomy, and neurochemistry of anxiety and stress has proceeded quite separately from behavioral and cognitive studies (see, e.g., Koob's chapter).

In the 1960s and 1970s, spurred on by an interest in human performance, a vast amount of work was done by cognitive psychologists on the effects of anxiety in normal subjects. Over the same period, there was a dramatic rise in the use of benzodiazepines to treat anxiety disorders. It is disappointing that there was little overlap in the anxiety research of psychopharmacologists, clinicians, and cognitive psychologists.

In some respects Eysenck's contribution is very similar to many of those that discuss how various manipulations impair some cognitive processes but spare others. Data are presented suggesting that the effects of anxiety are more pronounced on cognitive functions that are outside awareness (implicit memory processes) than they are on those of which subjects are aware (explicit processes). Furthermore, the affective tone of the stimuli being processed is critical in determining the effects of anxiety.

Leventhal begins his chapter by providing a summary of the traditional views of emotion, including cognitive arousal models, models based on facial expression, and feedback theories of emotions. He then develops a global theory of emotion that incorporates psychological, biological, developmental, and social perspectives. He considers emotion to be made up of functional modules that are hierarchically organized and, in various combinations, determine emotional behavior. Emphasizing the modular nature of emotion in terms of cognitive, expressive motor, and subjective processes, he views these component processes as partially independent but integrated in the generation of emotional reactions, reactions that can be productively studied from a developmental vantage point.

The functional significance of emotion is also considered. Emotion provides a means of communication with others, a theme considered by Darwin, and thus it is not surprising that Leventhal should discuss emotion from an evolutionary perspective. In addition, an understanding of emotion allows an appreciation of the internal environment and how that may organize and motivate behavior.

Leventhal argues that explanations of emotion that are solely biological or psychological will be of little value. One of the criteria for deciding whether some aspect of the psychology or biology of emotion is useful is whether in fact it is consistent with what is known in the other domain.

The distinction between factors that modulate as opposed to mediate cognition has been discussed by many researchers. Some have discussed, for example, the difference between intrinsic and extrinsic mechanisms (Krasne, 1978; Squire & Davis, 1981). In a sense, the mechanisms that *mediate* memory can be seen as those that are required for the establishment of a memory trace; the mechanisms that *modulate* memory are those that alter the strength of this trace. There is, perhaps, some analogy to the difference between a signal generator and an amplifier. The distinction has been discussed at various levels of analysis (see Churchland and Sejnowski's chapter in Part I). Behaviorally, factors such as motivation, mood, and arousal have been seen as having indirect, but powerful, effects on cognitive functions. In contrast, processing strategies, stimulus properties (such as modality and familiarity), and processing time have been considered to have direct effects. Similarly, at the neurochemical level, cholinergic neurons have been hypothesized to play a central and direct role in memory functions. The catecholamines, on the other hand, have been thought to affect memory indirectly, by altering arousal, mood, and motivation. We also know that many neuroanatomical systems are important in cognitive functions such as memory, and it has been suggested that some of these systems (e.g., the hippocampus) play a more direct role. Others (e.g., the amygdala) may affect cognitive functions indirectly, perhaps by altering sensitivity to reward; however, we should add that there is considerable overlap between the neurobiological substrates of memory and those of mood, arousal, and motivation.

Having attempted to define what we understand by modulation and mediation, we should also point out that this distinction is by no means clear. For example, McGaugh's chapter describes experiments in which treatments are given following acquisition and prior to retrieval. Because these treatments are not present at the time a memory trace appears to be formed (i.e., at the time of learning), they can be seen as modulating the strength of the memory. However, these same postacquisition treatments can also be seen as affecting the memory consolidation processes that are necessary for memory formation. In this sense, therefore, they are altering processes that mediate memory formation. Examples of postacquisition manipulations that have clear effects on memory are treatments that alter protein synthesis (e.g., Bennett, Rosenzweig, & Flood, 1972). The interested reader is referred to several reviews of this area (Weingartner & Parker, 1984). The body of research on retrograde amnesia also illustrates how treatments that follow acquisition can have profound effects on memory. Our decision to devote a part of this volume to the modulation of memory was not based on a position that processes mediating and modulating memory formation are clearly different. Instead, we wanted to highlight several areas of research that have generally been ignored in cognitive science but are important to consider in the study of memory.

As already pointed out, a number of areas in cognitive research have tended to ignore the modulation of cognition, and so it is not surprising that researchers in artificial intelligence and computational cognitive neuroscience have not considered arousal, motivation, and emotion. There have been attempts, however, to model the effects of anxiety on cognition using computer simulations (Blum, 1989). Perhaps a consideration of the modulatory factors will be an important feature of future cognitive systems.

We know that emotional states, motivational states, and arousal have a powerful effect on cognitive function. Why, therefore, have cognitive scientists considered them annoying artifacts to control and eliminate? The reasons are twofold. First, these terms are difficult to define clearly. Koob illustrates the problem in his discussion of stress and arousal. Second, emotional and motivational states are difficult to manipulate in a controlled fashion in humans, and yet operant psychologists have for many years successfully used various forms and programs of reinforcement to change human performance. These methods of altering motivation are not currently in use, but could easily be incorporated into cognitive research.

However, affect and cognition have been studied by psychologists interested in how emotion influences the perception and encoding of sensory information. One example of this research comes from the study of "perceptual defence," which was a particularly active area some three decades ago (see Erdelyi, 1974). This work has a number of parallels with the experiments described by Eysenck and showed that normal subjects processed threatening and nonthreatening stimuli differently with no apparent awareness.

Both Eysenck and Leventhal refer to another area of research in which psychologists have attended to the role of affect in cognition. This work examines how alterations in mood change the way subjects encode and retrieve information. In particular, affective states serve as contexts that determine how we organize information and the nature of what is accessible in memory.

While we have so far discussed how affect alters cognition, it is important to appreciate that cognition is a fundamental component in a number of theories of emotion. Perhaps the classic theory in this context is that of Cannon. Schacter and Singer (1962) logically extended the theory to show how cognitive factors can influence the conscious experience of emotion. Eysenck and Leventhal point out that even now it remains difficult to distinguish between the James-Lange and Cannon versions of the experience of emotion. In fact, the relative contributions of peripheral and central processes in the experience of emotion is a key theme in attempts to distinguish between these theories. This theme is discussed in Leventhal's chapter and is also a primary focus of McGaugh's discussion of the modulation of memory.

In Part II the contributors discuss the differentiated nature of cognition, showing how different manipulations could alter one aspect of cognition but not some other. Do the modulatory factors discussed in Part III exert specific or general effects on cognitive function? Eysenck's data suggest that implicit processes are affected more in anxiety states than are explicit ones. However, very little work has been done in addressing the specificity of the cognitive effects of alterations in arousal, motivation, and emotion. Such research is potentially very important, since it will have implications for theories of cognition and of emotion. Such information could also be helpful in considering methods of treating emotional disorders.

Finally, can what we know about cognition be used to treat patients with affective disorders? There is an enormous literature on the use of cognitive methods in the treatment of emotional disorders. Some of these cognitive treatments, including traditional psychotherapy, have been clinically effective (Elken et al., 1989). Moreover, psychodynamically oriented treatments are also cognitive in nature; their efficacy, however, remains to be demonstrated, and the theories on which they are based are not a part of contemporary cognitive science. A variety of cognitive techniques that manipulate self-image and self-efficacy have been used to treat patients with disturbed affect, such as depression. In general, these methods have not been successful in the absence of some form of pharmacotherapy. In contrast, probably the most effective treatment of an anxiety disorder, such as a simple phobia, is cognitive in nature (Marks, 1987). Graded exposure to phobic stimuli, systematic relaxation, and imaging techniques have all proved quite useful in treating phobic patients.

References

Bennett, E. L., Rosenzweig, M. R., & Flood, J. F. (1977). In S. Roberts, A. Lajtha, & W. H. Gispin (Eds.), *Mechanisms, regulation and special functions of protein synthesis in the brain*. Amsterdam: Elsevier.

Blum, G. S. (1989). A computer model for unconscious spread of anxiety-linked inhibition in cognitive networks. *Behavioral Science, 34*, 16–45.

Elken, I., Shea, M. T., Watkins, J. T., Imber, S. D., Sotsky, S. M., Collins, D. R., Glass, D. A., Pilkonis, P. A., Leber, W. R., Docherty, J. P., Fiester, S. J., & Parloff, M. B. (1989). National Institute of Mental Health treatment of depression collaborative research program. *Archives of General Psychiatry, 46*, 971–982.

Erdelyi, M. H. (1974). A new look at the new look: Perceptual defense and vigilance. *Psychological Review, 81*, 1–25.

Marks, I. M. (1987). *Fears, phobias and rituals*. New York: Oxford University Press.

Krasne, F. B. (1978). Extrinsic control of intrinsic neuronal plasticity: A hypothesis from work on simple systems. *Brain Research, 140*, 197–216.

Last, C. G., & Hersen, M. (1988). Handbook of anxiety disorders. Elmsford, NY: Pergamon Press.

Robbins, T. W. (1984). Cortical noradrenaline, attention and arousal. *Psychological Medicine, 14*, 13–21.

Schacter, S., & Singer, J. E. (1962). Cognitive, social and physiological determinants of emotional state. *Psychological Review, 69*, 379–399.

Squire, L. R., & Davis. H. P. (1981). The pharmacology of memory: A neurobiological perspective. *Annual Review of Pharmacology and Toxicology, 21*, 323–356.

Weingartner, H. J., & Parker, E. S. (1984). Memory consolidation . . . A psychobiology of cognition. Hillsdale, NJ: Lawrence Erlbaum.

IV

Clinical Perspectives

The contributors to Part IV are interested in clinical phenomena for a variety of reasons. Some use what we know about cognition and brain function to understand and treat diseases that manifest themselves as disturbances in cognitive function. Others study clinical phenomena, believing that these cast light on basic cognitive mechanisms. It is the interplay of these two perspectives that will be the focus of this final part of the volume.

We begin Part IV with a chapter by Lezak, who provides a clinical and psychometric perspective on the diagnosis of cognitive impairments in the clinic. The chapter following, by Milberg and Albert, explores how various cognitive systems function in association with one another. Milberg and Albert discuss in some detail the nature of the cognitive impairments observed in patients with localized lesions. A chronometric analysis of cognitive function in normals, used in conjunction with these clinical data, provides valuable information on the organization of cognitive functions in the brain. Next, Rapp and Caramazza focus on a very specific and well-defined aspect of cognitive functioning, the appreciation and production of language. Building on clinical findings from single case studies, they provide a detailed analysis of the ways in which language functioning might be impaired. In the chapter following, Siegel uses a conditioning model (derived from work discussed by various contributors in Part I) to describe mechanisms of drug tolerance and the development of strategies for treating another important clinical problem, drug addiction. Then, Gur and Gur offer an overview of the currently available techniques that provide an in vivo picture of brain function and discuss how these approaches might be useful in understanding the neurobiology of impaired and normal cognitive functioning. The final two contributions in Part IV consider the most common form of cognitive impairment, Alzheimer's disease. Palmer and Bowen review what we know about the neuropathology of this disorder, while Santucci et al. consider the different strategies that might be used to treat the behavioral cognitive dysfunctions associated with Alzheimer's disease.

19

Identifying Neuropsychological Deficits

MURIEL D. LEZAK

The Relevance of Cognition to Neuropsychological Assessment

The neuropsychological evaluation of organic brain dysfunction rests predominantly on cognitive activity for some very practical reasons: (1) We now have widely accepted and reasonably clear constructs for defining and describing the major categories of cognitive functions and many of their subsets. (2) Useful and readily applicable techniques are available for examining both discrete cognitive functions and complex cognitive behaviors; these include the host of mental ability tests, tests of specific cognitive functions, and laboratory measurement techniques. (3) Brain–behavior correlates for the major categories of cognitive functions and for many of their components have been reliably identified. (4) The expression of many specific kinds of cognitive disorders is now well known and referrable to brain lesion sites with known predictability. In short, the components of the brain–behavior relationship and their range of variability for most cognitive functions are well established. Psychometric tests and laboratory measures of cognitive functions, radiographic imaging, and neurophysiological techniques together constitute a two-way bridge, with many lanes making the conceptual links between brain substance and cognitive activity.

Other behavioral systems—specifically, the emotional/motivational complex and the executive functions (which have some metacognitive aspects) —have many fairly well-known brain correlates as well. However, their components are not as clearly identified nor are assessment techniques as available—or if available, as refined—as those we use to examine cognitive functions.

Brain correlates for emotional/motivational systems are much more complex and to that degree less well defined than they are for the cognitive functions: Interactions between neurotransmitters, other neurohumoral agents, past experience, and anatomic lesion sites appear to weigh more heavily and in more subtle ways in the behavioral expression of emotion than in that of cognition after brain damage. The parameters of these interactions are just beginning to be explored.

357

With respect to executive functions, some brain correlates are fairly well defined, but the range of variability remains obscure; other brain correlates are subject to much conjecture. The whole enterprise of establishing brain–behavior correlations for the executive functions is still limited by problems of definition and a lack of assessment tools with the reliability and predictive validity of those used in measuring cognitive functions.

Thus, neuropsychologists seeking to evaluate brain functions by psychological means turn to their familiar, well-established tests and guidelines for measuring cognitive functions even when the goal is an understanding of the whole brain.

Neuropsychological Methods for Evaluating the Brain's Status

Before the advent of techniques for visualizing brain structure and measuring the brain's physiological functioning, clinical neuropsychology was used primarily for diagnostic assistance. Many clinicians engaged in a test-focused search for definitive cognitive indicators of organicity (Lezak, 1988). These neuropsychological pioneers looked for signs of performance abnormalities associated with brain dysfunction, such as a rotated figure on a design copying task (e.g., Satz, 1966) or perseverative responses (e.g., Piotrowski, 1937). This approach to neuropsychodiagnosis followed in neurology's footsteps, for much of neurology's diagnostic accomplishments rested on the appreciation of the neuroanatomic correlates of specific indicators of deficit and their broader functional and physiological implications. While very useful for the diagnosing neurologist, the neurodiagnostic sign approach was of very limited utility in neuropsychology and of even less value for addressing behavioral issues.

Happily, quite early in the game, astute observers such as David Wechsler (1944) noted that brain damage tended to produce test score highs and lows with sufficient regularity to create patterns that are more or less associable with particular brain disorders and lesion sites. Knowledge of these patterns was soon applied to diagnostic problems (e.g., Reitan, 1955). Of greater significance for the future of clinical neuropsychology, pattern analysis enhanced appreciation of the variety and nature of neuropsychological impairments and opened up the involvement of neuropsychology in patient care, including planning, remediation, and rehabilitation.

Expanding knowledge of the complexities and subtleties of brain–behavior relationships led neuropsychology to abandon the quixotic search for behavioral signs of brain damage in favor of test score pattern analysis. From this background, techniques for comprehensive and practically meaningful behavior analysis have evolved, providing both theoretically and clinically relevant information and—where needed—diagnostically useful findings. Today, all systems of neuropsychological assessment incorporate some form of analysis of cognitive deficit patterns into their approach. Even systems that still purport to generate a single indicator of brain damage (actually, an estimate of the likelihood that the subject is brain damaged), such as the Halstead Reitan Battery (Boll, 1981) or some of its variants, are now accompanied by instructions—in readings, lecture courses, or manuals —on how to analyze the test performance patterns generated by that system to draw more sensitive diagnostic inferences or to develop a practical description of the patient's neuropsychological status.

Applications of Neuropsychological Assessment

The most usual cognitive parameters of many neuropathological entities and of many defined lesion sites are now well known (Damasio & Damasio, 1989; Heilman & Valenstein, 1985). This neuropsychological knowledge base is of sufficient repeatability, consistency, and constancy that neurology and psychiatry call upon it regularly to aid in answering slippery diagnostic questions that elude the neuroradiographic eye. The most frequent calls for diagnostic assistance from neuropsychology concern questions regarding the presence of diffuse brain damage, as occurs in mild head injury, exposure to certain neurotoxins, or early dementia. Neuropsychology's competence in making diagnostic discriminations rests largely on the identification of patterns of cognitive deficit or distortion.

Deficits must also be identified and described once the diagnosis has been documented, if the well-being of patients and their families is to be maximized. Knowing that a patient has a tumor or an area of infarct of such a size in such a place, or was unconscious for a specified length of time following a head injury or cardiac arrest, allows the examiner to make some general predictions regarding the course of the condition, the ultimate level of functioning, and deficit areas. Such knowledge, however, does not grant insight into the nature, severity, or practical implications of the patient's behavioral dysfunctions. For one thing, patients not infrequently surprise us with more or less impairment than the known lesion would seem to warrant. Moreover, even when the site and size of the brain damage has been mapped out and patient sidedness (hand, eye, etc.) seemingly conclusively ascertained, there will still be some patients who break our well-established rules and exhibit expressive deficits when receptive ones were anticipated, or language capacity or configural processing that appears to be organized on the "wrong" side of the brain.

But more than affecting the behavior of patients with anomalous neuropsychological presentations, the effects of brain damage on all except the most severely impaired persons will differ by virtue of level and kinds of premorbid skills and knowledge, attitudes and habits, the important demographic characteristics of age and sex, and minor idosyncrasies of brain organization (no two brains are exactly alike, see De Bleser, 1988; Ojemann, 1979). These individual differences, in turn, will have different and unique implications for the behavior of each patient: This will depend on his or her cultural level, support systems, physical competency, self and social expectations, and the interaction of all these variables with the patient's experience and interpretation of the brain's dysfunction, and with the neural system alterations occasioned by the brain damage and associated pathological changes. In short, only an individualized exploration of a patient's neuropsychological deficits can provide a definite understanding of his or her behavioral potential and limitations.

Neuropsychological Inference and Test Interpretation

The most obvious conclusions a neuropsychological examiner can draw from an array of examination data may be so taken for granted that the examiner loses sight of the inferential process leading to these conclusions. The examiner may not even be aware of the inferential process, having acquired a "cookbook" or predigested acquaintance with some predictive hypotheses relating (or purporting to relate) test performance patterns to brain sites and functions. Unfortunately, cookbook interpretations can lead to clinical malfeasance and fallacious research.

To use neuropsychological test findings meaningfully, to communicate about them effectively, and to design and evaluate studies using these data, their relationship to the concept of deficit must be clearly understood. Deficit measurement underlies all clinical and research applications of neuropsychological data; the measurement of deficit hinges on pattern analysis; and interpretation of the data entry into the pattern analysis paradigm rests on a series of assumptions.

Assumptions Underlying Deficit Measurement and Pattern Analysis

1. *Neuropsychological deficits involve the diminishment or loss of an ability, skill, function, or potential.*

Deficits represent a reduction in performance capacity or potential. It is rare that the underlying damage is of sufficient severity or located in such a vulnerable site that the functional capacity is totally lost. Neuropsychological deficit occurs in the form of more or less diminished or impaired abilities or performances relative to the patient's original potential or known performance level. Thus, deficit measurement deals with quantifying the amount of loss or impairment of cognitive functions, abilities, or skills.

If one takes any cognitive function into close consideration it becomes apparent that no matter how measurement techniques for that function handle the data the behavioral expression of that function actually distributes continuously. Memory, for example, in none of its aspects is an on–

off phenomenon. Visuospatial cleverness varies within the normal population: While impaired performances exhibit distortions rarely if ever normally seen, among visuospatially impaired persons one can discern degrees
of severity in their disabilities (e.g., see Benton, 1967). Similar observations
obtain in every other area of neuropsychological functioning and malfunc-
tioning: for example, attention, verbal skills, and mathematical abilities.

Deficits so profound that they amount to total loss of the function or
capacity are likely to involve the primary components of a functional
system. For example, global aphasias in which the patient has lost all
language skills and capacity are typically associated with destruction of the
primary speech center; cortical blindness occurs when the primary visual
cortex is destroyed (although even then some activity in the visual system
may be retained [see Weiskrantz, 1986]). When damage is this severe, deficit
measurement becomes irrelevant: Anyone can tell that the patient is
groping in full daylight or is unresponsive to verbal stimuli. Should there be
diagnostic doubt as to the authenticity of the deficit, only in those rare
instances when the neurologic or neuroradriographic examination cannot
supply definitive answers would one call for a neuropsychological examination for diagnostic purposes.

2. *Deficit measurement requires a value or standard for comparison.*

Deficit measurement requires the identification of a performance standard against which present performances can be compared. A critical issue
for the neuropsychological assessment of adults is ascertaining what scores
or values representing scores constitute the appropriate standard against
which an obtained score can be compared. This issue is not relevant to
psychological assessment, in which the examiner addresses such questions
as whether Alice is bright enough to take college preparatory courses, or
whether John's health preoccupations are of psychiatric proportions. There
the examiner's usual goal is to determine whether, in what ways, and by how
much the subject's performance differs from normal expectations. Comparisons between the subject's scores and those of a normative population
will answer these questions. In contrast, neuropsychological assessment
seeks to determine whether, in what ways, and by how much the patient's
present performance differs from what it was premorbidly—or at its best. A
first step in deficit measurement, therefore, is to ascertain or estimate the
patient's premorbid cognitive potential.

3. *Measurement of neuropsychological deficits relies upon three different kinds of
performance standards.*

In the normal adult population, excluding persons of very low intellectual capacity (of *borderline defective* caliber—i.e., − 1.3 S.D.—at best) or educationally deprived backgrounds, cognitive functions fall into three different
categories with respect to how performance levels are distributed. One of
these categories involves abilities that are normally distributed in the adult

population; one involves capacities and functions in which normal variations occur within a narrow performance range; and a third has to do with those abilities and skills acquired in the course of growing up and which are fully developed by adolescence if not earlier.

Three Kinds of Comparison Standards for Deficit Measurement

1. With respect to the distribution of performance levels in the adult population generally, the most commonly recognized category includes all skills and abilities as well as those complex neuropsychological functions that distribute normally (i.e., in the form of a bell-shaped distribution curve). The range of impaired functions or abilities in this category tends to be smaller than the range into which normal performances fall. Thus, the ability of neurologically intact adults in our culture to perform mathematical computations ranges from the dizzying heights of mental gymnastics achieved by theoretical mathematicians to seventh- or eighth-grade equivalents; the full range of abnormal, for adequately educated adults, goes from about sixth- or seventh-grade mathematics to finger counting.

The distinguishing feature of this category is that, like all other phenomena that distribute normally, the expression of these skills, abilities, and functions is multiply determined. Skills demonstrate the phenomenon well, as all learned skills are the interactive product of many variables: demographic factors, sensory and motor capacities, cognitive functions, and exposure to the appropriate stimuli and training. For example, the ability to write legibly involves—among numerous other variables—controlled and flexible hand movements linked to fine hand muscle coordination, years of motor practice (educational opportunity and support), knowledge of alphabet forms, good visual discrimination (at least during the formative years), the ability to transfer visual images into hand movements, and motivation to learn this skill (see Roeltgen, 1985). Complex cognitive abilities share in this distribution characteristic, for they, too, are multiply determined. Levels of accuracy of visuospatial judgments—which actually depend on learning, too, but learning that takes place implicitly in the course of development rather than explicitly in school (see Piaget, 1967)—also distribute normally within the normal population defined earlier.

When seeking an appropriate performance standard for normally distributed cognitive capacities, in most cases it is the potential that must be estimated rather than the actual level of functioning, as many adults do not fully realize their capacities, or do so in ways that are not readily documented. For example, the level of academic performance as indicated by school grades may underestimate a person's premorbid potential. This is particularly true of those who came from families in which educational prowess was neither valued nor encouraged. If all the examiner has to go on are high school grades, the ability level of educationally disinterested high school dropouts who entered the labor force at 16 or 17 and rose to positions of responsibility by virtue of their wits and acquired skills are likely to be undervalued and their actual deficits insufficiently appreciated. For persons with documented evidence of high intellectual competency, such as professionals or those with advanced degrees, their accomplishments represent an actual level of functioning that can serve as an acceptable performance standard, as it is much less likely that a PhD degree represents a significant underestimation of ability than does a set of poor grades from the second year of a rural high school.

Rarely, the academic record provides an overestimation of premorbid ability. Some families push and prod their children of mediocre ability with intensified teaching programs, tutors, and home drills, enabling them to make better grades and get higher achievement test scores than their abilities would normally warrant. And some people do drive themselves beyond their natural limits. Here, too, the examiner must tease out an appropriate estimate of potential from the available data.

In general, when dealing with normally distributed cognitive abilities and skills, establishment of a comparison standard rests on the evaluation of all relevant information about the patient. This includes *school documentation* (grades, test scores, academic achievement), *life history* (armed services record, job skills, and responsibilities), and *neuropsychological findings*. In this historical material the highest performance levels, representing those residual functions least affected by the patient's brain disorder, may offer the best estimate of the patient's premorbid abilities.

It should be obvious that the examiner must look to the indicators of highest mental competency for making judgments about an adult's premorbid ability. Few persons reach their full potential in any performance area, much less in all areas. Any given performance—whether it be a school grade or a job description—demonstrates that the individual can do at least that well. It may be that a given individual's potential is actually higher than his best performance would indicate. The best performance (or ideally, set of similarly high performances) provides only a minimum estimate of general ability: we cannot know what more that person would have accomplished by trying a little harder, having had better educational advantages, and so on.

Ascertaining an appropriate comparison standard for functions, abilities, and skills that distribute normally (i.e., in a bell-shaped curve) in the population at large is further complicated for persons whose premorbid functioning was exceptionally high or exceptionally low. For many adults of high premorbid cognitive competence, performances reflecting genuine deficits may look perfectly normal (i.e., fall within the *average* ability range, the middle 50% of a normal distribution) and not be recognized for what they are if a comparison standard is based on a normative population. Scores of persons whose best functioning levels were below the *average* range and who now perform at somewhat lower levels do not show the wide performance discrepancies that neuropsychologically impaired persons of *average* and better endowment may display, because the statistical structure of many tests makes them relatively insensitive to genuine performance discrepancies in the lower ability ranges. It is much easier to identify subtle deficits in bright than in dull people.

Enough information is available about most patients—from the school or occupational history, or from current examination data—to enable the examiner to make a reasonably appropriate estimate of premorbid potential. However, when a patient's accomplishments are unknown or have been mediocre at best, and brain damage is so widespread or severe as to have affected all aspects of cognitive functioning, the examiner may find it difficult to estimate a premorbid level for determining the relative severity of the patient's cognitive impairments. To overcome these obstacles, some ingenious techniques for estimating premorbid ability level have been devised. One set applies formulas to extract an estimate from demographic data—age, sex, amount of education (Karzmark, Heaton, Grant, & Matthews, 1985; Wilson, Rosenbaum, & Brown, 1979). Another approach to this problem is through somewhat indirect techniques for ascertaining the level of the premorbid reading vocabulary (Baddeley, Emslie, & Nimmo-Smith, 1988; Crawford, Parker, & Besson, 1988). This approach rests on the assumption that because cognitive abilities generally correlate most highly with vocabulary, vocabulary is the one knowledge base most likely to predict premorbid ability.

2. In contrast, functions that are minimally dependent on learning and experience and relatively unaffected by demographic variables for their presence and expression typically do not distribute normally within the normal adult population. From a neuropsychological point of view, the most prominent of this class of functions are attention and memory. For these functions, normal performances tend to occupy a relatively small range of their continua, whereas abnormal performances show up in long tails on the impaired side of the normal distribution. Thus, most neurologically intact adults between the ages of 16 and 60 can recall 12 or more words of a 15-word list upon hearing it repeated five times (Crawford, Stewart & Moore, 1989; Geffen et al., 1990). Abnormal performances by neurologically impaired patients have a much wider range, from 0 to 11 (Rosenberg, Ryan, & Prifitera, 1984). Similar range differences between tests performed by neurologically intact subjects and the abnormal scores of neurologically impaired patients show up when response speed is measured, as in tests involving attentional functions. Examples in this category could be given by tests sensitive to distractibility, visual tracking tests that involve capacities to focus and shift attention rapidly, and complex reaction-time measures. On all these tasks, normal performances tend to bunch up in a relatively small range at the fast end of the continuum with a relatively short and thin tail on the slow side. Abnormal scores spread out in the slow direction as far as the examiner's patience or lack of common sense (in allowing an abnormally slowed performance to drag on) will allow.

For these functions and abilities, normative population data provide the comparison standard for all adults. Thus, the comparison standard for this class of functions and abilities is similar to the one that obtains in clinical psychology: The subject's performance is compared with the population at large. Moreover, as in clinical psychology, the examiner must take into account age—and where possible, sex—in ascertaining the most suitable standard for comparing any given subject's performance. Education, too, contributes to performance efficiency on both memory and attentional tests (Lezak, 1983). Most probably this contribution takes place indirectly by increasing "test-wiseness" (i.e., with schooling comes sophistication about taking all kinds of tests) or reflecting the well-documented correlations of educational level with general health, and by inference, with stamina and vigor.

Interestingly, this difference between functions that distribute normally and those relatively independent of learning, experience, and demographic characteristics parallels the distinction some authorities make between *cortical* and *axial* functions. Those functions and abilities that distribute in a normal bell-shaped curve have important and dedicated cortical contributions such that lesions in the correlative cortical areas are sufficient to compromise them significantly and relatively discretely. Moreoever, these functions and abilities are distinguished by the fact that they are developed or enhanced by learning. Functions that do not distribute normally in the adult population—the axial functions—are likely not to be compromised by localized cortical lesions, nor does training or practice change them very much. Attentional deficits show up most often when damage is diffuse or widespread, or when it involves brainstem components. With a few exceptions, such as certain kinds of lateralized cortical lesions or frontal lobe damage, memory disorders are associated with damage to one or more components of the limbic system, which lie along or close to the neural axis and central structures of the brain.

3. Much adult cognitive activity involves well-learned skills acquired throughout infancy and childhood. Speech and language are probably the most obvious and omnipresent of these. By the time a younster reaches the teen years, an elaborate grammar and syntax have been acquired along with a more or less extensive vocabulary, clear enunciation, and culturally appropriate rhythm and tonal qualities of speech. Of all these aspects of speech and language, only the size of vocabulary distributes normally, so when neuropsychological deficits are suspected, vocabulary must be evaluated relative to an intraindividual comparison standard. All other aspects of speech and language are assumed to be fully intact in the adult. Should abnormalities appear in the structure of speech or in its rhythm or expressiveness, then deficits—relative to premorbid competency—may be inferred. For example, the agrammatic, telegraphic speech of a person with expressive aphasia indicates deficit on a neuropathological basis. No formally identified comparison standard is required to determine that agrammatic speech in an adult who once spoke well represents a neuropsychological deficit.

Deficits in many other activities come under this rubric: All adults can dress and groom themselves unaided (assuming adequte motor capacity), gesture appropriately, and know how to use common tools such as scissors, forks, combs, and hammers. Alterations or loss of these or similar kinds of skills represent significant neuropsychological deficits. Although the identification of neuropsychological deficit in these behaviors does not require elaborate inferential procedures, neuropsychological assessment may be useful in bringing these kinds of deficits to clinical attention.

The Clinical Use of Deficit Measurement

The appropriate conceptualization and clinical application of deficit measurement rests on an individualized approach to neuropsychological assessment. Each patient must be evaluated within a frame of reference afforded by that patient's background and present functioning. This requires the examiner to become acquainted with the patient's history, and to observe carefully how the patient acts and speaks and what is said. The examiner must not just record the scores the patient attains, but must pay attention to how the patient achieved them, knowing that dysfunctional patients often betray their higher premorbid cognitive competence than the scores would seem to warrant in a vocabulary, a problem-solving approach, or an acquaintance with concepts that is better than the level suggested by scores alone. On the other hand, the scores may represent residua of premorbid abilities, requiring the examiner to identify those that run higher than the patient's general level of competency as currently tested.

Exploring the possibility that the patient enjoyed higher premorbid levels of functioning than present performance on tests of normally distributed functions seems to indicate is only one examination responsibility. The examiner must look to normative criteria in evaluating memory and attentional functions, and also abilities with large response speed or physical competency components. Further, the examiner must be sensitive to the often subtle alterations in basic skills and knowledge that reflect neuropathological alterations in brain function.

Deficit measurement in neuropsychology thus requires a broad base of knowledge about cognitive functions, their statistical parameters, the tests used to measure them, and the variables—neuropathological, physiological, demographic, psychosocial, emotional—that can affect their expression. Acquiring the knowledge and experience base necessary for responsible evaluation of a person's neuropsychological status takes learning, training, and practice. Only knowledge, training, and guided experience will enable the clinician to know what to look for when doing deficit measurement, to become sensitively alert to historical markers and behavioral nuances, and to exercise good clinical judgment so that descriptions of the patient's neuropsychological status are relevant and useful.

References

Baddeley, A., Emslie, H., & Nimmo-Smith, M. I. (1988). Estimating premorbid intelligence. *Journal of Clinical and Experimental Neuropsychology, 10*, 326 (abstract).

Benton, A. L. (1967). Constructional apraxia and the minor hemisphere. *Confinia Neurologica, 29*, 1–16.

Boll, T. J. (1981). The Halstead-Reitan Neuropsychology Battery. In S. B. Filskov & T. J. Boll (Eds.), *Handbook of clinical neuropsychology*. New York: Wiley-Interscience.

Crawford, J. R., Parker, D. M., & Besson, J. A. O. (1988). Estimation of premorbid intelligence in organic conditions. *British Journal of Psychiatry, 153*, 178–181.

Crawford, J. R., Stewart, L. E., & Moore, J. W. (1989). Demonstration of savings on the AVLT and development of a parallel form. *Journal of Clinical and Experimental Neuropsychology, 11*, 975–981.

Damasio, H., & Damasio, A. R. (1989). *Lesion analysis in neuropsychology*. New York: Oxford University Press.

DeBleser, R. (1988). Localization of aphasia: Science or fiction. In G. Denes, C. Semenza, & P. Bisiacchi (Eds.), *Perspectives on cognitive neuropsychology*. East Sussex, England: Lawrence Erlbaum.

Geffen, G., Moar, K. J., O'Hanlon, A. P., & Clark, C. R. (1990). The Auditory Verbal Learning Test (Rey): Performance of 16 to 86 year olds of average intelligence. *The Clinical Neuropsychologist, 4*, 45–63.

Heilman, K. M., & Valenstein, E. (Eds.) (1985). *Clinical neuropsychology* (2nd ed.). New York: Oxford University Press.

Karzmark, P., Heaton, R. K., Grant, I., & Matthews, C. G. (1985). Use of demographic variables to predict Full Scale IQ: A replication and extension. *Journal of Clinical and Experimental Neuropsychology, 7*, 412–420.

Lezak, M. D. (1988). Neuropsychological tests and assessment techniques. In F. Boller, J. Grafman, G. Rizzolatti, & H. Goodglass (Eds.), *Handbook of neuropsychology* (pp. 47–68). Amsterdam: Elsevier.

Lezak, M. D. (1983). *Neuropsychological assessment* (2nd ed.). New York: Oxford University Press.

Ojemann, G. A. (1979). Individual variability in cortical localization of language. *Journal of Neurosurgery, 50*, 164–169.

Piaget, J. (1967). *Biologie et connaissance*. Paris: Gallimard.

Piotrowski, Z. (1937). The Rorschach inkblot method in organic disturbances of the central nervous system. *Journal of Nervous and Mental Disease, 86*, 525–537.

Reitan, R. M. (1955). Certain differential effects of left and right cerebral lesions in human adults. *Journal of Comparative and Physiological Psychology, 48*, 474–477.

Roeltgen, D. (1985). Agraphia. In K. M. Heilman & E. Valenstein (Eds.), *Clinical neuropsychology* (2nd ed.). New York: Oxford University Press.

Rosenberg, S. J., Ryan, J. J., & Prifitera, A. (1984). Rey Auditory-Verbal Learning Test performance of patients with and without memory impairment. *Journal of Clinical Psychology, 40*, 785–787.

Satz, P. A. (1966). A block rotation task: The application of multivariate and decision theory analysis for the prediction of organic brain disorder. *Psychological Monographs, 80*, 21 (Whole No. 629).

Wechsler, D. (1944). *The measurement of adult intelligence* (3rd ed.). Baltimore, MD: Williams and Wilkins.

Weiskrantz, L. (1986). *Blindsight*. Oxford, England: Oxford University Press.

Wilson, R. S., Rosenbaum, G., & Brown, G. (1979). The problem of premorbid intelligence in neuropsychological assessment. *Journal of Clinical Neuropsychology, 1*, 49–54.

20

The Speed of Constituent Mental Operations and Its Relationship to Neuronal Representation: An Hypothesis

WILLIAM MILBERG / MARILYN ALBERT

Although damage to the human cerebrum sometimes produces cognitive deficits that are highly predictive of the site of the causal lesion, in many cases deficits in cognition are not predictive of the site of the causal lesion, because they are only associated with diffuse or multiple lesions. The present chapter argues that the apparent variations in the "localizability" of cognitive functions are not random occurrences or merely artifacts of our methods of assessment. Rather, it is surmised that cognitive functions vary in the degree to which they may be localized because of the principles of neural representation that underlie them. It is hypothesized that the form in which a particular function is represented in the cerebral cortex is mandated by the intrinsic processing demands required by that function as well as by natural limitations imposed by biology. A number of possible "rules" will be examined that might be used to predict the "localizability" of a particular cognitive function, with the ultimate goal being the description of general principles of neural representation.

Lesion Localization and the Localization of Function

The clearest place to begin in the exposition of the principle of "localizability" is to consider differences in the apparent effect of lesions on the two cerebral hemispheres. There is a large body of clinical and experimental observation that suggests that the effects of lesions to the left hemisphere are more site specific than are the effects of lesions to the right hemisphere (Kertesz, 1983a). Sets of relatively focal lesions in the left hemisphere can produce substantial deficits to linguistic functions that in many cases are categorical and independent. For example, an ischemic lesion of the left frontal operculum that includes the third frontal gyrus and such underlying structures as the insula and the putamen (Mohr, Pessin, Finkelstein, Funkenstein, Duncan, & Davis, 1978) will frequently (in right-handed adults) result in an aphasia characterized by severely limited spontaneous speech (Levine & Sweet, 1982). Small variations in lesion placement, often within one or two centimeters, will determine whether these difficulties with speech output are accompanied by deficits in the use of grammatical information in both production and comprehension (Zurif, 1980), the ability to repeat (Benson, 1979a), and such attendant language skills as reading or writing.

Ischemic lesions to the left temporal lobe that include the supramarginal gyrus, the superior temporal gyrus and the Sylvian fissure (again in right-handed adults) will generally result in aphasic symptomatology characterized by poor auditory comprehension. Though the spontaneous speech of these patients may have a normal rhythm and flow, it is filled with numerous paraphasic errors or even jargon (Kertesz, 1983b). Again, depending on small variations in the placement of the lesion, patients may show variations in the ability to repeat, read, write, and name.

Although there are some differences in the number and designations of the diagnostic categories used, most nosological systems of aphasia describe more than a dozen combinations of language symptoms (affecting auditory input and output, vocabularies, repetition, reading, writing and naming, etc.) (Goodglass & Kaplan, 1972; Kertesz, 1979). In most cases, each symptom combination has been linked to a distinctive lesion site within the left cerebral hemisphere. The highly localization-specific, clinical-anatomical associations apparent with left hemisphere lesions have not been strictly limited to language symptoms. Other constellations of symptoms occurring with left hemisphere lesions include the various apraxias and agnosias (Bauer & Rubens, 1985; Geschwind, 1974). By contrast, deficits associated with the right hemisphere show less clustering into syndromes and are less clearly associated with a specific lesion location. For example, though the majority of evidence has emphasized the role of right posterior parietotemporal and occipital areas in constructional deficits, (e.g., drawing, block design), anterior portions of the right hemisphere have also been associated with deficits on such tasks (Kertesz, 1979). Similarly, disturbances in the perception of affect, the production of prosody, and the comprehension of metaphor, humor, and connotative semantic information have all been found following lesions to either the frontal, anterior temporal and parietal regions of the right cerebral hemisphere (Bowers, Bauer, Cosslett, & Heilman, 1985; Brownell, Potter, & Bihrle, 1986; Gardner, Brownell, Wapner, & Michelow, 1983; Ross, 1981). Even such clearly defined syndromes as hemispatial neglect can occur with lesions to almost any part of the dorsolateral surface of the right cerebral hemisphere (Heilman, Watson, Valenstein, & Damasio, 1983). The degree of localizability of functions also varies considerably within each hemisphere. Furthermore, some functions, syndromes, symptoms, or psychological test performance patterns do not appear to be particularly localized to either hemisphere. Naming as an indication of left hemisphere damage is a less localizing symptom than are deficits in auditory comprehension (e.g., Benson, 1979b), face recognition is more localizing than face discrimination (Damasio & Damasio, 1983), hemiparesis is more localizing than apraxia, and deficits on the Picture Arrangement subtest of the WAIS are more localizing than are deficits on the Block Design subtest of the WAIS (Milner, 1963; Newcombe, 1969).

As noted above, basic motor and sensory functions, certain aspects of linguistic performance (such as the production of fluent, grammatical sentences), and performance on a host of neuropsychological tasks appear to be predictive of the occurrence of lesions in specific cortical (and sometimes subcortical) areas. There may be some variability of the boundaries of these areas across individuals (Alexander, Naeser, & Palumbo, in press) but the likelihood of occurrence of these deficits decreases as the site of lesion is moved and seems to be relatively unaffected by lesion size (Naeser, 1983). In contrast, deficits in the generalized maintenance of attention or vigilance, the ability to perform verbal or nonverbal abstractions, certain aspects of pattern recognition (e.g., the ability to copy or draw), and certain other deficits in linguistic performance (e.g., naming) occur with lesions in a relatively wide variety of areas (Newcombe, 1969). As opposed to more focally represented functions, the likelihood and severity of these deficits do appear to vary as a function of lesion size.

Theories of Neural Representation

In recent years, theories of the neural representation of cognitive functions in the brain have increasingly led to a search for structures that behave as independent information-processing devices (Caplan, 1981; Caramazza, 1984; Coltheart, Patterson, & Marshall, 1980). The hope has been that the human cerebrum is organized as if it were a "cognitopic" mosaic (i.e., where each processing device is laid out in a nonoverlapping tessellation in the brain). Therefore, in principle at least, individual pieces could (through the agency of a lesion) be "removed." The usual assumption underlying this conceptual framework is that the elimination of the contribution of each neural information processing device will have a unique effect on behavior that should be predictable from theories of cognitive function that have been formulated independently of biological considerations (e.g., Caplan, 1981; Caramazza, 1984; Grodzinsky, 1986).

Although it is virtually axiomatic that networks consisting of large numbers of interconnected neurons underlie the execution of such complex mental operations as language, vision, and memory, the distribution of these neuronal networks varies considerably, making it improbable that the "cognitopic mosaic" can serve as a description of the neural representation of every function or mental operation. As the examples briefly outlined in the first section suggest, some mental operations indeed seem to be based on neurons clustered closely together within a relatively small volumetric space, whereas other mental operations seem to be based on neurons that are distributed over a relatively large volumetric space. The examples of comprehension, reading and writing, apraxia, and so on, are functional symptom categories that are certainly composed of more basic cognitive operations. Nevertheless, their localizability indicates that each has at least one such basic operation that is biologically determined and focally represented. Although some focally realized functions may be represented by spatially discrete neural structures, other functions, though apparently representative of functionally independent cognitive processes or even different cognitive domains, may actually share the same neural architecture, and may in principal not be neurally dissociable. This issue has been discussed in great detail by Kosslyn and his colleagues (Kosslyn & Van Kleeck, in press). It therefore seems likely that damage to neural architecture shared by different functions will result in an *association* of deficits (i.e., syndromes) rather than a *dissociation* of deficits. The best-known example of an association of deficits derived from different functional domains occurring with a relatively circumscribed lesion is Gerstmann's syndrome, which represents the association of abnormalities in writing, left-right discrimination, finger recognition, and arithmetic computation. The association of these disparate symptoms might indicate either that there is an actual sharing of a single basic functional component between the symptoms, or that several independent functional systems share the same architecture. The first possibility cannot be ruled out, but attempts to find a unifying component for the four symptoms of Gerstmann's syndrome have not been successful (Benton, 1977), making the second possibility more likely.

There are also numerous instances of associations of deficits that are representative of seemingly independent cognitive domains that only occur with multiple focal lesions, or with widely distributed neuronal cell loss. For example, patients with mild-to-moderate Alzheimer's disease typically show deficits in both drawing and proverb interpretation. However, these diffuse insults appear to spare many functions that are focally represented. Patients with advanced Alzheimer's disease almost never show the nonfluent speech or deficits in the production of grammatical sentences associated with ischemic lesions and rarely show deficits in basic motor or sensory functions until the final stages of disease, although a small vascular lesion of the motor strip is often suffcient to produce a permanent, severe hemiparesis in most right-handed adults.

The assertion that a unitary processing architecture can underlie numerous independent functions is by no means new (Freud, 1953; Hebb, 1949; Jackson, 1878; Lashley, 1950) or unique to the brain. The typical digital computer can sustain completely different types of programs (e.g., graphics, word processing, controlling an air conditioning/heating system) using the same set of transistors, diodes, and capacitors etched on a single silicon chip. Though these programs can sometimes run simultaneously, their access commands, editorial capabilities, and end products may be completely different from one another and incompatible. The possibility that a shared processing architecture may underlie some seemingly disparate mental operations has been virtually de-emphasized or ignored within the current zeitgeist of localized cognitopic mapping. Earlier theoreticians, including Jackson, Goldstein, Head, and Freud each presented varations of the argument that the entire brain participates in all higher mental operations. This contention was given its greatest theoretical force by Karl Lashley in the concepts of "mass action" and "equipotentiality." As Lashley wrote, "the same neurons that retain the memory traces of one experience must also participate in countless other activities." Although Donald Hebb believed that there was considerable specialization of structure within the brain, his theory of the "cell assembly" captured the idea that information is stored as a pattern of firing among a population of neurons, as opposed to a one-to-one correspondence to any particular member of that population. It is thus interesting to note that the theories of "mass action" and the "cell assembly" have provided the basis for a number of computer simulations of cognitive functions that use networks of neuronlike processing elements to perform such functions as complex pattern recognition and lexical access (McClelland, Rumelhart, & Hinton, 1986). However, most attempts to model cognitive operations using computer models of neural networks lend themselves to shared and unshared processing architectures, or make no commitments to the form of representation in which the function is actually realized.

Measurement Precision and Localizability

One must thus ask whether variations in"localizability" provide a clue to the form of representation of cognitive operations in the cerebral cortex and to the relationship between structure and function in the central nervous system; or whether the apparent degree to which functions may be localized is simply the result of the tools available to measure the functions in question. It could be argued, for example, that apparent variations in localizability are due to the fact that some functions are simply too poorly defined or difficult to observe, thus increasing the likelihood that measurement error will obscure the true focal representation of the underlying operation. This explanation is unlikely to be true because there are dramatic instances where the increased refinement of measures has led to poorer localization, and instances where poorly defined dimensions seem to be clearly localized. For example, Caplan (1988) has recently conducted an extensive study of the relationship among a number of measurements directly derived from linguistic theory. In this study a patient's competence with various aspects of sentence parsing was measured independently of his or her ability in phonological discrimination, memory, and the interpretation of semantic information. The most striking aspect of the data was the fact that most of the formally derived measures were not consistently predictive of lesion localization. For example, patients with right-frontal-lobe lesions were found to have scores similar to patients with left temporal lesions on some measures. Various lesions therefore resulted in associations and dissociations of symptoms that did not fit comfortably with linguistic theory. Caplan chose to pessimistically interpret this outcome as an indication of the unpredictability of the representation of all linguistic functions. It is more likely, however, that Caplan has demonstrated a critical feature of the manner in which processes are represented in the brain: Some aspects of language processing are focally represented and biologically determined, whereas other aspects of language, though dependent on the operation of these focally represented processes, are derivative of biologically determined processes and involve the brain as a whole. The fact that the linguistic measures correspond well to logically distinct categories formally derived from theories of language is no guarantee that those measures directly correspond to biologically determined and focally represented functions in the cerebral cortex.

In contrast, consider the difficult-to-define dimension of global verbal fluency, so often used in the evaluation of language ability (Goodlgass et al., 1972; Kertesz, 1979). Global fluency is one of the best predictors of lesion localization in the clinical aphasiological armamentarium (Benson, 1967). Though there have been attempts to break fluency down into quantifiable measures such as sentence length or words per minute, none of these measures are better predictors of lesion site than are the global ratings (Benson, 1967). Though fluency itself is not a variable of concern in most linguistic theories (fluency would most likely be relegated to the category of performance variables not relevant to linguistic competency), variations in fluency are highly correlated with other language symptoms, such as agrammatism and semantic comprehension, which map more easily into linguistically relevant processing categories. Global fluency (or some as yet undetermined aspect of fluency), though difficult to define, is a focally represented and biologically determined aspect of language performance that may be an important ingredient in linguistic competence.

Characteristics of Focally and Nonfocally Represented Functions

If measurement precision and theoretical construct validity is no guarantee that a function (or symptom) is going to be focally represented, are there any intrinsic characteristics of mental functions that determine whether they are going to be "localizable" and, by implication, focally represented and biologically determined? There are a number of characteristics that appear at first glance, to be descriptive of cognitive operations that are focally represented. However, as will be argued, very few of these characteristics represent actual biological constraints on the implementation of these operations. They are as follows:

Complexity

The most obvious dimension along which focally versus diffusely organized abilities might differ is that of complexity. The idea that processes requiring more intricate or more numerous neural calculations should require an increasing amount of processing space has intuitive appeal. The ability to access lexical items in the course of auditory comprehension (which is relatively localizable) appears to be more simple than the ability to comprehend proverbs or the ability to perform abstract category judgments; the latter two are abilities that show a significant decrement with lesions to many areas of the cortex (Lezak, 1983). This principle implies that the central nervous system follows some sort of capacity conservation principle, commiting varying amounts of neural volume as a function of the demands of the task at hand.

There are several reasons, however, why complexity alone is unlikely to account for differences in focal representation or localizability. First, the concept of complexity is sufficiently ambiguous to allow numerous counterexamples to direct localizability/complexity mapping. For example, the ability to copy geometric shapes is reduced by lesions to a wide variety of cortical areas, so that deficits in this ability (with a few exceptions) do not allow one to predict the locus of the causal lesion. The ability to maintain sequential word order, or use appropriate case, tense, and number markers in conjunction with nouns and verbs, is specifically reduced by lesions to the left frontal operculum. Which of the two functions is more complex? Using the criterion of the "Turing Machine" (i.e., a machine simulating an ability in a manner indistinguishable from a human), the former ability, which can be easily modeled using computer simulation, would be considered the simpler function. Yet the ability to generate (and understand) novel but grammatically correct sentences in a human language remains an only tenuously attained Holy Grail of artificial intelligence. In addition, because a function is diffusely represented does not mean that more neuronal units (or capacity) are involved. Regions of the brain vary greatly in neuronal density, so some cytoarchitectonic areas contain more neurons than others. For example, Brodmann's Area 44 (the equivalent of Broca's area) is nearly twice as dense as the immediately adjacent prefrontal Brodmann's Area 45 (Blinkov & Glazer, 1968). In principle at least, a focally represented function could use more neurons and require more capacity than a diffusely represented function.

Modularity

Fodor (1983) drew an important distinction between functions that could be considered "horizontal" or "domain specific"; "informationally encapsulated"; "cognitively impenetrable"; and hardwired and functions that are "vertical"; "domain general"; "informationally unencapsulated"; or "cognitively penetrable" with no specific corresponding neural architecture—the implication being that focal representation is a possible neural response to the need for some processes to be functionally isolated. Fodor likened these modular operations to hardwired, automatically performed "reflexes" that represent fixed, rule-bound pathways between stimulus (or input) and response (or output).

Vision, for example, would be considered a modular function par excellence. It depends on receptors and an information format that is uniquely adapted to process light energy (i.e., "domain specificity"); the underlying interpretive routines used to process such visual information as depth, texture, figure and ground, and so on, cannot under usual circumstances be used for any other forms of information and are not easily accessible to consciousness (i.e., encapsulated and impenetrable), and the visual system has highly specific and focally represented hardware (occipital cortex, superior colliculus, medial geniculate, etc.). The ability to perform basic arithmetic operations, though capitalizing on a number of modular systems such as language, would in itself be considered nonmodular. Arithmetic rules can be learned and evoked consciously, are not tied to any specific modality, and don't seem to have a single corresponding brain region.

Modular representation, however, has costs and benefits associated with it. The distinction (and attendant costs and benefits) between "automatic" and "controlled" processes maps somewhat onto the distinction between modular and nonmodular processes. The main benefits being efficiency (i.e., the function is performed with minimal cost to the rest of the system), accuracy, and speed. The costs of modular representation include inflexibility, lack of conscious control, and difficulty in error correction. Finally, the use of dedicated hardware carries along with it the danger of complete loss of function in the face of trauma or disease.

Unfortunately, most of these characteristics are secondary or emergent properties of cognitive functions that do not place necessary constraints on the form of underlying neural representations. Independent functions may certainly be implemented within a shared processing architecture yet remain "informationally encapsulated" and "domain specific." Most computers, for example, operate "wholistically" in the sense that the same central processing units and memory chips can be used for word processing, drawing pictures, and producing sounds. Imagine, if you will, the problem of deducing the structure of the underlying hardware of a computer by observing the effects of the destruction of a chip on various programs.

Functional independence of computer programs does not only refer to the fact that, for example, most word processing programs cannot be used to draw pictures and that most drawing programs cannot be used to write actual text. A graphics program may certainly be used to draw *pictures* of letters on a computer screen that are identical in appearance (e.g., font and size) to the letters typed with a word processing program. The pictures of letters may sometimes be combined with text in a word processing document; however, these pictures are ignored when any of the editing functions of the word processing program are used—only letters created by the word processing program are treated as "symbols" that can be moved, deleted, or altered as a class. A command may be entered to "change all the letter A's to the letter B." All the letters created by the word processor will be altered, though the "picture" of the letter A created by the drawing program will remain in A (unless a third program or command is used to specifically scan the picture of the letter and convert it into the symbolic letter).

Computer programs, or at least the by-products of those programs, can retain functional independence while sharing the same electronic architecture largely as a result of the physical properites of that architecture (Pylyshyn, 1986). A common serial computer performs its calculations with sufficient speed that different programs can alternately share the same physical computational structures. The computer achieves the functional independence of its programs by the specificity of commands needed to access its processing architecture. There is nothing inherently different in the physical structures or algorithms used to produce functionally independent computer programs. A thorough knowledge of a word processing manual (or even of the language used to produce the program) is unlikely to help one understand the effect of breaking a component inside the "black box" of the computer's central processing unit. Thus, although modularity may be a sufficient condition to produce focal representation, it is by no means a necessary condition. As was argued earlier, cognitive abilities may in principle be functionally independent, but may be implemented by shared neural structures.

Processing Speed as a Biological Constraint on Neural Representation

Processing speed is another alternative distinction between focally and diffusely represented functions. It can be argued that some functions, to be carried out effectively, must depend upon rapid processing time, whereas others can be effectively executed even when processing speed is reduced.

Given these considerations, it is hypothesized that the functions that need speed in their execution will be more focally represented than will functions without such speed requirements. Because synaptic transmission time has fixed upper limits, operations can be executed quickly only if they require few synapses or if the distance from one synapse to another is relatively short. Because complex operations such as those we are discussing undoubtedly involve extensive neural networks, the only way for a process to occur more quickly (if this is necessary) is if the distances traveled or the volume of tissue involved is reduced. For example, there are a large number of words, word endings, and word orderings that can be used to create intelligible grammatical sentences. Considering that the average English speaker produces comprehensible and syntactically correct sentences at the rate of 140 to 180 words per minute (Wingfield & Stein, 1989), or up to 126 milliseconds per syllable (Miller, Grosjean, & Lomanto, 1984), constituent word and sentence structure choices must be made extremely rapidly. By contrast, evoking the algorithms used in solving a basic arithmetic problem may take on the order of seconds even for a practiced normal adult.

If one examines this at a more hypothetical level, one can argue that the minimum requirement for a unit of information processing is that some rule-bound relationship (or function) between two variables must be calculated. The most likely determinant of speed needed for a unit of information processing would be the redundancy or complexity (nonlinearity) of the variables to be computed, as well as the real-time limits within which the computations must be made. Because of the biological limits set by synaptic transmission time, information will be processed more quickly over a short distance (or within a confined area) than over a long distance. It is suggested that the requirement for rapid information processing results in eventual localized control of function, because upper neural speed is a constant (or at least has fixed upper limits). The only way for a process to occur more quickly is to shorten the distances (or decrease the volume) over which the constituent information processing must occur. There should be a continuum of "localized control" related to the input/output metric. The greater the need for information processing speed, the greater the degree of local control. This claim is based completely on the limits imposed by "real-time" neural processing speed; however, it is likely that such things as the complexity of the informational translation will also affect the degree of localization. Hence, a single simple serial or linear translation of information will have a higher "localization" threshold than will a more complex parallel or nonlinear translation of information. A parallel processing routine may therefore be more "local" than a serial processing routine. In both cases, localization will be determined by the limits of real-time neural processing, though the complexity of the function in question will determine the time scale in which the information translation occurs.

At this point in time the actual data on the processing speeds of various consituent cognitive operations (with a few exceptions) are not available. Therefore, to assess whether this hypothesis has any intuitive basis in neuropsychology we informally surveyed a group of colleagues and asked them to rate the speed with which they thought a particular operation needed to be executed to be carried out effectively. We then asked them to rate the degree to which the functions listed were focally represented in the brain. Table 20.1 presents the results of this inquiry. A regression analysis between these two ratings indicated a strong relationship between the two estimates of the raters ($r = .88$). Thus, at least in the minds of our raters, there is a relationship between the necessary speed of execution of a cognitive operation and the focal representation of that function in the brain.

Table 20.1. Mean Constituent Speed and Localization Ratings of Selected Basic Cognitive Operations (N = 5 Raters)

	Operation	Rated speed[a]	Rated degree of local representation[b]
1.	Calculate locally defined syntactic rules	1	1
2.	Calculate globally defined syntactic rules	2	3
3.	Retrieve open class lexical item	2	2
4.	Retrieve closed class lexical item	1	1
5.	Name object to confrontation	3	3
6.	Retrieve cliche/stereotypy	3	3
7.	Comprehend single spoken lexical item	2	2
8.	Comprehend metaphor	4	4
9.	Comprehend joke	4	4
10.	Tell personal history	5	5
11.	Perform semantic judgment	3	3
12.	Automatically activate lexicon	1	1
13.	Convert a grapheme to phoneme	1	1
14.	Convert a grapheme to a lexical item	2	2
15.	Create primal visual sketch of object	1	1
16.	Recognize face	2	2
17.	Draw picture of object	4	4
18.	Parse a visual scene into figure ground	3	3
19.	Move fingers sequentially	1	1
20.	Track moving object with limb	1	2
21.	Memorize a novel list of words	4	4
22.	Imagery	4	4

[a]1 = fastest, 5 = slowest.

[b]1 = most focal, 5 = least focal.

Some Psychological and Biological Implications of the Constraints of Processing Speed

The theory concerning the link between processing speed and focal representation has several implications.

Consciousness and Modularity

In the contxt of the current discussion, consciousness may be conceptualized as a slow psychological operation that can potentially "track" and perhaps modify information translation in any modality. Consider that intra- and cross-modal attention switching (a basic boundary or delimiting condition of consciousness) occurs in seconds. The slower the rate of informational translation, the greater the likelihood that the process will be accessible to consciousness (and to modification through "consciousness"). Furthermore, if the translation is completed quickly enough, other functions (represented at a greater functional neural distance or occurring within a longer time scale) will simply not have time to influence the translation process, and likewise will not be directly influenced by the process. It is therefore suggested that certain operations may appear to occur "unconsciously" or "automatically" or to be "informationally encapsulated" and "cognitively impenetrable" as a result of the speed at which constituent information translation occurs. A function will appear to be "modular" when the speed of translation reaches some critical threshold (determined by attention switching or other limiting operations of consciousness).

Hardware and Software Modules

The concept of a processing "module" probably should be modified to distinguish between those modules that are hardwired and those that are not. It is proposed that the central nervous system uses both hardware and software modules (just as a computer can use a disk drive or a RAM disk). Hardware modules will tend to use the same local group of neurons to accomplish a particular task. Hardware modules tend not to be practice intensive (or are practice insensitive). Software modules are specific to individuals within a species, do not necessarily use the same local group of neurons within species (or they will use a more variable group of neurons than will hardware modules), and tend to be practice intensive (or practice sensitive). Both hardware and software modules have the usual characteristics of cognitive impenetrability, automaticity, and informational encapsulation. Because software modules are developed over time, they may tend to be more penetrable than hardware modules. Cognitive and motor skills (such as playing tennis or playing chess) are examples of software modules in the central nervous system.

Hardware modules should show increased focal representation developmentally (i.e., they should be correlated with age rather than practice). Evidence for this should be found in lesion studies and electrophysiological studies. Software modules should show increased focal representation with practice (i.e., they should be correlated with practice rather than with age). Evidence should be primarily electrophysiological (and more rarely come from lesions), because representation of software modules may not follow the usual vascular pathways and may show much more irregular distribution than do hardware modules. Software modules may depend on multiple neural systems, may develop their own special-purpose retrieval systems, and are not easily hurt with limited structural damage.

Equipotentiality

Except for special "hardware" needed for the initial transduction of information into a common "central" form, this formulation does not require that there be any inherent cellular differences between different parts of the brain subserving different functions (e.g., language versus vision). The specific areas of localization would be determined by the translation speed principle: functions would be localized in those areas that minimize total translation speed. Any development of specialized hardware (such as neurons with extra dendrites or extralong axons) would be in the service of the minimization of translation time, but should not be otherwise function specific. Hence, the cortical area for vision, for example, should be placed close to the input area for vision. This assumption also implies that any part of the brain could potentially subserve any function, if the pathway to that area allowed sufficiently fast translation to sustain the function in question. This suggests that the particular localization of a function is primarily determined by the location of input and output channels. As long as an area is close enough to the information translation channel to sustain the requisite speed of the translation, some recovery of function is possible. It would of course follow that if regions with slower translation are damaged, a function would be more recoverable than it would be in regions with fast translation, because a larger potential area could sustain the operation in the required time. Since brain areas tend to become "dedicated" with maturity, the likelihood of finding a sufficiently fast transaction route in an adult is less than it is in children. The hypothesis regarding speed of information translation therefore allows for equipotentiality, local representation, and a method of predicting recovery of function.

The Use of Neural Space

Fast information translation can be differentiated from slower translation by examining the relationship between the volumetric spread of activation and time. Fast, localized translation should be characterized by an initially slow spread of activation (over a given neural area), followed by increasing speed of spread as time goes on. Slow translation (that uses larger areas) would be characterized either by linearity or by initially fast spread of activation over time. The initially fast spread would indicate that no particular single area is involved in the underlying translation.

Imagine a sphere composed of neurons; the area within the sphere may be used in both fast and slow information translation. A wave of neuronal activation may propagate at a constant speed across this area, or in some nonlinear manner. Since the upper limit of this speed is biologically fixed (a constant), to obtain the results of a fast calculation quickly, the calculation must occur within a smaller volume. Since all neurons within this volume can potentially make the calculations needed, all neurons can potentially be activated. However, the critical information translation will have already been completed within the first few milliseconds of activation, not allowing the residual activation within the sphere to have any effect. A localized translation of information will be characterized by nonlinear activation, in which initially a small volume is activated, followed by a spread of activation that will asymptote as it approaches the basic biological speed limitations of the hardware. A truly nonlocal process will also be characterized by a nonlinear propagation of activation in which initial spread of activation is close to asymptotic speed, but slows down as activation is completed.

It also follows that neurophysiological measures of brain function that tap into information translation (e.g., evoked potentials and Positron Emission Tomography) may show a larger area of activation than is actually functional for a particular process: The translation will not necessarily tell you the form of the underlying propagation function or the form of representation underlying that function. Therefore, you must tap into the information translation "on-line."

Lesions to the System

The theory also predicts that the effect of a focal lesion will be much greater within the volume used in the "fast" translation (because the likelihood of disrupting the ongoing translation will be greater, since the spread of activation in the first few milliseconds of processing is confined to a small area) than in the "slow" translation (because greater neural area is covered in the first few milliseconds of activation). Slow functions should be more susceptible to "lesions" affecting the entire neural area in question. For the reasons noted above "hardware modules" are more likely to be affected by lesions than "software modules," but the principles of damage should be the same.

Summary and Conclusion

As our knowledge about the effects of brain lesions on well-defined psychological functions has grown, and neural network architecture is being used to develop formal computer models to simulate those functions, questions about the actual biological engineering of the nervous system can be posed with increasing specificity.

In this chapter we argued that a characteristic of cognitive operations that contributes to the form of cortical and subcortical representation is the speed at which that operation must be performed. While processing speed is certainly not likely to be the only characteristic of a cognitive function that contributes to the form in which it will be represented in the brain, it appears to be a possible determinant of the degree to which a function is subserved by clustered or dispersed neuronal networks. The hypothesis as outlined has implications for modularity, consciousness, recovery of function, and the neurophysiological measures one uses to record brain activity.

Acknowledgments

This work was supported by grants NS52685, RO1-AG03354 and PO1-AG04953 from the National Institute on Aging and NINCDS and Grant

097443765 from the Veterans Administration. We thank Gina McGlinchey-Berroth, Patrick Kilduff, Laura Grande, and Nancy Hebben for their valuable help with preparation of this manuscript.

Address reprint requests to: Dr. William Milberg, GRECC, VA Medical Center, 1400 VFW Parkway, West Roxbury, MA 02132.

References

Alexander, M. P., Naeser, M. A., & Palumbo, C. (in press). Broca's Area Aphasias: Aphasia after lesion including the frontal operculum. *Neurology*.

Bauer, R. M., & Rubens, A. B. (1985). Agnosia. In K. M. Heilman & E. Valenstein (Eds.), *Clinical neuropsychology*. New York: Oxford University Press.

Benson, D. F. (1967). Fluency in aphasia: Correlation with radioactive scan localization. *Cortex, 3*, 373–394.

Benson, D. F. (1979a). *Aphasia, alexia and agraphia*. New York: Churchill-Livingstone.

Benson, D. F. (1979b). Neurologic correlates of anomia. In H. Whitaker & H. Whitaker (Eds.), *Studies in neurolinguistics*. New York: Academic Press.

Benton, A. L. (1977). Reflections on the Gerstmann syndrome. *Brain and Language, 4*, 45–62.

Blinkov, S. M., & Glazer, I. (1968). *The human brain in figures and tables: A quantitative handbook*. New York: Basic Books.

Bowers, D., Bauer, R. M., Cosslett, H. B., & Heilman, K. H. (1985). Processing of faces by patients with unilateral hemisphere lesions. *Brain and Cognition, 4*, 258–272.

Brownell, H. H., Potter, H. H., & Bihrle, A. M. (1986). Inference deficits in right brain-damaged patients. *Brain and Language, 27*, 310–321.

Caplan, D. (1981). On the cerebral localization of linguistic functions: Logical and empirical issues surrounding deficit analysis and functional localization. *Brain and Language, 14*, 120–137.

Caplan, D. (1988). The biological basis for language. In F. J. Newmeyer (Ed.), *Language: Psychological and biological aspects* (pp. 237–255). New York: Cambridge University Press.

Caramazza, A. (1984). The logic of neuropsychological research and the problem of patient classification in aphasia. *Brain and Language, 21*, 9–20.

Coltheart, M., Patterson, K., & Marshall, J. C. (1980). *Deep dyslexia*. London: Routledge and Kegan Paul.

Damasio, A. R., & Damasio, H. (1983). Localization of lesions in achromatopsia and prosopagnosia. In A. Kertesz (Ed.), *Localization in neuropsychology*. New York: Academic Press.

Fodor, J. A. (1983). *Modularity of mind*. Cambridge, MA: MIT Press.

Freud, S. (1953). *On aphasia: A critical study*. New York: International Universities Press.

Gardner, H., Brownell, H. H., Wapner, W., & Michelow, D. (1983). Missing the point: The role of the right hemisphere in the processing of complex linguistic materials. In E. Perceman (Ed.), *Cognitive processing in the right hemisphere*. New York: Academic Press.

Geschwind, N. (1974). *Selected papers on language and the brain* (2nd ed.). Boston: D. Reidel.

Goodlgass, H., & Kaplan, E. (1972). *The assessment of aphasia and related disorders*. Philadelphia: Lea & Febiger.

Grodzinsky, Y. (1986). Language deficits in the theory of syntax. *Brain and Language, 27*, 135–139.

Hebb, D. O. (1949). *The organizaition of behavior*. New York: Wiley.

Heilman, K. M., Watson, R. T., Valenstein, E., & Damasio, A. R. (1983). Localization of lesions in neglect. In A. Kertesz (Ed.), *Localization in neuropsychology*. New York: Academic Press.

Jackson, J. H. (1878). On affections of speech from disease of the brain. *Brain, 1*, 304–330.

Kertesz, A. (1979). *Aphasia and associated disorders*. New York: Grune & Stratton.

Kertesz, A. (1983a). Issues in localization. In A. Kertesz (Ed.), *Localization in neuropsychology*. New York: Academic Press.

Kertesz, A. (1983b). Localization of lesions in Wernicke's aphasia. In A. Kertesz (Ed.), *Localization in neuropsychology*. New York: Academic Press.

Kosslyn, S. M., & Van Kleeck, M. (in press). *Broken brains and normal minds*.

Lashley, K. (1950 In search of the engram. *Physiological mechanisms in animal behavior*. New York: Academic Press.

Levine, D. M., & Sweet, E. (1982). The neuropathological basis of aphasia and its implications for the cerebral control of speech. In M. Arbib, D. Caplan, & J. Marshal (Eds.), *Neural models of language process* (pp. 299–326). New York: Academic Press.

Lezak, M. D. (1983). *Neuropsychological assessment*. New York: Oxford University Press.

McClelland, J. L., Rumelhart, D. E., & Hinton, G. E. (1986). The appeal of parallel distributed processing. In J. L. McClelland & D. E. Rumelhart (Eds.), *Parallel distributed processing* (pp. 3–44). Cambridge, MA: MIT Press.

Miller, J. L., Grosjean, F., & Lomanto, C. (1984). Articulation rate and its variability in spontaneous speech: A re-analysis and some implications. *Phonetica, 41*, 215–225.

Milner, B. (1963). Effects of different brain lesions on card sorting: The role of the frontal lobes. *Archives of Neurology, 9*, 90–100.

Mohr, J. P., Pessin, M. S., Finkelstein, S., Funkenstein, H. H., Duncan, C. W., & Davis, K. R. (1978). Broca aphasia: Patholotic and clinical. *Neurology, 28*, 311–324.

Naeser, M. A. (1983). CT scan lesion size and lesion locus in cortical and subcortical aphasias. In A. Kertesz (Ed.), *Localization in neuropsychology*. New York: Academic Press.

Newcombe, F. (1969). *Missile wounds of the brain*. Dublin: Oxford University Press.

Pylyshyn, Z. (1986). Cognitive science and the study of cognition and language. In E. C. Schwab & H. C. Nusbaum (Eds.), *Pattern recognition by humans and machines* (pp. 295–313). London: Academic Press.

Ross, E. D. (1981). The aprosodias: Functional-anatomical organization of affective components of language in the right hemisphere. *Archives of Neurology, 38*, 561–569.

Wingfield, A., & Stein, E. A. L. (1989). Modelling memory processes: Research and theory on memory and aging. In G. C. Gilmore, P. J. Whitehouse, & M. L. Wykle (Eds.), *Memory, aging and dementia*. New York: Springer Verlag.

Zurif, E. (1980). Language mechanisms: A neuropsychological perspective. *American Scientist, 68*, 305–311.

21

Cognitive Neuropsychology: From Impaired Performance to Normal Cognitive Structure

BRENDA C. RAPP / ALFONSO CARAMAZZA

In this chapter we will provide an overview of the cognitive neuro-psychological approach to questions of cognition. In order to do so, we will begin by placing the area within the broader context of cognitive science and neuroscience. We will then describe what we consider to be fundamental aspects of the methodology, types of evidence, contributions, and difficulties of research in cognitive neuropsychology. In our discussion we will draw primarily upon examples of research from our laboratory, but only because it is the work with which we are most familiar.

A Functional Architecture of Cognition

The fundamental working assumption in the cognitive sciences today is that there is a system of mental machinery that produces the patterns of performance that we observe when people engage in any and all tasks that require cognitive processing—tasks involving language, perception, memory, and so forth. That is, it is assumed that satisfactory explanations for the behavior we observe as people speak to neighbors, recognize old friends on the street, calculate the cost of their grocery lists, and so on, must make reference to the mental processes that are required to perform these skills. As a consequence, the goal of research in the cognitive sciences is to increase our understanding of the structure of this mental machinery or "functional architecture of cognition," which underlies and determines observed behavioral responses. The need for and the usefulness of such functional models of mental functions will become apparent from the example that follows.

The following are three different types of descriptions of the same language disorder observed in a patient subsequent to brain damage (Kohn, 1984). The first is a description of the patient's performance on a number of tasks, the second involves a description of the damaged neurological structures, and the third is formulated in terms of impaired cognitive, or functional, mechanisms.

Task performance

Conduction aphasia is typically characterized as a deficit in repetition that is combined with a relative sparing of spontaneous speech and comprehension.

Damaged neurological structures

[T]he disconnection which produces conduction aphasia arises from damage to a set of temporo-parietal connections. Given that the white matter fibers which connect the temporal and frontal areas make intermediate temporo-parietal and parieto-frontal connections, inferior parietal damage would at least partially disconnect the temporal and frontal areas.

Impaired cognitive mechanisms

[C]onduction aphasics possess a stable Working Memory and relatively intact Speech Monitoring. Pre-articulatory Programming . . . holds the greatest potential for characterizing the patterns of phonological breakdown exhibited by this syndrome. The postulated function of Pre-articulatory Programming involves the selection and sequencing of phonemic targets into a form which contains enough information for articulatory realization.

Even though each of these types of descriptions may be useful for certain purposes (ignoring for the purposes of this discussion the issue of their accuracy), we will argue that only the third offers an explanatory account that can provide the possibility of *prediction and generalization*. It is important to notice that, although this particular example makes reference to impaired performance, the arguments we will present regarding levels of description apply equally well to normal performance. That is, if one replaces "impaired performance" with "performance pattern X," one could describe the performance of unimpaired subjects at any of the three levels of description.

Description in Terms of Task Performance

The first, task-based, account informs us that the patient is impaired in repetition, although his spontaneous speech and comprehension are relatively normal. However, this description does not allow us to understand, for example, why this particular constellation of symptoms should occur or if the symptoms are related to one other. Most problematic is the fact that the description does not provide the basis for predicting whether or not the patient would be impaired on a task not included in the description, a task such as reading.

When we try to make a prediction of this sort, we find that, based on the knowledge that the patient is impaired in Task 1 (repetition), we have no reason to prefer the prediction that he will be impaired on Task 2 (reading) to the prediction that he will not. The problem is that repetition is a complex task, and the patient's difficulty with the task could have been the result of a number of rather different impairments, including difficulty hearing, difficulty remembering the words, difficulty recalling the way in which they are pronounced, and so on. Clearly, depending upon which of these, or other, deficits is responsible for poor repetition, we will have different predictions concerning reading, writing, and so forth. In order to generalize from Task 1 performance we would need to know, first, which process was responsible for poor Task 1 performance, and second, whether Task 2 requires the same process. Thus, even if we had observed this patient's performance on Tasks 3, 4, and 5 we still would not be in any better position to predict reading performance (Task 2). Generalization from one task to another requires an understanding of what it is that the tasks share. Therefore, in order to provide an explanation of performance that will allow for prediction and generalization, it seems that we must go beyond observed task performance and make reference to the underlying cognitive mechanisms employed in the given tasks. It becomes apparent that a description of a patient's performance on any number of tasks will not provide us with the necessary knowledge regarding the functional cognitive architecture.

Description in Terms of Damaged Neurological Structures

The second, neurologically based, account proves to be no more useful than the first in providing a basis for generalization. That is, knowledge that there has been disruption to temporal–frontal lobe connections does not provide the basis for predicting performance in reading or any other task. It would seem that an explanation of the observed performance cannot be based *solely* on knowledge of the brain structures involved in the patient's impairment, but instead requires an understanding of which cognitive functions are represented by the neurological structures and the effects that damage to those functions should have on the performance of the tasks of interest. Thus, in order to be able to predict task performance, a neurologically based description must make reference to a functional level theory.

This is not, however, meant to suggest that the development of a functional theory of cognition is in any way in opposition to, or a substitute for, work at the neurological level. Clearly a primary goal of cognitive science is to ultimately relate brain and cognition in such a way that a theory of the mind–brain can develop informed by data from all relevant domains. At present, however, the limitations in relating brain and cognition are due primarily to the fact that brain anatomy and cognitive architecture do not map onto one another in any transparent way (Teller, 1984). That is, it is not obvious, given a fact of neurophysiology, brain anatomy, or electrophysiology, what the psychological implications might be.

Thus, the brain anatomist or physiologist dedicated, for example, to mapping the interconnectivity of brain areas can work relatively oblivious of the advances of the cognitive scientist. Simiarly, the cognitive psychologist can work quite independently of the neurophysiologist. For instance, to the cognitive psychologist it is not apparent that there are neurological data that can be brought to bear upon issues of interest at the level of cognitive architecture, such as: Is syllable structure a part of the representation of a written word? Do the processes of reading and writing share any components? and so on. Just as these questions use the vocabulary of, and require data from, cognitive psychology, some questions about the brain require answers in the language of neurology. Ultimately, of course, the degree to which theories of mind and theories of brain will relate to one another will depend on the extent to which each of these is developed, and thus progress must continue at all levels. (See Mehler, Morton, & Jusczyk, 1984) for a more extensive discussion of these issues.)

Description in Terms of Impaired Functional Components

The third description, on the other hand, proposes a model of repetition that involves at least three processing mechanisms: speech monitoring, working memory, and prearticulatory programming components. The author suggests that the observed pattern of performance corresponds to damage to one of the three. Specifically, it is suggested that the performance of a patient who cannot repeat but can comprehend and speak spontaneously can be understood as resulting from damage to a prearticulatory programming mechanism whose function it is to select and sequence the representations to be used in pronunciation of words.

On the basis of this description we would predict impaired performance on all those tasks that require such a mechanism to be intact. In terms of the case at hand, if we had reason to think that this particular function was required for reading, then we would predict impaired reading performance.[1] With this type of description, prediction is possible on the basis of our understanding of the nature of the cognitive *function* that has been affected by neurological impairment.

The example we have considered allows us to see that if we wish to understand performance, the truly powerful unit of analysis is not at the level of the task or brain structures but rather at the level of the functional components of the cognitive system. Because of the explanatory power provided by functional theories of cognition, the primary concern of cognitive science is not with the development of an inventory of typical responses under particular stimulus conditions or on certain tasks; rather, the research question in cognitive science is typically: What observations can we gather that will allow us to choose the most accurate description of underlying psychological mechanisms from among the multitude of theoretically possible descriptions? The different fields of study in the cognitive sciences represent the use of different classes of observations and methodology to answer just these questions.

The Information-Processing Approach

The insight upon which explanatory accounts of cognitive processing are built is one which suggests that cognitive mechanisms can be usefully thought of as information-processing units (Neisser, 1967). The information-processing view focuses the analysis on the exchange or transformation of information that is required to perform a task. Clearly, however, the units of information exchange and transformation need not correspond to single tasks or single brain structures. As a consequence, a functional level of explanation will be, as was seen in the examples above, markedly different from those accounts that refer to tasks or neuroanatomy.

The information-processing metaphor has been used in all areas of cognitive science, from linguistics to cognitive neuroscience. Typically, within this framework, in order to explain behavior, one must specify: (1) the type and form of the information—or representation—that serves as input to a particular processing component or level; (2) the transformation of the information that the component performs; and (3) the form of the representation that results from this computation and serves as input to some other processing component.

For example, in visual science, certain theories of object recognition (see, for example, Barlow, 1982) propose that the absolute light intensities available at the retina serve as input to a level of processing that functions to emphasize the relative differences of dark and light. These representations of relative intensity differences, in turn, serve as the basis for localizing edges at certain orientations within the visual field. The description of these edges is then presumably used in further processing, which culminates in the recognition of an object.[2]

Thus, within the information-processing framework, a functional architecture refers to a description of the mental representations and processes that are involved in transforming the input to a cognitive system (e.g., the printed word in reading) to the output (e.g., pronunciation of the word). An understanding of the events that mediate between input and output is the general objective of cognitive science.

Cognitive Neuropsychology

Cognitive neuropsychology is the field of study that brings evidence obtained from the observation of brain-damaged performance to bear on questions of cognitive architecture. We will provide examples of the way in which observations of the performance of brain-damaged subjects can be used to develop claims about the structure of the representations and computations involved in cognitive processing.

Although it could have been the case that damage to the brain invariably resulted in a *general* impairment of cognitive functioning, it has been found, instead, that brain damage often results in the *selective* loss of psychological functions. For example, neurological damage can selectively affect reading but not writing, the recognition of objects but not speech, the ability to remember a list of items but not the ability to name objects, and so on. In terms of functional cognitive architecture, it can be said that brain damage can affect certain components while sparing others. This suggests that it can be assumed that patterns of dysfunction may reflect the componential structure of the cognitive system and that, as a consequence, the analysis of impaired performance in brain-damaged subjects may provide us with a unique opportunity to "look into" the cognitive machine and determine what some of the components are, how they work, how they are interrelated, and so on. The enterprise of cognitive neuropsychology is just that: to use patterns of performance of brain-damaged subjects in order to make inferences about what a normal cognitive system must be like to result, when damaged, in precisely the patterns of performance that have been observed. In terms of the pattern of repetition, comprehension, and speech performance described earlier one might ask: What must the underlying cognitive mechanisms be like in the normal system such that some pattern of damage to those mechanisms will result in the ability to speak spontaneously and comprehend spoken language and yet not be able to repeat? In this manner the patterns of impaired performance that are described in cognitive neuropsychological work provide information that serves to constrain the possible form that the underlying cognitive structure can take.

However, in order to use data from neurologically impaired patients, certain assumptions and methodological considerations are required (for a detailed discussion see Caramazza, 1986; McCloskey & Caramazza, 1988). Here we will review only the most fundamental of these—the *transparency assumption.*

In order to make inferences about normal cognitive structure from patterns of impaired performance we must assume (1) that the cognitive system of a brain-damaged patient was: normal before the damage and (2) that the lesion resulted in a relatively transparent modification of the normal system. By transparent we mean that the relationship between the pre- and postmorbid systems is fairly straightforward. That is, brain damage might result in impairment to preexisting components, processes, capacity, representational types, and so on, but it should not result in the de novo creation of cognitive operations forming a cognitive system that is fundamentally different from the normal premorbid one. Of course, this assumption could be wrong; if so, evidence obtained from different cases will be contradictory and will, furthermore, be inconsistent with results obtained from research based on unimpaired performance. Thus, the validity of this assumption will be evaluated by the successes of the approach as well as by the fact that we should see a convergence of results from research based on impaired performance and research with normal subjects.

Neuropsychological Evidence

Although different criteria can be used to motivate the existence of different components within the cognitive system, certainly one criterion is whether or not a processing component can be selectively disrupted by brain damage. If, however, brain damage were to result only in dissociations of function that were no finer than global cognitive systems, such that damage resulted only in the loss of entire systems such as language processing, calculation, music processing, and so on, then patterns of impaired performance would be of little value in determining the internal processing structure of these global systems. That is, the level of description that can be achieved based on selective patterns of impairment will be no greater than the specificity of the deficits. Fortunately, brain damage can selectively damage components at a relatively fine level, and patterns of impaired performance have been successfully used to inform theories regarding the internal structure of components within the various subdomains of cognition. Here we will briefly consider an example from the area of language processing—we will consider the case of how different patterns of selective impairment can provide evidence regarding the internal architecture of the lexical system—the system that is responsible for the processing of words.

The dominant view of the functional architecture of the lexical system is that it consists of an interconnected set of lexical components. A schematic representation of these components can be seen in Figure 21.1.

A major distinction is drawn between input and output lexical components; that is, lexical components involved in the comprehension and production of words, respectively. A second major distinction is drawn between the modality of input or output: phonological or orthographic. Thus, the orthographic input lexicon, which includes those mechanisms involved in the recognition of written words, is distinguished from the phonological input lexicon involved in the recognition of spoken words. These modality-specific input lexicons are further distinguished from their corresponding output lexicons, which consist of mechanisms involved in the production of written and spoken words. It is further assumed that the modality-specific lexical components are interconnected through a lexical-semantic system that stores the semantic representations for words. We will present a case with which we are familiar to serve as an example of how a complex pattern of impaired performance can contribute to the development and validation of models of this sort.

Patient R.G.B. (reported in Caramazza & Hills, 1990) was unimpaired in reading and in comprehending spoken words. However, the patient showed a clear dissociation between performance on tasks that required a spoken (phonological) output and performance on tasks that required a written

Orthographic input: Phonological input:

HOUSE   /haws/

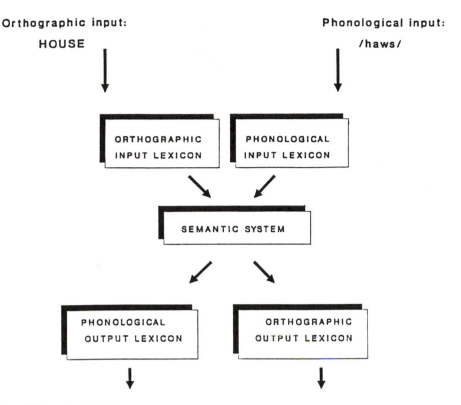

Figure 21.1 A schematic representation of the lexical system.

(orthographic) output; see Table 21.1. A large proportion of this patient's responses in tasks that require a spoken response consisted of semantic errors (e.g., the word the patient is supposed to produce is *pear* and the patient produces *banana*), whereas none of the responses in writing tasks were of this sort. Nonetheless, the patient knew the meaning of the words that he was unable to produce orally. Thus, when asked, immediately after

Table 21.1. The Distribution of R.G.B.'s Errors (Given in Proportion of Total Responses)

Task	Semantic errors	Other errors
Oral naming of pictures	36%	2%
Oral naming: tactile	36%	0%
Oral reading	32%	2%
Written naming: pictures	0%	6%
Writing to dictation	0%	6%
Auditory word-picture matching	0%	0%
Printed word-picture matching	0%	0%

producing an error, to define the stimulus word, R.G.B. produced definitions of the stimulus and not the response. For example, the oral reading
response he produced to the stimulus *records* was *radio*, but the definition he
produced was: "you play'em on a phonograph . . . can also mean notes you
take and keep;" or necklace (stimulus) → necktie (oral reading response) →
"you would wear, a woman would have around her neck . . . made of metal
. . . gold or silver (definition)" (see Basso, Taborelli, & Vignolo, 1978;
Michel, 1979 for reports of other patients with deficits in spoken but not
written naming).

A hypothesized language system must be structured such that it can be
damaged to produce just this pattern of performance. Based on the results
reported here, the following points can be made: (1) The dissociation
between spoken and written modalities of output supports a distinction
between phonological and orthographic output lexicons; (2) Even though
an auditorily presented word was consistently recognized and its meaning
accessed, the same word could not be consistently accessed for output. This
result provides strong support for the hypothesis that phonological information is represented or accessed separately for input and output processes. In the case of R.G.B., only phonological output processes were
affected.

A large number of other cases and different patterns of performance
converge on a lexical system that is distributed in the way depicted in Figure
21.1 (see Shallice, 1988, and Ellis & Young, 1988, for a review). In fact, the
architecture has been specified in considerably greater detail than we can
go into here.

We have begun with an example of how a selective pattern of impairment,
referred to as a *dissociation* in performance, can be used as the basis for
inference concerning the structure of a cognitive system. However, it is our
position that there is no fixed method that can be applied mechanically to
patterns of impaired performance in order to determine the inferences that
may justifiably be drawn from them about normal cognitive systems. Thus,
we would argue that patterns of dissociations in performance, association of
deficits, error patterns, and so on, are all useful sources of information.
This position contrasts markedly with that of Shallice (1988) and others
(Bub & Bub, 1988) who see a principled distinction between the use of
dissociations and other forms of evidence from impaired performance in
drawing inferences about cognitive processing. Their position is motivated
in large part by the view that "neuropsychological data may well prove to be
more important for sketching the broad outlines of the organisation of
subsystems than for testing the details of specific theories at the algorithmic
or mechanistic levels" (Shallice, 1988 p. 36). This position is based on the
belief that neurological damage is such that it can provide information that
will contribute only to a gross level of description of cognitive systems.
Clearly, however, this is an empirical question; there is no a priori reason to
suppose that only gross functional distinctions can be revealed through

analysis of the performance of brain-damaged subjects. In this context we might reconsider the case of R.G.B. The data obtained from this patient suggest a number of questions at a much finer grain of detail than those concerning the principal components of the lexical system: Why is it that *semantic* errors (*celery → lettuce*) result from damage to a *phonological* output component, although the patient understands the meaning of the words he names incorrectly? What might this observation suggest about the way phonological information is accessed and/or organized? There is no reason to believe that relevant hypotheses at this finer level of description cannot be formulated and tested with patients of this sort.

In order to illustrate the point that the utility of data from neurologically impaired patients cannot be narrowly defined, we will present two further cases, one that illustrates that an *associated pattern of deficits* may provide critical information in the development of a theory, and another in which *analyses of errors* that result from impaired performance are used to provide information about cognitive representations at a fine-grained level of detail.

We would point out that a pattern of associated deficits is sometimes *required* by a given cognitive model. For example, damage to a certain component should be reflected in impaired performance on all tasks that require that component. The discovery and documentation of this associated pattern of impaired performance on the predicted tasks would therefore constitute evidence in support of the model. (See Caramazza, 1986; McCloskey & Caramazza, 1988; Sokol & McCloskey, 1988, for a more extensive discussion, but also see Bub & Bub, 1988.)

For example, in the model presented in Figure 21.1 we have indicated that a single semantic system mediates between input and output modalities. In such an architecture, damage to the semantic system requires a comparable pattern of errors across all modalities of input and output. That is, a semantic deficit must result in an association of deficits on all tasks that require lexical processing.

Patient K.E. (reported in Hillis, Rapp, Romani, & Caramazza, 1990) showed just this pattern of impairment. K.E. made an equivalent proportion of semantic errors in writing to dictation, reading, picture naming, tactile naming, auditory and written word–picture matching, and written naming of pictures (see Table 21.2). Thus, K.E.'s performance provides precisely the association of deficits we would expect from a cognitive system structured like the one in Figure 21.1

K.E.'s performance was analyzed not only at a gross level, which determined that a comparable proportion of semantic errors was made across the input and output modalities, but also at a considerably fine-grained level of analysis. Thus, it was determined that K.E. had more difficulty with some semantic categories than with others; for example, there were more errors in the category of body parts than in the category of fruits. If semantic categories were rank-ordered from the best to worst performance

Table 21.2. The Distribution of K.E.'s Semantic Errors

Task	Semantic errors	Total errors
Auditory comprehension	42% (61/144)	42% (61/144)
Reading comprehension	36% (52/144)	36% (53/144)
Verbal naming of pictures	38% (108/288)	45% (130/288)
Oral reading	33% (96/288)	42% (122/288)
Written naming of pictures	32% (91/288)	46% (132/288)
Writing to dictation	24% (70/288)[a]	43% (123/288)
Tactile verbal naming	45% (21/47)	47% (21/47)
Tactile written naming	34% (16/47)	40% (19/47)

[a]Semantic errors were slightly less frequent and constituted a lower proportion of the total errors in writing tasks because K.E. also wrote a substantial number of nonword responses. Several of these responses appeared to be misspelled semantic errors; for example, beard → muschace but, since we could not be certain, were not classified as semantic errors.

in one modality, the rank order was preserved remarkably well across all of the modalities of input and output. For example, the categories of body parts, furniture and clothing were associated with the three of the four highest rates of semantic errors in all of the six tasks. The categories of fruits and vegetables were toward the lowest in rank order, regardless of modality, whereas transportation was uniformly in middle rank across tasks.[3]

The level of shared detail in this pattern of associated deficits provides a remarkably strong constraint on a theory of the underlying cognitive mechanisms. That is, whatever the model of cognitive processing one develops, it must account for the pattern of performance observed in K.E. It is not at all clear that a dissociation could provide any stronger constraints.

The second case illustrates the contribution of an *error analysis* to attempts to specify the structure of the orthographic representations that are used to guide the motor processes involved in writing a word. Logically these representations must, at the very least, include information about the identity of letters in a word and the order in which they occur (e.g., that the written word table has a *T* in first position, an A in second, etc.). If orthographic representations specify *only* the identity and the order of the letters that comprise a word, then, when these representations are damaged, we should expect errors to affect letter identities and/or order only. That is, damage to the graphemic representations should result in errors such as letter substitutions (*table* → *rable*), deletions (→ *tabl*), additions (→ *toable*), and transpositions (→ *talbe*). Concerning substitutions, if the identity of a particular letter is lost as a result of damage and if another is substituted in its place, we might expect the substituted letter to be chosen at random from among the letters of the alphabet.

This, however, has been shown not to be the case. Caramazza & Miceli (1990) and McCloskey, Goodman-Schulman, and Aliminosa (1988), have used data from Italian-speaking and English-speaking patients with hypothesized damage to orthographic representations to examine questions of orthographic representation. The writing errors of these patients indicate a remarkable consistency in substitution errors—vowels are substituted for vowels and consonants for consonants. Italian-speaking patient L.B. (Caramazza & Miceli, 1990), for example, produced 643 substitution errors in writing (e.g., *tavolo* → *tacolo*), 276 involving vowels and 367 involving consonants. In 640 of the 643 errors (99.5%) consonants were substituted for consonants and vowels for vowels. Based on this, and other evidence, Caramazza and Miceli inferred that L.B. had access to information about the consonant-vowel structure of the words even when letter identity information was unavailable. They suggested that the orthographic representations used in spelling specify not only letter identities and order but that they also include information concerning consonant-vowel structure. Thus, if the damaged representation of *tavolo* failed to specify the identity of the third letter but correctly specified that it was a consonant, this information could be used to produce a consonant in the third position.

$$\begin{array}{c} \text{C V C V C V} \\ \text{| | | | | |} \to \text{ta}\underline{c}\text{olo} \\ \text{T A \_ O L O} \end{array}$$

Regardless of whether this particular explanation provided for L.B.'s pattern of errors is correct, the point is that the pattern of vowel-vowel and consonant-consonant substitution would have been overlooked if one were "dissociation hunting." The pattern described here must be explained by a theory of the cognitive mechanisms involved in writing; by ignoring this level of detail we run the risk of missing extremely powerful constraints that can be placed upon our theories of processing.

In sum, these examples serve to illustrate that there are no a priori limitations on the kinds of evidence from brain damaged performance that may be useful. Very often the kind of evidence that will be most useful will depend on the degree of detail that has been achieved in a particular theory. If, for example, we have not delineated even the most basic processing components of a particular cognitive system, then gross dissociations in performance may be particularly revealing. In such a case we would not know what to make of subtle regularities or irregularities in performance. On the other hand, in the case where there is considerable knowledge regarding the principal components of a processing system, gross dissociations may prove to be irrelevant, whereas the details of error patterns may prove to be highly informative. Thus, the kinds of evidence we look for and are able to interpret will be closely related to the level at which our theories are articulated.

Problems Faced in Cognitive Neuropsychological Research

Because the goals of cognitive neuropsychology are just those of the other areas of cognitive science—to use data from subjects' performance as the basis for inference about the structure of cognitive mechanisms—some of the difficulties in this area are similar to the difficulties faced in all approaches to these questions. These general difficulties in cognitive science relate primarily to the complexity of making inferences that relate data to cognitive architecture. Other obstacles, however, either take a particular form in cognitive neuropsychology or are, in fact, peculiar to the types of observations and methodology required by work with brain-damaged patients. We shall present examples of difficulties generally faced by those doing research in cognitive science as well as examples of problems that are more specific to cognitive neuropsychological research.

A General Problem: The Complexity of Inference— or, Are the Conclusions Warranted?

Work in both cognitive psychology and cognitive neuropsychology uses observations of task performance to make inferences about underlying psychological mechanisms. As is the case with all inferences, these are built upon sets of assumptions. A major source of difficulty is the fact that it is not easy to make one's assumptions explicit and to seek independent confirmation of their validity (or, at least, of their reasonableness). Consequently, research is often subject to the criticism that empirical data exist that can be interpreted as undermining a particular assumption or that there are logical reasons to think that an assumption is not justified. Although, as already stated, this is a problem for research in cognitive science generally, here we will present an illustrative example involving research in cognitive neuropsychology.

Warrington, (1975), Warrington and Shallice (1979), and Warrington and McCarthy (1983), using data from neurologically impaired subjects, put forward the interesting proposal that the meaning of words is accessed from "general to specific," that "The precise meaning of a word may well be accessed ony as the end result of a process which involves the attaining of increasingly specific semantic representations" (Warrington and Shallice, 1979, p. 61). The claim is based on the observation of patients who could apparently demonstrate general knowledge of the category membership of words they could not otherwise identify or comprehend (Warrington, 1975). The patients were presented with pictures to name and with auditorily presented words that they had to define. For the items for which they could not provide a correct response, the patients were presented with "yes/no" questions that assessed knowledge of category membership as well as knowledge of item attributes (e.g., for animal words and pictures: Is it an animal? Is it a bird? Is it foreign? Is it bigger than a cat?; for objects: Is it an animal? Is it used indoors? Is it heavier than a telephone directory? Is it made of metal?) The results of this testing indicated that, with words that they could not otherwise identify or comprehend, the patients performed significantly above chance on both pictorial and auditory presentations with the most general questions; Is it an animal? However, their performance was at chance with respect to most of the other, more specific, questions.

Based primarily on this evidence the authors concluded (1) that superordinate (general, category membership) information and subordiate (item-specific) semantic information are represented autonomously such that access to superordinate information can occur without access to subordinate information, and (2) that in terms of the time course of access to semantic knowledge, superordinate information becomes available prior to subordinate information.

In Rapp and Caramazza (1989) we presented a number of arguments that suggested that the conclusions of general to specific access to word meaning were not warranted on the basis of the evidence provided by Warrington et al. (see also Patterson & Kay, 1982; Riddoch, Humphreys, Coltheart, & Funnell, 1988). Among other things, we pointed out that a basic assumption of the argument was not necessarily justified, thus casting doubt upon the entire inferential edifice. The following briefly summarizes the critical aspects of the inferential chain followed by Warrington, Shallice, and McCarthy.

The conclusions drawn by these authors were based on the following assumption (among others): Performance on a categorization task (questions such as: Is it an animal?) provides a "direct" measure of the *type* of semantic information available to the patients regarding a stimulus. This assumption served as a basis for the interpretation of the observations they reported: Patients are able to categorize items that they cannot name or identify. The following inferences were drawn: (1) The patients' ability to classify items that they cannot identify is evidence that the patients have accessed superordinate information for these items without having access to their subordinate or item-specific feature information, and (2) this, in turn, suggests a cognitive architecture in which superordinate information is represented independently of subordinate information, and further, it is consistent with the view that superordinate information is normally accessed prior to subordinate information.

Rapp and Caramazza (1988) argued that, if the basic assumption on which the argument rested was not justified, alternative interpretations of the data might follow. We suggested that, in fact, we can only assume that performance on the categorization task is a measure of whether or not *sufficient* semantic information is available to make the correct judgment. Thus correct performance does not necessarily reveal the *form* of the available information—whether it is of the superordinate or subordinate type. As a consequence, categorization performance may simply reflect the *amount* of semantic information available. On this basis, we suggested that, logically, accurate categorization could result from a situation where a semantic representation is only *partially* activated such that only a subset of the possible semantic information for a given concept is accessed. For example, if upon presentation of the word *ostrich* only parts of its semantic representation (e.g., flies, eats, runs, has feathers, or has a particular type of neck) become available, one could, without knowing that it is an ostrich, choose between its categories animal/transportation. On the other hand, questions such as, Is it larger than a cat? require that a particular semantic feature be available. Consequently, given a partially activated semantic representation, we would not be surprised to find superior performance on questions of superordinate category membership. Thus, if we do not assume that a categorization task assesses the availability of a certain type of information, then the interpretation given to these patients' performance— that superordinate information is available without accessing item-specific information—no longer follows, and the specific inferences regarding the cognitive architecture are also no longer justified.[4]

However, although we argued that the underlying assumption was not required, it may, nonetheless be true. Given the importance of the assumption made by Warrington et al. one would want to determine if there is any independent empirical evidence indicating that these patients, in fact, access primarily, or exclusively, superordinate information from items they are unable to identify.

In Rapp and Caramazza (1988) a patient similar to one referred to by Warrington et al. was used to seek such evidence. A task was devised based on the assumption that if the patient accessed only superordinate information from words, then words belonging to identical semantic categories should be confused more often than were words belonging to different categories. The results of this experiment indicated that this was not the case. Thus we could find no evidence to support the assumption that correct categorization indicates access to superordinate information.

In sum, Warrington et al. drew apparently not unreasonable conclusions from carefully observed patterns of performance; however, after having cast doubt upon basic assumptions of their argument, it is evident that there are reasons to question whether such conclusions were warranted. This is not an unusual situation, given that in cognitive science conclusions are normally drawn based on a number of assumptions for which empirical evidence is not always available or uncontroversial. Clearly this increases the difficulty involved in the interpretation of patterns of impaired and unimpaired performance.

A Game of Twenty Questions—or, The Reification of Dissociations in Cognitive Neuropsychology

Newell (1973), in a critique of work in cognitive psychology, argued that "you can't play 20 questions with nature and win" (p. 282). He criticized the tendency of work in cognitive psychology to be centered around the accumulation of evidence either for or against a particular dichotomy: nature/ nurture, peripheral/central, analog/digital, conscious/unconscious, single codes/multiple codes, and so on. Newell suggested that this approach is unproductive, not only because it is entirely possible that the answer is "both" (depending upon what part of the processing system we consider) but also because, in the attempt to discover evidence for or against one of the alternatives, evidence from other relevant questions tends to be ignored. Research tends to jump from one dichotomy to another with little integration of results across time and issues. In cognitive neuropsychology, perhaps because of the striking nature of the dissociations that are observed, there is a tendency for dissociations to function as dichotomies: A dissociation in performance is observed, names are assigned to the two elements of the dissociation, and other relevant evidence is not included in the explanation that is then provided. Further, because the dissociation has been labeled, the names are then referred to as though, by virtue of having been assigned to a phenomenon, they constitute an explanation.

An example of this problem can be seen in the work of Milberg, Blumstein, and Dworetzky (1987), who report a dissociation in performance in a group of neurologically impaired patients (clinically classified as Broca's aphasics on the basis of certain characteristics of their speech production and language comprehension).

The experimental task used to test these patients was an auditory lexical decision task. The general paradigm in this sort of task is as follows: a word (e.g., *beef*) or nonword (e.g., *helt*) is presented auditorily, and subjects must press a key as quickly as possible to indicate if the sequence of phonemes they heard corresponds to a word or not. With normal subjects, when a word is preceded by another word that is similar in meaning (e.g., *nurse/doctor*), the lexical decision to the second word is facilitated—it is faster than it would be if a neutral word had preceded it (e.g., *desk/doctor*). This effect is thought to occur because processing of the first word results in the activation of a semantic field within the semantic system that includes the meaning of the second word. Thus, the semantic representation of the second word is activated before the word itself actually appears. This activation will then facilitate the subsequent identification of the word when it is presented. (Tasks that produce facilitatory effects of this sort are called *priming tasks*.)

The results of the Milberg et al. experiment revealed that the group of so-called Broca's aphasics showed no facilitation on the task, although normals and other patient groups did. The authors went on to note that, in contrast to the impaired performance that they report on the *auditory lexical decision* task, this class of patients is thought, by many, to be unimpaired on tasks that require the *comprehension* of word meaning, such as categorization tasks. Thus, argued the authors, they had documented what appeared to be a dissociation between the ability of these patients to access semantic information normally on a categorization task and their inability to do so in the lexical decision task. The authors suggested that this dissociation could be understood as a selective impairment to *automatic* versus *controlled* access to word meaning. According to the authors, the lack of facilitation in the lexical decision task indicated damage to an automatic process—fast acting, of short duration and not under voluntary control—whereas unimpaired retrieval of semantic information in the categorization task indicated an intact controlled process—slower acting, under voluntary control, and influenced by expectancies and attentional demands.

In order for this conclusion to hold, the authors assumed, among many other things, that *the* aspect of processing that differentiates priming and categorization tasks is just this automatic/controlled dimension. However, if we examine the architectures of language processing that have been developed thus far, we can discover a number of other ways in which the two classes of tasks differ. These would need to be explored and ruled out before we could be confident that the automatic/controlled dimension was the relevant one.

Furthermore, one expects the dichotomy provided by these authors to provide an *explanation* of the performance of the patients on the experimental tasks. Although the dichotomy automatic/controlled is the focus of considerable research attention with unimpaired subjects, it is not clear that the labeling of the dissociation in performance on the two tasks as automatic/controlled constitutes an explanation of performance on either task. We would like, as Newell suggests, to know what the implications of such a distinction are in terms of the language processing architectures available to us: Do the terms *automatic* and *controlled* refer to different classes of semantic access mechanisms? If so, in what way do they function? What are the differences in function that distinguish them? Why should different semantic tasks engage different mechanisms?

Additionally, if there is assumed to be a causal relationship between the impairment responsible for the language characteristics that served as a basis for selecting the patients and the performance of the patients on the experimental tasks, one would expect that an explanation of the patients' performance should relate the two. In the case of Broca's aphasia, the characteristics that form the basis for classification tend to be the omission of function words and morphological markers in language production and impaired language comprehension (optionally). Therefore, in this case we would want to know how an impairment in automatic processing would result in just this pattern of impairment. For example, Why should a deficit in automatic processing result in the omission of function words and not of nouns? Thus, without knowing what is meant by, in this case, the labels *automatic* and *controlled* in terms of existing models, it is difficult to know what evidence can be brought to bear on the evaluation of such a proposal.

The reification of dichotomies is not a problem that afflicts cognitive neuropsychology exclusively. However, in cognitive neuropsychology this tendency is often reflected in the treatment of patterns of dissociations.

The Problem of Types and Groups

The problem of typing and grouping is one of the more complex problems faced more specifically by those who work within the cognitive neuropsychological approach and has been discussed extensively elsewhere (see Badecker & Carramazza, 1985). We will touch upon it here only as briefly as possible in order to provide the reader with an overall view of the difficulties involved.

At the heart of the problem is the diverse nature of neurological impairment—the fact that the damage suffered by different patients differs in various ways, although these differences can be major or relatively minor. Because of these differences, different cognitive functions may be affected in patients who otherwise share certain deficits. Thus, although two patients may have damage to the same component, one of the two patients may have damage to additional components as well. That is, one cannot safely assume that because patients A and B are impaired on Tasks 3, 4, and 5 and patient A is also impaired on Task 6, that patient B will also be impaired on Task 6. This would only be true if the mechanism responsible for damage to 3, 4, 5 and 6 in Patient A is the same mechanism responsible for damage to 3, 4, and 5 in patient B. The problem is, of course, that usually these are the very sorts of questions we are trying to explore and therefore we cannot assume that they are true. An example may help to illustrate this problem and the implications for grouping of patients that it raises.

In the Milberg et al. experiment, described earlier, the authors refer to two groups of patients selected on the basis of the characteristics of their performance in spoken production and language comprehension, what we shall call Task 1. Additionally, one group had been shown to be relatively intact on a semantic categorization task (Task 2), whereas the other group was impaired on an auditory priming task (Task 3). The authors assume that because the two groups were similar in terms of the characteristics of their language production and comprehension (Task 1), performance should be similar on Tasks 2 and 3, and therefore do not test both groups on both tasks. This assumption is fragile, because it is only true if the deficit responsible for the patient classification profile (Task 1) *requires* that all patients will be impaired on auditory priming (Task 2). Unless this is the case, the individual patients unimpaired on categorization (Task 3) may also be unimpaired on auditory priming (Task 2). If that were the case, then there would be no dissociation to speak of.

That is, we cannot assume that there is a patient of Type X who is identified by performance on Task 1 such that whatever is true of other patients of Type X on other tasks will also be true of this patient. This fact has serious implications for attempts to group patients as well as for the amount of testing that must be done with a patient—very little can be taken for granted, and it will be counterproductive to group patients and average across their performance.

It should be made clear, however, that the impediments to establishing patient types, and therefore to grouping patients, do not constitute a restriction on the relevance of neuropsychological data. Rather, these problems have methodological implications for how research can be carried out in this field. We would argue that the only way of dealing with the problem is by conducting single-patient studies (see Badecker & Caramazza, 1985; Caramazza, 1986a; but for a different view see Shallice, 1988, or Caplan, 1988). In the study of a single patient all the relevant parameters must be assessed for that patient, rather than being assumed from the performance of similar patients. In this manner questions regarding the nature of the underlying cognitive architecture can be effectively addressed. The results obtained from a single-patient study are comparable to the results obtained from a single experiment with unimpaired subjects. Theories can then be developed on the basis of a convergence of evidence from a number of such experiments.

Conclusions

Attempts to develop an explanatory account of cognitive processes are clearly fraught with a number of difficulties, some of which we have described here. It seems apparent to us, however, that these are obstacles that can be worked through or around. We have certainly not exhausted the list of dangers and problems, nor do we claim to be personally exempt from having committed some of the very mistakes that we have pointed out. Nonetheless, it seems to us that research in cognitive neuropsychology will be most productive when we are most familiar with not only the possibilities that the approach offers, but also the errors to which is most susceptible.

We have provided some examples that illustrate the ways in which data from neurologically impaired patients has been successfully used to develop models of cognition. We have made the point that the type of evidence that will be useful in this enterprise cannot be restricted a priori—any observation that provides a constraint on theory development is useful.

This work in cognitive model building has had, at the very least, two concrete results. First of all, it has greatly increased our understanding of the disorders that occur as a result of brain damage. Rather than being limited to understanding a disorder simply as one of impaired writing, for example, we can now specify that the damaged mechanism is, in a particular case, one required for holding an orthographic representation in memory while the motor commands for writing are being assembled. Clearly this increased understanding should also have an effect on theories and the methodology of rehabilitation. Second, the level of detail of our cognitive theories serves as a basis for an increasing specificity in theories of cognitive-neural mappings. Those interested in relating neurological structures and functions to psychological ones will have a richer and better articulated theory of the mind with which to work.

Besides these more practical applications of advances in this area, there is also, of course, simply the sheer satisfaction that is derived from developing better answers to age-old questions.

Acknowledgments

This work was made possible by the generous support of the Seaver Institute and NIH research grants NS22201 and DC00366 to The Johns Hopkins University. We gratefully acknowledge their support.

Notes

1. By choosing this particular example of a functional explanation we do not mean to endorse it. In fact we would argue that it is problematic not only because it is not sufficiently articulated but also because of the assumptions of patient homogeneity on which it is based. It is, nonetheless, an example of the *type* of explanation, that, because it makes reference to functional mechanisms, can potentially be useful in generating predictions and generalizations.

2. In this way projections from the retina through the peristriate cortex are assumed to represent a functionally hierarchical system: low-level analyzers converge on progressively higher, more abstract analyzers of visual information.

3. It is worth noting that the three categories that yielded the highest semantic (and total) error rates were from the group of categories whose members had the highest word frequency (frequency of occurrence in print). Furthermore, yet another high-frequency category, foods (excluding fruits and vegetables), elicited the fewest semantic errors in 5 of the 6 tasks, and only 1 semantic error in the remaining task. Therefore differences between categories, in this patient, cannot be attributed to word frequency.

4. Wherever the interpretation of tasks is required there are always assumptions regarding what performance on a particular task can be taken "as a measure of." Usually these assumptions carry certain implications that can be tested in order to increase our confidence in the appropriateness of the assumption. This sort of problem is not unique to the Warrington et al. case. It was pointed out in Hillis et al. (1990) that the ability to correctly mime the use of an item a patient is unable to name correctly has been taken by many as a measure of access to an *intact* semantic representation. Humphreys et al. (1988) test the validity of this particular assumption by examining more thoroughly the extent to which patient J.B. was able to access semantic information for items he could mime but not name. They found that, in fact, the assumption did not hold.

References

Badecker, W. &, Caramazza, A. (1985). On considerations of method and theory governing the use of clinical categories in neurolinguistics and cognitive neuropsychology: The case against agrammatism. *Cognition, 20*, 97–125.

Barlow, H. B. (1982). General principles: The senses considered as physical instruments. In H. B. Barlow & J. D. Mollon (Eds.), *The senses*. London: Cambridge University Press.

Basso, A., Taborelli, A., & Vignolo, L. A. (1978). Dissociated disorders of speaking and writing in aphasia. *Journal of Neurology, Neurosurgery and Psychiatry, 41*(6), 556.

Bub, F., & Bub, D. (1988). On methodology of single-case studies in cognitive neuropsychology. *Cognitive Neuropsychology, 5*(5), 565–589.

Caplan, D. (1988). On the role of group studies in neuropsychological and pathopsychological research. *Cognitive Neuropsychology, 5* (5), 535–548.

Caramazza, A. (1986). On drawing inferences about the structure of normal cognitive systems from the analysis of patterns of impaired performance: The case for single-patient studies. *Brain and Cognition, 5*, 41–66.

Caramazza, A., & Hillis, A. (1990). Where do semantic errors come from? *Cortex, 26*, 95–122.

Caramazza, A., & Miceli, G. (1990). The structure of graphemic representations. *Cognition, 37*, 243–297.

Ellis, A. W., & Young, A. W. (1988). *Human cognitive neuropsychology.* Hillsdale, NJ: Erlbaum.

Hillis, A., Rapp, B., Romani, C., & Caramazza, A. (1990). Selective impairment of semantics in lexical processing. *Cognitive Neuropsychology, 7*, 191–243.

Kohn, S. E. (1984). The nature of the phonological disorder in conduction aphasia. *Brain and Language, 23*, 97–115.

McCloskey, M., & Caramazza, A. (1988). Theory and methodology in cognitive neuropsychology: A response to our critics. *Cognitive Neuropsychology, 5*(5), 583–623.

McCloskey, M., Goodman-Schulman, R., & Aliminosa, D. (1988). *The structure of orthographic representations: Evidence from acquired dysgraphia.* Paper presented at the 29th annual meeting of the Psychonomic Society, Chicago.

Mehler, J., Morton, J., & Jusczyk, P. W. (1984). On reducing language to biology. *Cognitive Neuropsychology, 1*(1), 83–116.

Michel, F. (1979). Préservation du langage ecrit malgré un deficit majeur du langage oral. *Lyon Medical, 2441*, 141–149.

Milberg, W., Blumstein, S. E., & Dworetzky, B. (1987). Processing of lexical ambiguities in aphasia. *Brain and Language, 31*(1), 138–150.

Neisser, U. (1967). *Cognitive Psychology.* New York: Appleton-Century-Crofts.

Newell, A. (1973). You can't play 20 questions with nature and win: Projective comments on the papers of this symposium. In W. G. Chase (Ed.), *Visual information processing* (pp. 283–308). New York: Academic Press.

Patterson, K., & Kay, J. (1982). Letter-by-letter reading; Psychological descriptions of a neurological syndrome. *Quaterly Journal of Experimental Psychology, 34a*, 411–441.

Rapp, B., & Caramazza, A. (1989). General to specific access to word meaning: A claim reexamined. *Cognitive Neuropsychology, 6*, 251–272.

Riddoch, M. J., Humphreys, G. W., Coltheart, M., & Funnell, E. (1988). Semantic systems or system? Neuropsychological evidence re-examined. *Cognitive Neuropsychology, 5*(1), 3–25.

Shallice, T. (1988). *From neuropsychology to mental structure.* Cambridge, England: Cambridge University Press.

Sokol, S. M., & McCloskey, M. (1988). Levels of representation in verbal number production. *Applied Psycholinguistics, 9*, 267–281.

Teller, D. Y. (1984). Linking propositions. *Vision Research, 24*(10), 1233–1246.

Warrington, E. K. (1975). The selective impairment of semantic memory. *Quarterly Journal of Experimental Psychology, 27*, 635–657.

Warrington, E. K., & McCarthy, R. (1983). Category specific access dysphasia. *Brain, 106*, 859–878.

Warrington, E. K., & Shallice, T. (1979). Semantic access dyslexia. *Brain, 102*, 43–63.

22

Feedforward Processes in Drug Tolerance and Dependence

SHEPARD SIEGEL

The highly variable environment in which we live presents many threats to our existence, both external (e.g., temperature extremes) and internal (e.g., accumulation of toxic wastes produced by our cells). Survival requires that we resist these disturbances; temperature and body-fluid chemistry must fluctuate only within narrow limits. Deviations from optimal levels must be detected, and responses made to counter these deviations. The classic work of Claude Bernard (1878) emphasized the importance of maintaining a constant "*milieu intérieur*" in the face of vicissitudes posed by ever-changing exteroceptive and interoceptive stimulation. Walter Cannon (1932) extensively studied mechanisms of internal regulation, and propsed the term *homeostasis*. To a great extent, the subject matter of physiology is homeostasis: "For the animal organism the central problem of existence is that of maintaining the stability of its structure and function in the face of constant internal and external assaults" (Horrobin, 1970, p. 1).

One type of "assault" is provided by administration of a drug, and animals have a variety of homeostatic mechanisms to deal with many pharmacological insults. That is, the presence of the drug is detected and responses are made to attenuate the effect of the drug. Increasing knowledge in the area of neurochemistry has led to increasing appreciation of these feedback mechanisms in tolerance. For example, it is likely (e.g., Goldstein, 1976) that exogenous morphine initially supplements the activity of enkephalin. The response to morphine represents the combined effects of the drug and its endogenous ligand. If morphine is continually present, the excessive opioid activity elicits feedback processes that suppress the activity of enkephalin neurons. As a result of this suppression, the response to morphine decreases (or more morphine is needed to reestablish the initial effect of the drug). That is, tolerance is said to occur.

Feedback and Feedforward Systems

The above analysis of opiate tolerance illustrates the characteristics of a negative feedback system. The system contains a "misalignment detector" that detects deviations from optimal functioning (in this case, excessive opioid activity), and the output from this detector elicits a response (decreased enkephalin release) that attenuates the deviation. Thus, the situation is very much like the maintenance of room temperature by a thermostat: When a deviation from preset room temperature is detected (e.g., temperature falls), a response is initiated (heat production by the furnace) that attenuates the deviation.

Control systems in general, and biological control systems in particular, are often more complicated than this simple negative feedback loop: "There has been a growing awareness that feedback is not the only strategy used by the body to control physiological systems" (Houk, 1988, p. 97). Such awareness results from demonstrations that corrective action may be seen not only in response to a detected deviation, but also in response to the disturbance that elicits the deviation. For example, sophisticated temperature control systems may be sensitive not only to the temperature of the interior space being maintained but also to outside temperatures. With this additional capability, heat production can be initiated in anticipation of an impending fall in temperature. Such control systems are said to have "disturbance detectors," as well as misalignment detectors. The disturbance detector provides a mechanism of "warning" the system that stimulation of the misalignment detector is imminent. The advantages of disturbance detectors, acting in conjunction with misalignment detectors, are well known to control-system theorists (see Horrobin, 1970; Houk, 1980, 1988); the speed and smoothness of response of the controlled variable is enhanced:

For example, a thermostat could be made more effective if instead of a simple misalignment detector inside the house, there was in addition a disturbance detector on the outside wall. If the weather suddenly became cooler, this disturbance would be picked up by the outside thermometer before any alteration in inside temperature could take place. (Horrobin, 1970, p. 26)

Feedback refers to the process of control by misalignment detectors. *Feedforward* refers to the process of control by disturbance detectors. The contribution of feedforward to homeostatic regulation in many physiological systems (e.g., temperature maintenance, respiration during exercise) has been well documented (see Horrobin, 1970; Houk, 1980, 1988). It has recently become apparent that feedforward also makes an important contribution to the homeostatic processes mediating drug tolerance. That is, tolerance depends not only on feedback corrections made in response to drugs acting on central receptors but also on responses made in anticipation of such pharmacological stimulation.

Pavlovian Conditioning and Tolerance

"Feedforward means anticipation. It means responding, not to disturbances, but to stimuli that have been associated with disturbances in the past" (Toates, 1979, p. 99). The study of "responding . . . to stimuli that have been associated with disturbances in the past" is the study of Pavlovian conditioning; thus, an appreciation of feedforward regulation requires an appreciation of Pavlovian conditioning principles (see Houk 1980, 1988).

The Pavlovian Conditioning Situation

In the Pavlovian conditioning situation, a contingency is arranged between two stimuli; typically, one stimulus reliably predicts the occurrence of the second stimulus. Using the usual terminology, the second of these paired stimuli is termed the unconditional stimulus (UCS). The UCS, as the name implies, is selected because it elicits relevant activities from the outset (i.e., unconditionally), prior to any pairings. Responses elicited by the UCS are termed unconditional responses (UCRs). The stimulus signaling the presentation of the UCS is "neutral," (i.e., it elicits little relevant activity prior to its pairing with the UCS), and is termed the conditional stimulus (CS). The CS, as the name implies, becomes capable of eliciting new responses as a function of (i.e., conditional upon) its pairing with the unconditional stimulus.

In Pavlov's well-known conditioning research, the CS was a conveniently manipulated exteroceptive stimulus (bell, light, etc.), and the UCS was either food or orally injected dilute acid (both of which elicited a conveniently monitored salivary UCR). After a number of CS–UCS pairings, it was noted that the subject salivated not only in response to the UCS, but also in anticipation of the UCS (i.e., in response to the CS). The subject is then said to display a conditional response (CR).

Drugs as Unconditional Stimuli

A wide range of exteroceptive and interoceptive stimuli have been used in Pavlovian conditioning experiments (Razran, 1961). Drugs constitute a particularly interesting class of UCSs. After some number of drug administrations, with each administration being reliably signaled by a CS, pharmacological CRs can be observed in response to the CS. It was Pavlov who first demonstrated such pharmacological conditioning. He paired a tone with administration of apomorphine. The drug induced restlessness, salivation, and a "disposition to vomit." After several tone-apomorphine pairings, the tone alone "sufficed to produce all the active symptoms of the drug, only in a lesser degree" (Pavlov, 1927, p. 35).

Additional research by Krylov (reported by Pavlov, 1927, pp. 35–37) indicated that even if there is not an explicit CS (such as an auditory cue), naturally occurring predrug cues (opening the box containing the hypodermic syringe, cropping the fur, etc.) could serve as CSs. In Krylov's experiments, a dog was repeatedly injected with morphine, each injection eliciting a number of responses including copious salivation. After five or six such injections, it was observed that "the preliminaries of injection" (Pavlov, 1927, p. 35) elicited many morphinelike responses, including salivation.

THE PHARMACOLOGICAL CONDITIONAL RESPONSE. Most pharmacological conditioning research has been greatly influenced by Pavlov's theory of CR formation. According to this theory, the CR is a replica of the UCR, and, indeed, much drug conditioning work has demonstrated CRs that mimic the drug effect (Siegel, 1985; Stewart & Eikelboom, 1987). In contrast, in 1937 Subkov and Zilov reported that dogs with a history of epinephrine administration (each injection eliciting a tachycardiac response), displayed a conditional bradycardiac response. Subkov and Zilov cautioned against "the widely accepted view that the external modifications of the conditional reflex must always be identical with the response of the organism to the unconditional stimulus" (Subkov & Zilov, 1937, p. 296). Subsequent research has suggested that the characteristics of the pharmacological CR depend very much on the nature and mechanism of the drug effect (see Eikelboom & Stewart, 1982; Paletta & Wagner, 1986; Siegel, 1989). For many effects of many drugs, the CR is an anticipatory compensation for the drug effect. For example, the subject with a history of morphine administration (and its analgesic consequence) often displays a CR of hyperalgesia (Krank, 1987; Krank, Hinson, & Siegel, 1981; Siegel, 1975). Similar compensatory CRs have been reported with respect to the thermic (Siegel, 1978), locomotor (Mucha, Volkovsiks, & Kalant, 1981; Paletta & Wagner, 1986), behaviorally sedating (Hinson & Siegel, 1983), and gastrointestinal (Raffa & Porreca, 1986) effects of morphine. The CR seen with many nonopiate drugs is similarly opposite to the drug effect—e.g., atropine (Mulinos & Lieb, 1929), chlorpromazine (Pihl & Altman, 1971), amphetamine (Obál, 1966), methyl dopa (Korol & McLaughlin, 1976), lithium chloride (Domjan & Gillan, 1977), haloperidol (King, Schiff, & Bridger, 1978), ethanol (Lê, Poulos, & Cappell, 1979), and caffeine (Rozin, Reff, Mark, & Schull, 1984).

THE PHARMACOLOGICAL CONDITIONAL RESPONSE AND FEEDFORWARD CONTROL OF TOLERANCE. Drug-compensatory CRs would be expected to be a feature of normal drug administration procedures. In those cases in which the same drug is repeatedly administered, with discrete environmental stimuli signaling each drug administration, drug-compensatory CRs should function to increasingly attenuate the drug effect. A decreasing response to a drug, over the course of successive administrations, defines tolerance. Thus, feedforward processes (drug-compensatory conditional responses) augment feedback processes (drug-compensatory unconditional responses) in the control of tolerance.

Evidence for Feedforward Control of Tolerance

Feedforward control of tolerance is apparent in many demonstrations that tolerance is affected by experience with predrug cues, as well as by the drug.

Predrug Cues and Tolerance

On the basis of a feedforward analysis, cues associated with a drug elicit drug-compensatory CRs that contribute to tolerance; thus, tolerance should be more pronounced following drug administration in the presence of the usual predrug cues than in the presence of alternative cues. Results of a number of experiments by Mitchell and colleagues (e.g., Adams, Yeh, Woods, & Mitchell, 1969) did demonstrate that rats and people displayed the expected analgesia-tolerant response to the last of a series of morphine injections only if this final injection was presented in the context of the same environmental cues as the prior injections. Results of many subsequent experiments have confirmed and extended Mitchell's observations (reviewed by Siegel & MacRae, 1984). Environmental specificity of tolerance has also been demonstrated with many nonopiate drugs, including ethanol, pentobarbital, scopolamine, haloperidol, and several benzodiazepines (see Siegel, 1989).

Although environmental cues uniquely present at the time of drug administration are the most obvious type of predrug signals, there are many other potential cues for the effect of a drug: Ambient temperature (Kavaliers & Hirst, 1986), magnetic fields (Kavaliers & Ossenkopp, 1985), and a variety of cues induced by drugs (Greeley, Lê, Poulos, & Cappell, 1984; Siegel, 1988b) may all control the display of tolerance if they reliably signal a drug. Thus, pharmacological feedforward mechanisms may be engaged by any of a variety of exteroceptive or interoceptive stimuli present at the time of drug administration.

Other Evidence for Feedforward Control of Tolerance

If feedforward processes contribute to tolerance, it would be expected that nonpharmacological manipulations of the putative CS (cues present at the time of drug administration), known to attenuate Pavlovian conditioning, should similarly attenuate compensatory CR acquisition and thus tolerance. The results of many such manipulations have been assessed. Because much of these data are extensively reviewed elsewhere (Siegel, 1989), they are only briefly summarized here.

EXTINCTION OF TOLERANCE. The magnitude of established CRs is decreased by "extinction"—i.e., repeated presentations of the CS without the UCS. Similarly, tolerance to a variety of effects of many drugs is attenuated by repeated presentations of predrug cues without the drug. Such extinction has been demonstrated with respect to tolerance to the analgesic effect of morphine injected subcutaneously (e.g., Siegel, Sherman, & Mitchell, 1980) and directly into the ventricles of the brain (MacRae & Siegel, 1987). Furthermore, extinction has been demonstrated with respect to tolerance to amphetamine, midazolam (a short-acting benzodiazepine), and the synthetic polynucleotide, poly:IC (see Siegel, 1989).

Another procedure for extinguishing a CS–UCS association is to continue to present the CS and UCS, but in an unpaired manner (see Mackintosh, 1974). That is, the subject receives both conditioning stimuli, but the CS does not signal the UCS; rather, the CS is presented only during intervals between CS presentations. With this procedure, the CS signals the absence of the UCS. Fanselow and German (1982) reported that such unpaired presentations attenuate tolerance to the behaviorally sedating effect of morphine in rats. In this experiment, morphine was administered on a number of occasions in the presence of a distinctive environmental cue. When tolerance was established, continued presentation of the drug and cue, but in an explicitly unpaired manner, eliminated tolerance. That is, despite the fact that morphine-tolerant rats continued to receive morphine, tolerance was reversed if the continued morphine administrations were unpaired with the cue that was initially paired with the drug.

EXTERNAL INHIBITION OF TOLERANCE. Pavlov (1927) noted that presentation of a novel, extraneous stimulus disrupts the elicitation of established CRs. Such "external inhibition" of conditional responding has also been shown to eliminate tolerance to the hypothermic effect of ethanol (Siegel & Sdao-Jarvie, 1986) and the analgesic effect of morphine (Poulos, Hunt, & Cappell, 1988). That is, drug-experienced rats that normally display tolerance fail to do so when presented with an arbitrary novel stimulus.

RETARDATION OF THE DEVELOPMENT OF TOLERANCE. One technique for attenuating the development of a CS–UCS association is to present the CS alone prior to pairing it with the UCS (see Mackintosh, 1974). If feedforward processes contribute to tolerance, it would be expected that rats with extensive experience with drug administration cues prior to the time that these cues are paired with the drug should be relatively retarded in the acquisition of tolerance (compared to rats with minimal preexposure to these cues), despite the fact that the groups do not differ with respect to their histories of drug administration. Such an effect of CS preexposure has been demonstrated with respect to tolerance to the analgesic effect of morphine, and the immunostimulatory effect of poly:IC (see review by Siegel, 1989).

Another procedure for attenuating the development of a CS–UCS association is partial (rather than continuous) reinforcement. That is, if only a portion of the presentations of the CS are paired with the UCS, CR acquisition retarded (compared to the situation in which all presentations of the CS are paired with the UCS, see Mackintosh, 1974). On the basis of a feedforward analysis of tolerance, it would be expected that such partial reinforcement would retard the development of tolerance; a group in which only a portion of the presentation of drug administration cues are followed by the drug (i.e., a partial reinforcement group) should be slower to acquire tolerance than a group that never has exposure to drug-predictive cues without actually receiving the drug (i.e., a continuous reinforcement group), even when the two groups are equated with respect to all pharmacological parameters. Such findings have been reported with respect to tolerance to several effects of morphine (Krank, Hinson, & Siegel, 1984; Siegel, 1977, 1978).

OTHER EVIDENCE. In addition to the research just described, results of many other experiments have provided evidence that drug-anticipatory responses contribute to tolerance. For example, tolerance to morphine (Fanselow & German, 1982; Siegel, Hinson, & Krank, 1981) and pentobarbital (Hinson & Siegel, 1986) is subject to inhibitory learning. Morphine tolerance can also be manipulated by compound conditioning phenomena, such as "blocking" (Dafters, Hetherington, & McCartney, 1983) and "overshadowing" (Dafters & Bach, 1985; Walter & Riccio, 1983). A full discussion of these findings is beyond the scope of this chapter (for a fuller discussion, see Siegel, 1989), but it should be emphasized that many experiments, with a variety of drugs, support the contribution of drug-compensatory CRs to tolerance.

Feedforward Processes and Drug Withdrawal Symptoms

Feedback processes are elicited in response to the output of misalignment detectors. Since such misalignment-detector activity is often signaled, feedforward is a generally useful regulatory strategy that is used in combination with feedback. However, feedforward (like feedback) has disadvantages as well as advantages (Houk, 1980, 1988). Some drug-withdrawal symptoms are manifestations of disadvantages of feedforward regulation of drug effects.

Disadvantages of Feedforward

Under some circumstances, feedforward responses may be initiated erroneously; that is, a signal, although usually followed by systemic stimulation, may *not* be followed by such stimulation on a particular occasion. Consider such a state of affairs in the prototypical Pavlovian conditioning preparation, salivary conditioning, in which a bell is used to signal food:

Thus the controller for salivary secretion, which acts to keep the mouth appropriately moist, responds to the sounding of the bell after a period of training during which bell ringing is followed by the introduction of food into the mouth, the latter being a normal stimulus for salivation. In this example the conditioned reflex provides a feedforward mechanism that moistens the mouth in preparation for food that is about to arrive. It is clear that errors in performance (too much moisture) are bound to occur the first time the bell is not followed by presentation of food. The predictive value of the bell is good only as long as the environmental contingencies remain unchanged (Houk, 1980, p. 249).

Similarly, if "environmental contingencies" change, the feedforward response to impending pharmacological stimulation will result in "errors in performance": if the usual predrug cues are not followed by the usual pharmacological stimulation, the drug-compensatory CR will be elicited, but there will be no drug effect to be attenuated.

Withdrawal Symptoms as a Disadvantage of Feedforward Regulation

In discussing the role of compensatory CRs in withdrawal symptoms, it is important to make a distinction between the acute withdrawal reaction seen shortly after the initiation of abstinence (which typically lasts for days or, at most, weeks) and the apparently similar symptoms often noted after detoxification is presumably complete (see Hinson & Siegel, 1982). The acute withdrawal reaction is probably a manifestation of the disadvantages of feedback control (see Houk, 1980, 1988). The appearance of symptoms after a long drug-free period probably reflects the disadvantages of feedforward control; that is, it is likely that it is the anticipation of the drug, rather than the drug itself, that is responsible for these symptoms.

Consider the situation in which the addict expects a drug, but does not receive it; that is, no drug is available, but the addict is in an environment where he or she has frequently used drugs in the past, or it is the time of day when the drug is typically administered, or any of a variety of drug-associated stimuli occur. Research with animals demonstrates that presentation of cues previously associated with drug administration, but now not followed by the drug, results in the occurrence of drug-compensatory CRs. . . . In the situation in which the drug addict expects but does not receive the drug, it would be expected that drug-compensatory CRs would also occur. These CRs normally counter the pharmacological disruption of functioning which occurs when the anticipated drug is administered. However, since the expected drug is not forthcoming, the CRs may achieve expression as overt physiological reactions, e.g., yawning, running nose, watery eyes, sweating . . . or form the basis for the subjective experience of withdrawal sickness and craving. (Hinson & Siegel, 1982, p. 499).

Evidence That Feedforward Contributes to "Withdrawal Symptoms"

It has become increasingly apparent that just as feedforward contributes to tolerance (when the anticipated drug is administered), it also contributes to responses seen when the anticipated drug is not administered. That is, some drug "withdrawal symptoms" are, more accurately, drug "preparation symptoms."

ENVIRONMENTAL ELICITATION OF WITHDRAWAL DISTRESS. If feedforward contributes to "withdrawal symptoms," such symptoms should be especially pronounced in the presence of drug-associated cues. One way to evaluate the role of environmental cues in withdrawal distress is simply to ask addicts to recall the circumstances in which they suffer such distress. Several investigators have done just this, and have noted that both opiate addicts and alcoholics report that such distress is especially pronounced in the presence of drug-associated cues (see Biernacki, 1986; Siegel, 1988a). Similarly, clinicians have noted that opiate withdrawal symptoms are displayed when, during behavior therapy (even with long-detoxified former addicts), drugs are discussed or the paraphernalia of addiction (syringe and tourniquet) are viewed (see review by Siegel, 1988a). One's own personal experience may provide similar evidence of the importance of drug-associated cues in withdrawal distress and craving—environmental cues associated with smoking (or seeing others smoking, or talking about smoking) often elicit craving for a cigarette in individuals who have been addicted to nicotine.

Animals, experimentally addicted to morphine, also display environmentally elicited withdrawal symptoms. Consider Ternes's (1977) description of the behavior of monkeys that were repeatedly injected with morphine in the presence of an arbitrary auditory cue—tape-recorded music. This music became capable of eliciting withdrawal symptoms and relapse in these monkeys after a considerable period of abstinence: "After the animal had been weaned from the drug and maintained drug-free for several months, the experimenter again played the tape-recorded music and the animal showed the following signs: he became restless, had piloerection, yawned, became diruetic, showed rhinorrhea, and again sought out the drug injection" (Ternes, 1977, pp. 167–168).

There are several laboratory demonstrations of the ability of drug-associated cues to elicit withdrawal distress. For example, it has been noted that former addicts display physiological signs of narcotic withdrawal when they are exposed in the laboratory to heroin-related stimuli—e.g., they perform the "cooking up" ritual while being monitored by a polygraph, or when they see slides of opiate-related material (e.g., inserting a syringe into a vein), or a videotape depicting scenes of heroin preparation and administration (see Siegel, 1988a). Similar findings have been reported with respect to alcohol (see Siegel, 1987).

ENVIRONMENTAL CUES, RELAPSE, AND ABSTINENCE. On the basis of a feedforward analysis of anticipatory regulation, drug-compensatory responses, elicited by drug-predictive cues, elicit withdrawal distress. It follows that such withdrawal distress (and postdetoxification relapse to drug use) should be much less pronounced if the drug-experienced organism is not reexposed to drug-associated stimuli. Such findings have been reported both in epidemiological studies with humans (e.g., Maddux & Desmond, 1982; Robins, Helzer, & Davis, 175) and in experimental studies with animals (e.g., Hinson, Poulos, Thomas, & Cappell, 1986; Thompson & Ostlund, 1965) (see review by Siegel, 1988a).

Conclusions

Recent research has led to an increasing understanding of the homeostatic mechanisms responsible for the tolerance and withdrawal symptoms seen with many psychoactive drugs. Most of the research has concerned neurochemical responses, elicited by a pharmacological disturbance, that compensate for the disturbance. However, it is increasingly apparent that such compensatory responses can be elicited by stimuli that, in the past, have signaled the disturbance. That is, the feedback mechanisms that mediate tolerance and withdrawal symptoms are importantly influenced by feedforward mechanisms. Thus, a complete understanding of these pharmacological phenomena requires an appreciation of drug-anticipatory conditional responses, as well as drug-elicited unconditional responses. The study of conditional responding is the study of Pavlovian conditioning, and Pavlov was certainly aware of the importance of anticipatory responding for physiological regulation: "It is pretty evident that under natural conditions the normal animal must respond not only to stimuli which in themselves bring immediate benefit or harm, but also to other physical or chemical agencies—waves of light and the like—which in themselves only signal the approach of these stimuli" (Pavlov, 1927, p. 14).

References

Adams, W. J., Yeh, S. Y., Woods, L. A., & Mitchell, C. L. (1969). Drug-test interaction as a factor in the development of tolerance to the analgesic effect of morphine. *Journal of Pharmacology and Experimental Therapeutics*, *168*, 251–257.

Bernard, C. (1878). *Leçons sur les phénomènes de la vie communs aux animaux et aux végétaux.* Paris: Editions J-B Bailliere.

Biernacki, P. (1986). *Pathways from heroin addiction: Recovery without treatment.* Philadelphia: Temple University Press.

Cannon, W. B. (1932). *The wisdom of the body.* New York: Norton.

Dafters, R., & Bach, L. (1985). Absence of environment-specificity in morphine tolerance acquired in non-distinctive environments: Habituation or stimulus overshadowing? *Psychopharmacology*, *87*, 101–106.

Dafters, R., Hetherington, M., & McCartney, H. (1983). Blocking and sensory preconditioning effects in morphine analgesic tolerance: Support for a Pavlovian conditioning model of drug tolerance. *Quarterly Journal of Experimental Psychology*, *35B*, 1–11.

Domjan, M., & Gillan, D. J. (1977). After-effects of lithium-conditioned stimuli on consummatory behavior. *Journal of Experimental Psychology: Animal Behavior Processes*, *3*, 322–334.

Eikelboom, R., & Stewart, J. (1982). Conditioning of drug-induced physiological responses. *Psychological Reviews*, *89*, 507–528.

Fanselow, M. S., & German C. (1982). Explicitly unpaired delivery of morphine and the test situation: Extinction and retardation of tolerance to the suppressing effects of morphine on locomotor activity. *Behavioral and Neural Biology*, *35*, 231–241.

Goldstein, A. (1976). Opioid peptides (endorphins) in pituitary brain. *Science*, *193*, 1081–1086.

Greeley, J., Lê, D. A., Poulos, C. X., & Cappell, H. (1984). Alcohol is an effective cue in the conditional control of tolerance to ethanol. *Psychopharmacology*, *83*, 159–162.

Hinson, R. E., Poulos, C. X., Thomas, W., & Cappell, H. (1986). Pavlovian conditioning and addictive behavior: Relapse to oral self-administration of morphine. *Behavioral Neuroscience*, *100*, 368–375.

Hinson, R. E., & Siegel, S. (1982). Nonpharmacological bases of drug tolerance and dependence. *Journal of Psychosomatic Research*, *26*, 495–503.

Hinson, R. E., & Siegel, S. (1983). Anticipatory hyperactivity and tolerance to the narcotizing effect of morphine in the rat. *Behavioral Neuroscience, 97*, 759–767.

Hinson, R. E., & Siegel, S. (1986). Pavlovian inhibitory conditioning and tolerance to pentobarbital-induced hypothermia in rats. *Journal of Experimental Psychology: Animal Behavior Processes, 12*, 363–370.

Horrobin, D. F. (1970). *Principles of biological control.* Ayesbury, UK: Medical and Technical Publishing Company.

Houk, J. C. (1980). In V. B. Mountcastle (Ed.), *Medical physiology* (Vol. 1, pp. 246–267). St. Louis: Mosby.

Houk, J. C. (1988). Control strategies in physiological systems. *FASEB Journal, 2*, 97–107.

Kavaliers, M., & Hirst, M. (1986). Environmental specificity of tolerance to morphine-induced analgesia in a terrestial snail: Generalization of a behavioral model of tolerance. *Pharmacology, Biochemistry and Behavior, 25*, 1201–1206.

Kavaliers, M., & Ossenkopp. K.-P. (1985). Tolerance to morphine-induced analgesia in mice. *Life Sciences, 37*, 1125–1135.

King, J. J., Schiff, S. R., & Bridger, W. H. (1978). Haloperidol classical conditioning—paradoxical results. *Society for Neuroscience Abstracts, 4*, 495.

Korol, B. and McLaughlin, L. J. (1976). A homeostatic adaptive response to alpha-methyl-dopa in conscious dogs. *Pavlovian Journal of Biological Science, 11*, 67–75.

Krank, M. D. (1987). Conditioned hyperalgesia depends on the pain sensitivity measure. *Behavioral Neuroscience, 101*, 854–857.

Krank, M. D., Hinson, R. E., & Siegel, S. (1981). Conditional hyperalgesia is elicited by environmental signals of morphine. *Behavioral and Neural Biology, 32*, 148–157.

Krank, M. D., Hinson, R. E., & Siegel, S. (1984). Effect of partial reinforcement on tolerance to morphine-induced analgesia and weight loss in the rat. *Behavioral Neuroscience, 98*, 72–78.

Lê, A. D., Poulos, C. X., & Cappell, H. (1979). Conditioned tolerance to the hypothermic effects of ethyl alcohol. *Science, 206*, 1109–1110.

Mackintosh, N. J. (1974). *The psychology of animal learning.* London: Academic Press.

MacRae, J. R., & Siegel, S. (1987). Extinction of tolerance to the analgesic effect of morphine: Intracerbroventricular administration and effects of stress. *Behavioral Neuroscience, 101*, 790–796.

Maddux, J. F., & Desmond, D. P. (1982). Residence relocation inhibits opioid dependence. *Archives of General Psychiatry, 39*, 1313–1317.

Mucha, R. F., Volkovsiks, C., & Kalant, H. (1981). Conditioned increases in locomotor activity produced with morphine as an unconditioned stimulus, and the relation of conditioning to acute morphine effect and tolerance. *Journal of Comparative and Physiological Psychology, 95*, 351–362.

Mulinos, M. G., & Lieb, C. C. (1929). Pharmacology of learning. *American Journal of Physiology, 90*, 456–457.

Obàl, F. (1966). The fundamentals of the central nervous control of vegetative homeostasis. *Acta Physiologica Academiae Scientarum Hungaricae, 30*, 15–29.

Paletta, M. S., & Wagner, A. R. (1986). Development of context-specific tolerance to morphine: Support for a dual process interpretation. *Behavioral Neuroscience, 100*, 611–623.

Pavlov, I. P. (1927). *Conditioned reflexes* (G. V. Anrep, trans.). London: Oxford University Press.

Pihl, R. O., & Altman, J. (1971). An experimental analysis of the placebo effect. *Journal of Clinical Pharmacology, 11*, 91–95.

Poulos, C. X., Hunt, T., & Cappell, H. (1988). Tolerance to morphine analgesia is reduced by the novel addition or omission of an alcohol cue. *Psychopharmacology, 94*, 412–416.

Raffa, R. B., & Porreca, F. (1986). Evidence for a role of conditioning in the development of tolerance to morphine-induced inhibition of gastrointestinal transit in rats. *Neuroscience Letters, 67*, 229–232.

Razran, G. (1961). The observable unconscious and the inferable conscious in current Soviet psychophysiology: Interoceptive conditioning, semantic conditioning and the orienting reflex. *Psychological Review, 68*, 81–147.

Robins, L. N., Helzer, J. E., & Davis, D. H. (1975). Narcotic use in southeast Asia and afterwards. *Archives of General Psychiatry, 32*, 955–961.

Rozin, P., Reff, D., Mark, M., & Schull, J. (1984). Conditioned responses in human tolerance to caffeine. *Bulletin of the Psychonomic Society, 22*, 117–120.

Siegel, S. (1975). Evidence from rats that morphine tolerance is a learned response. *Journal of Comparative Physiology and Psychology, 89*, 498–506.

Siegel, S. (1977). Morphine tolerance acquisition as an associative process. *Journal of Experimental Psychology: Animal Behavior Processes, 3*, 1–13.

Siegel, S. (1978). Tolerance to the hypothermic effect of morphine in the rat is a learned response. *Journal of Comparative Physiology and Psychology, 92*, 1137–1147.

Siegel, S. (1985). Drug-anticipatory responses in animals. In L. White, B. Tursky, & G. Schwartz (Eds.), *Placebo: Theory, research, and mechanisms* (pp. 288–305). New York: Guilford Press.

Siegel, S. (1987). Pavlovian conditioning and ethanol tolerance. In K. O. Lindros, R. Ylikahri, and K. Kiianmaa (Eds.), *Advances in biomedical alcohol research* (pp. 25–36). Oxford, England: Pergamon Press. (Published as Supplement No. 1, *Alcohol and Alcoholism*, 1987.)

Siegel, S. (1988a). Drug anticipation and the treatment of dependence. In B. Ray (Ed.), *Learning factors in substance abuse* (pp. 1–24). (National Institute of Drug Abuse Research Monograph No. 84, Department of Health and Human Services Publication No. [ADM] 88-1576). Washington, DC: U. S. Government Printing Office.

Siegel, S. (1988b). State-dependent learning and morphine tolerance. *Behavioral Neuroscience, 102*, 228–232.

Siegel, S. (1989). Pharmacological conditioning of drug effects. In M. W. Emmett-Oglesby & A. J Goudie (Eds.), *Tolerance and sensitization to psychoactive drugs* (pp. 115–180). Clifton, NJ: Humana Press.

Siegel, S., Hinson, R. E., & Krank M. D. (1981). Morphine-induced attenuation of morphine tolerance. *Science, 212*, 1533–1534.

Siegel, S., & MacRae, J. (1984). Environmental specificity of tolerance. *Trends in Neurosciences, 7*, 140–143.

Siegel, S., & Sdao-Jarvie, K. (1986). Reversal of ethanol tolerance by a novel stimulus. *Psychopharmacology, 88*, 258–261.

Siegel, S., Sherman, J. E., & Mitchell, D. (1980). Extinction of morphine analgesia. *Learning and Motivation, 11*, 289–293.

Stewart, J., & Eikelboom, R. (1987). Conditioned drug effects. In L. L. Iverson, S. D. Iverson, & S. H. Snyder (Eds.), *Handbook of psychopharmacology, Vol. 9* (pp. 1–57). New York: Plenum Press.

Subkov, A. A., & Zilov, G. N. (1937). The role of conditioned reflex adaptation in the origin of hyperergic reactions. *Bulletin of Biological Medicine Experiments, 4*, 294–296.

Ternes, J. W. (1977). An opponent process theory of habitual behavior with special reference to smoking. In M. E. Jarvik, J. W. Cullen, E. R. Gritz, T. M. Vogt, and L. J. West (Eds.), *Research on smoking behavior* (pp. 157–182). (National Institute on Drug Abuse Research Monograph 17, DHEW Publication No. [ADM]79-581). Washington, DC: Superintendent of Documents, U. S. Government Printing Office.

Thompson, T., & Ostlund, W., Jr. (1965). Susceptibility to readdiction as a function of the addiction and withdrawal environments. *Journal of Comparative Physiology and Psychology, 60*, 388–392.

Toates, F. M. (1979). Homeostasis and drinking. *Behavioral and Brain Sciences, 2*, 95–139.

Walter, T. A., & Riccio, D. C. (1983). Overshadowing effects in the stimulus control of morphine analgesic tolerance. *Behavioral Neuroscience, 97*, 658–662.

23

The Impact of Neuroimaging on Human Neuropsychology

RUBEN C. GUR / RAQUEL E. GUR

We were asked by the editors to describe where the research in which we have been involved is leading, and how we envision its impact on the future of the neurosciences. To do that, we will: (1) summarize the methods we are using with the new technology for studying regional brain function, which will necessarily include only the essential information for evaluating the potential of these methods; (2) recount our experience using these methods in the study of behavioral dimensions; (3) describe an algorithm we have developed for analyzing behavioral data in systematic testing of neuropsychological hypotheses using neuroimaging technology; (4) comment on directions we perceive for future research with available technology, and speculate on the long-term potential of this research.

As will be obvious from this volume, the neurosciences have made tremendous progress in the recent past, and for someone interested in understanding neural substrates of behavioral dimensions there are now unprecedented techniques for measuring fundamental neuronal processes. To mention but some, molecular probes with monoclonal antibodies can yield information on the sources and extent of neuroanatomic changes, single-unit recordings can help localize brain regions activated by specific stimulus parameters, magnetic resonance imaging and spectroscopy can yield in vivo data on the state of brain anatomy and physiological indices, and autoradiographic techniques can assess local rates of metabolic activity or density and affinity of certain neuroreceptors. We will not review these methods here, but will concentrate on technology available for studies of brain–behavior relations in humans.

Neuroimaging Methods

Only some of the techniques for obtaining regional data on brain are applicable to humans. These techniques can be loosely termed *neuroimaging methods*, although this may be a misnomer for the class of techniques that yield data relevant to regional brain structure and function. The imaging of

417

these data is only one important way to display them, but it is of tremendous utility and heuristic value. We will divide neuroimaging methods into *neuroanatomic* and *neurophysiological* techniques, although the boundaries sometimes break. Thus, x-ray computed tomography (CT) is classified as an anatomic technique, but it can be enhanced (e.g., by administration of stable Xenon gas) to obtain physiological data (cerebral blood flow). Conversely, some essentially neurophysiologic methods (e.g., Xenon-133 clearance) can yield anatomically relevant data (percentage of tissue with fast clearance characteristics, which in normal brains can be assumed to reflect gray matter).

Neuroanatomic Methods

The two major techniques currently available for imaging brain anatomy are x-ray computed tomography (CT) and magnetic resonance imaging (MRI). MRI can display with exquisite detail brain structures and many lesions (Figure 23.1). These methods can be used for establishing links between brain regions and behavior. If a lesion is detectable on CT or MRI, then we can infer that at least some of the behavioral deficits have been caused by destruction of brain tissue, and hence this region is involved in

(a)   (b)

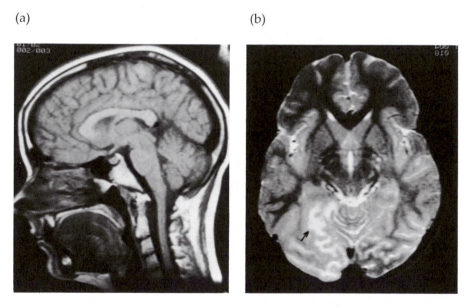

Figure 23.1 MRI: (a) A sagittal view of a normal brain. (b) An axial view of a cerebral infarct.

regulating the behavioral dimension showing impairment. The hazard is of being misled into thinking that a brain region is exclusively involved in regulating the impaired behavioral function, when it is only a component of a neural "network." The correlation between behavior and anatomic pathology is not optimal for studying networks.

MRI is superior to CT in many respects, particularly in brain areas near the skull or other bony tissue, where beam-hardening effects of bone distort the CT scans. Some lesions can be missed by CT and detected by MRI. On the other hand, abnormal signal intensity may occur in MRI scans of normal individuals, and the field still lacks standard scanning procedures and normative data. Very small lesions, or diffuse neuronal loss such as occurs in Alzheimer's disease, can have deleterious effects on behavior and yet be undetected by visual inspection of either CT or MRI scans. Computerized techniques for quantitating brain atrophy using tissue segmentation algorithms are being pursued, and these should prove more sensitive for detection of more subtle or diffuse anatomic changes.

Neurophysiological Methods

Behavior is affected not only by the anatomic integrity of the neural tissue, but also by the level, distribution, and dynamic features of neural activity. The capacity of brain regions to become active in response to task demands is likely to be of particular importance for establishing brain–behavior relations. In clinical populations, behavioral deficits can be more extensive than what is attributable to the death of neurons in regions showing anatomic destruction. These deficits can be caused by brain cells that are not dead but are either insufficiently active or too active. Some grave forms of brain dysfunction are caused by abnormalities in regional brain physiological activity in the absence of anatomic destruction. For example, epilepsy has severe behavioral manifestations and there is evidence for interictal deficits in cognitive and emotional functioning (Dodrill & Wilkus, 1976). Several techniques for physiological neuroimaging demonstrate abnormalities corresponding to an epileptogenic focus, yet CT or MRI scans are frequently uninformative or even normal (Gastaut, 1970).

Of the several physiological techniques that merit inclusion in the neuroimaging family, the oldest is electroencephalography (EEG), and the associated methods for recording of evoked potentials (EP). However, standardized methods to generate topographic display of regional EEG/EP measures have only recently been developed. BEAM (brain electric activity mapping) is an acronym which has been widely used, although it is a trade name and there are other systems, such as Bio-Logic's Brain Atlas (Bio-Logic, Inc.) and Lehmann's (1971), that follow a similar principle. The quantified EEG/EP

Bank: 1   Scale: 8
Freq: 2.50

Bio-logic®   File:A:SZFFT
Ctl.:
View:Top   Rec.:1   31.8
uV

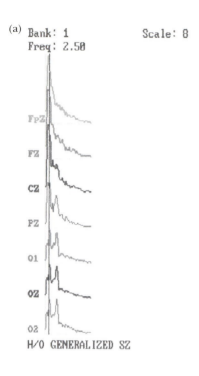

FpZ

FZ

CZ

PZ

O1

OZ

O2

H/O GENERALIZED SZ

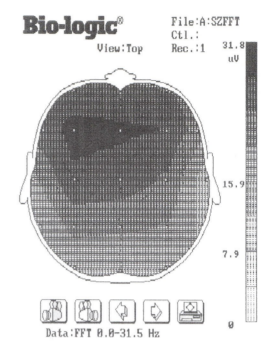

15.9

7.9

0

Data:FFT 0.0-31.5 Hz

(b)

Bank: 1   Scale: 16
Freq: 0.00

Bio-logic®   File:A:CVAFFT
Ctl.:
View:Top   Rec.:1   63.7
uV

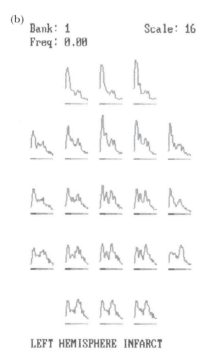

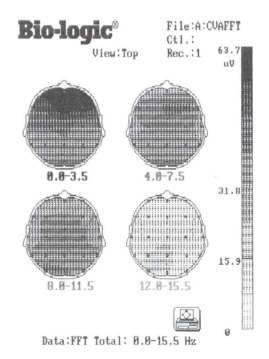

0.0-3.5   4.0-7.5

31.8

8.0-11.5   12.0-15.5

15.9

LEFT HEMISPHERE INFARCT

0

Data:FFT Total: 0.0-15.5 Hz

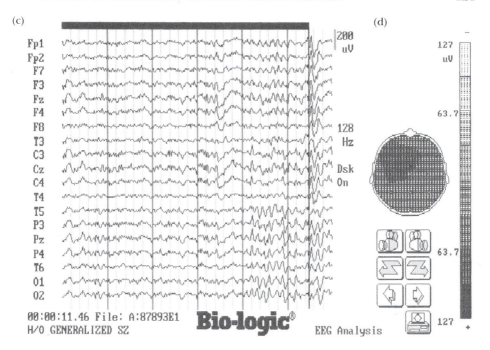

Figure 23.2 Presented are examples of topographic displays of electrophysiological data: (a) Integrated voltages of evoked responses for 12 latency windows following stimulus onset are mapped. In this case averaged evoked response amplitudes to a rare stimulus (oddball P300 paradigm) from a normal subject are shown. Note the posterior maxima in the 288- to 356-ms latency window. (b) Similar data to (a), but obtained from a schizophrenic patient. Note the overall decreased amplitude as compared with the normal subject, as well as the right-sided skew the positive potential has, especially at the 288- to 356-ms latency window. (c) In another example of a voltage map, the map displays the maxima of a spike obtained from an epileptic, which is indicated by the continuous EEG display. (d) This display is of EEG that has been subjected to frequency analysis, and the 12 integrated spectral bands are plotted topographically. In a patient with a right CVA, maximal amplitudes in the delta range (first three maps) are easily appreciated over the right hemisphere.

parameters are presented in gray to produce topographic displays such as is shown in Figure 23.2.

Isotopic techniques for imaging neurophysiology make use of the fact that active neurons have metabolic needs for oxygen and glucose, and that cerebral blood flow rates change in response to these needs. Measures obtained during the performance of cognitive tasks can help generate experimental data to delineate brain regions necessary for regulating specific cognitive processes. Such measures can also help identify regions of abnormal physiological activation associated with behavioral deficits.

The isotopic techniques for measuring cerebral metabolism and blood flow can be traced to the pioneering method of Kety and Schmidt (1948) for measuring whole-brain metabolism and blood flow. The technique used intracarotid injection of nitrous oxide, and measurement of arterial-venous differences in concentration yielded accurate and reproducible data on brain metabolism and blood flow. However, this technique is not only limited to providing whole-brain values, but also by its invasiveness.

Safe regional determinations were first made possible by the introduction of the Xenon-133 clearance techniques for measuring regional cerebral blood flow (rCBF). The highly diffusible Xenon-133 can be administered as gas mixed in air or in saline. Its clearance from the brain is measurable by scintillation detectors. The rate of clearance enables accurate quantitation of rCBF in the fast-clearing gray matter compartment, as well as calculation of mean flow of gray and white matter. Initial applications, using carotid injections (Olesen, Paulson, & Lassen, 1971), were invasive and only enabled measurements in one hemisphere at a time. Obrist, Thompson, Wang, and Wilkinson (1975) reported the Xenon-133 inhalation technique and presented models for quantifying rCBF with this noninvasive procedure. The technique permits simultaneous measurements from both hemispheres, and the number of brain regions that can be measured depends on the number of detectors. Initial studies were performed with up to 16 detectors,

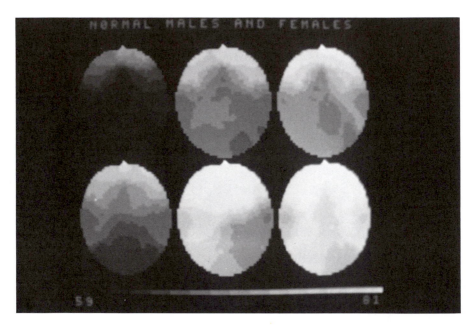

Figure 23.3 A topographic display of rCBF in a group of normal males (upper row) and females (lower row) during rest (first column), a verbal analogies task (middle column), and a spatial line orientation task (last column). Note the hyperfrontal pattern at rest, the higher CBF in females, and the increased CBF in the left for the verbal and the right for the spatial task.

8 over each hemisphere, but there are now commercially available systems with 32 detectors, and recently a system has been introduced that enables the placement of up to 254 detectors. The quantitative data can be displayed topographically, as is shown in Figure 23.3.

Note that rCBF is typically higher in the front of the brain. This "hyperfrontal" pattern has been observed routinely in normal subjects (Ingvar, 1979). The main limitation of the technique is that it is optimal for measuring rCBF on the brain surface near the skull, and thus limited to the study of cortical brain regions.

Positron emission tomography (PET) provides in vivo measures of biochemical and physiological processes with three-dimensional resolution. Selectively labeled radioisotopes are used to measure the rate of the biochemical process, and data are obtained using principles of computed tomography (Reivich et al., 1979). PET is used with radionuclides that decay through the emission of positrons. The radionuclides are usually given intravenously, and are taken up by tissue. Through the emission of a positron, they get rid of their energy and undergo the process of annihilation when the positron interacts with an electron. The two photons emitted from each annihilation travel in opposite directions, and the energy generated is measured by detector arrays. The coincidental counts are used to generate images reflecting the regional rate of radionuclide uptake. This information enables the calculation, depending on the specific radionuclide, of such varied physiological parameters as oxygen and glucose metabolism, blood flow, or receptor density and affinity of certain neurotransmitters. In order to relate this physiological information to anatomic regions of interest (ROI), an atlas of brain anatomy is required. Currently, such computerized "atlas" databases have been based on digitized images of sliced brains. We are near to being able to use customized atlas procedures based on MRI or x-ray CT scans. Multiple brain "slices" can be obtained with PET (Figure 23.4).

Another important application of PET is for assessment of the density

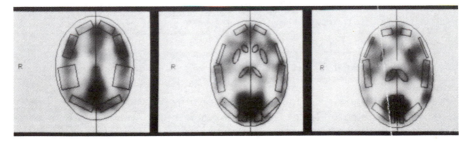

Figure 23.1   Placement of regions of interest. (*Source*: Gur et al., Regional brain function in schizophrenia. I. A positron emission tomography study, *Archives of General Psychiatry*, 1987, *44*, 119–125. Copyright © 1987 by the AMA. Used by permission.)

and affinity of neurotransmitters. Thus far, ligands have been developed for imaging D2 dopamine receptors (Wong et al., 1986; Farde, Hall, Ehrin, & Sedvall, 1986) and ligands for other receptors are currently being investigated. The potential of PET here is to provide the important link between anatomy and physiological activity that is mediated by excitatory and inhibitory effects of neurotransmitters. The investigation of this poorly understood link is, in our opinion, crucial.

Another technique more recently introduced for three-dimensional imaging of rCBF is single photon emission computed tomography (SPECT). The technique enables imaging of CBF using radionuclides, which, unlike positron emitters, do not require the availability of a dedicated cyclotron for their use. However, at present, reliable quantitation of rCBF with available radionuclides that can be safely administered is still very problematic, and much more work is required to make it applicable in systematic neurobehavioral research. SPECT can also be used for studying neurotransmitters, and ligands for imaging both D1 and D2 dopamine receptors have been described (Kung et al., 1989).

Behavioral Dimensions Studied with Neuroimaging

Mentation, as a product of the activity of brain regions, has metabolic costs. The regional distribution of activity is determined by neurotransmitter systems operating on neural networks, with regional specificity of receptor density and affinity. To understand brain regulation of behavior, it would help to investigate this entire chain. The application of neuroimaging techniques to the understanding of human behavior is still in its infancy. The sensitivity and limitations of these methods are being determined, and only crude dimensions of behavior have been examined. The research has proceeded in two directions. One line attempted to correlate normal variability in the neural measures with behavioral functions. The neuroimaging techniques can also be harnessed to the understanding of the neural substrates of abnormal behavior. Such clinical research will have payoffs for basic understanding of neurobehavioral issues as well. Here we describe some of the findings that may illustrate, if rudimentarily, some of the potential for future work.

Neuroanatomy and Behavior

Several studies have examined the relationship between cognitive dimensions and neuroanatomic data. For example, left hemispheric superiority in language processing has been related to CT evidence for asymmetry in the size of language-related structures (see Naeser, 1985, for review), and performance on cognitive tasks has been correlated with CT measures of atrophy. There seems to be a relationship between the extent and location of anatomic lesions and the severity and pattern of behavioral deficits (Raz, Raz, Yeo, Bigler, & Cullum, 1987). These findings suggest that neural substrates of behavioral dimensions can be studied by examining normal variability in neuroanatomic measures and by tracing patterns of behavioral deficits to regional destruction of neural tissue.

However, before much further progress can be achieved, there is a need for better normative data on control subjects and more precise quantitation of the CT and MRI results. The widely used technique for obtaining the neuroanatomic measures is to trace brain regions on the image and calculate the area using planimetry. A more rigorous method would be to apply segmentation algorithms to the data, and such procedures are currently being developed. Reliability and validity are yet to be established.

Neuroanatomic studies in patients with psychopathology have focused on schizophrenia. These studies reported increased ventricular size (higher ventricular/brain ratios) in schizophrenics, suggesting neuronal loss in mesial brain structures (Andreasen et al., 1986; Weinberger, Torrey, Neophytides, & Wyatt, 1979). However, most of these studies used CT, which is not optimal for assessing anatomic loss in brain regions close to the skull (because of the beam-hardening effects of bone described earlier). Another limitation of most studies is the use of planimetric measures of surfaces seen on the film scans. As noted earlier, segmentation of tissue can be done more precisely by algorithms applied to the actual pixel-by-pixel data. This was applied by Pfefferbaum et al. (1988), who used a semiautomated computerized approach for volumetric measurements. Diffuse cerebral atrophy distinguished schizophrenics from controls. Sulcal, more than ventricular enlargement, was implicated.

Neurophysiology and Behavior

The physiological neuroimaging techniques are more suitable for dynamic studies of brain–behavior relations. These techniques can be expected to give measures that are influenced not only by basal variability (among normal subjects and between patients and controls), but also by environmental stimuli, which can lead to changes in regional brain activity. This can be utilized to design experimental procedures, where the environmental factors are rigorously controlled, to test hypotheses on brain–behavior relations. Thus, subjects can be studied both at resting baseline states and while performing behavioral tasks.

When the techniques are applied to normal subjects, reliable patterns of physiological activity have been observed. For example, as seen in Figure 23.3, the Xenon-133 technique consistently shows "the hyperfrontal pattern" (Ingvar, 1979). PET shows higher glucose uptake in frontal regions, as well as in the posterior visual (calcarine) cortex (Figure 23.4). Repeated measures in the same normal individuals demonstrate excellent reliability for the Xenon-133 technique (Warach et al., 1987) and acceptable levels of reliability for PET indices (Gur et al., 1987).

Concerning the influence of stimuli, the techniques are sensitive to changes in brain activity produced by sensory stimulation and show specific effects for the visual, auditory and somatosenory modalities (see Reivich, Gur, & Alavi, 1983; Reivich, Alavi, & Gur, 1984, for review). Several studies have also examined the effects of cognitive tasks, consistently reporting increased metabolic activity during cognition compared to resting baseline (Leli et al., 1982; Risberg, & Ingvar, 1973). Regional specificity to task demands has also been documented. In our first experiment designed to test statistically the hypothesis of greater physiological activation in regions that serve specific cognitive functions, we studied a sample of normal right-handed young men. Measurements of rCBF were performed with the Xenon-133 inhalation technique at rest, and during the performance of a verbal analogies and a spatial gestalt-completion task (Ekstrom, French, Harman, & Dermen, 1976). We chose the verbal task because of evidence that performance of this task involves predominately left-hemispheric processing. By contrast, performance on the gestalt-completion task is impaired with right-hemispheric damage. Both tasks reliably increased CBF across regions. The increase was hemispherically asymmetric only for the verbal task, which produced greater increase in left hemispheric rCBF. Greater right-hemispheric increase was correlated with better performance on the spatial task, although for the entire sample the increase for this task was bilaterally symmetric (Gur & Reivich, 1980).

In a second experiment, the same verbal task and a different spatial task (line orientation) were administered to a sample of right- and left-handed males and females (Gur et al., 1982). Gender and handedness are dimensions of individual differences that have been shown to influence the direction and degree of hemispheric cognitive specialization (Levy & Reid, 1976). In this second rCBF study (Figure 23.5), females had higher rCBF and left-handers differed from right-handers in the pattern of hemispheric activation. Right-handers replicated the greater left-hemispheric increase for the verbal task and showed greater right-hemispheric increase with this spatial task. Left-handers did not show this pattern. This indicated that rCBF measured with the Xenon-133 technique is sensitive not only to the effects of cognitive effort on regional brain activity, but also to individual differences affecting the direction and degree of hemispheric specialization for cognitive function.

Subsequent studies with rCBF and PET have confirmed and extended these findings (Deutsch, Papanicolaou, Eisenberg, Loring, & Levin, 1986; Gur et al., 1983). In most cases, some of the results related to hypotheses that had been generated from earlier observations, but, as happens when new technology is applied, some of the results were unforeseen. For example, a PET study with the same verbal analogies and spatial line-orientation tasks just described, showed asymmetric effects in the expected temporoparietal

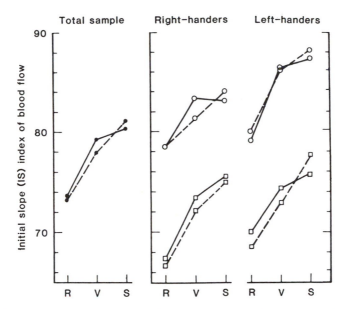

Figure 23.5 Initial slope (IS) index of blood flow to the left (———) and right (– – – –) hemispheres for the total sample and for right- and left-handed females (○) and right- and left-handed males (□) during performance of resting (R), verbal (V), and spatial tasks (S). (*Source*: Gur et al., Sex and handedness differences in cerebral blood flow during rest and cognitive activity, *Science*, 1982, *217*, 659–661. Copyright © 1982 by the AAAS. Used by permission.)

regions that had been implicated by clinical observations of brain damage. However, a less-established hypothesis proposed in 1972 by the British neuropsychologist Colwyn Trevarthen was also supported by the data. Trevarthen suggested that neural activation of lateralized cognitive processes could "spill over" to motor brain regions and produce orientation to the contralateral hemisphere (Trevarthen, 1972). Thus, when people have verbal thoughts they would orient themselves to the right, and when they have spatial cognitions they would orient themselves to the left. This hypothesis received some confirmation in studies examining the effects of verbal and spatial tasks on the direction of conjugate lateral eye movements while individuals were reflecting on the answers to test items. The PET study showed that the effects of the tasks were asymmetric not only in the temporoparietal regions, but also in regions controlling the orientation response (Gur et al., 1983).

Another behavioral dimension that received additional neurophysiological explanation from the new techniques is anxiety. A behavioral "law," traced to the results of an animal study by Yerkes and Dodson in 1908, posits a curvilinear, inverted-U relationship between anxiety and performance (Yerkes & Dodson, 1908). Performance is not very good when anxiety is extremely low, because the attentional and arousal components required for optimal performance are missing. At extremely high anxiety, however, performance also deteriorates. The brain mechanisms responsible for the operation of this law could not be studied without the availability of the neuroimaging techniques. Initial findings with PET and the Xenon-133 technique showed that this inverted-U relationship between anxiety and performance is paralleled by changes in cortical metabolism and blood flow. This suggests a neural mechanism that reduces cortical activity in high anxiety, perhaps reflecting a shift toward activation of subcortical regions which are more important in fight-or-flight situations.

Other PET and Xenon-133 studies have examined regional brain involvement in attention. These studies helped establish, for example, the existence of a network of brain regions that animal neuroanatomic studies (Mesulam, 1981) have implicated in attentional processing. We found that in humans these regions show greater right-hemispheric activation (Rosen, Gur, Reivich, Alavi, & Greenberg, 1981; Reivich et al., 1983). Further elaboration of attentional processes in relation to regional brain activity was elegantly carried out in a series of studies using PET measures of CBF (Posner, Petersen, Fox, & Raichle, 1988).

These early and preliminary studies seem to point to the possibility that neuroimaging techniques are sufficiently sensitive to detect the association between regional brain activity and behavior. They encourage more systematic work, using the neuroimaging techniques for testing hypothesis and generating new observations on brain–behavior relationships. This can be accomplished by manipulating behavioral dimensions in normal subjects and examining the effects of such manipulations on regional brain activity.

Physiological studies examining regional brain abnormalities in psychiatric populations have usually used resting conditions. Initial reports suggested diffusely reduced CBF and metabolism in schizophrenia (Mathew, Duncan, Weinman, & Barr, 1982; Mathew, Meyer, Francis, Schoolar, Weinman, & Mortel, 1981), and evidence for "hypofrontality" (Buchsbaum et al., 1982; Farkas, Wolf, Jaeger, Brodie, Christman, & Fowler, 1984).

The application of physiological neuroimaging techniques to psychiatric populations may be enhanced when combined with activation procedures. Since the behavioral abnormalities occur in response to environmental triggers, regional abnormalities in brain activity may be undetected at rest but become apparent in the pattern of changes in activity produced by task activation. This was seen even in a sample of patients with unilateral cerebral infarcts (Gur et al., 1987). Few studies have examined both resting and activated conditions in psychiatric populations. Gur et al. (1983) reported that resting rCBF was normal in medicated schizophrenics, but abnormalities were pronounced in the asymmetry of activation for the verbal analogies and the spatial line-orientation tasks. Schizophrenics failed to show the normal left-hemispheric activation for the verbal task and showed a "paradoxical" left-hemispheric activation for the spatial task. This supported Gur's (1979) hypothesis, based on behavioral studies, that schizo-phrenia is associated with left-hemispheric dysfunction and overactivation of the left, dysfunctional hemisphere. Further support for the hypothesis was obtained in a sample of unmedicated schizophrenics (Gur et al., 1985). In this group, resting CBF showed greater left-hemispheric values, and abnormal activation of the left hemisphere was associated with increased symptom severity. Weinberger, Berman, and Zec (1986), also using the Xenon-133 inhalation technique for measuring rCBF, reported that schizo-phrenics failed to show the normal increase in dorsolateral frontal activity during the performance of the Wisconsin Card Sorting Test (Heaton, 1981). Performance on this test is affected by frontal lobe damage.

There have been studies in other psychiatric disorders such as depression, obsessive-compulsive disorders, and anxiety disorders. Such studies could improve diagnostic accuracy and may help identify patterns of abnormal activation in specific forms of psychopathology. A potentially productive strategy would be to identify regional abnormalities in brain anatomy and physiological activity associated with focal and nonfocal brain disease, and relate the abnormal pattern to behavioral impairment.

A Method for Integrating Behavioral and Neuroimaging Data

A key issue for future work centers on the integration of multifaceted data on regional brain function. In addition to the anatomic information on the integrity of brain regions, such as is now available from x-ray CT and from MRI scans, the PET and Xenon-133 techniques provide information on regional brain physiological activity, and topographic EEG techniques com-plement our armamentarium of data on regional brain function with infor-mation on electric activity. Imaging technology provides means for integrat-ing such multifaceted data in a comprehensible manner, but theory is needed to guide our understanding of how the activity of the brain is related to behavior. Now the challenge is to integrate behavioral theories on regional brain function with the anatomic and physiological neuroimaging data.

Theories on brain regulation of human behavior have been tested in clinical populations by correlating behavioral deficits with clinical signs and postmortem findings. Neuropsychological testing in patients with brain disease provides measures of specific behavioral functions (e.g., mem-ory, learning, praxis). In the process of formulating theories on brain regulation of behavior, the pattern of scores is used to test hypotheses on regional brain involvement for specific behavioral dimensions. The pattern of deficits is used in clinical practice to implicate brain regions putatively affected by a disease (Gur et al., 1985; Gur, Trivedi, Saykin, & Gur, 1988; Gur, Saykin, Blonder, & Gur, 1988).

The process of testing neurobehavioral theories can be helped by quan-tification of theoretical statements concerning regional brain involvement in the regulation of behavioral dimensions. We have proposed an algorithm that applies such a quantification to standard neuropsychological test scores (Trivedi & Gur, 1987, 1989). The algorithm yields a value for each brain region, which reflects expectations that the region is neurally compromised, given the pattern of neuropsychological scores. These regional values can be examined statistically to test the behavioral hypotheses against clinical data and other neuroimaging data. They can also be presented topographically using standard procedures for translating numbers into a gray scale or a color scale. This can facilitate comprehension of the spatial distribution of implicated regions. Initial testing of the algorithm in clinical cases and populations was encouraging (Gur, Saykin, Blonder, & Gur, 1988; Gur, Trivedi, Saykin, & Gur, 1988). There was consistency between the "behavioral images" and the location of lesion in patients with unilateral cerebral infarcts (Figure 23.6).

The topographic displays showed correspondence with clinical and CT data, and were congruent with the clinical interpretation of the neuropsychological data. Note that in both cases, the behavioral image suggested that regions considerably larger than the CT lesion were behaviorally "hypofunctional," including a contralaterally homotopic region. This was not detected by the clinical evaluation, and may reflect inadequacy in the "spatial resolution" of the behavioral images. However, it could also be a true behavioral manifestation of remote physiological effects of focal lesions. As noted earlier, areas larger than the anatomic lesions have shown metabolic suppression in PET studies of local cerebral glucose metabolism using the 18-F-FDG technique (Kuhl et al., 1980; Kushner et al., 1984) and in studies of regional cerebral blood flow using the Xenon-133 clearance technique (Gur et al., 1987). But comparing behavioral images with PET and CT scans, it would be possible to evaluate which behavioral effects are accompanied by remote physiological suppressions.

Directions for Future Research and Long-Term Potential

The impact of neuroimaging technology is only beginning to surface. Some techniques are still in the process of evolving toward optimal resolution and standard scanning procedures. Normative data are lacking, and only embryonic steps have been taken to identify neural networks regulating human behavior. Although already several brain diseases can be reliably detected by neuroimaging, the diagnostic utility of the techniques for the vast majority of neurological and for all psychiatric diseases is still unestablished. It is clear that no one technique will likely provide all the answers. Each has its relative advantages and limitations and yields data that are complementary but overlapping in part. Several immediate steps are necessary to harness this technology in the service of improved theoretical insight on normal brain–behavior functions and in the diagnosis and treatment of psychopathology.

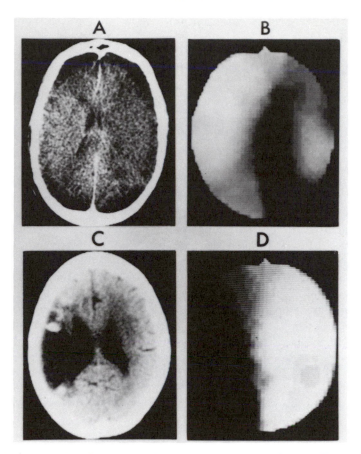

Figure 23.6 CT scans and corresponding behavioral images of two patients with cerebral infarcts, one in the right hemisphere (2A and B for CT and behavioral image, respectively) and one in the left (2C and D). The CT scans were reversed so that the left hemisphere is to the viewer's left. (*Source*: Gur et al., "Behavioral imaging"—A procedure for analysis and display of neuropsychological test scores: I. Construction of algorithm and initial clinical evaluation, *Neuropsychiatry, Neuropsychology, and Behavioral Neurology*, 1988, *1*, 53–60. Used by permission.)

Applying these techniques to patients with destructive brain lesions—such as are produced by stroke, head trauma, or brain tumors—could help in determining the extent and topographical distribution of anatomic brain damage and abnormal (suppressed or abnormally increased) metabolic activation capacity. It has already been mentioned that metabolism is suppressed even in regions that are anatomically intact when evaluated by techniques for assessing regional brain anatomy (CT and MRI). This in itself could explain why so frequently patients show behavioral deficits that are more pervasive than what can be explained by regions of anatomic involvement. However, in addition to suppression of baseline physiological activity, behavioral deficits can stem from failure of brain regions to become activated in response to task demands. This appears to be the case in psychiatric populations, where resting baseline CBF and metabolism are normal, but disturbances emerge in the pattern of regional activation for cognitive tasks. However, this may also occur in populations with focal brain lesions. For example, when we applied the Xenon-133 technique to patients with unilateral stroke, we found that the extent of asymmetric physiological suppression was largest when right-hemispheric stroke patients were performing a spatial task and, conversely, when left-hemispheric stroke patients were engaged in a verbal task. Combining standard behavioral activation procedures with neuroimaging studies of regional brain physiological activity will, we believe, become a useful tool for hypothesis testing as well as for the diagnosis and evaluation of treatment effects.

The ability of PET to image the distribution of neurotransmitters can be combined with its measurements of topographical distribution of metabolism to understand how neurotransmitter systems regulate behavior. It would be revealing, for instance, to compare Parkinson's disease with Alzheimer's disease. Dopamine deficiency seems responsible for the symptoms of the former, and some evidence suggests involvement of cholinergic pathways in the latter. Neuroimaging of neurotransmitter density and affinity as well as the distribution of physiological activity in these populations can be compared to schizophrenia, where there is evidence for increased dopamine (Wong, Wagner, & Tune, 1986). Such studies can progress in a complementary fashion, using brain disease with specific pathology in comparison to disorders of behavior where the brain pathology is unclear, and the metabolic studies could help identify regional brain involvement. For example, studies of reading disability and other learning disabilities could help test hypotheses on regional brain dysfunction that underlie these disorders.

But looking at pathology and deficit is only one side of the coin. Normal individuals vary considerably in behavior and abilities, and studying the relation between this variability and regional brain physiology can shed light on the neural underpinnngs of this variability. For example, exceptionally talented individuals in the sciences, art, and the humanities could be evaluated. Again, this should be done at rest and during behavioral "challenge" procedures. How does regional brain activity of a mathematician differ from that of an architect? Which brain regions does a mathematician activate when attempting to develop a calculus for a particular problem? Which brain regions does a lawyer activate when preparing or presenting a case? Could we train individuals to activate appropriate brain regions and thereby enhance their abilities? The technology permits us to begin answering such questions.

Thus far, the cognitive dimension has received the greatest attention, particularly in the examination of verbal and spatial tasks in relation to hemispheric activation. But emotion and conative or motivational factors could also be studied with the techniques. The studies on anxiety and attention and affect described earlier can be extended and refined to other dimensions within those examined (such as additional emotions), and new dimensions can be explored (such as pleasure/pain). Extension to populations representing a range of normal and abnormal behavior could be informative in understanding regional brain involvement in the regulation of behavior and could be of considerable relevance to the diagnosis and treatment of dysfunction. The key, in our opinion, is to integrate data from both anatomic and physiological methods. Such integration is difficult, and will constrain progress in theory and application.

References

Andreasen, N. C., Nasrallah, H. A., Dunn, V., Olson, S. C., Grove, W. M., Ehrhard, J. C., Coffman, J. A., & Crossett, J. H. (1986). Structural abnormalities in the frontal system in schizophrenia: A magnetic resonance imaging study. *Archives of General Psychiatry, 43*, 136–144.

Buschbaum, M. S., Ingvar, D. H., Kessler, R., Waters, R. N., Cappelletti, J., Van Kammen, D. P., King, C., Johnson, J. L., Manning, R. G., Flynn, R. W., Mann, L. S., Bunney, W. E., & Sokoloff, L. (1982). Cerebral glucography with positron tomography. *Archives of General Psychiatry, 39*, 251–259.

Deutsch, G., Papanicolaou, A. C., Eisenberg, H. M., Loring, D. W., & Levin, H. S. (1986). CBF gradient changes elicited by visual stimulation and visual memory tasks. *Neuropsychologia, 24*, 283–287.

Dodrill, C. B., & Wilkus, R. J. (1976). Relationships between intelligence and electroencephalographic epileptiform activity in adult epileptics. *Neurology, 26*, 525–531.

Ekstrom, R. B., French, J. W., Harman, H. H., & Dermen D. (1976). *Manual for kit of factor-referenced cognitive tests.* Princeton, NJ: Educational Testing Service.

Farde, L., Hall, H., Ehrin, E., & Sedvall, G. (1986). Quantitative analysis of D2 dopamine receptor binding in the living human brain by PET. *Science, 231*, 258–261.

Farkas, T., Wolf, A. P., Jaeger, J., Brodie, J. D., Christman, D. R., & Fowler, J. S. (1984). Regional brain glucose metabolism in chronic schizophrenia. *Archives of General Psychiatry, 41*, 293–300.

Gastaut H. (1970). Clinical and electroencephalographic classification of epileptic seizure. *Epilepsia, 11*, 102–113.

Gur, R. C., Gur, R. E., Rosen, A. D., Warach, S., Alavi, A., Greenberg, J., Reivich, M. (1983). A cognitive-motor network demonstrated by positron emission tomography. *Neuropsychologia, 21*, 601–606.

Gur, R. C., Gur, R. E., Silver, F. L., Obrist, W. D., Skolnick, B. E., Kushner, M., Hurtig, H. I., & Reivich, M. (1987). Regional cerebral blood flow in stroke: Hemispheric effects of cognitive activity. *Stroke, 18*, 776–780.

Gur, R. C., Hungerbuhler, J. P., Younkin, D., Rose, A. D., Skolnick, B. E., & Reivich, M. (1982). Sex and handedness differences in cerebral blood flow during rest and cognitive activity. *Science, 217*, 659–661.

Gur, R. C., & Reivich, M. (1980). Cognitive task effects on hemispheric blood flow in humans: Evidence for individual differences in hemispheric activation. *Brain and Language, 9*, 78–92.

Gur, R. C., Saykin, A. J., Blonder, L. X., & Gur, R. E. (1988). Behavioral imaging": II Application of the quantitative algorithm to hypothesis testing in a population of hemiparkinsonian patients. *Neuropsychiatry, Neuropsychology, and Behavioral Neurology, 1,* 87–96.

Gur, R. C., Trivedi, S. S., Saykin, A. J., & Gur, R. E. (1988). "Behavioral imaging"—A procedure for analysis and display of neuropsychological test scores: I. Construction of algorithm and initial clinical evaluation. *Neuropsychiatry, Neuropsychology, and Behavioral Neurology, 1,* 53–60.

Gur, R. C., Trivedi, S. S., Saykin, A. J., Resnick, S. M., Malamut, B. L., & Gur, R. E. (1985). Behavioral imaging. *Journal of Clinical and Experimental Neuropsychology, 7,* 633.

Gur, R. E. (1979). Cognitive concomitants of hemispheric dysfunction in schizophrenia. *Archives of General Psychiatry, 36,* 269–274.

Gur, R. E., Gur, R. C., Skolnick, B. E., Caroff, S., Obrist, W. D., Resnick, S., & Reivich, M. (1985). Brain function in psychiatric disorders: III. Regional cerebral blood flow in unmedicated schizophrenics. *Archives of General Psychiatry, 42,* 329–334.

Gur, R. E., Resnick, S. M., Gur, R. C., Alavi, A., Caroff, S., Kushner, M., & Reivich, M. (1987). Regional brain function in schizophrenia. I. Repeated evaluation with positron emission tomography. *Archives of General Psychiatry, 44,* 126–129.

Gur, R. E., Skolnick, B. E., Gur, R. C., Caroff, S., Rieger, W., Obrist, W. D., Younkin, D., & Reivich, M. (1983). Brain function in psychiatric disorders: I. Regional cerebral blood flow in medicated schizophrenics. *Archives of General Psychiatry, 40,* 1250–1254.

Heaton, R. K. (1981). *Wisconsin Card Sorting Test manual.* Odessa, FL: Psychological Assessment Resources.

Ingvar, D. H. (1979). "Hyperfrontal" distribution of the cerebral grey matter flow in resting wakefulness: On the functional anatomy of the conscious state. *Acta Neurologica Scandinavica, 60,* 12–25.

Kety, S. S., & Schmidt, C. F. (1948). The nitrous oxide method for the quantitative determination of cerebral blood flow in man: Theory, procedure and normal values. *Journal of Clinical Investigation, 27,* 476–483.

Kuhl, D. E., Phelps, M. E., Kowell, A. P., Metter, E. J., Selin, C., & Winter, J. (1980). Effects of stroke on local cerebral metabolism and perfusion: Mapping by emission computed tomography of 18 FDG and 13 NH3. *Annals of Neurology, 8,* 47–60.

Kung, H. F., Pan, S., Kung, M. P., Billings, J., Kasliwal, R., Reilley, J., & Alavi, A. (1989). In vitro and in vivo evaluation of [I-123] IBZM: A potential CNS D-2 dopamine receptor imaging agent. *Journal of Nuclear Medicine, 30,* 88–92.

Kushner, M., Alavi, A., Reivich, M., Dann, R., Burke, A., & Robinson, G. (1984). Contralateral cerebellar hypometabolism following cerebral insult: A positron emission tomographic study. *Annals of Neurology, 15,* 425–434.

Lehmann, D. (1971). Multichannel topography of human alpha EEG fields. *Electroencephalography and Clinical Neurophysiology, 31,* 439–449.

Leli, D. A., Hannay, H. G., Falgout, J. C., Wilson, E. M., Wills, E. L., Kathol, C. R., & Halsey, J. H. (1982). Focal changes in cerebral blood flow produced by a test of right-left discrimination. *Brain Cognition, 1,* 206–223.

Levy, J., & Reid, M. (1976). Variations in writing posture and cerebral organization. *Science, 194,* 337–339.

Mathew, R. J., Duncan, G. C., Weinman, M. L., & Barr, D. L. (1982). Regional cerebral blood flow in schizophrenia. *Archives of General Psychiatry, 39,* 1121–1124.

Mathew, R. J., Meyer, J. S., Francis, D. J., Schoolar, J. C., Weinman, M., Mortel, K. F. (1981). Regional cerebral blood flow in schizophrenia: A preliminary report. *American Journal of Psychiatry, 138,* 112–113.

Mesulam, M. M. (1981). A cortical network for directed attention and unilateral neglect. *Annals of Neurology, 10,* 209–325.

Naeser, M. (1985). Quantitative approaches to computerized tomography in behavioral neurology. In M. M. Mesulam (Ed.), *Principles of behavioral neurology* (pp. 363–383). Philadelphia: F. A. Davis.

Obrist, W. D., Thompson, H. K., Wang, H. S. & Wilkinson, W. E. (1975). Regional cerebral blood flow estimated by 133-Xenon inhalation. *Stroke, 6*, 245–256.

Olesen, J., Paulson, O. B., Lassen, N. A. (1971). Regional cerebral blood flow in man determined by the initial slope of the clearance of intra-arterially injected 133Xe. *Stroke, 2*, 519–540.

Pfefferbaum, A., Zipursky, R. B., Lim, K. O., Zatz, L. M., Stahl, S. M., & Jernigan, T. L. (1988). Computed tomographic evidence for generalized sulcal and ventricular enlargement in schizophrenia. *Archives of General Psychiatry, 45*, 633–640.

Posner, M. I., Petersen, S. E., Fox, P. T., & Raichle, M. E. (1988). Localization of cognitive operations in the human brain. *Science, 240*, 1627–1631.

Raz, N., Raz, S., Yeo, R. A., Bigler, E. D., & Cullum, C. M. (1987). Relationship between cognitive and morphological asymmetries in Alzheimer's dementia: A CT study. *International Journal of Neuroscience, 35*, 235–243.

Reivich, M., Alavi, A., & Gur, R. C. (1984). Positron emission tomographic studies of perceptual tasks. *Annals of Neurology, 15*, 61–65.

Reivich, M., Gur, R. C., & Alavi, A. (1983). Positron emission tomographic studies of sensory stimuli, cognitive processes, and anxiety. *Human Neurobiology, 2*, 25–33.

Reivich, M., Kuhl, D., Wolf, A. P., Greenberg, J., Phelps, M., Ido, T., Casella, V., Fowler, J., Hoffman, E., Alavi, A., Som, P., & Sokoloff, L. (1979). The 18-F-fluorodeoxyglucose method for the measurement of local cerebral glucose utilization in man. *Circulation Research, 44*, 127–137.

Risberg, J., & Ingvar, D. H. (1973). Patterns of activation in the grey matter of the dominant hemisphere during memorizing and reasoning. *Brain, 96*, 737–756.

Rosen, A., Gur, R. C., Reivich, M., Alavi, A., & Greenberg, J. (1981). Task-related arousal and local glucose metabolism in the human brain. *International Neuropsychological Society Bulletin*.

Trevarthen, C. (1972). In J. Cernacvek & F. Podivinsky (Eds.), *Cerebral interhemispheric relations*. Bratislava: Publishing House Slovak Academy Sciences.

Trivedi, S. S., & Gur, R. C. (1987). Computer graphics for neuropsychological data. *Proceedings of the National Computer Graphics Association, 3*, 22–32.

Trivedi, S. S., & Gur, R. C. (1989). Topographic mapping of cerebral blood flow and behavior. *Computers in Biology and Medicine, 19*, 219–229.

Warach, A., Gur, R. C., Gur, R. E., Skolnick, B. E., Obrist, W. D., & Reivich, M. (1987). The reproducibility of the Xe-133 inhalation technique in resting studies: Task order and sex related effects in healthy young adults. *Journal of Cerebral Blood Flow and Metabolism, 7*, 702–708.

Weinberger, D. R., Berman, K. F., & Zec, R. F. (1986). Physiological dysfunction of dorsolateral prefrontal cortex in schizophrenia: I. Regional cerebral blood flow evidence. *Archives of General Psychiatry, 43*, 114–124.

Weinberger, D. R., & Torrey, E. F., Neophytides, A. N., & Wyatt, R. J. (1979). Structural abnormalities in the cerebral cortex of chronic schizophrenic patients. *Archives of General Psychiatry, 36*, 935–939.

Wong, D. F., Gjedde, A., Wagner, H. N., Jr., Dannals, R. F., Douglass, K. H., Links, J. M., & Kuhar, M. J. (1986). Quantification of neuroreceptors in the living brain. II. Inhibition studies of receptor density and affinity. *Journal of Cerebral Blood Flow and Metabolism, 6*, 147–153.

Wong, D. F., Wagner, H. N., Jr., & Tune, L. E. (1986). Positron emission tomography reveals elevated D2 dopamine receptors in drug-naive schizophrenics. *Science, 234*, 1558–1563.

Yerkes, R. M., & Dodson, J. D. (1908). The relation of strength of stimulus to rapidity of habit-formation. *Journal of Comparative Neurology and Psychology, 18*, 458–482.

24

Neurochemical Approaches to Cognitive Disorders

ALAN M. PALMER / DAVID M. BOWEN

Evidence is presented here to indicate that higher mental functions, in particular memory, may be studied in terms of the neurochemistry of patients with dementia. It is well known that progressive dementia represents a common cause of impaired mental function, particularly in the elderly.There are a variety of causes, but it is most commonly associated with Alzheimer's disease (AD). By itself, AD accounts for approximately half of all cases of dementia irrespective of age of onset (Bowen, 1981). The disease is characterized at the microscopic level by the presence of large numbers of neurofibrillary tangles (abnormal intraneuronal inclusions) and senile plaques (clusters of dystrophic neurites and glial processes). These changes are particularly prominent in the neocortex and hippocampus, and the abnormal proteins associated with these structures are the subject of detailed ongoing study (Bowen & Procter, 1991). Genetic and molecular biological approaches to AD have focused on the relationship between chromosome 21 and the protein constituents of the senile plaque, but the relative importance of this is unclear in the pathogenesis of "sporadic" AD, the common form of the disease (Bowen, Beyreuther et al., 1988; Finch & Davies, 1988; Muller-Hill & Beyreuther, 1989; Pericak-Vance et al., 1988; Selkoe, 1989).

AD and its associated neuropathology was first described at the turn of the century (see Bowen, 1981), but it has been the more recent development of neurochemistry that has permitted the identification of selectively affected neurons. The AD brain tissue usually studied has evidence of no more than slight cerebrovascular disease. Brains from demented patients showing both cerebrovascular disease and quite intense plaque and tangle formation (Diagnositc Division 0:3 of Corsellis, 1962) account for approximately one fifth of all examples of dementia of late onset (Bowen, 1981). These have increased activity of a hypertensive agent-related peptide-bond hydrolyzing enzyme. Apart from this, "mixed" and AD subjects show similar neurochemical changes (Bowen & Davison, 1986), which suggests that the pathological process of AD is the major cause of dementia. Most neurochemical studies have utilized tissue obtained post-mortem, though a small number of studies (mainly from this laboratory) have also examined

436

antemortem tissue. Antemortem data has also been obtained by in vivo imaging techniques and ex vivo examinations of cerebrospinal fluid. It is generally accepted that abnormal proteins (e.g., beta-amyloid and ubiquitinated products) have a role in the pathogenesis of AD, but it is likely that abnormal neurotransmitter function is more amenable to therapeutic intervention. The present review therefore considers transmitters in detail and seeks to link these with proposed pathogenic mechanisms, which themselves may be "triggered" by an environmental factor (such as a toxin or infectious agent; e.g., Bowen, 1981; Pearson & Powell, 1989). Emphasis is laid on the neocortex because of the selective (evolutionary) increase in the size of this region in Man (Bowen & Davison, 1978). Moreover, limbic structures are difficult to study due to their small size (but see Cotman, Geddes, Monahan, & Anderson, 1987; Greenamyre, Penney, D'Amato, & Young, 1987; Hyman, Van Hoesen, & Damasio, 1987).

Postmortem Studies of AD

Problems Associated with Postmortem Neurochemical Studies

The prospect of obtainng meaningful data from diseased postmortem brain may be questioned as parameters under scrutiny may be influenced by factors such as patient age, sex, drug history, immediate premortem status (sudden death or prolonged coma) and postmortem delay. Interpretation of postmortem data therefore requires fastidious consideration of these factors in order to separate changes that are due to the brain disease from those that occur as a result of epiphenomena (Anon, 1977; Cheetham, Crompton, Katona, Parker, & Horton, 1988; Palmer, Lowe, Francis, & Bowen, 1988; Procter, Palmer, et al., 1988; Rossor et al., 1982; Spokes, 1979).

Tissue atrophy is another factor that may confound interpretation of data. This is because the practice of expressing results relative to unit mass (e.g., per mg of total protein) does not make allowance for any reduction in volume of brain structure. Shrinkage of some, but not other, components may lead to an apparent increase in unaffected components. Perhaps more important, should losses occur either equally in all cellular components or in a major component such as pyramidal neurons, the apparent loss in neurochemical measures may be an underestimate (possibly not even detected), even in severely atrophied tissue. Only a few neurochemical studies have attempted to make allowance for this factor (reviewed by Bowen & Davison, 1986; see also Najlerahim & Bowen, 1988; Palmer & Bowen, 1985; Palmer, Stratmann, Procter, & Bowen, 1988; Procter, Lowe et al., 1988). Thus the hyptertensive-agent-related enzyme (angiotensin-converting enzyme) may be altered in AD (Bowen, 1981; Bowen, White et al., 1979; see also Arregui, Perry, Rossor, & Tomlinson, 1982; Bowen, Bayreuther et al., 1988).

It is not possible to distinguish reliably whether a reduced concentration of a marker is due to loss of the neurons that normally possess it, or to an

alteration in the rate of its turnover, which may be a secondary phenomenon. The sodium-dependent uptake and calcium-dependent release of
transmitter (depolarization evoked) provide important markers for the
study of brains from experimental animals. However, because both of these
processes require the maintenance of membrane potentials, they cannot be
examined post-mortem unless tissue is obtained shortly after death and
assayed immediately or carefully frozen in isotonic media (Bowen, Sims,
Lee, & Marek, 1982; Dodd, Hambley, & Hardy, 1988; but see Procter, Palmer,
et al., 1988). Postmortem changes also influence the activity of specific
marker enzymes such as dopamine-beta-hydroxylase and tryptophan hydroxylase. Finally, it is difficult to identify changes occurring early in AD by
examination of tissue obtained post-mortem, where the disease has usually
run its full course (see Palmer, Francis, Bowen et al., 1987).

Cortical Neurons

The cortex may contain as many as 90% of the brain's neurons (Jones &
Hendry, 1988), subdivided into two types: pyramidal and nonpyramidal.

EXCITATORY AMINO ACIDS (PYRAMIDAL NEURONS). Pyramidal cells are the largest
and most abundant neuron type in the cerebral cortex (Winfield, Gatter, &
Powell, 1980). They send axons to other cortical areas (corticocortical and
commissural projections) as well as to a variety of subcortical structures
(corticofugal projections). The neurotransmitter(s) associated with these
neurons is considered to be an excitatory amino acid (EAA), principally
aspartate and glutamate (see Palmer, Hutson, Lowe, & Bowen, 1989). A loss
of EAAergic pyramidal neurons from the hippocampus in AD is quite well
established (Hyman et al., 1987) and related neurons in the amygdala are
probably also affected (Francis, Pearson et al., 1987).

Loss of nerve cells throughout all areas of cortex has long been the
impression of neuropathological examination. However, the quantitative
measurement of cell numbers in the cerebral cortex is complicated by the
13 to 18% reduction in brain volume (Hubbard & Anderson, 1981; Miller,
Alston, & Corsellis, 1981). Earlier studies suggested that nerve cell numbers
were unaltered (but see Colon, 1973). More recent observations, using
image analyzers that make allowance for atrophy by counting cells in
columns of cerebral cortex rather than in individual fields, indicate that
large (presumably pyramidal) neurons are depleted by 23 to 46% (Hubbard
& Anderson, 1985; Mountjoy, Roth, Evans, & Evans, 1983; Terry, Peck, De
Teresa, 1981). This work has been confirmed and extended by counts of the
number of pyramidal neurons in cortical layers III and V (Mann, Yates, &
Marcyniuk, 1985).

Pyramidal cells are probably intimately involved in the formation of
neuritic plaques and neurofibrillary tangles. The abundance of pyramidal

cells makes it likely that they make a substantial contribution to the degenerating neurites that partially constitute senile plaques. Moreover, the neurofibrillary tangles accumulate in the cytoplasm of neurons, which generally resemble pyramidal cells in size, shape, and dendritic morphology. Although the presence of plaques and tangles in AD cortex has been known since Alzheimer's original report, the precise elements of cortical circuitry compromised by these pathological changes have remained obscure. Recent quantitative data on the distribution of plaques and tangles (Lewis, Campbell, Terry, & Morrison, 1987; Pearson, Esiri, Hiorns, Wilcock & Powell, 1985; Rogers & Morrison, 1985) suggest that AD may result, at least in part, from a loss of structural and functional integrity of corticocortical association projection fibers.

The integrity of pyramidal neurons has been difficult to assess biochemically because of the ubiquitous distribution and many functions of EAA's. However, postmortem studies of both the uptake and binding of radiolabeled D-aspartic acid suggests that loss of some terminals of EAAergic neurons occurs, though the magnitude and distribution of change, the extent of influence of artifact and epiphenomena and the location of the cell bodies is not yet clear (Cowburn, Hardy, Roberts, & Briggs, 1988; Cross, Slater, Candy, Perry, & Perry, 1987; Hardy et al., 1987; McGeer, Singh, & McGeer, 1987; Palmer, Procter, Stratman, & Bowen, 1986; Procter, Lowe et al., 1988; Procter, Palmer et al., 1988; Simpson, Royston et al., 1988).

GAMMA AMINOBUTYRIC ACID (GABA). The nonpyramidal neurons of the cortex are, broadly speaking the intrinsic local circuit neurons. Various subclasses of nonpyramidal cells have been described on the basis of the appearance of the cell body and dendrites and include spiny and nonspiny and bipolar and multipolar. Immunohistochemical studies have indicated that the majority of nonpyramidal interneurons stain with antisera against GABA or its biosynthetic enzyme glutamic acid decarboxylase (GAD). Thus, GABA constitutes a major inhibitory transmitter system in the cortex, accounting for as many as 30% of all cortical neurons (Jones, 1986, but see Fisher, Buchwald, Hull, & Levine, 1988).

Large reductions in cortical GAD activity in AD were originally published alongside some of the first reports of diminished choline acetyltransferase (ChAT) activity (Bowen et al., 1976a; Davies & Maloney, 1976; Perry et al., 1977). However, detailed studies of the brains of humans and experimental animals suggested that loss of GAD activity was attributable to the terminal hypoxia associated with protracted death (Bowen, Goodhardt et al., 1976; Bowen, Smith, White, & Davison 1976a, 1976b; Bowen et al., 1977; Spillane et al., 1977). No change in GAD activity was found in a recent study were AD and control subjects were carefully matched for agonal state (Reinikinen et al., 1988). Surprisingly, concentrations of GABA in subcortical structures have not been shown to be similarly affected (Spokes, 1979), although GABA concentrations in the parietal cortex have recently been shown to be affected by agonal state (Lowe et al., 1988). There is evidence of reduced GABA concentration in the temporal, frontal, parietal and occipital lobes (reviewed by Lowe et al., 1988). The reductions found in most studies are less substantial and widespread than those reported by Ellison, Beal, Mazurek, Bird, & Martin (1986) who assayed subjects of similar age but included only pathologically severe examples of the disease. Large and widespread reductions in uptake sites of GABAergic neurons have also been reported (Hardy, Cowburn, Barton, Reynolds, Dodd, et al., 1987). However, because this was based upon active uptake determinations in tissue that had been frozen, thawed and subfractionated, it is difficult to exclude the possibility that inappropriate preparations were produced in disease-affected tissue. Indeed, Simpson, Cross, Slater, & Deakin (1988) find preservation of this uptake site (assayed using a ligand binding technique) in all regions examined, except the temporal cortex.

NEUROPEPTIDES. The cortical neuropeptides are released upon depolarization and seem to play a role in slow chemical signaling. They appear to be localized in one broad class of interneuron, forming approximately 5 to 10% of the total cortical cell population. Current information indicates that virtually all of these interneurons are GABAergic (see Jones, 1986; Jones & Hendry, 1988). Neuropeptides have been demonstrated to be stable in postmortem tissue, and so they have been extensively studied in AD (for reviews see Allen et al., 1984; Beal & Martin, 1986; Beal et al., 1985; Davies et al., 1980; Perry et al., 1981b; Rossor & Iversen, 1986; Rossor et al., 1980).

Despite relatively high concentrations in cerebral cortex, cholecystokinin, vasoactive intestinal polypeptide, and neuropeptide Y are unaffected. This may be adduced as evidence of relative sparing of the neurons that contain these peptides. However, such an interpretation needs to be made with some caution, because tissue shrinkage consequent upon cell loss may mask reductions in a chemical marker when expressed relative to tissue protein. The normal concentrations of vasoactive intestinal peptide are of particular interest in view of the reported coexistence between this peptide and ChAT activity within a population of nonpyramidal bipolar neurons (Eckenstein & Baughman, 1984). Such coexistence is not found in ascending cholinergic projections, so the sparing of the peptide in AD may also indicate a sparing of intrinsic cholinergic neurons. Galanin exists in human basal forebrain cholinergic neurons, but its "overactivity" may accentuate cholinergic dysfunction (see Chan-Palay, 1988). The localization of corticotropin-releasing factor in the cerebral cortex is not clear but it has been reported reduced (Bissettee, Reynolds, Kilts, Widerlov, & Nemeroff, 1985; De Souza, Whitehouse, Kuhor, Price, & Vale, 1986), whereas galanin appears unaltered (Beal et al., 1988).

Various molecular forms of somatostatin have been detected in rat and human brain. Many studies have demonstrated reduced somatostatinlike immunoreactivity (SLIR), which appears to be attributable to loss of a variety of the forms. Changes have been found to be greater in studies where only subjects displaying severe histopathology were examined (Beal et al., 1985; Davies et al., 1980; Davies et al., 1982) than in those where no such selection criteria was employed (Ferrier et al., 1983; Francis, Bowen et al., 1987; Lowe et al., 1988; Reinikainen, Reikkinen, Jolkkonen, Kosma, & Suininer, 1987; Rossor et al., 1980; Tamminga, Foster, Fedio, Bird, & Chase, 1987). A difference in selection criteria may also explain why Beal et al. (1986) found reduced concentrations of neuropeptide Y whereas other groups have reported no change (see Allen et al., 1984; Dawbarn, Rosser, Mountjoy, Roth, & Emson, 1986). A proportion of SLIR neurons in human cerebral cortex also stain for neuropeptide Y, so it is difficult to reconcile loss of SLIR with the sparing of neuropeptide Y.

Corticopetal Neurons

ACETYLCHOLINE. DeKosky and Bass (1985), in their seminal review, pointed out that Alfred Pope's 1964 finding of a lower acetylcholinesterase activity in AD brains provided an early indication that the cholinergic system may be affected in this condition. However, it was not until over a decade later, when changes in ChAT activity were demonstrated in the neocortex, that an appreciation of the potential importance of cholinergic dysfunction began to develop. Thus, on the basis of the suggestion that cholinergic dysfunction is the major contributing cause of the dementia of AD, the hope emerged that, just as for Parkinson's disease, palliative replacement therapy might prove effective. However, in spite of over 80 published attempts to enhance brain cholinergic function, such treatment has met with only modest clinical results at best (see Collerton, 1986, for review).

Evidence for presynaptic cholinergic dysfunction in the cerebral cortex in AD has now been obtained in all postmortem studies that have measured ChAT activity together with studies of radiolabeled choline uptake (Rylett, Ball, & Colhoun, 1983; see also Bowen, Smith, White, Flack et al., 1987; Bowen, Smith, White, Goodhardt et al., 1977) and acetylcholine content and release (Nordberg, Adem, Nilsson, & Winblad, 1987; Richter, Perry, & Tomlinson, 1980). These neurochemical data are supported by a myriad of histological studies, which have reported loss of cholinergic perikarya from the medial forebrain nucleus basalis of Meynert (for reviews see Arendt & Bigl, 1986; Perry 1986; Whitehouse et al., 1982; Wilcock, Esiri, Bowen, & Smith, 1983). Reduced ChAT activity is also a feature of "senile dementia of Lewy body type" (E. K. Perry, personal communication; Perry, Irving, Blessed, Perry, & Fairbairn, 1989) but preliminary data suggest this is not a feature of dementia of frontal lobe type (see Gustafson, 1987; Neary, Snowden, Northen, & Goulding, 1988). Frontal cortex (obtained at diagnostic craniotomy) from three examples (similar to patients of Neary et al., 1988) have been assayed for ChAT (mean $\pm$ SD of 10.2 ± 3.2 nmol of acetylcholine formed/min/100 mg protein, Bowen and Neary unpublished data; control value is $8.3 + 2.4$ nmol/min/100 mg, Bowen, Benton, Spillane, Smith, & Allen, 1982).

SEROTONIN. Serotonergic neurons, like cholinergic cells, innervate large areas of cerebral cortex from discrete extracortical (raphe) nuclei. However, serotonergic activity has been less thoroughly investigated in AD, largely because of the difficulty of measuring the activity of tryptophan hydroxylase in postmortem brain. With the exception of studies of the serotonin (5-HT) carrier (Bowen et al., 1983; D'Amato et al., 1987; Perry, Marshall, & Blessed, 1983) all estimates of serotonergic neurons in postmortem samples have relied upon determination of the concentrations of 5-HT and its major metabolite, 5-hydroxyindoleacetic acid (5-HIAA; Arai, Kusarka, & Iizuka, 1984; Bowen et al., 1979; 1983; Cross et al., 1983; D'Amato et al., 1987; Francis et al., 1985; Gottfries et al., 1983; Palmer, Wilcock, Esiri, Francis, & Bowen, 1987, Reinikainen et al., 1988), and these confirm earlier Swedish studies (see Bowen, 1981). The 5-HT content in the neocortex from AD subjects has in general been found to be reduced, whereas 5-HIAA was unaltered except for two reports of a reduced content of this metabolite (Cross et al., 1983; Reinikainen et al., 1988). This discrepancy may be related to postmortem delays, which were shorter in the latter studies than in studies of large groups of samples (Gottfries et al., 1983; Palmer, Wilcock et al., 1987c; Reynolds et al., 1984). There is some evidence from the histopathological measurements to indicate that intrinsic cortical change and serotonergic denervation are related, because significant negative correlations were formed between tangle counts and 5-HIAA content in both frontal and temporal cortex of AD subjects (Palmer, Wilcock et al., 1987).

Neurofibrillary tangles and neuronal loss occur in the dorsal raphe nucleus, but it is unclear what proportion of cells are affected and whether the affected cells relate topographically to areas of pronounced neocortical damage (Curcio & Kemper, 1984; Mann, Yates, & Marcyniuk, 1984; Yamamoto & Hirano, 1985).

CATECHOLAMINES. The study of catecholaminergic neurons has been hampered by the postmortem instability of the enzymes responsible for the synthesis of dopamine (DA) and noradrenaline (NA). Apart from two studies of dopamine-beta-hydroxylase activity post-mortem—one indicating reduced activity (Perry, Blessed et al., 1981), the other finding no change (Davies & Maloney, 1976)—biochemical studies of the cerebral cortex have focused upon the determination of the concentrations of DA and NA and their principal metabolites, homovanillic acid (HVA) and 3-methoxy-4-hydroxyphenylglycol (MHPG) respectively. The concentration of DA has consistently been found not to be reduced (Arai et al., 1984; Gottfries et al., 1983; Iversen et al., 1984; Palmer, Wilcock et al., 1987) and an unaltered concentration of the minor metabolite of DA, dihydroxyphenylacetic acid (DOPAC) was found in the single study to examine this compound (Palmer, Wilcock et al., 1987). HVA concentrations are also reported reduced in some regions (Gottfries et al., 1983; Palmer, Wilcock et al., 1987), but elevated in others (Gottfries et al., 1983; Palmer, Wilcock et al., 1987). Reduced concentration of NA have generally been reported (Arai et al., 1984; Gottfries et al., 1983; Iversen et al., 1984; Palmer, Wilcock et al., 1987; but see D'Amato et al., 1987) whereas concentrations of MHPG have been found to be reduced, unaltered, or even elevated (Arai et al., 1984; Cross et al., 1983; Gottfries et al., 1983; Palmer et al., 1987; Reinikainen et al., 1988). This may be a reflection of the postmortem accumulation of MHPG (Palmer, Lowe, Francis, & Bowen, 1988) which together with the high turnover rate and low concentrations of catecholamines in the cortex make questionable the validity of determination of tissue concentrations post-mortem. Moreover, since oxygen is a cofactor of both tyrosine hydroxylase and dopamine-beta-hydroxylase, the terminal hypoxia usually associated with AD may also be partly responsible for some of the observed changes. Noradrenergic changes in the cortex may be associated with neuronal loss from the brainstem noradrenergic nucleus, the locus ceruleus (e.g., Iversen et al., 1984; Mann et al., 1984; Marcyniuk, Mann, & Yates, 1986). However, there is not an obvious relationship between cell loss from this nucleus and the decline in dopamine-beta-hydroxylase activity in the cortex (Perry, Blessed et al., 1981). Dopamingeric cell bodies within the substantia nigra are generally preserved in both structure and number but there is evidence of reduced cell densities in the ventral tegmental area (Mann, Yates, & Marcyriuk, 1987).

Neurotransmitter Recognition Sites

Ligand binding studies in AD have indicated that many neurotransmitter recognition sites are *unaltered* in the cerebral cortex. This includes those for catecholamines (alpha$_1$, alpha$_2$, and beta$_1$ and beta$_2$ subtypes), opiates (mu and kappa-delta subtypes), cholecystokinin, adenosine, histamine, and GABA (Bowen et al., 1979, 1983; Cross, Crow, Ferrier, & Johnson, 1986; Cross, Crow, Johnson et al., 1984; D'Amato et al., 1987; Davies & Verth, 1978; Hays & Paul, 1982; Lang & Henke, 1983; Shimohama, Taniguchi, Fujinara, & Kamyama, 1987). Assay of muscarinic receptor subtypes shows the high affinity pirenzapine binding site (M$_1$ receptor), the major postsynaptic receptor, to be apparently preserved (Caulfield, Straughan, Cross, Crow, & Birdsall, 1982; Mash, Flynn, & Potter, 1985; Smith et al., 1988). This suggests that a suitable target for excess acetylcholine is available but successful therapy with a cholinomimetic would help prove that this receptor is functionally intact (see Young, Kish, & Warsh, 1988). M$_2$ muscarinic receptors and nicotinic receptor sites (both of which are present in smaller numbers and may be primarily presynaptic) are reduced in the disease (Caulfield et al., 1982; Mash et al., 1985, Flynn & Mash, 1986; Nordberg et al., 1987; Perry, Perry, et al., 1987; Smith et al., 1988; Whitehouse et al., 1986). Further studies on the location and physiological role of these receptors are necessary to assess the significance of these losses.

Although altered populations of EAA recognition sites have been reported in the neostriatum and the hippocampus (Pearce et al., 1984; Cotman et al., 1987; Greenamyre et al., 1987; Maragos, Chu, Young, D'Amato, & Penney, 1987), the neocortex has shown few alterations (Cowburn et al., 1988; Cowburn, Hardy, Briggs, & Roberts, 1989; D'Amato et al., 1987; Mouradian, Contreras, Monahan, & Chase, 1988; Procter, Sterling, Stratman, Cross, & Bowen, 1989; Procter, Wong, Stratman, Lowe, & Bowen 1989). Lower glutamate binding in cortical layers 1 and 2 and reduced glycine modulation of the N-methyl-D-aspartate (NMDA) receptor have been described (see Procter, Sterling et al., 1989; Procter, Wong et al., 1989; the reduced binding of [^{3}H]-TCP to poorly washed membranes may also reflect this latter change; see Simpson, Royston et al., 1988). Reduced binding of radiolabeled somatostatin (Beal et al., 1985) and corticotropin-releasing factor (DeSouza et al., 1986) have been reported in the cerebral cortex together with loss of DA D_2 receptors from the neostriatum, possibly reflecting loss of EAA nerve terminals from this region (Cross, Crow, Ferrier, Johnson, & Markaikis, 1984; Rinne, Sako, Paljora, Molsa, & Rinne, 1986; see Pearce et al., 1984). The most consistent abnormality is that of recognition sites for 5-HT. Binding of [^{3}H] 5-HT (to the 5-HT$_1$ site) has been shown to be significantly reduced in the areas of cortex examined (Cross, Crow, Ferrier et al., 1984, 1986; Cross, Crow, Johnson et al., 1984; Perry et al., 1984). Lower binding of [^{3}H] ketanserin (to the 5-HT$_2$ site) has also been described (Cross, Crow, Johnson et al., 1984; Perry et al., 1984; Procter, Lowe et al., 1988; Reynolds et al., 1984). The binding of [^{3}H] lysergic acid diethylamide (to both 5-HT$_1$ and 5-HT$_2$ sites) is reported significantly reduced in all studies, including the only one to make allowance for tissue atrophy by examination of the entire temporal lobe (Bowen et al., 1979, 1983). Further study of 5-HT$_{1A}$ receptors as a marker of pyramidal cells is required (Davies, Deisz, Prince, & Proutka, 1987; Middlemiss, Palmer, Edel, & Bowen, 1986; Procter, Middlemiss, & Bowen, 1988; Sprouse & Aghajanian, 1988). The reductions in binding to the 5-HT$_{1A}$ recognition site in a new series (Table 24.1) are

Table 24.1. 5-HT$_{1A}$ Recognition Site in the Neocortex

Region	Alzheimer's disease ($n = 6$)	Control ($n = 11$)
Frontal lobe		
Pars opercularis	9.3 ± 5.1	15.2 ± 7.0
Orbital gyrus	$8.9 \pm 3.1*$	18.0 ± 6.3
Temporal pole	$7.8 \pm 3.0\dagger$	24.6 ± 8.5
Parietal lobule (superior)	$3.4 \pm 2.0\dagger$	14.5 ± 6.4

Note: Fmoles/mg protein ($\pm$ SD), assayed using 1 nM of radioligand according to Procter, Middlemiss, & Bowen (1988). Controls were age-matched to AD (mean of 82, 73–89 years).

*$p < 0.02$, $\dagger p < 0.005$ (Two-tailed Mann Whitney U test). Procter, in preparation.

apparently more widespread than has been reported previously in younger subjects. However, this is based on data expressed per unit mass and brain tissue loss is greatest in the youngest AD patients. (Bowen et al., 1979; Hubbard & Anderson, 1981).

Antemortem Studies of AD

Postmortem tissue represents the end stages of AD. Thus, assessment of neurochemical pathology in living patients has an important part to play in the understanding of the pathophysiology of AD.

Cerebrospinal Fluid and Improved Diagnosis

Progressive dementia of unknown cause is most commonly diagnosed as AD, but biopsy studies of presenile patients have demonstrated a sizable minority (30 to 35%) with no specific histological abnormalities, described as having "undiagnosed dementia" (Neary, Snowden, Mann, et al., 1986; Sim, Turner, & Smith 1966). Abnormalities in cerebral transmission described earlier suggest that analysis of neurotransmitters or metabolites in spinal fluid may provide a means of both increasing the diagnostic accuracy of AD and monitoring the course of the disease (see Growdon & Corkin, 1983).There is not a good spinal fluid marker of the cholinergic lesion, largely because of the rapid enzymatic hydrolysis of acetylcholine to form choline and acetyl-coenzyme A. However, acetylcholinesterase activity has been proposed to be of some diagnostic value, particularly when expressed relative to butyrylcholinesterase activity (Appleyard et al., 1987; Huff, Maire, Growdon, Corkin, & Wurtman, 1986). In contrast to choline, metabolites of monoamine neurotransmitters are not actively taken up into parent neurites, but are instead transported from the extracellular fluid into spinal fluid. This permits indirect assessment of NA, 5-HT, and DA neurotransmitter function in the brain by analysis of the concentration of their major metabolites (MHPG, 5-HIAA, and HVA) in spinal fluid. However, the sites of origin of these metabolites in lumbar fluid is not entirely clear (Wood, 1980).

Gottfries, Gottfries, & Roos (1969) found cerebrospinal fluid concentrations of 5-HIAA and HVA to be reduced by 50% in presenile patients. A deficiency of such studies is that the diagnosis of AD has not been verified by neuropathological examination, so a few examples of Pick's disease, dementia of frontal lobe type and "dementia of the Lewy body type" (Perry et al., 1989, but see Lennox et al., 1989) have probably "contaminated" most series. This laboratory has utilized spinal fluid from patients where the clinical diagnosis of AD was neuropathologically confirmed post-mortem or antemortem. The concentration of HVA was reduced (Figure 24.1; Palmer et al., 1984) and is thought to reflect dysfunction of the EAAergic corticostriatal pathway (Pearce et al., 1984; Procter, Palmer et al., 1988). The data fail to provide a diagnostic marker, because values for AD patients do not differ from those for patients with undiagnosed dementia. This was also the finding for SLIR content (Francis et al., 1984). Now there is an intensive search for a protein-derived marker.

Imaging Techniques

The emergence of neuroimaging technology has made it possible to study brain biochemistry during life (Gur, this volume; McGeer, 1986; Maragos et al., 1987; Neary et al., 1987, Procter, Lowe et al., 1988). Some of the considerations that apply to postmortem studies are also relevant to neuroimaging studies. In particular, tissue atrophy is a complicating factor in positron

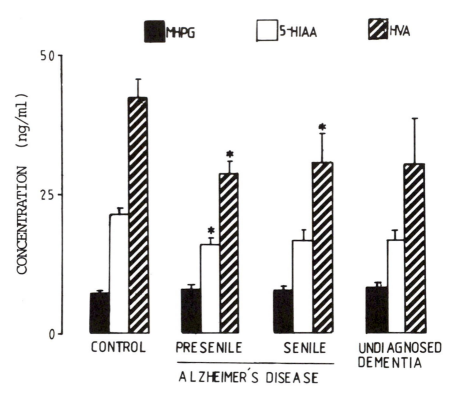

Figure 24.1 Monoamine metabolites in the CSF of patients with dementia. The number of control and presenile Alzheimer patients are 24 to 26 and 28 to 35, respectively. The senile Alzheimer group (>70 years) contains 9 or 10 patients, and the group without Alzheimer histology ("undiagnosed dementia") contains 5 to 7 patients. Bars represent SEM. *Results significantly different from control ($p < 0.05$, ANOVA-LSD; Palmer, 1987).

emission tomography (PET) studies, for which it is particularly difficult to make allowance. Shrinkage and loss of pyramidal neurons in temporal-parietal regions is likely to be the major cause of cerebral atrophy in AD (Najlerahim & Bowen, 1988). A focus of apparent glucose hypometabolism has been found consistently in the temporal and parietal lobes by PET. Measurements by this technique are sensitive to atrophy (Chawluk et al., 1987), and hypometabolism is not seen in vitro (Sims, Bowen, & Davison, 1981; Sims, Finegan, Blass, Bowen, & Neary, 1987). The scanning results may therefore actually index (at least in part) the early shrinkage (or even loss) of pyramidal cells (Procter, Palmer et al., 1988). Specific markers of pyramidal neurons that may be of use in detecting foci of degeneration have not yet been clearly identified, although quite a wide range of receptor ligands are under study (Mazière & Mazière, 1991).

Tissue Analysis

Small amounts of AD brain tissues, removed for diagnostic purposes (see Neary, Snowden, Bowen, Sims, Mann, & Benton, 1986; Neary, Snowden, Mann et al., 1986; Neary, Snowden, Sims, Mann, Yates et al., 1986), have also been used for biochemical analyses (see Bowen, Smith, White, Goodhardt et al., 1977), an approach pioneered by Korey, Scheinberg, Terry, & Stein (1961). The few samples offered to the authors (39, between 1976 and 1987) originated from seven centers. The material has, however, contributed to knowledge of the biochemistry of AD. First, it has helped to separate disease-related change from the artifact and epiphenomena normally associated with postmortem tissue. Second, it has allowed much firmer conclusions to be drawn about neuronal integrity by permitting assessment of a variety of chemical markers for a single neuronal type. Third, it has provided a clear picture of the neuronal changes that occur early in the disease, and, together with postmortem data, changes that occur at a later stage. Finally, antemortem tissue can provide an opportunity to relate neurochemical measures to specific behavioral changes, as such an inquiry is very difficult in postmortem tissue because of the preponderance of severely affected subjects and the variable and often prolonged duration between clinical assessment and death.

Unlike most other transmitters, the concentrations of aspartate and glutamate do not provide a good index of the neurotransmitter pool of these substances, so the reduced concentration of glutamate (Korey et al., 1961 and Table 24.2, but see Perry, Yong, Bergeron, Hansen, & Jones, 1987) and the elevated concentration of aspartate (Table 24.2) may reflect abnormal energy metabolism in AD. The unaltered release of EAAs is surprising in view of the substantial body of postmortem data indicating dysfunction of pyramidal neurons. However, the data is based on static incubations in media with no inhibitor of EAA uaptake, so determinations may have been complicated by reuptake, particularly in AD if energy metabolism is com-

Table 24.2. Summary of Neurotransmitter Measures Assessed Antemortem in Patients with Alzheimer's Disease

	AD as Percentage of Control	
	Temporal cortex	Frontal cortex
EAAs		
Aspartate release	NS[1]	—
Glutamate release	NS[1]	—
Aspartate concentration	124[2]	NS[2]
Glutamate concentration	86[2]	NS[2]
GABA		
Release	NS[1]	—
Concentration	NS[3]	145[3]
GAD activity	NS[4]	—
Somatostatin		
SLIR release	—	NS[5]
SLIR concentration	NS[5]	NS[5]
Acetylcholine		
Synthesis	41[6]	47[6]
ChAT activity	35[6]	36[6]
Choline uptake	57[7]	—
5-HT		
Release	—	51[6]
Concentration	31[6]	NS[6]
5-HIAA concentration	44[6]	63[6]
Uptake	39[6]	—
NA		
Release	—	NS[8]
Concentration	32[8]	—
MHPG concentration	NS[8]	—
Uptake	47[8]	—
DA		
Release	—	NS[8]
Concentration	NS[8]	NS[8]
DOPAC concentration	NS[8]	NS[8]
HVA concentration	NS[8]	NS[8]

Note: Percentages are given only for values that were significantly different from control; NS, not significant; —, not determined. Superscript numbers indicate reference [1]Smith, Bowen, Sims, Neary, & Davison, 1983; [2]Procter, Palmer et al., 1988; [3]Lowe et al., 1988; [4]Spillane et al., 1977; [5]Francis, Bowen et al., 1987; [6]Palmer, Francis, Benton et al., 1987; [7]Sims et al., 1983; [8]Palmer, Francis, Bowen et al., 1987.

promised. In contrast to the losses observerd in postmortem tissue, measures of GABAergic neurons and of somatostatin-releasing cells were unaltered. This suggests that either the changes seen post-mortem are artifactual or the measures are affected only at a relatively late stage of the disease. The latter explanation does seem more likely in view of the more pronounced reduction in the concentration of both GABA and SLIR in severely affected postmortem tissue and anatomical data indicating that GABA and somatostatin coexist in the same neuron (Jones & Hendry, 1988).

Although postmortem findings generated the cholinergic hypothesis of AD and stimulated much experimental effort across the world, the central question of whether cholinergic function is impaired in AD would have remained equivocal without access to antemortem tissue (Sims et al., 1980). This is because ChAT does not limit the rate of acetylcholine synthesis under normal circumstances. Unfractionated preparations containing synaptosomes have been used to demonstrate reduced acetylcholine synthesis and "high affinity" choline uptake in AD; the ChAT activity of cell free homogenates displayed a similar change (Table 24.2). The concordance of these findings strongly suggests loss of cholinergic varicosities from the cortex. Similarly, postmortem data indicating noradrenergic and serotonergic denervation has been corroborated by antemortem studies, where the release and uptake of NA and 5-HT have been measured in addition to their concentrations. Postmortem data indicating a significant relationship between tangle counts and the concentration of 5-HIAA and a sparing of dopaminergic varicosities in the cortex has also been confirmed in antemortem tissue (Table 24.2; Palmer, Francis, Benton et al., 1987).

Clinical Neurophysiology

The technique of sensory event–related potentials appears to be an in vivo method of assessing the integrity of pyramidal cells (see Procter & Bowen, 1988). The measurement of the late components of the visual evoked response may provide a method for assessing the effect of drugs free of the confounding effects of cerebral atrophy and the complexity of neuropsychological testing. The demonstration of significant clinical effects would of course be necessary eventually, but this approach may help establish whether a major clinical trial was warranted.

Topics of Speculation and Debate

The genetics of AD has entered a most controversial phase and the concepts of a neurofibrillary-tangle-specific epitope and a hematogenous origin of senile plaque amyloid are also under intense debate. The function of the B-amyloid precursor protein is unknown, although there are two leading possibilities (Marx, 1989). One is that it is an intercellular adhesion molecule, the other is that it is involved in the maintenance of neuronal growth or survival, a function also attributed to the NMDA receptor complex. Both may be localized to the membranes of excitatory amino-acid-releasing pyramidal cells in the cortex.

Pyramidal Neurons and Nerve Growth Factor

Specific biochemical indices of subpopulations of cortical pyramidal neurons have not been established for use in either postmortem or in vivo imaging studies. This is because little is known of the biochemical changes induced by loss of pyramidal cells giving rise to corticofugal, commissural, and corticocortical projections. Results could also be difficult to interpret due to the complexity of GABA (Cheetham et al., 1988) as well as that of NMDA receptors and the possibility that not all constituent recognition sites are equally affected.

No technique has yet been developed for selectively destroying pyramidal neurons in the cerebral cortex of laboratory animals. Hence, inferences drawn here about the possible role of pyramidal neurons in the AD are based on imperfect markers. The losses of cortical gray matter may reflect shrinkage and fallout of pyramidal neurons, as these constitute the major neuronal type, whereas reduced binding to the 5-HT$_{1A}$ site may index loss of a subpopulation of these cells (Davies et al., 1987). Nonetheless, the fact that the disease is associated with alterations in several other potential biochemical markers of cortical pyramidal neurons (Table 24.3) and that structural changes in the cerebral cortex are demonstrated by many other techniques, indicate a role for these cells in pathogenesis. The possibility of pyramidal neurons in the cortex being primarily affected, leading to retrograde degeneration of cholinergic afferents (Pearson et al., 1985) is strengthened by the finding (Ayer-LeLievre, Olson, Ebendal, Seiger, & Persson, 1988) in hippocampal pyramidal neurons of a gene coding for a

Table 24.3. Excitatory Amino-Acid Neurotransmission in the Neocortex from Alzheimer Subjects Some 3 Years (Neurosurgical Samples) and 6–10 Years (Autopsy Material) after Onset

	Early stage	Late stage
Glutamate, content		
Total	deceased	deceased*
Extracellular	unchanged	nd
Aspartate, content		
Total	increased	unchanged
Extracellular	unchanged	nd
Glutaminase activity	unchanged	decreased†
D-aspartate uptake	nd	decreased*
D-aspartate "binding"	nd	decreased†
NMDA receptor binding		
Cation channel	nd	unchanged
Glutamate site	nd	unchanged
Glycine site	decreased	decreased
"Zinc site"	nd	unchanged
Kainate receptor binding	nd	unchanged
Quisqualate receptor binding	nd	unchanged‡

Note: Decrease/increase, identifies significant changes; nd, not determined; * specimens obtained within 3 h of death; † subject to unexplained variability or artifact; ‡ hippocampus (see Cotman, Geddes, Monahan, & Anderson, 1987; Cowburn, Hardy, Briggs, & Roberts, 1989; Cowburn, Hardy, Roberts & Briggs, 1988; Procter, Lowe et al., 1988; Procter, Palmer et al., 1988; Procter, Wong et al., 1989; Procter, Sterling et al., 1989; Steele, Palmer, & Bowen, in preparation).

protein cholinergic growth factor (NGF). Discussion of NGF replacement therapy is becoming fashionable (Ad Hoc Working Group, 1989) but its clinical efficacy remains to be established, as does any effect of NGF on the viability or metabolism of cortical pyramidal neurons.

Excitatory Amino Acids

Neurochemical and histological evidence described earlier support the hypothesis that there is an abnormality of EAA neurotransmission in AD. What remains most unclear is its contribution to pathogenesis and symptomatology. Inhibition of the NMDA receptor complex appears to inhibit learning in rats (Morris, Anderson, Lynch, & Baudry, 1986), so in AD the reduced influence of glycine modulation of this complex may have an analogous effect. The excitotoxic role of EAAs is well known and has been implicated in the pathogenesis of AD (Mattson, 1988). However, evidence is emerging to suggest trophic effects at low micromolar concentrations (Balazs, Jorgensen, & Hack, 1988; Brewer & Cotman, 1988; Moran & Patel, 1989). The factors that determine the relative balance of toxic and trophic effects are not known. Similarly unknown is the relationship, if any, between the NMDA receptor complex and B-amyloid precursor protein, as well as the role of toxic and trophic factors in the production of the histological and protein abnormalities of AD. Glycine, by virtue of its modulatory effect on the NMDA-receptor complex (Kemp et al., 1988), may be one such factor. The cortical pathophysiology of the strychnine-insensitive recognition site for this amino-acid warrants further study, together with the possibility that two anatomically distinct forms of the NMDA complex exist (Monaghan, Olvermann, Nguyen, Watkins, & Cotman, 1988).

Pharmacological Treatment

Although cholinergic therapy in general has been remarkably unsuccessful, treatment of AD patients with 1,2,3,4-tetrahydro-9-aminoacridine (tacrine), has been reported to elicit striking improvement in everyday behavior as well as performance in formal tests of memory and orientation (Summers, Majouski, Marsha, Tachiki, & Kling, 1986). Moreover, systemic administration of tacrine has been found to influence brain concentrations of monoamine transmitters and metabolites and to inhibit "high affinity" uptake of choline. In vitro findings indicate that, in the low micromolar range, tacrine modulates monoaminergic transmission and inhibits acetylcholinesterase activity (see Bowen, Steel, Lowe, & Palmer, 1989; De Souza, Robinson, Cross, & Green, 1989).

Many transmitter abnormalities have been described in postmortem AD brains, but correction of cholinergic dysfunction is probably essential for the treatment of cognitive decline (Lowe et al., 1988). However, Palmer, Stratman et al. (1988) have argued that alterations in monoamines are related to changes in noncognitive behavior, which often determines the need for institutional care (see also Zweig et al., 1988). Thus, the range of tacrine's effects on transmitter systems suggests that the reported improvement it caused in everyday behavior may be partly due to a correction of monoaminergic abnormalities, whereas the cholinergic effects of the drug may account for the cognitive improvement. Tacrine also interacts with presynaptic and postsynaptic aspects of EAA transmitter functions, but the clinical significance of this is not clear (Albin, Young, & Penney, 1988; Steele, Palmer, Lowe, & Bowen, 1989).

Increased Protein Degradation?

In the early 1970s, the action of free radicals, activation of lysosomes, and accumulation of degenerative products such as lipofuscin and neurofibrillary tangles, were considered particularly relevant to brain aging (see Bowen & Davison, 1986). Recent findings with novel peptide bond-hydrolyzing enzymes, cholinesterase staining and antiubiquitin immunocytochemistry suggest that protein degradation needs to be studied in AD (see Lennox et al., 1989; Marx, 1989; Mesulam & Geula, 1988; Small & Simpson, 1988; Taylor, Bartlett, & Greenfield, 1988), for amino-acid sequences of the amyloid precursor protein as well as NGF and its "prohormone," see Muller-Hill & Beyreuther, 1988; Ullrich, Gray, Berman, & Dall, 1983). Figure 24.2 presents the hypothesis that the pathology of the disease spreads through the brain in a pattern related to increased protein degradation, occurring as a result of a toxin/infectious-agent–induced reduction in the pH value within and around an anatomically defined series of cortical and subcortical connections. Another mechanism proposed for increased protein degradation is proteolytic activity associated with cholinesterase-rich pyramidal neurones. (See Mesulam & Geula, 1988.)

Cognitive Neurochemistry

Relating brain transmitters to the behavioral changes of AD has become a major goal of cognitive neurochemistry. Postmortem studies provided an indication of vulnerable transmitter systems that have been studied further using neurosurgical samples. Neurochemical measures assessed antemortem in this laboratory have the advantage that the disease has not yet run its full course and that there is a fixed and short duration (less than 2 weeks) between neuropsychological assessment and the removal of tissue (Neary, Snowden, Bowen, Sims, Mann, Benton et al., 1986). A rating of the magnitude of dementia was obtained from patients on the basis of their performance on a number of tests that assessed the extent of the following clinical domains: memory, perceptuo-spatial abilities, and language. The rating correlated with acetylcholine synthesis but not with ChAT activity

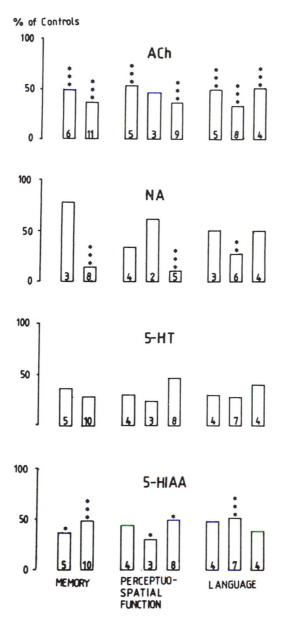

% of Controls

Figure 24.2   NA, 5-HT, and 5-HIAA concentrations and acetylcholine (ACh) synthesis in the temporal cortex of patients with Alzheimer's disease in relation to impairments of memory, perceptuospatial function and language. Histobars represent mean values of subgroups of Alzheimer patients as a percentage of control values. Memory impairment was rated as moderate (first histobar) and severe (second histobar); perceptuospatial dysfunction was absent/mild, moderate, and severe; and language impairment was absent, mild, and moderate/severe. The number of patients in each subgroup is indicated inside the histobars. Subgroups significantly different from the control are indicated by asterisks, $^{*}P < 0.05$, $^{**}p < 0.02$, $^{***}p < 0.01$ from controls (Kruskal-Wallis analysis of variance followed by the Mann-Whitney U test; Palmer, 1987).

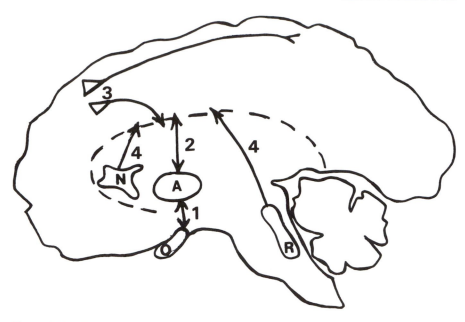

Figure 24.3   Diagrammatic representation of a possible sequence of events (1–4) in AD leading to the loss of cholinergic neurons from the nucleus basalis of Meynert (N). Retrograde degeneration of ascending fibers (4; also from the cholinergic septal-hippocampal pathway) may occur as a consequence of reduced output of neurotrophic factors by the cortex (e.g., Ayer-LeLievre et al., 1988; Dawbarn, Allen, & Semenenko, 1988) secondary to abnormal (e.g., tangle, possibly granulovacuolar degeneration/Hirano body/Lewy body–containing) or loss of corticocortical, corticofugal (3), and hippocampal pyramidal (5-HT1A-bearing excitatory amino acid-releasing) cells inferred from early (Gottfries, Gottfries, & Roos, 1969; Korey, Scheinberg, Terry, & Stein, 1961) and recent data (see text). A critical change in these cells is ascribed here to a toxin/infectious agent–(Bowen, 1981; Pearson & Powell, 1989) induced alteration in energy metabolism (Procter, Francis, et al., 1988; Saraiva et al., 1985; Sims, Bowen, & Davison, 1981; Sims, Finegan et al., 1987; Terry, 1978) leading to a reduced pH value within and around pyramidal cells and stabilization of neutral-pH labile peptide-bond hydrolases (e.g., Bowen & Davison, 1973). Owing to the protracted course of the disease, only a small portion of the brain may show these changes at any given moment (see Gottfries, Kjallquist, Ponten, Ross, & Sundberg, 1974). It has been suggested that these changes begin in olfactory areas of the (1, 2; 0, olfactory bulb; A, amygdala; Pearson & Powell, 1989). A similar mechanism may also account for the loss of noradrenergic and serotonergic neurons from brain stem nuclei (R). (After Middlemiss, Bowen, & Palmer, 1986.)

(Francis et al., 1985; Neary, Snowden, Mann et al., 1986). Both acetylcholine synthesis and NA concentration tended to be lower in patients with severe memory impairment than in those with moderate impairment (Figure 24.3). Acetylcholine synthesis also appeared to decline with increasing impairment of perceptuo-spatial function. In comparison with controls, the most marked reduction was that of NA in subgroups of patients with severe impairments of memory and of perceptospatial function. Mood disturbance (mainly anxiety) appeared to be a feature of a number of patients but did not influence any of the neurochemical parameters measured (Palmer, Francis, Benton et al., 1987).

Loss of pyramidal neurons may be a critical change in AD since atrophy-corrected counts of pyramidal cells in cortical layers III and V were also found to correlate with the magnitude of dementia (Neary, Snowden, Mann et al., 1986). The significant association between PET-determined ^{18}F-deoxyglucose uptake and behavior (Cutler et al., 1985; Miller et al., 1987) may also reflect a relationship between pyramidal cell dysfunction and the clinical symptoms of AD. This lends weight to the view that loss of functional and structural integrity of pyramidal neurons comprising corticocortical projection systems leads to a global cortical disconnection syndrome, where each cortical regions function in isolation (Morrison et al., 1986; see also Hyman et al., 1987, for discussion of the hippocampus). Greenamyre and Young (1989) conclude that: "Given our current understanding of EAA pharmacology, it is unlikely that EAA agonist or antagonist therapy will be of benefit." Modulation of the glycine site on the NMDA receptor complex may produce clinically beneficial results, possibly with minimal side effects—e.g., use of the antibiotic D-cycloserine is indicated (Bowen, 1990). The success of such an approach will presumably depend upon the extent of neuronal loss in subjects receiving such treatment.

Acknowledgments

Data on 5-HT$_{1A}$ binding and the "zinc site" are from PhD theses of Andrew Procter and Janet Steele, being prepared. The help of many colleagues in the collection and classification of specimens is gratefully acknowledged. The research was supported by the Alzheimer's Disease Society, Astra Alab, The Brain Research Trust ("Miriam Marks Department") and the Medical Research Council.

Note

Evidence linking EAA-releasing pyramidal neurones, senile plaque amyloid, and early changes in AD is emerging. β-amyloid protein (Wright et al., *Nature*, *349*, 653–654, 1991) is reported to be toxic to cultured pyramidal neurones (Yankner et al., *Proceedings of the National Academy of Sciences USA*, *87*, 9020–9023, 1990), in particular in the presence of EAAs (Koh et al., *Brain Research*, *533*, 315–320, 1990), and early loss of EAA-releasing neurones seems to be a feature of AD (Lowe et al., *Neuroscience*, *38*, 571–578, 1990; Kesslak et al., *Neurology*, *41*, 51–54, 1991). Desferrioxamine, reported to slow decline in activity of daily living, may reduce, in sequence, free iron (Crapper, McLachlan et al., *Lancet*, *337*, 1304–1308, 1991), free radical

formation, membrane damage, and abnormal proteolysis of amyloid precursor protein, thus reducing β-amyloid protein concentration. D-cycloserine by virtue of its partial agonist property (Chessell et al., *Brain Research*,
submitted) may simultaneously arrest progressive deterioration, by preventing the postulated excitotoxic damage (above and Greenamyre & Young,
1989), and improve mental performance.

References

Ad hoc working group on nerve growth factor and Alzheimer's disease (1989). Potential
    use of nerve growth factor to treat Alzheimer's disease. *Science, 243,* 11.
Albin, R. L., Young, A. B., & Penney, J. B. (1988). Tetrahydro-9-aminoacridine (THA)
    interacts with the phencyclidine (PCP) receptor site. *Neuroscience Letters, 88,* 303–
    317.
Allen, J. M., Ferrier, I. N., Roberts, G. W., Cross, A. J., Adrian, T. E., Crow, T. J., & Bloom, S.
    R. (1984). Elevation of neuropeptide Y (NPY) in substantia innominata in Al-
    zheimer type dementia. *Journal of the Neurological Sciences, 64,* 325–331.
Anonymous. (1977). Cholinergic involvement in senile dementia. *Lancet, i,* 408.
Appleyard, M. E., Smith, A. D., Berman, P., Wilcock, G. K., Esiri, M. M., Neary, D., &
    Bowen, D. M. (1987). Cholinesterase activities in cerebrospinal fluid of patients
    with senile dementia of Alzheimer's type. *Brain, 110,* 1309–1322.
Arai, H., Kusarka, K., & Iizuka, R. (1984). Changes of biogenic amines and their metabo-
    lites in postmortem brains from patients with Alzheimer-type dementia. *Journal of
    Neurochemistry, 43,* 388–393.
Arendt, T., & Bigl, V. (1986). Alzheimer's plaques and cortical cholinergic innervation.
    Neurology, 17, 277–279.
Arregui, A., Perry, E. K., Rossor, M., & Tomlinson, B. (1982). Angiotensin converting
    enzyme in Alzheimer's disease: Increased activity in caudate nucleus and cortical
    areas. *Journal of Neurochemistry, 38,* 1490–1492.
Ayer-LeLievre, C., Olson, L., Ebendal, T., Seiger, A., & Persson H. (1988). Expression of the
    B-nerve growth factor gene in hippocampal neurons. *Science, 240,* 1339–1341.
Balazs, R., Jorgensen, O. S., & Hack, N. (1988). N-methyl-D-aspartate promotes the survival
    of cerebellar granule cells in culture. *Neuroscience, 27,* 437–451.
Beal, M. F., Clevens, R. A., Chattha, G. K., MacGarvey, M. U., Mazurek, M. F., & Gabriel, S.
    M. (1988). Galanin-like immunoreactivity is unchanged in Alzheimer's disease and
    Parkinson's disease dementia cerebral cortex. *Journal of Neurochemistry, 51,* 1935–
    1941.
Beal, M. F., & Martin, J. B. (1986). Neuropeptides in neurological disease. *Annals of
    Neurology, 20,* 567–565.
Beal, M. F., Mazarek, M. F., Geetinder, K. C., Sevendsen, B. S., Bird, E. D., & Martin, J. B.
    (1986). Neuropeptide Y immunoreactivity is reduced in cerebral cortex in Al-
    zheimer's disease. *Annals of Neurology, 20,* 282–288.
Beal, M. F., Mazarek, M. F., Tran, V. T., Chatta, G., Bird, E. D., & Martin, J. B. (1985).
    Reduced numbers of somatostatin receptors in its cerebral cortex in Alzheimer's
    disease. *Science, 224,* 284–291.
Bissettee, G., Reynolds, G. P., Kilts, C. D., Widerlov, E., & Nemeroff, C. B. (1985).
    Corticotropin-releasing factor-like immunoreactivity in senile dementia of the
    Alzheimer type. *Journal of the American Medical Association, 254,* 3067–3069.
Bowen, D. M. (1981). Alzheimer's disease. In A. N. Davison & R. H. S. Thompson (Eds.),
    The molecular basis of neuropathology (pp. 649–665). London: Edward Arnold.
Bowen, D. M. (1990). Treatment of Alzheimer's disease: Molecular pathology versus
    neurotransmitter, based therapy. *British Journal of Psychiatry, 157,* 327–330.

Bowen, D. M., Allen, S. J., Benton, J. S., Goodhardt, M. J., Haan, E. A., Palmer, A. M., Simon, N., Smith, C. C. T., Spillane, J. A., Esiri, M. M., Neary, D., Snowden, J. S., Wilcock, G. K., & Davison, A. N. (1983). Biochemical assessment of serotonergic and cholinergic dysfunction and cerebral atrophy in Alzheimer's disease. *Journal of Neurochemistry*, *41*, 266–272.

Bowen, D. M., Benton, J. S. Spillane, J. A., Smith, C. C. T., & Allen, S. J. (1982). Choline acetyltransferase activity and histopathology of frontal neocortex from biopsies of demented patients. *Journal of Neurological Sciences*, *57*, 191–202.

Bowen, D. M., Beyreuther, K., Cross, A. J., Davies, P., Diringer, H., Goldgaber, Hock, F. J., Khachaturian, Z. S., Kurz, A. F., Masters, C. L., Multhaup, G., Price, D. L., & Saper, C. B. (1988). Cell injury: Molecular biology and genetic basis. In A. S. Henderson & J. H. Henderson (Eds.), *Etiology of dementia of Alzheimer type* (pp. 165–176). Chichester, England: Wiley.

Bowen, D. M., & Davison, A. N. (1973). Cathepsin A in human brain and spleen. *Bichemistry Journal*, *131* 417–419.

Bowen, D. M., & Davison, A. N. (1978). Biochemical changes in the normal aging brain and in dementia. In B. Isaacs (Ed.), *Recent advances in geriatric medicine* (pp. 41–60). Edinburgh: Churchill Livingstone.

Bowen, D. M., & Davison, A. N. (1986). Biochemical studies of nerve cells and energy metabolism in Alzheimer's disease. *British Medical Bulletin*, *42*: 75–80.

Bowen, D. M., & Procter, A. W. (1991). The neurochemistry of the major dementias of old age and indications for future drug treatment. In R. Jacoby & C. Oppenheimer (Eds.), *Textbook of psychiatry in the elderly*. London: Oxford University Press, in press.

Bowen, D. M., Sims, N. R., Lee, K. A. D., & Marek, K. L. (1982). Acetylcholine synthesis and glucose oxidation are preserved in human brain obtained shortly after death. *Neuroscience Letters*, *31* 195–199.

Bowen, D. M., Goodhardt, M. J., Strong, A. J., Smith, C. B., White, P., Branston, N. M., Symon, L., & Davison, A. N. (1976). Biochemical indices of brain structure, function and "hypoxia" in cortex from baboons with middle artery occlusion. *Brain Research*, *117*, 503–507.

Bowen, D. M., Smith, C. B., White, P., & Davison, A. N. (1976). Neurotransmitter-related enzymes and indices of hypoxia in senile dementia and other abiotrophies. *Brain*, *99*, 459–496.

Bowen, D. M., Smith, C. B., White, P., Flack, R. H. A., Carrasco, L. H., Gedye, J. L., & Davison, A. N. (1977). Chemical pathology of the organic dementias: II Quantitative estimation of cellular changes in postmortem brains. *Brain*, *100*, 427–447.

Bowen, D. M., Smith, C. B., White, P., Goodhardt, M. J., Spillane, J. A., Flack, R. H. A., & Davison, A. N. (1977). Chemical pathology of the organic dementias: I Validity of biochemical measurements in human postmortem brain specimens. *Brain*, *100*, 397–426.

Bowen, D. M., Steel, J. E., Lowe, S. L., & Palmer, A. M. (1989). Tacrine in relation to amino acid transmitters in Alzheimer's disease. In R. J. Wurtman, Corkins & Growdon, J. H., (Eds.), *Alzheimer's disease* (pp. 91–96). New York: Raven Press.

Bowen, D. M., White, P., Spillane, J. A., Goodhardt, M. J., Curzon, G., Iwangoff, P., Meyer-Ruge, W., & Davison, A. N. (1979). Accelerated aging or selective neuronal loss as an important cause of dementia? *Lancet*, *i*, 11–14.

Brewer, G. J., & Cotman, C. W. (1988). NMDA promotes branching, MK801 stimulates elongation of dentate granule neurons. *Society of Neuroscience, Abstract*, *14*, 115.

Caulfield, M. P., Straughan, D. S., Cross, A. J., Crow, T., & Birdsall, N. J. (1982). Cortical muscarinic receptor subtypes in Alzheimer's disease. *Lancet*, *ii*, 1277.

Chan-Palay, V. (1988). Galanin hyperinnervates surviving neurons of the human basal nucelus of Meynert in dementia of Alzheimer's and Parkinson's disease: A hypothesis for the role of galanin in accentuating cholinergic dysfunction in dementia. *Journal of Comparative Neurology*, *173*, 543–557.

Chawluk, J. B., Alavi, A., Dann, R., Hartig, H. I., Bais, S., Kushner, M., Zimmerman, R. A., & Reivich, M. J. (1987). Poistron emission tomography in aging and dementia: Effect of cerebral atrophy. *Journal of Nuclear Medicine, 28,* 431–437.

Cheetham, S. C., Crompton, M. R., Katona, L. E., Parker, S. J., & Horton, R. W. (1988). Brain GABA$_A$/benzodiazepine binding sites and glutamic decarboxylase activity in depressed suicide victims. *Brain Research, 460,* 114–123.

Collerton, D. (1986). Cholinergic function and intellectual decline in Alzheimer's disease. *Neuroscience, 19,* 1–28.

Colon, E. J. (1973). The cerebral cortex in presenile dementia: A quantitative study. *Acta Neuropathologica, 23,* 281–290.

Corsellis, J. A. N. (1962). Mental illness and the ageing brain. Maudsley monograph No. 9. London: Oxford University Press.

Cotman, C. W., Geddes, J. W., Monahan, D. T., & Anderson, K. J. (1987). Excitatory amino acid receptors in Alzheimer' disease. In P. Davies & C. E. Finch (Eds.), *Molecular neuropathology of aging* (pp. 67–84). (Banbury report 27). Cold Spring Harbor, NY, Laboratory.

Cowburn, R. F., Hardy, J. A., Briggs, R. S., & Roberts, P. J. (1989). Characteristics, density and distribution of kainate receptors in normal and Alzheimer's diseased human brain. *Journal of Neurochemistry, 52,* 140–147.

Cowburn, R., Hardy, J., Roberts, P., & Briggs, R. (1988). Presynaptic and postsynaptic glutamatergic function in Alzheimer's disease. *Brain Research, 52,* 403–407.

Cross, A. J., Crow, T. J., Ferrier, I. N. & Johnson, J. A. (1986). The selectivity of the reduction of serotonin S2 receptors in Alzheimer-type dementia *Neurobiology of Aging, 7,* 3–8.

Cross, A. J., Crow, J. J., Ferrier, I. W., Johnson, J. A., Bloom, S. R., & Corsellis, J. A. N. (1984). Serotonin receptor changes in dementia of the Alzheimer type. *Journal of Neurochemistry, 43,* 1574–1582.

Cross, A. J., Crow, T. J., Ferrier, I. N., Johnson, J. A., & Markakis, D. (1984). Striatal dopamine receptors in Alzheimer-type dementia. *Neuroscience Letters, 52,* 1–6.

Cross, A. J., Crow, T. J., Johnson, J. A., Joseph, M. H., Perry, E. K., Perry, R. H., Blessed, G., & Tomlinson, B. E. (1983). Monoamine metabolism in senile dementia of Alzheimer type. *Journal of the Neurological Sciences, 60,* 383–392.

Cross, A. J., Crow, T. J., Johnson, J. A. Perry, E. K., Perry, R. H., Blessed, G., & Tomlinson, B. E. (1984). Studies on neurotransmitter receptor systems in neocortex and hippocampus in senile dementia Alzheimer type. *Journal of the Neurological Sciences, 64,* 109–117.

Cross, A. J., Slater, P., Candy, J. M., Perry, E. K., & Perry, R. H. (1987). Glutamate deficits in Alzheimer's disease. *Journal of Neurology, Neurosurgery and Psychiatry, 50,* 357–358.

Curcio, C. A., & Kemper, T. (1984). Nucleus Raphe Dorsalis in dementia of the Alzheimer type: Neurofibrillary changes and neuronal packing density. *Journal of Neuropathology and Experimental Neurology, 43,* 359–368.

Cutler, N. R., Haxby, J. V., Duara, R., Grady, C. L., Kay, A. D., Kessler, R. M., Sandaram, M., & Rappoport, S. I. (1985). Clinical history, brain metabolism and neuropsychological function in Alzheimer's disease. *Annals of Neurology, 18,* 298–309.

D'Amato, R. J., Zweig, R. M., Whitehouse, P. J., Wenk, G. C., Singer, H. S., Mayeux, R., Price, D. L., & Snyder, S. H. (1987). Aminergic systems in Alzheimer's disease and Parkinson's disease. *Annals of Neurology, 22,* 229–236.

Davies, M. F., Deisz, R. A., Prince, D. A., & Peroutka, S. J. (1987). Two distinct effects of 5-hydroxtryptamine on single cortical neurons. *Brain Research, 423,* 347–352.

Davies, P., Katzman, R., & Terry, R. D. (1980). Reduced somatostatin-like immunoreactivity in cerebral cortex from areas of Alzheimer disease and Alzheimer senile dementia. *Nature, 288,* 274–280.

Davies, P., & Maloney, A. J. P. (1976). Selective loss of central cholinergic neurons in Alzheimer's disease. *Lancet, i,* 1403.

Davies, P., & Verth, A. H. (1978). Regional distribution of muscarinic acetylcholine receptor in normal and Alzheimer type dementia brains. *Brain Research, 138*, 385–392.

Dawbarn, D., Rossor, M. W., Mountjoy, C. Q., Roth, M., & Emson, P. C. (1986). Decreased somatostatin immunoreactivity but not neuropeptide Y immunoreactivity in cortex in senile dementia of Alzheimer type. *Neuroscience Letters, 70*, 154–159.

DeKosky, S., & Bass, N. H. (1985). Biochemistry of senile dementia. In A. Lajtha (Ed.), *Handbook of neurochemistry* Vol 10, pp 617–649. New York: Plenum Publishing Corp.

De Souza, R. J., Robinson, T. N., Cross, A. J., & Green, R. (1989). The effects of tetra-hydroaminoacridine of endogenous 5-hydroxytryptamine and dopamine release from rat brain slices. *British Journal of Pharmacology, 96*, 352.

De Souza, E. B., Whitehouse, P. J., Kuhor, M. J., Price, D. C., & Vale, W. W. (1986). Reciprocal changes in corticotropin-releasing factor (CRF)-like immunoreactivity and CRF receptors in cerebral cortex of Alzheimer's disease. *Nature, 319*, 593–545.

Dodd, P. R., Hambley, J. W., Cowburn, R. F., & Hardy, J. A. (1988). A comparison of methodologies for the study of functional transmitter neurochemistry in human brain. *Journal of Neurochemistry, 50*, 1333–1345.

Ellison, D. W., Beal, M. F., Mazurek, M. F., Bird, E. D., & Martin, J. B., (1986). A postmortem study of amino-acid neurotransmitters in Alzheimer's disease. *Annals of Neurology, 20*, 616–621.

Ferrier, I. W., Cross, A. J., Johnson, J. A., Roberts, G. W., Corsellis, J. A. N., Lee, Y. C., O'Shaughinessy, D., Adrim, T. E., McGregor, A. J., Baratese-Hamilton, & Bloom, S. B. (1983). *Journal of the Sciences, 62*, 154–170.

Finch, C. E., & Davies, P. (1988). *The molecular biology of Alzheimer's disease*. Cold Spring Harbor, NY, Laboratory.

Fisher, R. S., Buchwald, N. A., Hull, C. D., & Levine, M. S. (1988) GABAergic basal forebrain neurons project to the neocortex: The localisation of glutamic acid debarboxylase and choline acetyltransferase in feline corticopetal neurons. *Journal of Comparative Neurology, 272*, 489–502.

Flynn, D. D., & Mash, D. C. (1986). Characterization of C.[3H] nicotine binding in human cerebral cortex: Comparison between Alzheimer's disease and normal. *Journal of Neurochemistry, 47*, 1948–1954.

Francis, P. T., Bowen, D. M., Lowe, S. L., Neary, D., Mann, D. M. A., & Snowden, J. S. (1987). Somatostatin content and release measured in cerebral biopsies from demented patients. *Journal of the Neurological Sciences, 78*, 1–16.

Francis, P. T., Bowen, D. M., Neary, D., Palo, J., Wikstrom, J., & Olney, J. (1984). Somatostatin-like immunoreactivity in lumber cerebrospinal fluid from neurohistologically examined demented patients. *Neurobiology of Aging, 5*, 183–186.

Francis, P. T., Palmer, A. M., Sims, N. R., Bowen, D. M., Davison, A. N., Esiri, M. M., Neary, D., Snowden, J. S., & Wilcock, G. K. (1985). Neurochemical studies of early-onset Alzheimer's disease: Possible influence on treatment. *New England Journal of Medicine, 313*, 7–11.

Francis, P. T., Pearson, R. C. A., Lowe, S. L., Neal, J. N., Stephens, P. H., Powell, T. P. S., & Bowen, D. M. (1987). The dementia of Alzheimer's disease: An update. *Journal of Neurology, Neurosurgery and Psychiatry, 50*, 242–243.

Gottfries, C. G., Adolfsson, R., Aquilonius, S. M., Carlsson, A., Eckernas, S. A., Nordberg, L., Oreland, L., Svennerholm, L., Wiberg, A., & Winblad, B. (1983). Biochemical changes in dementia disorders of Alzheimer type (AD/SD AT). *Neurobiology of Aging, 4*, 261–271.

Gottfries, C. G., Gottfries, I., & Roos, B. E. (1969). Homovanillic acid and 5-hydroxyindoleacetic acid in the cerebrospinal fluid of patients with senile dementia, pre-senile dementia and Parkinsonism. *Journal of Neurochemistry, 16*, 1341–1345.

Gottfries, C. G., Kjallquist, A., Ponten, U., Roos, B. F, & Sundberg, G. (1974). Cerebrospinal fluid pH and monoamine and glucolytic metabolities in Alzheimer's disease. *British Journal of Psychiatry, 124*, 280–287.

Greenamyre, T. J., Penney, J. B., D'Amato, C. J., & Young, A. B. (1987). Dementia of the

Alzheimer type: Changes in hippocampus L-[3H] glutamate binding. *Journal of Neurochemistry, 48,* 543–551.

Greenamyre, T. J. & Young, A. B. (in press). Excitatory amino acids and Alzheimer's disease. *Neurobiology of Aging.*

Growdon, J. H., & Corkin, S. (1983). Cerebrospinal fluid measures in Alzheimer's disease. In T. Crook, S. Ferris, & R. Bartus (Eds.), *Assessment in geriatric psychopharmacology* (pp. 251–263). Mark Powley Associates Inc.

Gustafson, L. (1987). Frontal lobe degeneration of non-Alzheimer type II. Clinical picture of differential diagnosis. *Archives of Gerontology and Geriatrics, 6,* 209–223.

Hardy, J., Cowburn, R., Barton, A., Reynolds, C., Dodd, P., Wester, P., O'Carroll, A. M., Lofdahl, E., & Winblad, B. (1987). A disorder of cortical GABAergic innervation in Alzheimer's disease. *Neuroscience Letters, 73,* 192–196.

Hardy, J.,Cowburn, R., Barton, A., Reynolds, G., Lofdahl, E., O'Carroll, A. M., Wester, P., & Winblad, B. (1987). Region-specific loss of glutamate innervation in Alzheimer's disease. *Neuroscience Letters, 73,* 77–80.

Hays, S. E., & Paul, T. M. (1982). CCK receptors and neurological disease. *Life Sciences, 31,* 319–322.

Hubbard, B. M., & Anderson, J. M. (1981). A quantitative study of cerebral atrophy in old age senile dementia. *Journal of the Neurological Sciences, 50,* 135–145.

Hubbard, B. M., & Anderson, J. M. (1985). Age-related variation in the neuron content of the cerebral cortex in senile dementia of Alzheimer type. *Neuropathology and Applied Neurobiology, 11,* 369–382.

Huff, F. J., Maire, J.-C., Growdon, J. H., Corkin, S., & Wurtman, R. J. (1986). Cholinesterases in cerebrospinal fluid: Correlations with clinical measures in Alzheimer's disease. *Journal of the Neurological Sciences, 72,* 121–129.

Hyman, B. T., Van Hoesen, G. W., & Damasio, A. R. (1987). Alzheimer's disease: Glutamate depletion in the hippocampal perferant pathway zone. *Annals of Neurology, 22,* 37–40.

Iversen, L. L., Rossor, M. W., Reynolds, G. P., Hills, R., Roth, M., Mountjoy, C. Q., Foote, J. H., Morrison, J. H., & Bloom, F. E. (1984). Loss of pigmented dopamine-B-hydroxylase positive cells from locus ceruleus in senile dementia of Alzheimer's type. *Neuroscience Letters, 34,* 95–100.

Jones, E. G. (1986). Neurotransmitters in the cerebral cortex. *Journal of Neurosurgery, 65,* 135–153.

Jones, E. G., & Hendry, S. H. C. (1988). Expression of neuronal diversity in the central nervous system in E. G. Jones (Ed.), *Molecular biology of the human brain* (pp. 3–11). New York: Alan R. Liss.

Kemp, J. A., Foster, A. C., Leeson, P. D., Priestley, T., Tidgett, R., Iversen, L. L., & Woodruff, G. N. (1988). 7-chlorokynurenic acid is a selective antagonist at the glycine modulatory site of the N-methyl-D-aspartate receptor complex. *Proceedings of the National Academy of Sciences, (U.S.A.), 85,* 6547–6550.

Korey, S. R., Scheinberg, L., Terry, R., & Stein, A. (1961). Studies in presenile dlementia. *Transactions of the American Neurological Association, 86,* 99–102.

Lang, W., & Henke, H. (1983). Cholinergic receptor binding autoradiography in brains of nonneurological and senile dementia of Alzheimer type patients. *Brain Research, 267,* 271–280.

Lennox, G., Lowe, J., Byrne, E. J., Landon, M., Mayer, R. J., & Godwin-Austen, R. B. (1989). Diffuse Lewy body disease. *Lancet, i,* 323–324.

Lewis, D. A., Campbell, M. J., Terry, R. D., & Morrison, J. H. (1987). Laminar and regional distributions of neurofibrillary tangles and neuritic plaques in Alzheimer's disease: A quantitative study of visual and auditory cortices. *Journal of Neuroscience, 7,* 1799–1816.

Lowe, S. L., Francis, P. T., Procter, A. W., Palmer, A. M., Davison, A. N., & Bowen, D. M. (1988). Gamma-aminobutyric acid concentration in brain tissue at two stages of Alzheimer's disease. *Brain, 111,* 785–799.

Mann, D. M. A., Yates, P. O., & Marcyniuk, B. (1984). Monoaminergic neurotransmitter systems in senile dementia of Alzheimer type. *Clinical Neuropathology*, *3*, 199–205.

Mann, D. M. A., Yates, P. O., & Marcyniuk, B. (1985). Alzheimer's presenile dementia, senile dementia of Alzheimer's type and Down's syndrome in middle age for an age related continuum of pathological change. *Neuropathology and Applied Neurobiology*, *10*, 185–207.

Mann, D. M. A., Yates, P. O., & Marcyniuk, B. (1987). Dopaminergic neurotransmitter systems in Alzheimer's disease and Down's syndrome at middle age. *Journal of Neurology, Neurosurgery, and Psychiatry*, *50*, 341–344.

Maragos, W. F., Chu, D. C. M., Young, A. B., D'Amato, C. J., & Penney, J. B. (1987). Loss of hippocampus 3H TCP binding in Alzheimer's disease. *Neuroscience Letters*, *74*, 371–376.

Marcyniuk, B., Mann, D. M. A., & Yates, P. O. (1986). Loss of nerve cells from the locus ceruleus in Alzheimer's disease is topographically arranged. *Neuroscience Letters*, *64*, 247–252.

Marx, J. L. (1989). Brain protein yields clues to Alzheimer's disease. *Science*, *243*, 1664–1666.

Mash, D. C., Flynn, D. D., & Potter, L. T. (1985). Loss of M_2 muscarinic receptors in the cerebral cortex in Alzheimer's disease and experimental cholinergic denervation. *Science*, *228*, 1115–1117.

Mattson, M. P. (1988). Neurotransmitters in the regulation of neuronal cytoarchitecture. *Brain Research Reviews*, *13*, 179–212.

Mazière, B., & Mazière, M. (1991). Positron emission tomography studies of brain receptors. *Fundamental and Clinical Pharmacology*, *5*, 61–91.

McGeer, P. L. (1986). Brain imaging in Alzheimer's disease. *British Medical Bulletin*, *42*, 24–28.

McGeer, E. G., Singh, E. A., McGeer, P. L. (1987). Sodium-dependent glutamate binding in senile dementia. *Neurobiology of Aging*, *8*, 219–223.

Mesulam, M. M., & Geula, C. (1988). Acetylcholinesterase-rich pyramidal neurons in the human neocortex and hippocampus: Absence at birth, development during the life span, and dissolution in Alzheimer's disease. *Annals of Neurology*, *24*, 765–773.

Middlemiss, D. N., Palmer, A. M., Edel, N., & Bowen, D. M. (1986). Binding of the novel serotonin against 8-hydroxy-2-(di-n-propylamino) tetralin in normal Alzheimer brain. *Journal of Neurochemistry*, *46*, 993–996.

Miller, A. K. H., Alston, R. C., & Corsellis, J. A. N. (1981). Variations with age in the volumes of grey and white matter in the cerebral hemispheres of man: Measurements with an image analyzer. *Neuropathology and Applied Neurobiology*, *6*, 119–132.

Miller, J. D., de Leon, M. J., Ferris, S. H., Kluger, A., George, A. E. Reisberg, B., Sachs, H. J., & Wolf, A. R. (1987). Abnormal temporal lobe response in Alzheimer's disease during cognitive processing as measured by C2-deoxy-D-glucose and PET. *Journal of Cerebral Blood Flow and Metabolism*, *7*, 248–251.

Monaghan, D. T., Olverman, H. J., Nguyen, L., Watkins, J. C., & Cotman, C. W. (1988). Two classes of N-methyl-D-aspartate recognition sites: Differential distribution and differential regulation by glycine. *Proceedings of the National Academy of Sciences (U.S.A.)*, *85*, 9836–9840.

Moran, J., & Patel, A. J. (in press). Stimulation of the N-methyl-D-aspartate receptor promotes the biochemical differentiation of cerebellar granule neurons and not astrocytes. *Brain Research*.

Morris, R. G. M., Anderson, E., Lynch, G., & Baudry, M. (1986). Selective impairment of learning and blockade of long-term potentiation by an N-methyl-D-aspartate receptor antagonist AP-5. *Nature*, *319*, 774–776.

Morrison, J. H., Scherr, S., Lewis, D. L., Campbell, M. J., Bloom, F. E., Rogers, J., & Benioff, R. (1986). The laminar and regional distribution of neocortical somatostatin and neuritic plaques: Implications for Alzheimer's disease as a global neocortical disconnection syndrome. In A. B. Scheibel & S. T. Vincent (Eds.), *The biological substrates of Alzheimer's disease* (p. 115). New York: Academic Press.

Mountjoy, C. Q., Roth, M., Evans, N. J. R., & Evans, H. M. (1983). Cortical neuronal counts in normal elderly controls and demented patients. *Neurobiology of Aging, 4,* 1–11.

Mouradian, M. M., Contreras, P. C., Monahan, J. B., & Chase, T. N. (1988). [3H] MK-801 binding in Alzheimer's disease. *Neuroscience Letters, 93,* 225–230.

Muller-Hill, B., & Beyreuther, K. (1989). Molecular biology of Alzheimer's disease. *Annual Review of Biochemistry, 58,* 287–307.

Najlerahim, A., & Bowen, D. M. (1988). Biochemical measurements in Alzheimer's disease reveal a necessity for improved neuroimaging techniques to study metabolism. *Biochemistry Journal, 251,* 305–308.

Neary, D., Snowden, J. S., Bowen, D. M., Sims, N. R., Mann, D. M. A., Benton, J. S., Norten, B., & Yates, P. O. (1986). Neuropsychological syndromes in presenile dementia due to cerebral atrophy. *Journal of Neurology, Neurosurgery and Psychiatry, 49,* 163–174.

Neary, D., Snowden, J.S., Bowen, D. M., Sims, N. R., Mann, D. M. A., Yates, P. O., & Davison, A. N. (1986). Cerebral biopsy in the investigation of presenile dementia due to cerebral atrophy. *Journal of Neurology, Neurosurgery and Psychiatry, 49,* 157–162.

Neary, D., Snowden, J. S., Mann, D. M. A., Bowen, D. M., Sims, N. R., Northen, B., (1986). Alzheimer's disease: A correlative study. *Journal of Neurology, Neurosurgery and Psychiatry, 49,* 229–237.

Neary D. Snowden, J. S., Northen, B., & Goulding, P. (1988). Dementia of frontal type. *Journal of Neurology, Neuroscience, and Psychiatry, 51,* 353–361.

Neary D., Snowden, J. S., Shields, R. A., Barjan, A. W. I., Northen, B., MacDermott, N., Prescott, M. C., & Testa, H. J. (1987). Single photon emission tomography using 99mTc-MM-PAO in the investigation of dementia. *Journal of Neurology, Neuroscience and Psychiatry, 50,* 1101–1109.

Nordberg, A., Adem, A., Nilsson, L., & Winblad, B. (1987). Cholinergic deficits in CNS and peripheral nonneuronal tissue in Alzheimer's disease. In M. J. Dowdall & J. N. Hawthorne (Eds.), *Cellular and molecular basis of cholinergic function* (pp. 858–868). Chichester, England: Ellis Horwood.

Palmer, A. M., & Bowen, D. M. (1985). 5-Hydroxyindoleacetic acid and homovanillic acid in the cerebrospinal fluid and caudate nucleus of histologically verified example of Alzheimer's disease. *Biochemical Society Transactions, 13,* 167–168.

Palmer, A. M., Francis, P. T. Benton, J. S., Sims, N. R., Mann, D. M. A., Neary, D., Snowden, J. S., & Bowen, D. M. (1987). Presynaptic serotonergic dysfunction in patients with Alzheimer's disease. *Journal of Neurochemistry, 48,* 8–15.

Palmer, A. M., Francis, P. T., Bowen, D. M., Benton, J. S., Neary, D., Mann, D. M. A., & Snowden, J. S. (1987). Catecholaminergic neurons assessed antemortem in Alzheimer's disease. *Brain Research, 414,* 365–375.

Palmer, A. M., Hutson, P. H., Lowe, S. L., & Bowen, D. M. (1989). Extracellular concentrations of aspartate and glutamate in rat neostriatum following chemical stimulation of frontal cortex. *Experimental Brain Research, 75,* 659–663.

Palmer, A. M., Lowe, S. I., Francis, P. T. & Bowen, D. M. (1988). Are postmortem biochemical studies of human brain worthwhile? *Biochemical Society Transactions, 16,* 472–475.

Palmer, A. M., Procter, A. W., Stratmann, G. C., & Bowen, D. M., (1986). Excitatory amino acid-releasing and cholinergic neurons in Alzheimer's disease. *Neuroscience Letters, 66,* 199–204.

Palmer, A. M., Sims, N. R., Bowen, D. M., Neary, D., Palo, J., Wikstrom, J., & Davison, A. N. (1984). Monoamine metabolite concentrations in lumber cerebrospinal fluid of patients with histologically verified Alzheimer's dementia. *Journal of Neurosurgery and Psychiatry, 47,* 481–484.

Palmer, A. M., Stratmann, G. C., Procter, A. W., & Bowen, D. M. (1988). Possible neurotransmitter basis of behavioural changes in Alzheimer's disease. *Annals of Neurology, 23,* 616–620.

Palmer, A. M., Wilcock, G. K., Esiri, M. M., Francis, P. T., & Bowen, D. M. (1987). Monoaminergic innervation of the frontal and temporal lobes in Alzheimer's disease. *Brain Research, 401,* 231–238.

Pearce, B. R., Palmer, A. M., Bowen, D. M., Wilcock, G. K., Esiri, M. M., & Davison, A. N. (1984). Neurotransmitter dysfunction and atrophy of the caudate nucleus in Alzheimer's disease. *Neurochemical Pathology, 2,* 221–232.

Pearson, R. C. A., Esiri, M. M., Hiorns, R. W., Wilcock, G. K., & Powell, T. P. S. (1985). Anatomical correlates of the distribution of the pathological changes in the neocortex in Alzheimer's disease. *Proceedings of the National Academy of Sciences (U.S.A.), 82,* 4531–4534.

Pearson, R. C. A., & Powell, J. P. S. (1989). The neuroanatomy of Alzheimer's disease. *Reviews of Neuroscience, 2,* 101–123.

Pericak-Vance, M. A., Yamaoka, L. H., Haynes, C. S., Speer, M. C., Haines, J. L., Gaskell, P. C., Hung, W. Y. Clark, C. M., Heyman, A. L., Trofatter, J. A., Eisenmenger, J. P., Gilbert, J. R., Lee, J. E., Alberts, M. J., Dawson, D. V., Bartlett, R. J., Earl, N. L., Siddique, T., Vance, J. M., Coneally, P. M., & Roses, A. D. (1988). Genetic linkage studies in Alzheimer's disease families. *Experimental Neurology, 102,* 271–279.

Perry, E. K., (1986). The cholinergic hypothesis—ten years on. *British Medical Bulletin, 42,* 63–69.

Perry, E. K., Blessed, G., Tomlinson, B. E., Perry, R. H., Crow, T. H., Cross, A. J., Dockray, G. J., Dimaline, R., & Arregui, A. (1981). Neurochemical activities in human temporal lobe released to aging and Alzheimer-type changes. *Neurobiology of Aging, 2,* 215–256.

Perry, R. H., Dockrae, G. J., Dimaline, R., Perry, E. K., Blessed, G., & Tomlinson, B. E. (1981). Neuropeptides in Alzheimer's disease, depression and schizophrenia. *Journal of the Neurological Sciences, 51,* 465–472.

Perry, E. K., Gibson, P. H., Blessed, G., Perry, R. H., & Tomlinson, B. E. (1977). Neurotransmitter enzyme abnormalities in senile dementia. *Journal of the Neurological Sciences, 34,* 247–265.

Perry, R. H., Irving, D., Blessed, G., Perry, E. K., & Fairbairn, A. F. (1989). Clinically and neuropathologically distinct form of dementia in the elderly. *Lancet, i,* 166.

Perry, E. K., Marshall, E. F., & Blessed, G. (1983). Decreased imipramine binding in the brains of patients with depression. *British Journal of Psychiatry, 142,* 188–192.

Perry, E. H., Perry, R. H., Candy, J. M., Fairbairn, A. F., Blessed, G., Dick, D. J., & Tomlinson, B. E. (1984). Cortical serotonin S2 receptor binding abnormalities in patients with Alzheimer's disease. *Neuroscience Letters, 51,* 353–357.

Perry, E. K., Perry, R. H., Smith, C. J., Dick, D. J., Candy, J. M., Edwardson, J. A., Fairbairn, A., & Blessed, G. (1987). Nicotinic receptor abnormalities in Alzheimer's and Parkinson's diseases. *Journal of Neurology, Neurosurgery and Psychiatry, 50,* 806–809.

Perry, T. L., Yong, V. W., Bergeron, C., Hansen, S., & Jones, K. (1987). Amino-acids, glutathione, and glutathione transferase activity in the brains of patients with Alzheimer's disease. *Annals of Neurology, 21,* 331–336.

Pierotti, A. R., Harmar, A. J., Simpson, J., & Yates, C. M. (1986). High molecular-weight forms of somatostatin are reduced in Alzheimer's disease and Downs syndrome. *Neuroscience Letters, 63,* 141–146.

Procter, A. W., & Bowen, D. M. (1988). Beta-carbolines for Alzheimer's disease?—More evidence, a test of efficacy and some precautions. *Trends in Neuroscience, 11,* 208–209.

Procter, A. W., Lowe, S. L., Palmer, A. M., Francis. P. T., Esiri, M. M., Stratmann, G. C., Najlerahim, A., Patel, A., Hunt, A., & Bowen, D. M. (1988). Topographical distribution of neurochemical changes in Alzheimer's disease. *Journal of the Neurological Sciences, 84,* 125–140.

Procter, A. W., Middlemiss, D. N., & Bowen, D. M. (1988). Selective loss of serotonin recognition sites in the parietal cortex in Alzheimer's disease. *International Journal of Geriatrics and Psychiatry, 3,* 37–44.

Procter, A. W., Palmer, A. M., Francis, P. T., Lowe, S. L., Neary, D., Murphy, E., Doshi, R., & Bowen, D. M. (1988). Evidence of glutamatergic denervation and possible abnormal metabolism in Alzheimer's disease. *Journal of Neurochemistry, 50*, 790–802.

Procter, A. W., Sterling, J. H., Stratmann, G. C., Cross, A. J., & Bowen, D. M. (1989). Loss of glycine-dependent radiological binding to the N-methyl-D-aspartate-phencyclidine receptor complex in patients with Alzheimer's disease. *Neuroscience Letters, 101*, 62–66.

Procter, A. W., Wong, E. H. F., Stratmann, G. C., Lowe, S. C., & Bowen, D. M. (1989). Reduced glycine stimulation of [³H] MK801 binding in Alzheimer's disease. *Journal of Neurochemistry, 53*, 698–704.

Reinikainen, K. J., Riekkinen, P. J., Jolkkonen, J., Kosma, V. M., & Suininer, H. (1987). Decreased somatostatin-like immunoreactivity in cerebral cortex and cerebrospinal fluid in Alzheimer's disease. *Brain Research, 402*, 103–108.

Reinikainen, K. J., Paljarvi, L., Huuskonen, M., Soininen, H., Laakso, M. & Riekkinen, P. J. (1988). A post mortem study of noradrenergic, serotonergic and GABAergic neurons in Alzheimer's disease. *Journal of the Neurological Sciences, 84*, 101–116.

Reynolds, G. P., Arnold, L., Rossor, M. W., Iversen, L. L., Mountjoy, L. Q., & Roth, M. (1984). Reduced binding of [³H] ketanserin to cortical 5HT₂ receptors in senile dementia of Alzheimer type. *Neuroscience Letters, 44*, 47–51.

Richter, J. A., Perry, E. K., & Tomlinson, B. E. (1980). Acetylcholine and choline levels in postmortem human tissue: Preliminary observations in Alzheimer's disease. *Life Sciences, 26*, 1683–1689.

Rinne, J. O., Sako, E., Paljora, L., Molsa, P. K., & Rinne, U. K. (1986). Brain dopamine D-2 receptors in senile dementia. *Journal of Neurology Transactions, 65*, 51–62.

Rogers, J., & Morrison, J. H. (1985). Quantitative morphology and regional laminar distributions of senile plaques in Alzheimer's disease. *Neuroscience, 5*, 2801–2808.

Rossor, M. N., Emson, P. C., Mountjoy, C. Q., Roth, M., & Iversen, L. L. (1980). Reduced amounts of immunoreactive somatostatin in the temporal cortex in senile dementia of Alzheimer type. *Neuroscience Letters, 20*, 373–377.

Rossor, N. M., Garrett, W., Johnson, A. L., Mountjoy, C. Q., Roth, M., & Iversen, L. L. (1982). A postmortem study of cholinergic and GABA systems in senile dementia. *Brain, 105*, 315–330.

Rossor, M. N., & Iversen, L. L. (1986). Non-cholinergic neurotransmitter abnormalities in Alzheimer's disease. *British Medical Bulletin, 42*, 70–74.

Rylett, R. J., Ball, M. J., & Colhoun, E. H. (1983). Evidence for high affinity choline transport in synaptosomes prepared from hippocampus of patients with Alzheimer's disease. *Brain Research, 289*, 169–175.

Selkoe, D. (1989). Biochemistry of altered brain proteins in Alzheimer's disease. *Annual Review of Neuroscience, 12*, 1–7.

Shimohama, S., Taniguchi, T., Fujinara, M., & Kameyama, M. (1987). Changes in B-adrenergic receptor subtypes in Alzheimer type dementia. *Journal of Neurochemistry, 48*, 1215–1221.

Sim, M., Turner, E., & Smith, W. T. (1966). Cerebral biopsy in the investigation of presenile dementia clinical aspects. *British Journal of Psychiatry, 112*, 119–125.

Simpson, M. D. C., Cross, A. J., Slater, P., & Deakin, J. F. W. (1988). Loss of cortical GABA uptake sites in Alzheimer's disease. *Journal of Neurological Transactions, 71*, 219–226.

Simpson, M. D. C., Royston, M. C., Deakin, J. F. W., Cross, A. J., Mann, D. M. A., & Slater, P. (1988). Regional changes in [³H]D-aspartate and [³H]TCP binding sites in Alzheimer's disease brains. *Brain Research, 462*, 76–82.

Sims, N. R., Bowen, D. M., Allen, S. T., Smith, C. C. T., Neary, D., Thomas, D. J., & Davison, A. N. (1983). Presynaptic cholinergic dysfunction in patients with dementia. *Journal of Neurochemistry, 40*, 503–509.

Sims, N. R., Bowen, D. M., & Davison, A. N. (1981). [¹⁴C] acetylcholine synthesis and [¹⁴C] carbon dioxide production from [U-¹⁴C] glucose by tissue prisms from human neocortex. *Biochemical Journal, 196*, 867–876.

Sims, N. R., Bowen, D. M., Smith, C. C. T., Flack, R. H. A., Davison, A. N., Snowden, (copy unreadable) Glucose metabolism and acetylcholine synthesis in relation to neuronal activity in Alzheimer's disease. *Lancet, i*, 333–336.

Sims, N. R., Finegan, J. M., Blass, J. P., Bowen, D. M., & Neary, D. (1987). Mitochondrial function in brain tissue in primary degenerative dementia. *Brain Research, 436*, 30–38.

Small, D. H., & Simpson, R. J. (1988). Acetylcholinesterase undergoes autolysis to generate trypsin-like activity. *Neuroscience Letters, 89*, 223–228.

Smith, C. J., Perry, E. K., Perry, R. H., Candy, J. M., Johnson, M., Bonham, J. R., Dick, D. J., Fairbairn, A., Blessed, G., & Birdsall, N. J. M. (1988). Muscarinic cholinergic receptor subtypes in hippocampus in human cognitive disorders. *Journal of Neurochemistry, 50*, 547–856.

Spillane, J. A., White, P., Goodhart, M. J., Flack, R. H. A., Bowen, D. M., & Davison, A. N. (1977). Selective vulnerability of neurons in organic dementia. *Nature, 266*, 3558–3560.

Spokes, E. G. S. (1979). An analysis of factors influencing measurements of dopaine, noradrenaline, glutamate decarboxylase, and choline acetyltransferase in human postmortem brain. *Brain, 102*, 333–346.

Sprouse, J. S., & Aghajanian, G. K. (1988). Responses of hippocampal pyramidal cells to putative $5\text{-}HT_{1A}$ and $5HT_{1B}$ agonists: A comparative study with dorsal raphe neurons. *Neuropharmacology, 27*, 707–715.

Steele, J. E., Palmer, A. M., Lowe, S. L., & Bowen, D. M. (1990). The influence of tetrahydro-9-aminoacridine on release of excitatory amino acids. *Clinical Neuropharmacology, 13*, 58–66.

Summers, W. K., Majouski, L. V., Marsh, G. M., Tachiki, K., & Kling, A. (1986). Oral tetrahydroaminoacridine in long-term treatment of senile dementia, Alzheimer type. *New England Journal of Medicine, 315*, 1241–1245.

Tamminga, C. A., Foster, N. L., Fedio, P., Bird, E. D., & Chase, T. N. (1987). Alzheimer's disease: Low cerebral somatostatin levels correlate with impaired cognitive function and cortical metabolism. *Neurology, 37*, 161–165.

Taylor, S. J., Bartlett, M. J., & Greenfield, S. A. (1988). Release of acetylcholinesterase within the guinea pig substantia nigra: Effects of 5-hydroxytryptamine and amphetamine. *Neuropharmacology, 27*, 507–514.

Terry, R. D., Peck, A., De Teresa, R., Scheschter, M. D., & Horoupian, D. S. (1981). Some morphometric aspects of the brain in senile dementia of the Alzheimer type. *Annals of Neurology, 10*, 184–192.

Ullrich, A., Gray, A., Berman, C., & Dall, T. J. (1983). Human B-nerve growth factor gene sequence highly homologous to that of mouse. *Nature, 303*, 821–825.

Vincent, S. R., Johannson, U., & Hokfelt, T. (1982). Neuropeptide coexistence in human neurons. *Nature, 298*, 65–67.

Whitehouse, P. J., Marinu, A. M., Antuono, P. G., Lomenstein, J. T., Coyle, J. T., Price, D. L., & Kellar, K. J. (1986). Nicotine acetylcholine binding sites in Alzheimer's disease *Brain Research, 371*, 146–151.

Whitehouse, P. J., Price, D. L., Strable, R. G., Clark, A. W., Coyle, J. T., & Delong, M. R. (1982). Alzheimer's disease and senile dementia: Loss of neurons in the basal forebrain. *Science, 215*, 1237–1239.

Wilcock, G. K., Esiri, M. M., Bowen, D. M., & Smith, C. C. T. (1983). The nucleus basalis in Alzheimer's disease: Cell counts and cortical biochemistry. *Neuropathology and Applied Neurobiology, 9*, 175–179.

Winfield, D. A., Gatter, K. C., & Powell, T. P. S. (1980). An electron microscopic study of the types and proportions of neurons in its cortex of motor and visual areas of the cat and rat. *Brain, 103*, 245–258.

Wood, J. H. (1980). Sites of origin and cerebrospinal fluid concentration gradients. In J. H. Wood (Ed.), *Neurobiology of cerebrospinal fluid* (pp. 53–62). New York: Plenum Publishing Corp.

Yamamoto, T., & Hirano, A. (1985). Nucleus raphe dorsalis in Alzheimer's disease; Neu-
    rofibrillary tangles and loss of large neurons. *Annals of Neurology, 17*, 573–577.
Young, L. T., Kish, S. J., Li, P. P., & Warsh, J. J. (1988). Decreased brain [^{3}H] inositol 1,4,5-
    trisphosphate binding in Alzheimer's Disease. *Neuroscience Letters, 94*, 198–202.
Zweig, R. M., Ross, C. A., Hedreen, J. C., Steele, C., Cardillo, J. E., Whitehouse, P. J.,
    Folstein, M. F., & Price, D. L. (1988). The neuropathology of aminergic nuclei in
    Alzheimer's disease. *Annals of Neurology, 24*, 233–242.

25

Approaches to the Treatment of Alzheimer's Disease

ANTHONY C. SANTUCCI / VAHRAM HAROUTUNIAN
LINDA M. BIERER / KENNETH L. DAVIS

Alzheimer's disease (AD), a progressive neurodegenerative disease, is estimated to affect as many as 5% of the population of the United States above age 65 (Katzman, 1976; Plum, 1979; Terry & Katzman, 1983). At present, AD represents a major mental health problem for which there exists no known treatment. With projections of significant increases in the elderly population (over 51 million by the year 2030, according to the U.S. Census Bureau), the threat of AD approaching epidemic proportions is very real. If this health crisis and its effects are to be avoided, it is imperative that effective therapeutic strategies be developed now. This chapter outlines some of the approaches that have been taken to alleviate the impairments of AD and suggests areas of research that may aid in the ultimate development of an effective treatment strategy.

Clinical Pharmacology

Until quite recently attempts at developing treatment strategies for AD consisted of administering a number of drugs with little if any scientific rationale, and not surprisingly, even less efficacy. Vasodilators and nootropics were the classes of compounds from which therapeutic agents were chosen. However, as our understanding of the pathophysiology of AD has increased, it has become clear that AD is not a condition that primarily affects either cerebral vascular circulation or glucose utilization. Hence, there is no reason to suppose, nor any evidence to suggest, that increasing the profusion of cerebral tissue or enhancing metabolic activity would in any way diminish the cognitive symptoms of AD.

Between 1976 and 1977 it became possible, for the first time, to conceptualize a rational approach to the treatment of AD. A series of studies from the United Kingdom were published unequivocally establishing that AD was not a nonselective neurodegenerative process, but had a surprising predilection for cholinergic neurotransmission (Davies & Maloney, 1976; Perry, Gibson, Blessed, Perry, & Tomlinson, 1977; Perry, Perry, Blessed, &

467

Tomlinson, 1977; Perry, Perry, Gibson, Blessed, & Tomlinson, 1977). It was subsequently demonstrated that the magnitude of cholinergic deficits, specifically decreased choline acetyltransferase activity, correlated well with the degree of intellectual impairment when indexed shortly before death, further strengthening the link between impaired cholinergic tranmission and AD (Perry et al., 1981; Perry et al., 1978). Although it was subsequently to be determined that other neurochemical deficits are also associated with AD (see Perry, 1987, for a summary), the discovery of the preeminence of the cholinergic deficit made it possible to conceive an approach to the treatment of AD, namely, the reversal of the cholinergic deficiency. Hence, cholinomimetics, or drugs that increase cholinergic activity, were tested in AD.

Presynaptic Agents

There are numerous pharmacological approaches to increasing cholingergic activity. One approach is to administer agents that act at presynaptic sites and facilitate acetylcholine (ACh) synthesis and/or release. For example, investigators have attempted to increase ACh availability by increasing the amounts of the precursor choline or dietary lecithin. However, increasing amounts of precursor has not been shown to alter spontaneous release of ACh (Bierkamper & Goldberg, 1979), because the rate-limiting step in the synthesis of ACh appears to be, under most conditions, a sodium-dependent, high-affinity transport system that brings choline into the nerve terminal (Simon & Kuhar, 1975). Thus, there is little reason to suppose that an increase in the amount of precursor would facilitate cholinergic function and lead to cognitive improvements in AD patients. Indeed, as a survey of the literature indicates, clinical trials with choline or lecithin have almost ubiquitously shown no significant cognitive improvement in symptoms of AD patients (Bartus, Dean, Beer, & Lippa, 1982).

One presynaptic agent that has been shown to increase ACh availability is 4-aminopyridine (4-AP). 4-aminopyridine is generally thought to facilitate the release of ACh and some other neurotransmitters by facilitating an increase in transmembrane calcium influx (Lundh & Thesleff, 1977; Vizi, van Dijk, & Foldes, 1977). A study examining the mnemomimetic effects of 4-AP in animals indicated an enhancement of 72-hour retention of a passive avoidance response when rats were injected with low doses (0.05 and 0.1 mg/kg) of 4-AP immediately after training (Haroutunian, Barnes, & Davis, 1985). Findings of Wessling and Agoston (1984) complement these animal data by indicating that 4-AP produced memory enhancement in AD patients when the drug was given over a period of 6 weeks. These data encourage future studies examining the mnemomimetic effect of 4-AP in AD patients.

Although reasonable at face value, presynaptic strategies have the difficulty of requiring an ever-decreasing number of neurons to synthesize and release more ACh than they might under normal conditions. Furthermore, because animal studies indicate that aged cholinergic neurons already have difficulty in the synthesis and release of ACh (Decker, 1987), the surviving cholinergic cells may experience further damage when required to "over-function" in response to presynaptic agents. Thus the extensive cholinergic cell loss often seen in AD patients (Davies & Maloney, 1976; Perry, Gibson et al., 1977; Perry, Perry, Blessed et al., 1977; Perry, Perry, Gibson et al., 1977), coupled with limited overlapping projections (Coyle, McKinney, Johnson, & Hedreen, 1983) and possible functional abnormalities (Decker, 1987), may preclude surviving cells from compensating functionally. This suggests that presynaptic strategies may not be clinically useful, especially in the latter stages of AD at a time when the neurodegenerative process has resulted in the greatest amount of cell loss.

Synaptic Agents

The increase in ACh produced by synaptic agents is secondary to the inhibition of cholinesterase, the enzyme that degrades ACh. Thus, by preventing the breakdown of ACh, synaptic agents increase the half-life of the neurotransmitter in the synapic cleft. Like the presynaptic strategy, this approach also suffers from the likelihood that the efficacy of the synaptic agent is still dependent on the integrity of the cholinergic neuron. However, this strategy has the advantage of preserving what is probably the important temporal and phasic relationship between pre- and post-synaptic elements of cholinergic transmission.

The prototypical synaptic agent has been the cholinesterase inhibitor, physostigmine, an agent most extensively studied in the treatment of AD. Clinical investigations using both intravenous (Blackwood & Christie, 1986; Christie, Shering, Ferguson, & Glen, 1981; Davis & Mohs, 1982) and oral (Beller, Overall, & Swann, 1985; Mohs et al., 1985; Thal, Fuld, Masur, & Sharpless, 1983) preparations have been conducted in AD patients. These studies have typically used a methodology for the administration of these agents that takes into account the considerable variability in patient response. Due to differences in the hydrolysis of physostigmine, the status of the cholinergic compounds, and sensitivity to the effect of cholinergic compounds, a careful dose titration phase (i.e., identifying an individual's "best dose") must precede any test of physostigmine or related compounds.

The general consensus that derives from studies of physostigmine is that a substantial subgroup of patients have a small, but clinically meaningful, improvement following physostigmine administration (see Hollander, Mohs, & Davis, 1986; Mohs & Davis, 1987, for summaries of findings). This improvement in cognitive function certainly does not happen in all patients, nor is it comparable to the restoration of motor function observed in Parkinson's disease patients following replacement therapy with L-Dopa. Nevertheless, synaptic agents seem clearly superior to vasodilators and nootropics. Indeed, as a consequence of the positive results reported with a longer-acting cholinesterase inhibitor, tetrahydroaminoacridine (THA) (Summers, Majovski, Marsh, Tachiki, & Kling, 1986), a 17-center trial is now underway across the United States to test the efficacy of this agent. Data recently derived from the administration of THA indicate that even better effects than those that occur from physostigmine may be possible. An important difference between physostigmine and THA, which may help account for the greater efficacy of the latter, is that THA may be less prone to hydrolysis and may give more predictable and sustained blood levels than physostigmine. But these results should be considered preliminary at best, especially because recent data indicate that THA administration may result in liver toxicity.

Enhancement of cognitive function following oral physostigmine in some, but not all, AD patients and the variability in the degree of enhancement have provided impetus to find a biochemical measurement that might differentiate treatment responders from nonresponders. Two potential candidates have been implicated. One is nocturnal plasma cortisol levels. A strong correlation between symptom improvement and increases in cortisol has been reported (Mohs et al., 1985). Since cortisol levels increase with central cholinergic stimulation (Davis et al., 1982), such a correlation suggests an association of symtpom improvement with enhanced cholinergic activity. The other potential "marker" of treatment responders is cholinesterase inhibition in the cerebrospinal fluid (CSF) in response to physostigmine administration. Increasing cholinesterase inhibition in the CSF was found to correlate with memory improvement during physostigmine treatment (Thal et al., 1983). Either of these biochemical indices may prove to be a critical screen in determining whether or not AD patients will respond to physostigmine therapy.

Postsynaptic Agents

Stimulation of postsynaptic receptors by means of cholinergic agonists has also been examined in the treatment of AD. This strategy has the advantage of being independent of the cholinergic neuron and thus unaffected by the continued degeneration of the cholinergic neuron. However, application of postsynaptic agents in treatment of AD assumes that tonic actions of drugs on receptors will be equivalent to the phasic action of the endogenous neurotransmitter ACh, which may or may not be true. Furthermore, this approach is made even more problematic by the complexity of cholinergic receptors; cholinergic receptors can be subdivided into M1, M2, M3, M4, M5, nicotinic, phosphatidyl inositol linked, and cyclic GMP linked. The differential distribution and probable functions of all these receptor subtypes are likely to have profound implications for drug development and remain unresolved to date.

Results obtained from clinical trials examining cholinergic agonists are equivocal. For example, when the muscarinic agonist, arecoline, was administered subcutaneously to AD patients, a transient improvement in memory was demonstrated (Christie et al., 1981). In contrast, oxotremorine, a muscarinic agonist that has a half-life of several hours, failed to produce memory enhancement in AD patients while inducing side effects, primarily depression and gastrointestinal distress, sufficient to warrant discontinuance of the study (Davis et al., 1987). Finally, an orally active cholinergic agonist, RS86, has yielded some improvement in cognition (cited in Mohs & Davis, 1987; Wettstein & Spiegel, 1984), but as with physostigmine, the degree of improvement was moderate. In short, it is safe to say that some agonists may produce a transient improvement in AD patients but that the consistent efficacy of any one agent has not been established.

Outcome Measures

Regardless of the pharmacological strategy employed, the development of instruments to measure cognitive and behavioral improvements is essential. In this regard, a number of symptom rating scales have been developed for the assessment of dementia (see *Psychopharmacology Bulletin*, 1988, *24*[4]). For example, the Alzheimer's Disease Assessment Scale (ADAS) was designed for use in geriatric psychopharmacological studies and measures the most frequently noted cognitive deficits in AD—including memory loss, disorientation, apraxia, and aphasia—as well as noncognitive behavioral symptoms such as mood state and the ability to perform activities of daily living. The ADAS is appropriate for use in subjects with mild to moderate dementia and can be administered in both in- and outpatient settings. Several forms of this instrument have been developed in order to minimize learning effects in studies in which this test is administered repeatedly (Mohs & Cohen, 1988; Mohs, Rossen, & Davis, 1983).

The usefulness of the ADAS has recently been demonstrated in two independent studies, one with oral physostigmine (Mohs et al., 1985), the other with the muscarinic agonist RS86 (Hollander et al., 1987). Results from these investigations indicated that improvements on the ADAS robustly correlated with the effects of these cholinergic compounds to elevate nocturnal cortisol. As mentioned earlier, because cortisol secretion is enhanced by central cholinergic stimulation (Davis et al., 1982), the elevation in cortisol secretion noted in response to physostigmine and RS86 may be interpreted to reflect central cholinomimetic activity. Therefore, the fact that improvements in ADAS scores correlate with cortisol secretion suggests that the ADAS may provide a sensitive clinical outcome measure of cognitive and behavioral changes produced by pharmacological treatments.

While the ADAS is sensitive to the deterioration in cognitive and noncognitive behaviors that may be anticipated during the course of the progression of dementing illness, this assessment scale is not sensitive to change at the end stages of AD. In order to measure the effectiveness of pharmacological agents administered to patients in the terminal stages of dementia, new scales must be developed with the capacity to reflect treatment-induced improvements and/or the retardation of symptom progression. One such attempt has been a modification of the Clinical Dementia Rating suggested by Heyman and his asociates (Heyman et al., 1987). Moreover, as treatment strategies aimed at preventing the onset or initial development of symptoms in AD patients are devised, assessment instruments that are both capable of reflecting the early stages of dementia and sensitive to change in very mild symptomatology need to be created.

Animal Models

The pace of the development of cholinomimetic agents for the treatment of AD has, by necessity, been slow. The recruitment of elderly people who can safely participate in complicated pharmacological trials has contributed significantly to this problem. Consequently, there has been a tremendous incentive to develop an "animal model" of AD. Development of an appropriate animal model of AD would not only broaden our understanding of those neural processes responsible for normal cognitive functioning but would also help identify those pathophysiological processes responsible for AD and would allow evaluation of potential therapeutic treatments. In short, development of an adequate "animal model" would provide vital information necessary for the understanding and treatment of AD (Olton & Wenk, 1987).

Behavioral Pharmacology

The most widely used approach to developing an animal model of AD is to attempt to mirror the invariable cholinergic deficit of the disease by producing lesions of the nucleus basalis of Meynert (nbM), or septal nucleus. These lesions produce a hypocholinergic state in the cortex and hippocampus, respectively, similar to what is observed in the brains of AD patients. Although a variety of methods have been employed, administration of the excitotoxin ibotenic acid directly into these cholinergic brain nuclei has been often used. When tested on a variety of experimental tasks such as passive (Haroutunian, Kanof, & Davis, 1985; Santucci, Kanof, & Haroutunian, 1989) and active avoidance (Flicker, Dean, Watkins, Fisher, & Bartus, 1983), a radial arm maze (Bartus et al., 1985), brightness discrimination and its reversal (Santucci & Haroutunian, 1989), and other tasks (see Olton & Wenk, 1987), these lesioned animals exhibit cognitive deficits centering around the inability to learn and retain new information, but are otherwise normal. If however, cholinomimetic agents (presynaptic, postsynaptic, or synaptic) are administered to these cholinergic-lesioned animals, performance is inevitably normalized (Dokla & Thal, 1988; Haroutunian et al., 1985; Mandel & Thal, 1988; Murray & Fibiger, 1985; 1986; Ridley, Murray,

Johnson, & Baker, 1986; Santucci et al., 1989). A case in point is provided by the results of an experiment examining the efficacy of physostigmine in nbM-lesioned rats (Haroutunian et al., 1985). Results from this study indicated that lesion-induced retention deficits were alleviated when rats, tested 72 hours after passive avoidance training, received posttraining injections of 0.06 mg/kg of physostigmine.

The robustness with which the cognitive deficits of hypocholinergic animals are improved by cholinomimetic drugs raises the question of why these same agents are not more efficacious in AD patients. This question has a potentially straightforward answer. Although deficits in brain acetylcholine in AD are virtually ubiquitous, AD is more than simply a cholinergic deficit. Deficiencies in the neuropeptides somatostatin (Beal et al., 1985; Davies, Katzman, & Terry, 1980; Rossor, Emson, Mountjoy, Roth, & Iversen, 1980) and corticotropin-releasing factor (CRF) (Bissette, Reynolds, Kilts, Widerlov, & Nemeroff, 1985; De Souza, Whitehouse, Kuhar, Price, & Vale, 1986; Powers et al., 1987) appear to be relatively constant across Alzheimer's patients. Additionally, variable but frequently encountered abnormalities include deficiencies in norepinephrine (Cross et al., 1981, 1983; Crow et al., 1984; Perry et al., 1981) and serotonin (Cross et al., 1983; Crow et al., 1984), which are probably secondary to degeneration in the locus coeruleus and median raphe nucleus, respectively. Thus, a more appropriate animal model of AD would add these deficits to the cholinergic abnormality.

Such "multiple lesions" studies have been initiated in our laboratory. For example, with the use of cysteamine, it has been possible to deplete selectively central somatostatin stores (Haroutunian, Mantin, Campbell, Tsuboyama, & Davis, 1987; Sagar et al., 1982). In one study (Haroutunian, Kanof, & Davis, 1989) cysteamine-induced somatostatin deficiency neither worsened the cognitive deficit produced by an nbM lesion nor affected the efficacy of a cholinergic agent to reverse the memory deficits of hypocholinergic animals. It thus appears that somatostatin deficits neither exacerbate those cognitive impairments produced by lesions of the cholinergic system nor degrade the ability of cholinomimetics to reverse these impairments.

The pattern of results is different, however, when noradrenergic lesions are combined with cholinergic lesions. Although a depletion of cortical norepinephrine of approximately 90% does not further impair the cognitive performance of animals with a cholinergic nucelus basalis abnormality, this combination of lesions completely eliminates the ability of cholinomimetics to restore animals to normal levels of memory functioning (Haroutunian, Kanof, Tsuboyama, & Davis, 1990). Thus, the addition of a noradrenergic deficit to a cholinergic (nbM) lesion blocks the beneficial effect of cholinomimetic treatment.

Obviously these results have clear implications for the treatment of AD. If a rational "first-generation" approach to the treatment of AD was the administration of cholinomimetics, a rational "second-generation" approach would be to combine cholinomimetics with adrenergic agents. This combined pharmacological approach was taken by Haroutunian, Kanof, Tsuboyama et al. (1990) in their examination of the effects of cholinergic/ noradrenergic lesions in rats. Data from this experiment indicated that exceedingly low doses of clonidine, a noradrenergic alpha-2 agonist, in conjunction with physostigmine, improved the retention performance of animals with combined lesions to a level comparable to animals with intact noradrenergic and cholinergic systems. Hence, clinicians can look forward to experimental studies in which adrenergic agents will be combined with cholinergic agents in those patients who have a minimal response to cho-linomimetics.

Neurotrophic Factors

Although there is some reason, based on animal studies, to be enthusiastic about the course of pharmacological investigations in AD, this approach is still based on reversing a series of deficits that are the sequelae of a degenerative process. It is possible to conceive of an approach that might either end or reverse that degeneration. Agents that may serve as potential therapeutic candidates in this regard are those compounds collectively known as neurotrophics (e.g., nerve growth factor—NGF, epidermal growth factor—EGF, gangliosides, etc.). The most studied neurotrophic agent is NGF. Although NGF has traditionally been associated with peripheral sym-pathetic and sensory neurons, recent research indicates that NGF may contribute significantly to the growth and maintenance of central nervous system cells, with extraordinary specificity for cholinergic neurons (Gnahn, Hefti, Heumann, Schwab, & Thoenen, 1983; Honegger & Lenoir, 1982; Martinez, Dreyfus, Jonkait, & Black, 1985; Mobley, Rutkowski, Tennekoon, Buchanan, & Johnston, 1985; Schwab, Otten, Agid, & Thoenen, 1979; Seiler & Schwab, 1984). Furthermore, other data indicate that exogenous admin-istration of NGF can facilitate neurochemical and behavioral restoration of function following cholinergic-depleting CNS lesions. For instance, partial recovery of cortical choline acetyltransferase and acetylcholinesterase defi-cits (assayed 6 weeks post lesion) has been reported when nbM-lesioned rats were periodically infused with NGF over a 2-week postoperative period (Haroutunian, Kanof, & Davis, 1986). Nerve growth factor, when admin-istered over a period of 4 weeks, has also been reported to attenuate the retrograde degeneration of transected cholinergic septohippocampal neu-rons (Hefti, 1986). Finally, repeated NGF injections have been demonstrated to produce behavioral recovery of function when animals with partial transections of the septohippocampal pathway were trained in a radial arm maze (Will & Hefti, 1985). Undoubtedly the methodologies associated with NGF administration and related compounds will improve over the next few years. As knowledge regarding this group of compounds grows, it is conceiv-able that their use could be extended to patients with AD.

Tissue Transplantation

Although tissue transplantation (graphing) techniques have been employed in amphibians, fish, and birds since the end of the nineteenth century, it has been only during the last two decades that similar procedures have been applied successfully to the central nervous systems of mammalian species (Gash, Collier, & Sladek, 1985). During the 1980s, neural transplantation has received considerable attention from brain scientists (Gash, 1984). This is evidenced by the fact that neural grafts have been employed in alleviating such diverse deficits as motor impairments produced by lesions of the dopaminergic system (Bohn, Cupit, Marciano, & Gash, 1987; Perlow et al., 1979), neuroendocrine deficiencies (Gash, Sladek, & Sladek, 1980; Gibson et al., 1984), learning impairments following cortical lesions (Labbe, Firl, Mufson, & Stein, 1983; Stein, Labbe, Attella, & Rakowsky, 1985), and thalamic anatomical abnormalities produced by lesions of the developing posterior cortex (Haun & Cunningham, 1984). The dramatic increase in the experimental analysis of neural grafting has led to the preliminary use of this technique in alleviating impairments of Parkinson's disease in humans (Madrazo et al., 1987; but see Sladek & Shoulson, 1988).

The potential therapeutic value of neural grafts has also been explored with regard to AD as indicated by the extensive literature documenting the effect of grafts into cholinergic-lesioned animals (see Gash et al., 1985). These studies establish that neurochemical, histopathological, and behavioral sequelae of a lesion-induced cholinergic deficiency can be reversed by fetal ventral forebrain cell implants. In one of the first studies of this kind, Dunnett and his colleagues reported that embryonic septal grafts implanted into the hippocampus of fimbria/fornix-lesioned rats reversed lesion-induced T-maze alternation deficits (Dunnett, Low, Iversen, Stenevi, & Bjorklund, 1982). Restitution of performance in this study was reported to accompany restoration of acetylcholinesterase staining pattern in the denervated hippocampus. Since this initial report, others have confirmed this graft-induced improvement effect in cholinergic-lesioned animals and have shown that recovery is most dramatic when the implanted tissue is integrated with the host hippocampus (Kimble, Breiller, & Stickrod, 1986).

Like the septohippocampal system, grafts of fetal origin have been reported to reverse impairments produced by lesions to the basal forebrain–frontal cortical system (e.g., Arendash, Strong, & Mouton, 1985). Two recent studies have indicated that, in rats, memory and neurochemical deficits produced by unilateral lesions of the nbM were ameliorated when fetal grafts of the cholinergic ventral forebrain were placed into the denervated neocortex (Dunnett et al., 1985; Fine, Dunnett, Bjorklund, & Iversen, 1985). Data from our own laboratory have extended these findings by demonstrating the efficacy of fetal ventral forebrain implants to reverse some of the memory and neurochemical deficits produced by bilateral lesions of the nbM (Santucci, Haroutunian, Gluck, & Davis, 1991).

Although these preliminary results with animals are encouraging, the application of fetal cell transplants as a therapeutic strategy for the treatment of AD awaits the resolution of at least two main issues: (1) the moral issue regarding the source of the donor tissue and (2) the utility of the "transplantation" strategy in general. With regard to the first point, recent research has indicated that cultured neuroblastoma cells can survive for a prolonged period of time when grafted into hippocampi of monkeys and may serve well as donor tissue (Gash, Notter, Okawara, Kraus, & Joynt, 1986). If it is subsequently shown that they are capable of reversing a variety of neurological and behavioral impairments, cultured cells may provide a practical alternative to fetal cell implants and thus eliminate the problem of "source of donor." Therefore, research aimed at creating cultured cell lines may prove to be quite useful in developing an ultimate treatment for AD. The second issue is more troublesome. The extensiveness of the histopathological abnormalities in AD (eg., multiple neurotransmitter involvement, wide neuroanatomical deficiencies, etc.) questions the ultimate utility of a "transplantation" approach to the treatment of AD. However, this should not prevent researchers from investigating neural transplants as a potential therapy for AD. Since its inception, neural grafting has been a field filled with scientific surprises and exploding in data, and its progress should be eagerly followed.

Genetics

The ultimate goal of any program on the therapeutics of AD is to prevent the disease. Prevention requires some understanding of the physiology, or at the very least, identification of risk factors. Two factors have been repeatedly shown to be associated with increased risk for AD: age, and a family history of progressive dementia. In fact, genetics may play a particularly large role in the development of AD. Fifty percent of first-degree relatives of probands with AD develop a progressive dementia by approximately age 90 (Breitner & Folstein, 1984; Mohs, Breitner, Silverman, & Davis, 1987). These data suggest that cases of AD previously referred to as "sporadic" may actually be familial cases and that potential victims in previous generations did not live long enough for the phenotypic expression of the Alzheimer's gene or genes.

These data point toward a molecular genetic approach to identify the etiology of AD and ultimately prevent AD. The abnormal histopathology, which is to say the plagues and tangles seen in the Alzheimer's brain, are a clue to the molecular genetic abnormalities that may underlie the disease. Proteins uniquely present in the Alzheimer's brain, or present in far greater quantities, can be traced back to the chromosome on which their gene is located. This information can be applied to the analysis of various families with a particular loading for AD to determine if the candidate gene cosegregates with the expression of AD. One of the most promising possibilities is for the preamyloid protein. This protein appears to derive from a gene located on chromosome 21 (St. George-Hyslop et al., 1987). Trisomy 21, or Down's syndrome, inevitably results in the histopathological (Ball & Nuttall, 1981; Crapper, Dalton, Skopitz, Scott, & Hachinshi, 1975; Ellis, McCulloch, & Corley, 1974) and neurochemical (Yates, Ritchie, Simpson, Maloney, & Gordon, 1981; Yates, Simpson, Maloney, Gordon, & Reid, 1980) changes of AD in those Down's individuals who survive into their fifth decade of life. Thus there is considerable interest in determining what may be the precise relationship between the preamyloid gene and the development of AD. If a definitive connection is made between the preamyloid gene and AD, or for that matter between any gene and AD, it is then conceivable that "genetic treatments" may be developed wherein the defective genetic material be corrected or compensated.

Admittedly, the realization of "genetic treatments" is still a number of years away. However, the rate at which such treatments may be developed can be significantly increased by the use of a much more appropriate animal model of AD using transgenic techniques. Because it is clear that genetics play a role in AD, development of such an animal model would allow investigators to examine the pathophysiology of AD and explore potential therapeutic treatments in an animal system that better reflects the human situation.

Whatever the molecular basis of AD, and the genes responsible for its familial expression, the key question from the perspective of prevention is why do some individuals manifest the disease in their early 50s and others not until their late 80s. Factors leading to the expression of the gene, when ultimately elucidated, hold the promise of prevention. Given the extraordinary human and economic cost of AD, it is therefore not surprising that this approach will receive substantial attention in the years to come.

Conclusion

This chapter outlined various past strategies that have been employed for the treatment of AD, and discussed approaches that investigators are currently pursuing. This review was not intended to be exhaustive but instead was aimed at highlighting those preclinical and clinical lines of research that hold promise for development of an effective therapy and ultimate cure for AD. It is clear that AD is a multisymptom disease that is most likely mediated by various neurochemical deficiencies, among them abnormalities of cholinergic transmission. It is also clear however that augmentation of the cholinergic system alone is not nearly an entirely effective therapeutic strategy. Animal lesion studies suggest that, for some patients, combining cholinergic and adrenergic agents may prove to be more efficacious than cholinergic therapy alone. Preliminary studies based on such a combined drug treatment strategy are, in fact, in the initial stages of investigation in our clinical laboratory. Finally, the ultimate goal of any research program dealing with the therapeutics of AD is to prevent the development and/or halt the progression of the disease to a clinically meaningful degree. Toward this end, the use of neurotrophic factors to retard or prevent the degeneration of brain cells and the use of molecular biological strategies to identify and prevent the expression of the gene(s) responsible for AD are two areas that have the potential of contributing significantly to the treatment of this disorder. Although a cure for AD is not imminent, we are encouraged by the recent progress that has been made and are confident that the lines of research presently being pursued will continue to contribute to the understanding and treatment of AD.

References

Arendash, G. W., Strong, P. N., & Mouton, P. R. (1985). Intracerebral transplantation of cholinergic neurons in a new animal model for Alzheimer's disease. In J. T. Hutton & A. D. Kenny (Eds.), *Senile dementia of the Alzheimer type* (pp. 351–376). New York: Alan R. Liss.

Ball, M. J., & Nuttall, K. (1981). Topography of neurofibrillary tangles and granulovacuoles in hippocampi of patients with Down's syndrome: Quantitative comparison with normal aging and Alzheimer's disease. *Neuropathology and Applied Neurobiology, 7*, 13–20.

Bartus, R. T., Dean, R. L., Beer, B., & Lippa, A. S. (1982). The cholinergic hypothesis of geriatric memory dysfunction. *Science, 217*, 408–417.

Bartus, R. T., Flicker, C., Dean, R. L., Pontecorvo, M., Figueiredo, J. C., & Fisher, S. K. (1985). Selective memory loss following nucleus basalis lesions: Long term behavioral recovery despite persistent cholinergic deficiencies. *Pharmacology, Biochemistry and Behavior, 23*, 125–135.

Beal, M. F., Mazurek, M. F., Tran, V. T., Chattha, G., Bird, E. D., & Martin, J. B. (1985). Reduced numbers of somatostatin receptors in the cerebral cortex in Alzheimer's disease. *Science, 229*, 289–291.

Beller, S. A., Overall, J. E., & Swann, A. C. (1985). Efficacy of oral physostigmine in primary degenerative dementia: A double-blind study of response to different dose level. *Psychopharmacology, 87*, 147–151.

Bierkamper, G. G., & Goldberg, A. M. (1979). The effect of choline on the release of acetylcholine from the neuromuscular junction. In A. Barbeau, J. H. Growden, & R. J. Wurtman (Eds.), *Nutrition and the brain* (pp. 243–251). New York: Raven Press.

Bissette, G., Reynolds, G. P., Kilts, C. D., Widerlov, E., & Nemeroff, C. B. (1985). Corticotropin-releasing factor (CRF)-like immunoreactivity in senile dementia of the Alzheimer's type. *Journal of the American Medical Association, 254*, 3067–3069.

Blackwood, D. H. R., & Christie, J. E. (1986). The effects of physostigmine on memory and auditory P300 in Alzheimer-type dementia. *Biological Psychiatry, 21*, 557–560.

Bohn, M. C., Cupit, L., Marciano, F., & Gash, D. M. (1987). Adrenal medulla grafts enhance recovery of striatal dopaminergic fibers. *Science, 237*, 913–915.

Breitner, J. C. S., & Folstein, M. F. (1984). Familial Alzheimer's dementia: A prevalent disorder with specific clinical features. *Psychological Medicine, 14*, 63–80.

Christie, J. E., Shering, A., Ferguson, J., & Glen, A. I. M. (1981). Physostigmine and arecoline: Effects on intravenous infusions in Alzheimer presenile dementia. *British Journal of Psychiatry, 138*, 45–50.

Coyle, J. T., McKinney, M., Johnson, M.V., & Hedreen, J. C. (1983). Synaptic neurochemistry of the basal forebrain cholinergic projections. *Psychopharmacology Bulletin, 19,* 441–447.

Crapper, D. R., Dalton, A. J., Skopitz, M., Scott, J. W., & Hachinshi, V. C. (1975). Alzheimer's degeneration in Down's syndrome—electrophysiological alterations and histopathological findings. *Archives of Neurology, 32,* 618–623.

Cross, A. J., Crow, T. J., Johnson, J. A., Joseph, M. H., Perry, E. K., Perry, R. H., Blessed, G., & Tomlinson, B. E. (1983). Monoamine metabolism in senile dementia of Alzheimer type. *Journal of the Neurological Sciences, 60,* 383–392.

Cross, A. J., Crow, T. J., Perry, E. K., Perry, R. H., Blessed, G., & Tomlinson, B. E. (1981). Reduced dopamine-beta-hydroxylase activity in Alzheimer's disease. *British Medical Journal, 282,* 93–94.

Crow, T. J., Cross, A. J., Cooper, S. J., Deakin, J. F. W., Ferrier, I. N., Johnson, J. A., Joseph, M. H., Owen, F., Poulter, M., Lofthouse, R., et al. (1984). Neurotransmitter receptors and monoamine metabolites in the brains of patients with Alzheimer type dementia and depression, and suicides. *Neuropharmacology, 23*(12B), 1561–1563.

Davies, P., Katzman, R., & Terry, R. D. (1980). Reduced somatostatin-like immunoreactivity in cerebral cortex from cases of Alzheimer's disease and Alzheimer senile dementia. *Nature (London), 288,* 279–280.

Davies, P., & Maloney, A. J. F. (1976). Selective loss of central cholinergic neurons in Alzheimer's disease. *Lancet, ii,* 1403.

Davis, B. M., Brown, G. M., Miller, M., Friessen, H.G., Kastin, A. J., & Davis, K. L. (1982). Effects of cholinergic stimulation on pituitary-hormone release. *Psychoneuroendocrinology, 7,* 347–354.

Davis, K. L., Hollander, E., Davidson, M., Davis, B. M., Mohs, R. C., & Horvath, T. B. (1987). Induction of depression with oxotremorine in patients with Alzheimer's disease. *American Journal of Psychiatry, 144,* 468–471.

Davis, K. L., & Mohs, R. C. (1982). Enhancement of memory process in Alzheimner's disease with multiple-dose intravenous physostigmine. *American Journal of Psychiatry, 139,* 1421–1424.

Decker, M. W. (1987). The effects of aging on hippocampal and cortical projections of the forebrain of cholinergic system. *Brain Research Reviews, 12,* 423–438.

De Souza, E. B., Whitehouse, P. J., Kuhar, M. J., Price, D. L. & Vale, W. W. (1986). Reciprocal changes in corticotropin-releasing factor (CRF)-like immunoreactivity and CRF receptors in cerebral cortex of Alzheimer's disease. *Nature, 319,* 593–595.

Dokla, C. P. J., & Thal, L. J. (1988). Effect of cholinesterase inhibitors on Morris water task behavior following lesions of the nucleus basalis magnocellularis. *Behavioral Neuroscience, 102,* 861–871.

Dunnett, S. B., Low, W. C., Iversen, S. D., Stenevi, U., & Bjorklund, A. (1982). Septal transplants restore maze learning in rats with fornix-fimbria lesions. *Brain Research, 251,* 335–348.

Dunnett, S. B., Toniol, G., Fine, A., Ryan, C. N., Bjorklund, A., & Iversen, S. D. (1985). Transplantation of embryonic ventral forebrain neurons to the neocortex of rats with lesions of nucleus basalis magnocellularis—II. Sensorimotor and learning impairments. *Neuroscience, 16,* 787–797.

Ellis, W. G., McCulloch, J. R., & Corley, C. L. (1974). Presenile dementia in Down's syndrome—ultrastructural identity with Alzheimer's disease. *Neurology, 24,* 101–106.

Fine, A., Dunnett, S. B., Bjorklund, A., & Iversen, S. D. (1985). Cholinergic ventral forebrain grafts into the neocortex improve passive avoidance memory in a rat model of Alzheimer's disease. *Proceedings of the National Academy of Sciences (U.S.A.), 82,* 5227–5229.

Flicker, C., Dean, R. L., Watkins, D, L., Fisher, S. K., & Bartus, R. T. (1983). Behavioral and neurochemical effects following neurotoxic lesions of a major cholinergic input to the cerebral cortex in the rat. *Pharmacology, Biochemistry and Behavior, 18,* 973–981.

Gash, D. M. (1984). Neural transplants in mammals. In J. R. Sladek & D. M. Gash (Eds.), *Neural transplants* (pp. 1–12). New York: Plenum Publishing Corp.

Gash, D. M., Collier, T. J., & Sladek, J. R. (1985). Neural transplantation: A review of recent developments and potential applications to the aged brain. *Neurobiology of Aging, 6,* 131–150.

Gash, D. M., Notter, M. F. D., Okawara, S. H., Kraus, A. L., & Joynt, R. J. (1986). Amitotic neuroblastoma cells used for neural implants in monkeys. *Science, 233,* 1420–1422.

Gash, D. M., Sladek, J. R., & Sladek, C. D. (1980). Functional development of grafted vasopressin neurons. *Science, 210,* 1367–1369.

Gibson, M. J., Krieger, D. T., Charlton, H. M., Zimmerman, E. A., Silverman, A. J., & Perlow, M. J. (1984). Mating and pregnancy can occur in genetically hypogonadal mice with preoptic area brain grafts. *Science, 225,* 949–951.

Gnahn, H., Hefti, F., Heumann, R., Schwab, M. E., & Thoenen, H. (1983). NGF-mediated increase of choline acetyltransferase (ChAT) in the neonatal rat forebrain: Evidence for a physiological role of NGF in the brain? *Developmental Brain Research, 9,* 45–52.

Haroutunian, V., Barnes, E., & Davis, K. L. (1985). Cholinergic modulation of memory in rats. *Psychopharmacology, 87,* 266–271.

Haroutunian, V., Kanof, P. D., & Davis, K. L. (1985). Pharmacological alleviation of cholinergic lesion induced memory deficits in rats. *Life Sciences, 37,* 945–952.

Haroutunian, V., Kanof, P. D., & Davis, K. L. (1986). Partial reversal of lesion-induced deficits in cortical cholinergic markers by nerve growth factor. *Brain Research, 386,* 397–399.

Haroutunian, V., Kanof, P. D., & Davis, K. L. (1989). Interactions of forebrain cholinergic and somatostatinergic systems in the rat. *Brain Research, 496,* 98–104.

Haroutunian, V., Kanof, P. D., Tsuboyama, G. K., & Davis, K. L. (1990). Restoration of cholinomimetic activity by clonidine in cholinergic plus noradrenergic lesioned rats. *Brain Research, 507,* 261–266.

Haroutunian, V., Mantin, R., Campbell, G. A., Tsuboyama, G. K., & Davis, K. L. (1987). Cysteamine-induced depletion of central immunoactivity: Effects of behavior, learning, memory, and brain neurochemistry. *Brain Research, 403,* 234–242.

Haun, F., & Cunningham, T. J. (1984). Cortical transplants reveal CNS trophic interactions in situ. *Developmental Brain Research, 15,* 290–294.

Hefti, F. (1986). Nerve growth factor promotes survival of septal cholinergic neurons after fimbrial transections. *Journal of Neuroscience, 6,* 2155–2162.

Heyman, A., Wilkinson, W. E., Hurwitz, B. J., Helms, M. J., Haynes, C. S., Utley, C. M., & Gwyther, L. P. (1987). Early-onset Alzheimer's disease: Clinical predictors of institutionalization and death. *Neurology, 37,* 980–984.

Hollander, E., Davidson, M., Mohs, R. C., Horvath, T. B., Davis, B. M., Zemishlany, Z., & Davis, K. L. (1987). RS86 in the treatment of Alzheimer's disease: Cognitive and biological effects. *Biological Psychiatry, 22,* 1067–1078.

Hollander, E., Mohs, R. C., & Davis, K. L. (1986). Cholinergic approaches to the treatment of Alzheimer's disease. *British Medical Bulletin, 42,* 97–100.

Honegger, P., & Lenoir, D. (1982). Nerve growth factor (NGF) stimulation of cholinergic telencephalic neuron in aggregating cell cultures. *Developmental Brain Research, 3,* 229–238.

Katzman, R. (1976). The prevalence and malignancy of Alzheimer disease. *Archives of Neurology, 33,* 217–218.

Kimble, D. P., Breiller, R., & Stickrod, G. (1986). Fetal brain implants improve maze performance in hippocampal-lesioned rats. *Brain Research, 363,* 358–363.

Labbe, R., Firl, A., Mufson, E. J., & Stein, D. G. (1983). Fetal brain transplants: Reduction of cognitive deficits in rats with frontal cortex lesions. *Science, 221,* 470–472.

Lundh, H., & Thesleff, S. (1977). The mode of action of 4-aminopyridine and guanidine on transmitter release from motor nerve terminals. *European Journal of Pharmacology, 42,* 411–412.

Madrazo, I., Drucker-Colin, R., Diaz, V., Martinez-Mata, J., Torres, C., & Becerril, J. J. (1987). Open microsurgical autograft of adrenal medulla to the right caudate nucleus in two patients with intractable Parkinson's disease. *New England Journal of Medicine, 316*, 831–834.

Mandel, R. J., & Thal, L. J. (1988). Physostigmine improves water maze performance following nucleus basalis magnocellularis lesions in rats. *Psychopharmacology, 96*, 421–425.

Martinez, H., Dreyfus, C. F., Jonkait, G. M., & Black, I. B. (1985). Nerve growth factor promotes cholinergic development in brain striatal cultures. *Proceedings of the National Academy of Sciences (U.S.A.), 82*, 7777–7781.

Mobley, W. C., Rutkowski, J. L., Tennekoon, G. I., Buchanan, K., & Johnston, M. V. (1985). Choline acetyltransferase activity in striatum of neonatal rats increased by nerve growth factor. *Science, 229*, 284–287.

Mohs, R. C., Breitner, J. C. S., Silverman, J. M., & Davis, K. L. (1987). Alzheimer's disease: Morbid risk among first-degree relative approximates 50% by 90 years of age. *Archives of General Psychiatry, 44*, 405–408.

Mohs, R. C., & Cohen, L. (1988). Alzheimer's disease assessment scale (ADAS). *Psychopharmacology Bulletin, 24*, 627–628.

Mohs, R. C., Davis, B. M., Johns, C. A., Mathe, A. A., Greenwald, B. S., Horvath, T. B., & Davis, K. L. (1985). Oral physostigmine treatment of patients with Alzheimer's disease. *American Journal of Psychiatry, 142*, 28–33.

Mohs, R. C., & Davis, K. L. (1987). The experimental pharmacology of Alzheimer's disease and related dementias. In H. Y. Meltzer (Ed.), *Psychopharmacology: The third generation of progress* (pp. 921–928). New York: Raven Press.

Mohs, R. C., Rosen, W. G., & Davis, K. L. (1983). The Alzheimer's disease assessment scale: An instrument for assessing treatment efficacy. *Psychopharmacology Bulletin, 19*, 448–450.

Murray, C. L., & Fibiger, H. C. (1985). Learning and memory deficits after lesions of the nucleus basalis magnocellularis: Reversal by physostigmine. *Neuroscience, 14*, 1025–1032.

Murray, C. L., & Fibiger, H. C. (1986). Pilocarpine and physostigmine attenuates spatial memory impairments produced by lesions of the nucleus basalis magnocellularis. *Behavioral Neuroscience, 100*, 23–32.

Olton, D. S., & Wenk, G. L. (1987). Dementia: Animal models of the cognitive impairments produced by degeneration of the basal forebrain cholinergic system. In H. Y. Meltzer (Ed.), *Psychopharmacology: The third generation of progress* (pp. 941–953). New York: Raven Press.

Perlow, M. F., Freed, W. F., Hoffer, B. J., Seiger, A., Olson, L., & Wyatt, R. J. (1979). Brain grafts reduce motor abnormalities produced by destruction of nigrostriatal dopamine system. *Science, 204*, 643–647.

Perry, E. K. (1987). Cortical neurotransmitter chemistry in Alzheimer's disease. In H. Y. Meltzer (Ed.), *Psychopharmacology: The third generation of progress* (pp. 887–895). New York: Raven Press.

Perry, E. K., Blessed, G., Tomlinson, B. E., Perry, R. H., Crow, T. J., Cross, A. J., Dockray, G. J., Dimalaine, R., & Arregui, A. (1981). Neurochemical activities in human temporal lobe related to aging and Alzheimer-type changes. *Neurobiology of Aging, 2*, 251–256.

Perry, E. K., Gibson, P. H., Blessed, G., Perry, R. H., & Tomlinson, B. E. (1977). Neurotransmitter enzyme abnormalities in senile dementia. *Journal of the Neurological Sciences, 34*, 247–265.

Perry, E. K., Perry, R. H., Blessed, G., & Tomlinson, B. E. (1977). Necropsy evidence of central cholinergic deficits in senile dementia. *Lancet, i*, 189.

Perry, E. K., Perry, R. H., Gibson, P. H., Blessed, G., & Tomlinson, B. E. (1977). A cholinergic connection between normal aging and senile dementia in the human hippocampus. *Neuroscience Letters, 6*, 85–89.

Perry, E. K., Tomlinson, B. E., Blessed, G., Bergman, P. H., Gibson, P. H., & Perry, R. H. (1978). Correlation of cholinergic abnormalities with senile plaques and mental test scores in senile dementia. *British Medical Journal, 2*, 1427–1429.

Plum, F. (1979). Dementia, an approaching epidemic. *Nature, 279*, 372–373.

Powers, R. E., Walker, L. C., De Souza, E. B., Vale, W. W., Struble, R. G., Whitehouse, P. J., & Price, D. L. (1987). Immunohistochemical study of neurons containing corticotropin-releasing factor in Alzheimer's disease. *Synapse, 1*, 405–410.

Ridley, R. M., Murray, T. K., Johnson, J. A., & Baker, H. F. (1986). Learning impairment following lesion of the basal nucleus of Meynert in the marmoset: Modification by cholinergic drugs. *Brain Research, 376*, 108–116.

Rossor, M. N., Emson, P. C., Mountjoy, C. Q., Roth, M., & Iversen, L. L. (1980). Reduced amounts of immunoreactive somatostatin in the temporal cortex in senile dementia of Alzheimer's type. *Neuroscience Letters, 20*, 373–377.

Sagar, S. M., Landry, D., Millard, W. J., Badger, T. M., Arnold, M. A., & Martin, J. B. (1982). Depletion of somatostatin-like immunoreactivity in the rat central nervous system by cysteamine. *Journal of Neuroscience, 2*, 225–231.

St. George-Hyslop, P. H., Tanzi, R. E., Polinsky, R. J., Haines, J. L., Nee, L., Watkins, P. C., Myers, R. H., Feldman, R. G., Pollen, D., Drachman, D., et al. (1987). The genetic defect causing familial Alzheimer's disease maps on chromosome 21. *Science, 235*, 885–890.

Santucci, A. C., & Haroutunian, V. (1989). Nucleus basalis lesions impair memory in rats trained on nonspatial and spatial discrimination tasks. *Physiology and Behavior, 45*, 1025–1031.

Santucci, A. C., Kanof, P. D., & Haroutunian, V. (1989). Effect of physostigmine on memory consolidation and retrieval processes in intact and nucleus basalis lesioned rats. *Psychopharmacology, 99*, 70–74.

Santucci, A. C., Kanof, P. D., & Haroutunian, V. (1991). Fetal transplant-induced restoration of spatial memory in rats with lesions of the nucleus basalis of Meynert. *Journal of Neural Transplantation and Plasticity, 2*, 65–74.

Schwab, M. E., Otten, U., Agid. Y., & Thoenen, H. (1979). Nerve growth factor (NGF) in the rat CNS: Absence of specific retrograde axonal transport and tyrosine hydroxylase induction in locus coeruleus and substantia nigra. *Brain Research, 168*, 473–483.

Seiler, M., & Schwab, M. E. (1984). Specific retrograde transport of nerve growth factor (NGF) from neocortex to nucleus basalis in the rat. *Brain Research, 300*, 33–39.

Simon, J. R., & Kuhar, M. J. (1975). Impulse-flow regulation in high affinity choline uptake in brain cholinergic nerve terminals. *Nature, 255*, 162–163.

Sladek, J. R., & Shoulson, I. (1988). Neural transplantation: A call for patience rather than patients. *Science, 240*, 1386–1388.

Stein, D. G., Labbe, R., Attella, M. J., & Rakowsky, H. A. (1985). Fetal brain tissue transplants reduce visual deficits in adult rats with bilateral lesions of the occipital cortex. *Behavioral and Neural Biology, 44*, 266–277.

Summers, W. K., Majovski, V., Marsh, G. M., Tachiki, K., & Kling, A. (1986). Oral tetrahydroaminoacridine in long-term treatment of senile dementia, Alzheimer's type. *New England Journal of Medicine, 315*, 1241–1245.

Terry, R. D., & Katzman, R. (1983). Senile dementia of the Alzheimer's type. *Annals of Neurology, 14*, 497–506.

Thal, L. J., Fuld, P. A., Masur, D. M., & Sharpless, N. S. (1983). Oral physostigmine and lecithin improve memory in Alzheimer's disease. *Annals of Neurology, 13*, 491–496.

Vizi, E. S., van Dijk, J., & Foldes, F. F. (1977). The effect of 4-aminopyridine on acetylcholine release. *Journal of Neural Transmission, 41*, 265–274.

Wessling, H., & Agoston, S. (1984). Effects of 4-aminopyridine in elderly patients with Alzheimer's disease. *New England Journal of Medicine, 310*, 988–989.

Wettstein, A., & Spiegel, R. (1984). Clinical trials with the cholinergic drug RS 86 in Alzheimer's disease (AD) and senile dementia of the Alzheimer's type (SDAT). *Psychopharmacology, 84*, 572–573.

Will, B., & Hefti, F. (1985). Behavioral and neurochemical effects of chronic intraventricular injections of nerve growth factor in adult rats with fimbria lesions. *Brain Research*, *17*, 17–24.

Yates, C. M., Ritchie, I. M., Simpson, J., Maloney, A. F. J., & Gordon, A. (1981). Noradrenaline in Alzheimer-type dementia and Down syndrome. *Lancet*, *ii*, 39–40.

Yates, C. M., Simpson, J., Maloney, A. F. J., Gordon, A., & Reid, A. H. (1980). Alzheimer-like cholinergic deficiency in Down syndrome. *Lancet*, *ii*, 979.

Commentary on Part IV

In Chapter 19, Lezak gives a rationale for the use of analytic neuropsychological instruments to diagnose cognitive deficits. The challenge for a clinical neuropsychologist is to gather, organize, and synthesize a wealth of behavioral data available on a patient, which must then be used to determine whether the patient is impaired and, if so, what determines the impaired functioning. Lezak describes some of the difficulties involved in making clinical evaluations. For example, in determining the presence and nature of a deficit, one frequent problem is relating the current performance of a patient to his or her previous history. Neuropsychologists often begin by using standardized psychometric tests of skills and abilities, for which a large normative data base is available. By evaluating patients using such standardized instruments, one often assembles a profile that reveals islands of relatively spared functioning and other areas of relatively impaired functioning. The challenge for the clinician is to make appropriate inferences about the functional significance of such a profile for a given patient.

There are many practical questions that a clinical neuropsychologist is asked to answer. They all concern an evaluation of the extent of a cognitive dysfunction, how it might be expressed, or the determinants of the dysfunction. Is an individual with severely compromised reasoning ability and communication skills competent to stand trial? Should a child with a learning disability be taught in a different setting than other children? Does the nature of a learning disability as delineated by the pattern of neuropsychological deficits and residual capacities provide etiological indications (e.g., exposure to an environmental toxin, such as lead, during development)? Is a cognitively impaired parent capable of providing appropriate care for his or her children? Can a patient with significant impairments in reasoning and judgment survive outside a protective environment and, if so, what support would be necessary to make this possible? These are some of the questions that may rely heavily, if not exclusively, on normative data. The clinician's skill is in the integration of such data across a wide number of domains in the context of the patient's life history and present circumstances.

484

Lezak, however, points out a number of limitations in the applicability of normative data to an evaluation of the patient's performance. Normative data are not going to be helpful in determining whether a person who is incapable of playing the piano has a deficit; for most humans this is not considered an impairment. But if the individual was once a concert pianist, then clearly it is noteworthy. This behavior, which might be considered a deficit, may occur for a number of very different reasons. The patient may be depressed and no longer interested in playing the piano. Alternatively, a specific brain lesion may have altered the subject's ability to read music, hear particular sounds, or perform complex motor sequences. Normative data, an identification of qualitative indicators of deficit, and the patient's psychosocial history combine to provide a meaningful description of the patient's neuropsychological status. This description, in turn, may contribute to a diagnostic conclusion, may enable the patient and such interested persons as family members and clinical staff to plan for patient care, and may form the basis for treatment strategies that manage or remediate cognitive deficits.

Milberg and Albert discuss the evidence for and against the localization of various cognitive functions. Rather than assume that the modular organization of the brain reflects the modular organization of a function, they ask the question: Is there anything about the neural implementation of a particular function that one can deduce from the psychological demands inherent in that function? Their emphasis is slightly different from that of most cognitive neuropsychologists, who assume that separate neural structures directly correspond to the boxes drawn by psychological theory. Milberg and Albert argue that these boxes may not always exist in discrete neural locations. They begin by considering the effect of lesions in the left and right hemispheres and point out that lesions in the left hemisphere seem to produce more predictable and discrete changes in cognition. They note that different cognitive domains can share the same neural architecture and that therefore it is not surprising that damage to one brain area can produce deficits in a number of different domains. For example, patients with Gerstmann's syndrome, which results from a lesion of the left parieto-occipital region, exhibit deficits in writing, left-right discrimination, finger recognition, and arithmetic computation. The failure to find a unifying component to these associated deficits makes it likely that functionally independent systems share a common architecture. Milberg and Albert point out that this idea has been de-emphasized in current neuropsychology. Their study of the association of cognitive functions complements the analysis of Weiskrantz on cognitive dissociations in Part II.

Milberg and Albert also consider how speed of processing can be used to determine whether functions are localized or distributed. They argue that functions that require rapid execution are more localized than slower processes. Support for this idea comes from a variety of sources, including their own demonstration of a correlation between the estimated speed of various functions and the rated degree of localized representation of that function in the brain. For example, the comprehension of a joke is a slow process probably diffusely represented in the brain, whereas the ability to convert a grapheme to a phoneme is very rapid and highly localizable. Milberg and Albert's discussion has close parallels with that of Moscovitch and Umilta in Part II on modularity. However, Moscovitch and Umilta discuss more "what" is implemented, whereas Milberg and Albert consider "how" it is implemented (see Part I). Milberg and Albert also discuss the difference between hardware and software modules, a distinction that is often difficult to state clearly. Although the debate over what is hard and what is soft may be stimulating, it may be no more resolvable than the debate over the parallel distinction between nature and nurture.

Rapp and Caramazza provide a framework for considering a functional architecture of cognition. They argue that the most powerful unit of analysis for understanding performance is at the level of the functional components that make up some cognitive system. Therefore, a careful analysis of the functional components of a task will be considerably more useful in understanding how some cognitive system works than will the study of potential neuroanatomic substrates. Rapp and Caramazza see neurobiological approaches as complementary to their functional analysis of behavior but note that attempts to map brain anatomy and cognitive architecture have not yet been particularly successful. They provide evidence that the performance of brain-damaged patients allows an examination of the functional components of cognition.

Like Milberg and Albert, Rapp and Caramazza, believe that an analysis of impairments that are associated with one another can be as valuable as demonstrations of functional dissociations in helping define cognitive architecture. For example, the performance of their patient K. E. across a wide range of tasks was used to support the presence of a single semantic system. This patient made semantic errors in writing, comprehension, reading, and tactile naming. These errors were comparable across modalities of input or output. An analysis of the types of errors that patients make can be a useful source of information about cognitive mechanisms. Finally, the authors stress the value of single-case studies for understanding cognitive mechanisms. In particular, they argue that grouping patients according to the site of a lesion, or their performance on one or more tasks, has a number of pitfalls. The assumption that a lesion will produce identical effects in all patients is highly questionable. Because superficially similar performance on a number of tasks may result from different underlying impairments, the fact that two patients perform similarly on a number of tasks does not provide a sufficient basis for assuming that the patients are members of a homogeneous group or type. In short, a detailed analysis of the cognitive functioning in a single patient can tell us an enormous amount about cognition. This approach using single-subject designs with additional subjects serving as replications has been used extensively by psychophysicists and clinical neurologists.

Siegel argues convincingly for the parallels between the development of tolerance to a drug effect and associative learning as described by Pavlovian conditioning. He begins with an analysis of associative learning, which he describes in terms of feedback systems. The basic concept behind his feedback and feedforward processes is homeostasis. He postulates the presence of certain "misalignment detectors" that register deviations from the steady state. The system then builds mechanisms to predict these deviations (i.e., learns). How this takes place is discussed in terms of conditioning, and the formulation developed is neither behavior nor drug specific. For example, Siegel is able to discuss the tolerance to morphine-induced analgesia using the same model he uses to discuss ethanol-induced hypothermia. Furthermore, the model can be applied directly to an important clinical problem—the treatment of drug dependence. For instance, it predicts that withdrawal symptoms will be more severe in contexts associated with drug administration. Therefore, it may be important to extinguish the conditioned response mediating this effect by exposing addicts to drug-associated cues under clinical supervision. It may be difficult to do this in the exact setting in which drugs such as opiates are self-administered. However, the graded presentation of drug paraphernalia to patients without actual drug administration may be a valuable part of any drug treatment program. Siegel's chapter raises the issue of how broadly we wish to define learning or the plasticity of some system. For example, does the response of the immune system to some biological stimulus also constitute an example of learning and memory?

Siegel presents a high-level (psychological) account of drug tolerance without the use of neurobiology, but a number of the drug responses that Siegel discusses have been well characterized neurobiologically. For example, the analgesic response to morphine is mediated by mu-opiate receptors (Neil & Terenius, 1986). Moreover, various biological explanations have been proposed to account for the development of tolerance to morphine's effects (e.g., Nestler, Erdos, Terwilliger, Duman, & Tallman, 1989). As in Part I, we wish to emphasize that different levels of analysis of phenomena such as drug tolerance should not be considered inconsistent with one another and hopefully will be integrated in the coming years.

In Chapter 23, Gur and Gur give an overview of the techniques currently available for providing an in vivo picture of the brain. Some techniques give information about brain structure, whereas others provide a picture of brain function. The neuroimaging techniques have been around for decades, and the technology has been built on our knowledge of computer science and of the neuroanatomy and neurophysiology of the brain. It is very understandable why so many researchers are drawn to these methods, which allow an in vivo representation of biochemical and physiological activity in three-dimensional space. The imaging techniques can provide a direct link between behavior and brain activity in both normal and impaired individuals.

The most frequently used imaging systems are those involving electrical recording from the scalp of activity that originates in various areas of the brain. These methods are now being commonly used in cognitive studies and, because they are noninvasive, have the advantage of allowing multiple measurements in the same subject.

By using radiolabeled isotopes, biochemical and physiological activity in the brain can be mapped using positron emission tomography (PET). In recent years the developers of imaging programs have emphasized pharmacokinetic and hard- and software engineering. For this reason, the tools available and the resolution of the method have improved considerably. Diagnostic category has been the most common treatment variable in studies using these techniques; however, it might be argued that, because relatively little attention has been paid to what the subject is doing at the time of testing, these techniques have not had a major impact on the cognitive aspects of neuroscience. Many of the behavioral results described by Gur and Gur essentially confirm data obtained using different methods. Yet if imaging methods are used in conjunction with carefully designed tasks for isolating components of cognition (e.g., like those described by Rapp and Caramazza), then they will be able to provide cognitive neuroscientists with powerful tools for exploring cognition.

Not surprisingly, imaging techniques have often been used to examine brain pathology in various neuropsychiatric disorders such as Alzheimer's disease and schizophrenia, and more recently in disturbances in affect such as anxiety disorders. Here, too, the data have been of modest value and have tended to confirm what we already knew about these disorders from other sources rather than provide new insights. In order to reap the maximum benefit from the techniques described by Gur and Gur, it will be important to integrate the data from the various in vivo methods. It would be useful to compare in the same subjects data from PET, evoked responses, MRI, and other techniques. It will also be important to compare the data obtained from in vivo studies with those obtained in vitro. Finally, as we noted earlier, none of the techniques alone will make a major contribution to cognitive neuroscience without a systematic examination of behavior.

The final two chapters in the section consider Alzheimer's disease (AD), which is perhaps the most common and dramatic clinical example of cognitive failure. Over the last decade, a great deal has been learned of the neuropathology of this disease, and Palmer and Bowen provide an overview of this research. What should be clear to the reader is that Alzheimer's disease does not affect only one brain system. A little over a decade ago the key features of the neuroanatomy and neurochemistry of the disease appeared to be the presence of fibrillary tangles and plaques, and a cholinergic deficit. The chapter by Palmer and Bowen illustrates that many more systems are changed in Alzheimer patients.

The greatest amount of information has come from postmortem studies of the brains of patients with Alzheimer's disease. Palmer and Bowen begin by pointing out that there are a number of difficulties in making use of such data; nevertheless, many valuable findings have been reported. There is nerve loss from the cortex. Of the two types of cortical neurons, it is the pyramidal cells that appear to be more involved in the disease. Moreover, it may be changes in these cells that are responsible for the formation of plaques and tangles. Changes in the functioning of various neurochemical systems have also been reported. For example, alterations in GABAergic, catecholaminergic, and in particular serotonergic neurons have all been noted. As already pointed out, changes in the cholinergic system have received the most extensive attention. Finally, alterations in the function of various neuropeptides, such as somatostatin, have also been observed.

Palmer and Bowen also discuss methods other than postmortem studies for obtaining information on the neurochemistry of Alzheimer's disease. Brain biopsies and cerebrospinal fluid obtained from patients can also be used to obtain convergent descriptions of the disease. These data may be particularly important as our ability to diagnose and treat the disease is improved. The response of patients to various drugs can be a valuable source of information as well. For example, patients may improve following treatment with some specific neuropharmacological agent. Alternatively, their sensitization to the effects of some other drug (e.g., the cholinergic antagonist scopolamine) may give information about the integrity of some neurotransmitter system (such as the cholinergic system, see Sunderland et al., 1987).

To make effective use of the neurobiological data that Palmer and Bowen systematically provide, it would be most useful to compare the changes seen in AD with those observed in other disorders such as Huntington's disease and Korsakoff's disease. Such studies are likely to enhance our understanding of the neurochemistry of cognition. The choice of groups should be based on the cognitive features of the various disorders.

In the final chapter of this section, Santucci, Haroutunian, Bierer, and Davis discuss the strategies available for the treatment of Alzheimer's disease, some of which are developed from neuropathological findings like those described by Palmer and Bowen. Santucci et al. describe four different approaches to treating AD, beginning with pharmacological treatments. Not surprisingly, many of the drug trials conducted for the treatment of AD use agents that affect the cholinergic nervous system. Such drugs include those that increase presynaptic neuronal functioning (such as choline), agents that increase the availability of acetylcholine in the synapse (such as acetylcholinesterase inhibitors—e.g., tetrahydroaminoacridine), and agents that stimulate the postsynaptic neuron (such as arecoline). To date, none of these have been particularly effective in reversing the cognitive deficits of the disease or altering the course of the disease. In addition, a variety of other drugs have also been tested. For example, nootropics have been widely tested with little scientific rationale and have shown little (if any) efficacy. Our current knowledge of the postmortem neuropathology of AD suggests that reversing the cognitive impairments associated with disease by pharmacological means is not realistic, although results obtained by studying living patients lead to a less pessimistic conclusion (Bowen, 1990). A multitransmitter approach may be more appropriate than attempts that alter the functional status of a single neurotransmitter system.

The second approach argues that, since Alzheimer's disease is a degnerative disorder, treatments that alter the production of nerve growth factors might have therapeutic value. While the administration of nerve growth factors to animals has produced promising data, the technology for using these agents as a treatment option in humans has not yet been developed.

A third strategy involves the use of tissue transplantation, which has been used in lower animals for a long time. In the instance of Parkinson's disease (Madrazo et al., 1987), the technology has been applied to humans. It has not yet been used in humans with Alzheimer's disease. Santucci et al. point out that research is still needed to solve problems associated with donor–recipient compatibility. One solution involves the development of cultured cell lines that can be used for transplantation.

The final approach involves the prevention, rather than the treatment, of Alzheimer's disease. A knowledge of the genetics of the disease and risk factors may make it possible to intervene in individuals with a high risk for developing the disease. As already noted, pharmacological challenges may be helpful in identifying such individuals.

The chapters by Palmer and Bowen and by Santucci et al. focus little attention on behavior. This raises the question of whether a detailed behavioral analysis of Alzheimer's patients is necessary to diagnose and develop strategies for treating the disease. We return to this point in the following discussion.

We shall focus on two themes in the remainder of this discussion. The first considers whether our knowledge of cognition at any level of analysis has enhanced our understanding of clinical phenomena. Specifically, has our knowledge of cognition helped in categorizing patients, in predicting the course of their disease, and in treating them? Second, we consider in more detail whether studying clinical phenomena can enhance our knowledge of cognitive neuroscience.

How do we diagnose cognitive impairments? Prototypic examples of impaired cognition are those seen in dementia associated with Alzheimer's disease, amnesia in Korsakoff's disease, impairments in attention in attention deficit disorder, the loss of specific knowledge associated with highly discrete brain lesions (e.g., Warrington & McCarthy, 1987), specific avoidance behaviors seen in simple phobias, and dissociations in memory seen in multiple personality disorder (Putnam, 1989). All these disorders are diagnosed primarily on the basis of behavior, and diagnostic formulations include an analysis of past as well as current behavior. In other cases, such as Huntington's disease, a diagnosis may be made primarily on the basis of a biological determinant, such as autosomal dominant genetic transmission, along with symptom expression.

The behavioral methods used in diagnosis are based on both the psychometric tradition and a detailed clinical evaluation such as that described by Lezak. These methods are not generally derived from the latest models or findings in experimental psychology. The types of dissociations and models described by the contributors in Part II, and by both Milberg and Albert and Rapp and Caramazza in this section have so far had little impact on diagnosis. However, these models are likely to influence evaluation procedures in the future. Just as none of the current cognitive models form a basis for diagnosing disordered cognition, there is no standard neurobiological procedure currently available for establishing a diagnosis of a neuropsychiatric disorder such as depression or a neurological disorder such as Alzheimer's disease. However, Gur and Gur, Palmer and Bowen, and Santucci et al. do discuss some of the many procedures that may ultimately be useful in diagnosing Alzheimer's disease and other neuropsychiatric disorders.

In fact, there are parallels between the approach described by Rapp and Caramazza to characterize their subjects and the general clinical approach for diagnosing a patient. Both are interested in what processes might account for a failure in cognitive function. To be able to function as an effective clinician, one must have some model or scheme that can account for the alternative ways that cognition may be altered. For example, in trying to understand why a patient has a memory impairment, the clinician must be able to evaluate the likelihood that impaired functioning may be due to one or more of a number of factors. These might include nighttime sleeping medication, the presence of depression, a previous history of impaired functioning due to poor education, and various stressors such as loss of employment or death of a spouse. Such diagnostic issues illustrate the importance of considering factors that modulate cognitive function (see Part III).

At one level, diagnosis may be seen as a method of classifying cognitive disorders in a way rather similar to the classification of cognitive processes discussed in Part II. Both require a detailed and organized description of behavior. As we noted in Part II, classification can take place without a consideration of mechanism, although this focus has played an increasing role in the evolution of diagnostic schemes as we have learned more about brain–behavior relationships. For example, DSM-III-R (American Psychiatric Association, 1987) has a greater emphasis on mechanism than DSM-I. An appreciation of mechanism is clearly important for developing treatments.

Part of the difficulty in diagnosing neuropsychiatric disorders is that patients categorized under a single diagnosis (e.g., Alzheimer's disease) may in fact represent a heterogeneous population. That is, there may be different subtypes of the disease with different neurobiological and behavioral characteristics. Interdisciplinary study in which neuropathological, bio-

chemical, and genetic analyses are added to the detailed analysis of behavior should allow us to identify and characterize different forms of these
diseases. Perhaps a genetic marker, an altered level of a neurotransmitter
metabolite, or neuropathological findings available from a brain biopsy
will be useful in further classifying psychiatric disorders and will contribute
to the development of later versions of DSM-III-R (American Psychiatric Association, 1987).

An early diagnosis of a disorder is often difficult, although it may be very
important for successful treatment. In particular, a number of the treatments for Alzheimer's disease discussed by Santucci et al. might slow the
course of the disease and therefore have greater benefit the earlier the patient is diagnosed.

The diagnosis and treatment of cognitive disorders are intimately connected. Treatments are based on accurate clinical diagnoses, and of course,
that is one of the primary reasons for making a diagnosis in the first place. It
should be clear from the different contributions to this volume that a variety of approaches might prove useful in treating cognitive impairments.
Part I discussed in detail how accounts of cognitive behavior could be
provided at levels ranging from the molecular to the psychological. These
different levels of analysis have led to the development of different treatments for cognitive impairments.

In Table A we have selected a variety of disorders that illustrate how treatments derived from some of these different levels of analysis can be useful.
The examples were chosen to reflect disorders whose treatment ranges
from exclusively biological manipulations through disorders that are
treated with both biological and behavioral methods to disorders most
effectively treated by behavioral means. It should be noted, as was discussed
in Part I, that a complete description of a disorder at one level of analysis
does not imply that the disorder cannot also be completely described at another level.

We wish to emphasize that Table A is intended to be illustrative—it represents our interpretation of the strategies we expect to be the most effective
for treating the disorders listed in the coming years. In some cases it reflects
current clinical practice, where this has been shown to have a major impact
on a disorder. In other cases current treatments are minimally effective, and
our rating is based primarily on future prospects. We have also included in
the table an approximation of the extent to which each disorder is associated with a localized neurological dysfunction (after Milberg and Albert)
and a similar approximation of whether the behavioral impairment is general or highly specific. The rating scale we used for characterizing treatments ranges from 1, which corresponds to situations in which treatment is
almost exclusively biological (e.g., drug treatments, brain tissue transplantation), to 5, which corresponds to situations in which the treatment is almost
exclusively behavioral.

Table A. Treatments, Neuropathologies, and Behaviors
Associated with Various Clinical Disorders

Diagnosis	Treatment[a]	Neuropathology[b]	Abnormal Behavior[c]
Alzheimer's disease	1	2	1
Huntington's disease	1	1	1
Korsakoff's disease	1	3	3
Schizophrenia	2	?	2
Aphasia	2	5	5
Attention deficit disorder	3	?	3
Drug abuse	3	?	4
Generalized anxiety disorder	3	?	3
Simple phobia	4	?	5
Multiple personality disorder	5	?	4

[a]Ranging from 1 = exclusively biological to 5 = exclusively behavioral.
[b]Ranging from 1 = very diffuse to 5 = very localized pathology.
[c]Ranging from 1 = very varied to 5 = very focused abnormal behavior.

Alzheimer's disease and Huntington's disease are examples of biologically determined diseases. For this reason the strategies being used to treat
these diseases are biologically based. As was discussed earlier, current clinical trials involve biologically based treatments that do not provide cures, although they may have an impact on some symnptoms and may slow the
course of these diseases. However, the most common clinical treatment in
current use involves the behavioral management of the patient. This reflects the sad reality of our current state of knowledge in the development
of biological treatments.

Korsakoff's disease and Parkinson's disease also have clear and well-
understood neuropathologies, and there is some appreciation of their
etiologies. In the case of Parkinson's disease, the treatment of the disorder
has revolved around the finding of drastic reductions in dopamine concentrations in the substantia nigra of Parkinsonian patients. This finding
has prompted the use of dopamine precursors to reverse the neurotransmitter deficiency. While this treatment has been useful in temporarily helping
Parkinsonian patients, it has not cured the disease. The relative success of altering the function of a single neurotransmitter in Parkinson's disease,
however, has had a major impact on scientific thought about other disorders. For example, the dopamine theory of schizophrenia and the cholinergic hypothesis in Alzheimer's disease have been the sources of almost
all treatment strategies, but their yield might be considered disappointing.
It might be argued that over the last 30 years there have been few, if any,
breakthroughs in clinical psychopharmacology because we have tended to
think in terms of one neurotransmitter as the key to each disease. After
reading the chapter by Palmer and Bowen, which outlines the diverse nature of the neuropathology of this disease, one would expect the effects of

cholinergic manipulation in Alzheimer's disease to be limited. Besides the use of the dopamine precursor L-DOPA, another biological manipulation, brain tissue transplantation, has begun to be used to treat Parkinson's disease. The logic of this approach is to replace a part of the brain that has been identified as being impaired. One might speculate that such an approach may eventually be used to treat disorders with more focused lesions, perhaps like those described by Warrington, and McCarthy (1987) or by Rapp and Caramazza. However, at present, such treatments are not available; instead, patients often receive behavioral training that helps them to relearn skills and develop new strategies to compensate for their disabilities.

We have already mentioned the dopamine hypothesis of schizophrenia, and in fact the administration of dopamine antagonists is the most widely used treatment of this disease. These drugs have had a major impact on the time patients spend hospitalized and have effectively attenuated a number of the psychotic symptoms associated with the disease. However, as in Parkinson's disease and Alzheimer's, drugs have not proved to be cures. We do not wish to imply that a cure for a disease is the only worthwhile goal. A drug (such as insulin) can be very valuable in treating a disorder (such as diabetes) successfully without curing it. However, we feel that the management of schizophrenia is currently much less successful than that of diabetes.

Numerous factors contribute to the development of attention deficit disorder and generalized anxiety disorder. Both biological and behavioral determinants have been proposed, and for this reason, both behavioral and biological manipulations have been useful in treating these disorders. Similarly, as was noted in the discussion of Siegel's chapter, drug abuse has been discussed at both biological and psychological levels of analysis, and treatments have been developed from both perspectives.

As we discussed at the end of Part III, simple phobias appear to be most successfully treated behaviorally. Most theorists would also argue that the development of simple phobias is most easily understood at a behavioral level of analysis. This is also the case for multiple personality disorder. Here a key symptom is memory dissociation of autobiographical experience. These dissociations occur as a response to trauma, such as sexual abuse in early childhood. Traditional psychotherapeutic methods are used to treat these patients with the aim of integrating the different personalities.

So far we have considered methods of treating cognitive dysfunctions in patient populations. Another potentially useful application of our knowledge of cognitive neuroscience would be to enhance the cognitive functions of nonpatient populations. Such subjects might exhibit some chronic impairment in cognitive function (as seen in the elderly) or a relatively short-lived impairment like that seen under stress or sleep deprivation conditions, or they may be unimpaired normals. Certainly, there has been enormous investment in the pharmaceutical industry in developing cognitive enhancers, particularly for target populations such as the nominally healthy elderly.

Having considered the contribution of cognitive neuroscience to our understanding of clinical phenomena, we now examine whether clinical phenomena have taught us anything about the psychobiology of cognition. We begin by considering the value of single-patient studies. Rapp and Caramazza argue persuasively for the value of such studies. Historically, there are many examples where the study of a single patient has yielded an enormous amount of information about cognition. Classic examples include Broca's (1861) studies of language, the emotional changes that accompanied the accident of Phineas Gage (Harlow, 1865), Fritsch and Hitzig's (1870) studies of motor function, Milner's (1966) studies of memory in H.M., and Luria's (1973) clinical studies. Just as the study of patients with highly specific cognitive impairments has significantly added to our knowledge of cognitive neuroscience, studies of other individuals with striking abilities may also prove useful. Examples include Luria's (1968) mnemonist and some of the remarkable idiots savants (Scheerer, Rothmann, & Goldstein, 1945; Horwitz, Kestenbaum, Person, & Jarvik, 1965; Viscott, 1970). We have, perhaps, learned more about cognition from single-case studies than from studying groups of patients. In fact, the study of groups of patients has often been built on the careful and detailed analysis of single patients. For example, Alzheimer used such an analysis of a progressive dementing patient to subsequently define the disease that bears his name. While it is a clinically important issue to be able to predict how patients as a group (i.e., with a given diagnosis) will respond to some treatment, Rapp and Caramazza point out the value of considering individual data and some potential shortcomings of group studies in developing theories of underlying cognitive structure. It may also be noted that in experimental psychology, the single subject has served as the foundation of various theories of sensory processing, perception, attention, and learning and memory (e.g., Ebbinghaus, 1885; Wundt, 1874).

Clinical populations have been a rich source of information about brain–behavior relationships. It has been argued that brain abnormalities observed in a detailed postmortem analysis of a patient may be related to the behavioral impairments observed in the patient. If a series of patients show similar abnormalities and similar behavioral impairments, then the brain–behavior link is strengthened. There are many examples of such analyses dating back to Broca (1861). In contrast, the study of the behavior of nominally normal subjects together with postmortem analyses has been of value only in providing normative data against which to compare data from impaired subjects. The study of brain–behavior relationships in normals, however, is now proceeding more rapidly with the development of the techniques described by Gur and Gur.

In addition to providing cognitive scientists with new insights, clinical data have also produced findings that confirm models and theories derived from basic experimental studies of cognition. Experimental studies in normals often produce findings much less striking than those seen in patients. It may be noted that even those working with artificial neuronal networks are interested in clinical phenomena. For example, McClelland and Rumelhart (1986) have attempted to model clinically observed amnesia using parallel distributed networks.

In conclusion, the study of basic cognitive neuroscience has helped our understanding of clinical phenomena. Likewise, the study of clinical phenomena has greatly added to our knowledge of cognitive neuroscience. The demonstration of clear reproducible dissociations and associations in cognition from the study of single patients has been recognized as extraordinarily valuable to cognitive scientists. While there is considerable evidence of integration in the development of treatments of cognitive dysfunction, a greater interaction between basic and clinical scientists might be particularly useful in the development of more powerful diagnostic schemes.

References

American Psychiatric Association (1987). *Diagnostic and statistical manual of mental disorders (DSH-III-R)*. (3rd ed., rev.). Washington, DC: American Psychiatric Association.

Bowen, D. M. (1990). Treatment of Alzheimer's disease: Molecular pathology versus neurotransmitter based therapy. *British Journal of Psychiatry, 157*, 327–330.

Broca, P. (1861). Remarques sur le siège de la faculté du langage articule, suives d'une observation d'aphémie. *Bulletin et Mémoires de la Societé anatomique de Paris, 2*, 330–357.

Ebbinghaus, H. (1885). *Uber das Gedachtnis* Leipzig: Dunker. H. Ruyer & C. E. Bussenuis (Trans.). *Memory*. New York: Teachers College Press, 1913.

Fritsch, G., & Hitzig, E. (1870). Ueber die elektrische Erregbarkeit des Grosshirns. Reichert und Du Bois—*Reymond's Arch. fur Anatomie und Physiologie*, 300–332.

Harlow (1865). Phineas Gage. *New England Journal of Medicine, 2*, 327–346.

Horwitz, W. A., Kestenbaum, C., Person, E., & Jarvik, L. (1965). Identical twin—"idiot savants"—calendar calculators. *American Journal of Psychiatry, 121*, 1075–1079.

Luria, A. R. (1968). *The mind of a mnemonist*. Cambridge, MA: Harvard University Press.

Luria, A. R. (1973). *The working brain*. New York: Basic Books.

Madrazo, I., Drucker-Colin, R., Diaza, V., Martinez-Mata, J., Torres, C., & Becerril, J. J. (1987). Open microsurgical autograft of adrenal medulla to the right caudate nucleus in two patients with intractable Parkinson's disease. *New England Journal of Medicine, 316*, 831–834.

McClelland, J. L., & Rumelhart, D. E. (1986). Amnesia and distributed memory. In J. L. McClelland & D. E. Rumelhart (Eds.), *Parallel distributed processing* (Vol. 2 pp. 503–527). Cambridge, MA: MIT Press.

Milner, B. (1966). Amnesia following operation on the temporal lobes. In C. M. W. Whitty & O. L. Zangwill (Eds.), *Amnesia*. London: Butterworth.

Neil, A., & Terenius, L. (1986). Receptor mechanisms for nociception. *International Anesthesiology Clinics, 24*(2), 1–15.

Nestler, E. J., Erdos, J. J., Terwilliger, R., Duman, R. S., & Tallman, J. F. (1989). Regulation of G-proteins by chronic morphine in the rat locus coeruleus. *Brain Research, 476*, 230–239.

Putnam, F. W. (1989). Diagnosis and treatment of multiple personality disorder. New York: Guilford Press.

Scheerer, M., Rothmann, E., & Goldstein, K. (1945). A case of "idiot savant": An experimental study of personality organization. *Psychology Monographs*, *269*, 1–61.

Sunderland, T., Tariot, P., Cohen, R. M., Weingartner, H., Muller, E. A., & Murphy, D. L. (1987). Anticholinergic sensitivity in patients with dementia of an Alzheimer type and age-matched controls. *Archives of General Psychiatry*, *44*, 418–426.

Viscott, D. S. (1970). A musical idiot savant. *Psychiatry*, *33*, 494–515.

Warrington, E. K., & McCarthy, R. A. (1987). Categories of knowledge. Further fractionations and an attempted integration. *Brain*, *110*, 1273–1296.

Wundt, W. (1874). *Grundzuger der physiologischen Psychologie*. Leipzig: Engelmann. E. Titchner (Trans.). New York: Macmillan. 1904. 2.

Concluding Remarks

We hope the reader has obtained some sense of the disciplines that together provide a picture of the psychobiology of cognition. All the contributors, as representatives of these disciplines, have presented positions that have added to our understanding of cognitive neuroscience. They present exciting findings that highlight how much has been learned in recent years about brain–behavior relationships.

It should also be clear to the reader that it remains difficult to integrate all this research to form a single coherent picture of the mind. Attaining this ultimate goal of cognitive neuroscience will require continued efforts at interdisciplinary communication. Many argue that it will never be possible to reduce mental states to neurobiological states or, conversely, to deduce mental states from neurobiological states. Churchland (1986) provides an extensive discussion of these positions. We believe, however, as do most of the contributors, that cognition will eventually be understood in terms of neurobiological determinants. Even if this agenda is never completely fulfilled, many fascinating observations about brain and behavior have emerged and will continue to emerge in the attempt.

Most researchers in cognitive neuroscience would readily agree on some of the key questions that remain to be answered. How do we transform sensory information into experience? How do we learn and remember? How do we process feelings and emotions? To what extent does awareness or consciousness of an experience modify our appreciation and memory of the experience? In order to answer these questions, we need both conceptual schemes for formulating experiments and the technical support that allows relevant data to be collected. However, one of the problems in having available a growing number of powerful tools for exploring brain structure and brain processes is that it is easy to end up with assay-driven research, where measurement can be so extensive that we are overwhelmed with data but have no system for organizing it. Clearly, well-developed conceptual schemes, such as those provided by a number of the contributors, are crucial for progress in cognitive neuroscience.

Our knowledge of cognitive neuroscience has been, and will increasingly be, applied to improving our society. In Part IV methods of improving impaired cognitive functioning were discussed at some length. If one considers the incidence of cognitive impairments in the population, cognitive neuroscience could have a major impact in this area.

We might speculate that in the future cognitive neuroscience will also have an influence in many nonclinical areas. In Part I the interaction between neuroscience and artificial intelligence was discussed. We expect this interaction to grow in the coming years. Other areas that might benefit from progress in cognitive neuroscience include education, human engineering, and industrial psychology. When a similar volume is written twenty years from now, a final section might describe the application of cognitive neuroscience to some of these areas.

Reference

Churchland, P.S. (1986). *Neurophilosophy*. Cambridge, MA: MIT Press.

Index